Marine Algae in Pharmaceutical Science

Marine Algae in Pharmaceutical Science

Editors
Heinz A. Hoppe · Tore Levring
Yukio Tanaka

W DE G

Walter de Gruyter · Berlin · New York 1979

Editors
Heinz A. Hoppe
Finkenstraße 5
D-2150 Buxtehude
Germany

Professor Tore Levring
University of Gothenburg
Marine Botanical Institute
S-413 19-Gothenburg/Sweden

Dr. Yukio Tanaka
St. Mary's Hospital
3830 Lacombe Avenue
Montreal, Que. H3T 1UB
Canada

CIP-Kurztitelaufnahme der Deutschen Bibliothek

Marine algae in pharmaceutical science / ed.
Heinz A. Hoppe . . . – Berlin, New York : de Gruyter, 1979.
ISBN 3-11-007375-7

NE: Hoppe, Heinz August [Hrsg.]

Library of Congress Cataloging in Publication Data FEB 8 1979

Marine algae in pharmaceutical science.

Includes articles from a special symposium organized in connection with the 9th International Seaweed Symposium, held in Santa Barbara, Calif., 1977.
Bibliography: p.
Includes indexes.
1. Marine algae – Therapeutic use – Congresses.
2. Marine pharmacology – Congresses. I. Hoppe, Heinz August, 1907 – II. Levring, Tore, 1913 – III. Tanaka, Yukio, 1929 – IV. International Seaweed Symposium, 9th, Santa Barbara, Calif., 1977.
RS165.A45M37 615'.329'392 79-4419
ISBN 3-11-00-7375-7

 Printing: Karl Gerike, Berlin. – Binding: Lüderitz & Bauer, Buchgewerbe GmbH, Berlin. – Printed in Germany

PREFACE

In connection with the 9th International Seaweed Symposium held in Santa Barbara, California in August 1977, a special symposium was organized to discuss "Marine Algae in Pharmaceutical Sciences". International Seaweed Symposia have always been a forum for those interested in any aspect of marine algae, whether their taxonomy, morphology, ecology, physiology, cultivation, chemistry, biochemistry or application. The last-mentioned in particular, being the fruit of all other aspects of algal sciences, is of immense depth.

One of the many applications of algae that has enjoyed growing interest in recent years is their usefulness and potential as sources of pharmaceutically and medically important substances. Apparently we have scarcely begun to tap this enormous treasure inherited from Nature, and it seemed to us that a special gathering of those involved in this particular field could further stimulate and expand existing interest.

This monograph, comprising articles by symposium participants as well as other experts, is a single-volume survey of current knowledge in this field. It is hoped that at future Seaweed Symposia follow-up meetings in this and other fields will be arranged with the same purpose as this special symposium, and with equal success.

We are most grateful to our Santa Barbara hosts, who kindly agreed to this venture and without whom our original plans could never have been realized. The continuing interest and insight of the publishing house Walter de Gruyter & Co., Berlin and New York, in algal sciences is the essential nutrient to our endeavors, and we are particularly grateful for their support. We would also like to express our sincere appreciation to the speakers and audience of the symposium, and especially to the contributors to this monograph.

Heinz A. Hoppe — Tore Levring — Yukio Tanaka

Contents

Part III. Articles on Selected Algae and Algal Products

Contributors

Ahing, S. T., Department of Dentistry, St. Mary's Hospital, 3830 Lacombe Avenue, Montreal, Quebec H3T IM5, Canada

Aktin, E., University of Istanbul, Department of Neurology, Faculty of Medicine, Istanbul, Turkey

Andersen, K. K., Department of Biochemistry, University of New Hampshire, Durham, New Hampshire 03824, USA

Aubert, M., Institut National de la Santé et de la Recherche Médicale, C.E.R.B.O.M., 1, Avenue Jean-Lorrain, 06300 Nice, France

Aubert, J., Institut National de la Santé et de la Recherche Médicale, C.E.R.B.O.M., 1, Avenue Jean-Lorrain, 06300 Nice, France

Bhattacharjee, S. S., McGill University, Department of Microbiology & Immunology, 3775 University Street, Montreal, Canada

Bonotto, S., Département of Radiobiology, C.E.N. - S.C.K., B-21400 Mol, Belgium

Chapman, V. J., Department of Botany, University of Auckland, Auckland, New Zealand

Czapke, K., Morski Instytut Rybacki al Zjednoczenia, Gdynia, Poland

Deig, E. F., University of California, Berkeley, Naval Biosciences Laboratory, Naval Supply Center, Oakland, California 94625, USA

Der Marderosian, A., The Philadelphia College of Pharmacy and Science, 43rd Street, Woodland Avenue & Kingsessing Mall, Philadelphia, Pennsylvania 19104, USA

Diaz-Pifferer, M., Department of Marine Sciences, University of Puerto Rico, Mayaguez, Puerto Rico 00708, USA

Ehresmann, D. W., University of California, Berkeley, Naval Biosciences Laboratory, Naval Supply Center, Oakland, California,94625, USA

Fenical, W., Institute of Marine Resources, Scripps Institution of Oceanography, La Jolla, California 94625, USA

Feochari, K., Clinical Investigation Service, St. Mary's Hospital, 3830 Lacombe Avenue, Montreal, Quebec H3T IM5,Canada

Gauthier, M., Institut National de la Santé et de la Recherche Médicale, C.E.R.B.O.M., 1. Avenue Jean-Lorrain, 06300 Nice, France

Glombitza, K.-W., Institut für Pharmazeutische Biologie der Universität Bonn, Nussallee 6, D-5300 Bonn, BRD

Gnassia-Barelli, M., Laboratoire de Physique et Chimie marines, La Darse, F-06230 Villefrache-sur-Mer, France

Gopal, B. V., Cellulose Products of India Ltd, Shenbaga Building, RA/9 Poonga Street, Thirunagar, Madurai-625006, India

Gueho, E., Laboratoire de Cryptogamie, 8 Avenue Rockefeller, F-69373 Lyon Cedex 2, France

Güler, E., University of Istanbul, Department of Pharmacy and Technology, Faculty of Pharmacy, Istanbul, Turkey

Güven, K. C., University of Istanbul, Department of Pharmacy and Technology, Faculty of Pharmacy, Istanbul, Turkey

Hager, L. P., University of Illinois, Urbana, Illinois, USA

Hatch, M. T., University of California, Berkeley, Naval Biosciences Laboratory, Naval Supply Center, Oakland, California 94625, USA

Hébert, M., Department of Dentistry, St. Mary's Hospital, 3830 Lacombe Avenue, Montreal, Quebec H3T IM5, Canada

Hoppe, H. A., Finkenstrasse 5, D-2150 Buxtehude, BRD

Ikawa, M., Department of Biochemistry, University of New Hampshire, Durham, New Hampshire 03824, USA

Kneifel, H., Abteilung für Algenforschung und Algentechnologie der Gesellschaft für Strahlen- und Umweltforschung mbH., Bunsen-Kirchhoff-Strasse 13, D-4600 Dortmund, BRD

Knight, L. C., Washington University, School of Medicine, The Edward Mallinckrodt Institute of Radiology, 510 South Kingshighway, St. Louis, Missouri 63110, USA

Koyuncuoǧlu, H., University of Istanbul, Department of Pharmacology, Faculty of Medicine, Istanbul, Turkey

Levring, T., Marine Botanical Institute, Carl Skottsbergs Gata 22, S-413 19 Göteborg, Sweden

McConnell, O. J., Institute of Marine Resources, Scripps Institution of Oceanography, La Jolla, California 92093, USA

McElvany, K. D., Washington University, School of Medicine, The Edward Mallinckrodt Institute of Radiology, 510 South Kingshighway, St. Louis, Missouri 63110, USA

Michanek, G., Marine Botanical Institute, Carl Skottsbergs Gata 22, S-413 19 Göteborg, Sweden

Misra, A., Department of Botany, L. N. Mithila University, Darbhanga, 846004 India

Mshigeni, K. E., Department of Botany, University of Dar es Salaam, P. O. Box 35060, Dar es Salaam, Tanzania

Mziray, W. R., Department of Botany, University of Dar es Salaam, P. O. Box 35060, Dar es Salaam, Tanzania

Nisizawa, K., Department of Fisheries, College of Agriculture and Veterinary Medicine, Nihon University, Shimouma-3, Setagaya, Tokyo 154, Japan

Noguchi, T., Departments of Chemistry and Biochemistry, University of New Hampshire, Durham, New Hampshire 03824, USA

Okamoto, K., Department of Chemistry, Kyoto University of Education, Fushimi-ku, Kyoto, 612, Japan

Okasaki, M., Department of Chemistry, Kyoto University of Education, Fushimi-ku, Kyoto, 612, Japan

Otsuka, K., Department of Chemistry, Kyoto University of Education, Fushimi-ku, Kyoto, 612, Japan

Pesando, D., Laboratoire de Physique et Chimie marines, La Darse, F-06230 Villefranche-sur-Mer, France

Pohl, P., Institut für Pharmazeutische Biologie der Universität Kiel, Grasweg 9, D-2300 Kiel, BRD

Sinha, R., Department of Botany, L. N. Mithila University, Darbhanga, 846004 India

Siuda, J. F., University of Pittsburgh, Pittsburgh, Pa., USA

Soliman, F. M., Department of Pharmacognocy, College of Pharmacy, University of Rhode Island, Kingston, Rhode Island 02881, USA

Stara, J. F., The United States Environmental Protection Agency, Cincinatti, Ohio 45237, USA

Tanaka, Y., 10 Surry Road, Dollard des Ormeaux, Quebec H3B 2C3, Canada

Theiler, R., University of Illinois, Urbana, Illinois, USA

Tsuchiya, Y., Kitasato University, School of Fisheries Sciences, Sankriku-Machi, Kesen-Gun, Iwate Prefecture 022-01, Japan

Uebel, J. J., Department of Biochemistry, University of New Hampshire, Durham, New Hampshire 03824, USA

Weber, A., Institut für Allgemeine Botanik, Universität Hamburg, Jungiusstraße 6-8, D-2000 Hamburg 36, BRD

Welch, M. J., Washington University, School of Medicine, The Edward Mallinckrodt Institute of Radiology, 510 South Kingshighway, St. Louis, Missouri 63110, USA

Wettern, M., Institut für Allgemeine Botanik, Universität Hamburg, Jungiusstrasse 6-8, D-2000 Hamburg 36, BRD

Yamamoto, T., Department of Chemistry, Kyoto University of Education, Fushimi-ku, Kyoto, 612, Japan

Yaphe, W., McGill University, Department of Microbiology & Immunology, 3775 University Street, Montreal, Canada

Youngken Jr., H. W., College of Pharmacy, Office of the Dean, University of Rhode Island, Kingston, Rhode Island 02881, USA

Zurheide, F., Institut für Pharmazeutische Biologie der Universität Kiel, Grasweg 9, D-2300 Kiel, BRD

Part I. General Reviews

THE VEGETATION IN THE SEA

Tore Levring
Marine Botanical Institute, University of Göteborg, Carl Skottsbergs Gata 22, S-413 19 Göteborg, Sweden

The plant life in the sea is rich. The often enormous quantities of seaweeds found on a beach after a storm may serve to prove this point . Marine vegetation consists mainly of algae, but also a few phanerogams are found. These belong to the genera *Zostera*, *Ruppia*, *Cymodocea*, *Posedonia* and others. In suitable localities they often occur in great quantities and form green meadows on loose bottoms in shallow waters. These "sea grasses" are locally of great importance to the animal life there.

The vegetation can be divided biologically into two large groups:

1) Benthos, i.e. the attached forms.
2) Plankton (phytoplankton), free living and mainly unicellular forms.

This classification is not by any means systematic but purely biological. However, benthos and phytoplankton belong on the whole to different taxonomic groups. The Chlorophyta (green algae) is the main exception as this group contains both benthic and planktonic forms. As only benthic algae have been of interest so far as raw material for technical use, we will only deal with this group in this brief review. Phytoplankton is, however, the main primary producer of the sea and is as such of fundamental importance to marine life.

Marine Algae in Pharmaceutical Science
edited by H. A. Hoppe, T. Levring, Y. Tanaka

Like all other plants marine algae are dependent on and influenced by different environmental factors: substratum, medium, temperature and light. They differ highly from land plants in that they are surrounded by water and not by air. The lower limit of benthic vegetation is mainly determined by the reduction of light, i. e. the radiant energy. This lower limit varies: in coastal areas of northwestern Europe it is about 30 metres, in tropical waters or in the Mediterranean, which have very clear water, about 100 metres or occasionally 200 metres.

ENVIRONMENTAL FACTORS

The Substratum

The chemical composition of the substratum is normally of no significance to algal vegetation but a more or less hard bottom is essential. Contrary to terrestrial plants algae have no true roots by means of which to draw mineral salts from the substratum. They take their nutrients directly from the surrounding medium, the sea water. The substratum, therefore, need only be suitable for germination and allow the plants to cling it. It may therefore consist of different kinds of stones, wood or even a glass jar. Soft bottoms are often without vegetation. The seagrasses are exceptions. They often form green meadows on muddy bottoms. A few tropical algae (i. e. *Avrainvillea*, *Caulerpa*, *Halimeda*, *Penicillus*, *Rhizocephalus*, *Turbinaria*, *Udotea*), all forms which take hold in sand (or on coral reefs) with their root-like excrescence and rhizoids represent another group, which does not require a hard substratum.

Apart from these there are also numerous forms, perhaps the majority, which grow on other plants as epiphytes or in a few cases even as parasites. They are mostly several decimeters to a few centimeters high or even microscopic in size. Some of the epiphytic forms appear to demand certain species of plants as substratum, whereas in other cases the species of the host plant seems to be of no significance.

The Medium

Sea water is an aqueous solution containing a great number of chemical elements. The main ions are sodium, magnesium, chloride and sulphate. There are also dissolved gases (carbon dioxide, oxygen) and organic compounds in the solution. The salinity of ocean water amounts to approximately 36‰, in some areas slightly more. In littoral areas and inland seas the salinity may be lower, as, for instance in the Baltic, where it gradually becomes as low as 3‰ or even less.

The osmotic pressure of the sea water is determined by its salinity and amounts normally to about 25 Atm. Cells of plants living in this kind of medium must have a somewhat higher osmotic pressure. The sensitiveness of various species to a lower or higher salinity differs greatly and is, of course, of great importance for spreading and distribution. The phenomenon of this so called *osmotic resistance* was investigated particularly by Biebl (1962), Höffler (1931), Hoffmann (1929), and Kylin (1938). It is a striking fact that the Chlorophyta (green algae) are less susceptible than the Rhodophyta (red algae). Biebl also points out the larger range of osmotic resistance in surface algae as compared to the algae living in deeper water.

With regard to their susceptibility to a varying salinity algae can be subdivided into four groups:

Euryhaline species - of little susceptibility

Stenohaline species - live in a certain narrow range of salinity

Hyperhaline species - require a high salinity

Hypohaline species - to be found in brackish water.

Although the salinity varies in some regions, the proportions of the more important elements dissolved in the water remain constant. The major ions are shown on Tabl. 1. Most of the elements necessary for plant growth are always to be found in sufficient quantity in sea water. The content of phosphate and nitrate (and ammonia), however, shows great variation owing to plant and animal life.

Table 1.

Principle Ions of Sea Water
(Percentage composition of ions)

	%		%
Na^+	30.4	Cl^-	55.2
Mg^{++}	3.7	SO_4^{--}	7.7
Ca^{++}	1.16	Br^-	0.19
K^+	1.1	H_3BO_3	0.07
Sr^{++}	0.04	HCO_3^- and CO_3^{--}	0.35
		Minor constituents 0.02 - 0.03	

Apart from the uptake of different elements associated with the formation of enzymes etc., certain ions - often as inorganic compounds - are accumulated in the cells to a surprisingly large extent. The function of these accumulations is often obscure.

Potassium is often accumulated to an extent 20 to 30 times greater than its concentration in sea water. Several marine algae store nitrate in a similar way. The ancient use of many seaweeds as manure is at least partly connected to these facts.

Very peculiar is the accumulation of iodine (up to 30,000 times) in many marine algae, i.e. *Laminaria*, and the storage of sulfuric acid in *Desmarestia*.

Carbon dioxide is an essential agent in the photosynthetic process. Sea water contains carbon dioxide as bicarbonate and carbonate ions in a rather complicated system combined with hydrogene ions:

$$H_2O + CO_2 \text{ (dissolved)} \rightleftharpoons H_2CO_3 \quad HCO_3^- + H^+ \rightleftharpoons CO_3^{--} + 2H^+$$

In this process the atmospheric carbon dioxide is in nearly all cases in equilibrium with the carbonic acid system in the surface layer of the water. This system is also dependent on salinity, temperature and pH value. Normally, the pH of sea water is 8.2 (8.1 - 8.3). With increasing depth it falls to about 8.0.

There are also dissolved organic compounds in sea water. Some of these substances, however, seem to play no major role as nutrients. Others, which especially occur close to the coast, act as growth stimulators, and are probably of great importance. According to modern investigations the following chemical groups have been found in sea water: carbohydrates (i.e. polysaccharides and lipides) nitrogen-containing compounds such as proteins, polypeptides, amino acids, and humus acids (Dyrssen 1966, Kalle 1962). It is also known that many algae probably the majority - require certain vitamins - cobolamin (B_{12}), thiamine (B_1), biotin, and others - or mixtures of them for normal development. Such growth substances are also found in natural sea water (Burkholder 1959). They derive mainly from bacterial activities. Several growth factors like ascorbic acid and niacin are also produced by the marine algae themselves, like *Ascophyllum*, *Fucus*, and *Laminaria* (Jensen 1964, Larsen 1958).

The colour of water, especially in coastal areas, depends to some extent on the so-called y e l l o w s u b - s t a n c e (Kalle 1938, Jerlov 1976). According to recent investigations (Kalle 1966) this substance seems to be formed in sea water from carbohydrates in the presence of certain amino acids.

Temperature

The temperature of the water is of great importance for the development, the growth and the geographical distribution of marine algae. The annual variation in temperature differs greatly in different regions. In tropical waters these variations are always slight and amount to only a few degrees. In contrast, there are littoral regions, such as the Atlantic coast of North America from Cape Hatteras to Newfoundland, or the coasts of the Kattegat and the Skagerrak in Europe, where the variations can amount to 18°to 20°C or more. In the Mediterranean the variations lie between 17° and 14°C. Generally the variations in temperature are greater in coastal regions than in the open sea. The variations in temperature are naturally most marked at the surface; this is particularly true of sections of the sea more or less sealed off, such as the Kattegat, the Skagerrak and the Baltic Sea.

Not only are annual mean temperatures of various regions important for algal growth and geographical distribution but also the maximum and the minimum temperatures are decisive factors.

With respect to the phytoplankton the oceanic regions are divided into four large groups (Table 2), which also can be used for the benthic vegetation (Steemann Nielsen 1944).

Table 2.
Distribution of Phytoplankton and Benthos

Temperature in degrees centigrade	Group	Corresponds to
$>10^{\circ}$	Oligothermic (cold region)	I Arctic
$0-20^{\circ}$	Mesothermic α (cold temperate region)	II Boreal
$14-24^{\circ}$	Mesothermic β (warm temperate region)	III Subtropical
$<24^{\circ}$	Polythermic (warm region)	IV Tropical

The algae are divided into two groups according to their susceptibility to variations in temperature:

1) *stenothermal species* - with a low resistance to changes in temperature (range of tolerance 10 degrees centigrade)

2) *eurythermal species* - with a high resistance to changes in temperature.

Water Movements

Water movements in the sea are mainly caused by tides, winds, changes of temperature, and the inflow of fresh water, and result in vertical and horizontal currents, waves and intermixing between water masses.

The tide in the open sea never amounts to more than 1 metre. The tidal wave is enlarged when it reaches areas of lower depth. In bays where special conditions become effective, differences in the tide levels can be registered, e.g. in the central parts of the bay between Brittany and Normandy in France (St. Malo and Mont Saint Michel) approximately 12 metres, and in the Bay of Fundy on the inner coast of Nova Scotia in Canada 21 metres. In open coastal areas the differences in the tide levels are approximately 1.5 to 3 metres.

In some areas like the Mediterranean, the Kattegat and the Skagerrak there is practically no tide or it amounts to some 10 - 30 cm.

The regular change in the water level plays an important role in the vertical distribution of the vegetation in the tidal zone. Tidal currents - and of course other currents too - can often lead to turbulence in the water having a balancing effect on water temperatures. Turbulence may also cause water that is lacking in nutrient salts to be mixed with water rich in them. In general, conditions of growth are favorable for benthos as well as for plankton in areas with such water movements.

Submarine Light

The significance of submarine illumination to the marine vegetation is a question of great importance. Irradiance is reduced by absorption and scattering when passing through the water. This attenuation is different for different wavelengths and is also highly dependent on the presence of dissolved and suspended substances in the sea water. Thus the colour of the submarine illumination below 5-10 metres in normal coastal waters, which are more or less turbid, is greenish. In areas with clear oceanic water it is bluish instead.

As submarine radiant enery is absorbed by the different pigments of algae and used for their photosynthesis its total value and its spectral composition is of great importance.

In coastal areas with more or less turbid water (i.e. north-western Europe) it is well known that the Chlorophyta are found in shallow water - down to about 5 metres. The Rhodophyta are found at all depths but predominate in deeper water. The Phaeophyta hold an intermediate position. This is in good accordance with the fact that the Chlorophyta are able to use red and blue light about equally well, but not green light. They are therefore not able to exist in purely greenish illumination. Greenish light however is an excellent source of radiant energy for Rhodophyta. As the Phaeophyta hold the intermediate position it could be said that the two last divisions are less dependent on the spectral composition of the radiant energy than the Chlorophyta.

In areas with clear, oceanic water, where the transmittance is much higher, algal growth can be expected in greater depths than in coastal areas of the type mentioned above. An extensive algal vegetation has been reported in depths down to 100 metres or even lower at Bermuda, situated in the extremely clear waters of the Gulf Stream in the North Atlantic and in the Caribbean area (Taylor 1960, 1961). The number of Chlorophyta in this deep water vegetation is reported to be considerable. Similar reports are available from the Mediterranean (Levring 1958), and Madeira (Levring 1974). As the irradiance below 10-15 metres in such areas belongs to the bluish part of the spectrum, the occurrence of green algae is hardly a surprise, as they are able to make use of the blue light for their photosynthesis (Fig. 1).

In Arctic and Antarctic waters algae have been found growing at great depths. Lund (1957) traced several species down to a depth of 80 metres on the coast of Greenland. Wilce (1967) has found a well developed algal vegetation under the ice down to at least 50 metres (occasionally even below 100 m) on the west coast of Greenland and arctic Canada. According to Zaneveld (1966) benthic algae have been found in the Ross Sea (Antarc-

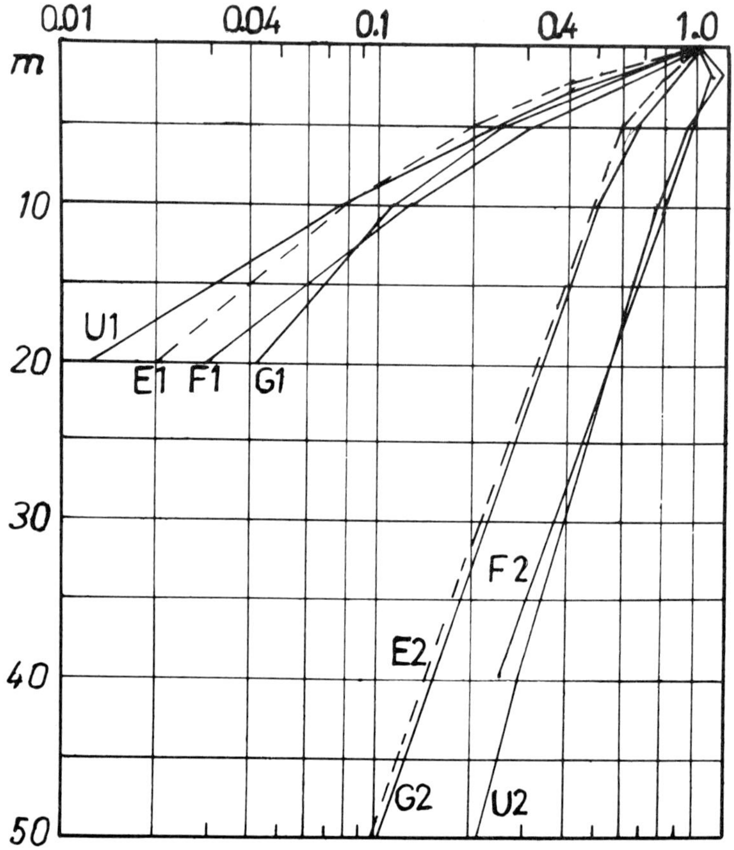

Fig. 1. Attenuance of radiant energy and photosynthesis at different depths at 2 stations off the coast of North Carolina. 1, coastal waters; 2, Gulf Stream. U = *Ulva lactuca*, F = *Fucus vesiculosus*, G = *Gracilaria foliifera*. E = radiant energy.

tica) down to the remarkable depth of 150-300 metres. No satisfactory explanation has been given for the occurrence of these algae in habitats where - at least for the greater part of the year - no light is present.

The relation between submarine light and the vertical distribution of algal vegetation may be summarized in the following way: There are different types of algae which are adapted to light of different intensities. For the constitution of these types the various pigments play an important role. They make the utilization of the different spectral compositions possible. (Fig. 2). In other words, these types are not only adapted to different light intensities, but also to different spectral compositions of the radiant energy (Levring 1969 a,b).

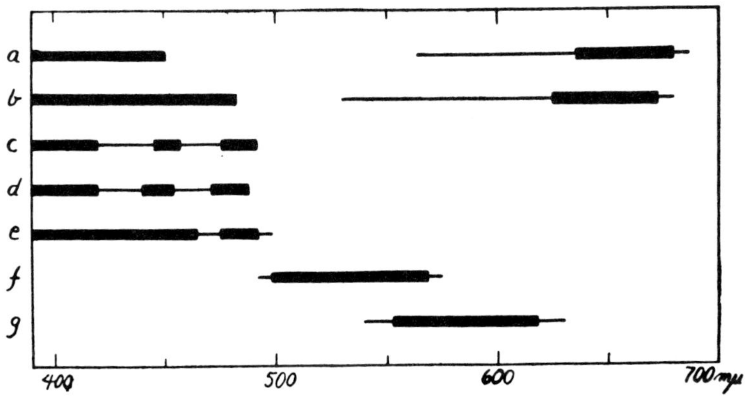

Fig. 2. Schematic representation of the absorption of different pigments of the algae: Chlorophyll *a* (a) and *b* (b), carotene (c), xanthophyll (d), fucoxanthin (e), phycoerytrin (f) and phycocyanin (g) (from Levring).

THE BENTHIC VEGETATION

Growth and Periodicity

Photosynthetic activity is, of course, of decisive importance for the growth of algae. It is well known that the majority of the Chlorophyta are found in shallow water of its coastal areas with greenish submarine illumination, as they are restricted to the use of red and blue light for their photosynthesis. In areas with very clear water with bluish illumination various Chlorophyta - naturally dim light forms - have been found in depths down to 200 metres. Due to the accessory pigments of Phaeophyta - fucoxanthin - and Rhodophyta - phycoerythrin - representatives of these two divisions are less dependent on the spectral compositions of the submarine illumination. Very remarkable is the discovery of algae in great depths (80 - 300 metres) under the ice in Arctic and Antarctic areas mentioned above.

Growth can roughly be regarded as the photosynthetic surplus:

Photosynthesis - Respiration = Growth

Photosynthesis as well as respiration are dependent on temperature. However, with rising temperature the rate of respiration increases more than photosynthesis. This explains the well-known phenomenon that algae in temperate regions have their great growth period in the spring. At that time of the year the water is still cold and consequently the rate of respiration is low. Light conditions, on the other hand, are so favourable that they permit an almost maximum degree of photosynthesis, which, of course, results in a large photosynthetic surplus. In many areas, perhaps mainly where differences in temperature of the surface water are considerable, i.e. in northern Europe, it is well known that extensive communities of certain algae (*Ulothrix*, *Urospora*, *Monostroma*, *Porphyra*, etc.) develop for a few months in the spring or early summer. The development of such seasonal communities is certainly favoured by the balance between radiant energy and water temperature mentioned above. This fact may also, at least partly, explain the existence of the very heavy algal growth in many cold water areas of the world.

Kylin (1907) distinguishes three different main types of periodicity:

1) Forms that are vegetative and fertile all the year round (e.g. *Hildenbrandia*).

2) Forms with vegetative development during the whole year, but the period of fertility limited in some cases to the summer (*Fucus vesiculosus*, *Polysiphonia nigrescens*) and in most others to winter and spring (e.g. *Fucus serratus*).

3) Forms showing vegetative growth and fertility only during a limited period. The photosynthetically active, vegetative organs are shed after the termination of the growth period (e.g. *Desmarestia aculeata*, *Polysiphonia elongata*).

Conditions with only slight annual variations in temperature result in a less pronounced periodicity.

A photoperiodic reaction may also be important for the development and life cycle of several marine algae. The existence of this phenomenon has been proved in some cases. Further investigations in this field will certainly give valuable information about the seasonal variation of algal vegetation.

Geographical Distribution

Although floristic studies of algal vegetation from various parts of the world have been carried out during the last century, our present knowledge of the different floral regions is still in many respects incomplete and fragmentary. This applies especially to quantitative estimations of the algal vegetation. A commendable attempt to gather all uptodate facts in this particular field has recently been made by Michanek (1975). It appears that our knowledgefor many areas of the world is good, but for others incomplete. For the estimation of the potential yields of marine algae in different geographical areas such knowledge is very important especially for questions connected with the commercial use of seaweeds.

The composition of the algal flora of a certain region cannot be explained by climatic or hydrographic factors only. Historic factors have also to be considered. The algae are plants of extremely high age. It is therefore necessary to bear in mind also former connections of coastal areas, which have vanished today. The distribution of different species along a more or less continous coast-line, as, for exemple, on the west coast of Europe, is in the first place undoubtedly due to temperature conditions. In the classic work by Börgesen and Jónsson (1905) on the distribution of the algae in northern Europe and the Arctic Ocean the different species were divided into five groups according to their distribution: arctic, subarctic, boreal-arctic, cold boreal, and warm boreal. This system has generally been accepted and used by other algologists, and it has also been completed and enlarged in different ways. (Tab. 3.) The boundaries between the various groups are not clearly marked but more or less overlapping, so that they cannot always be regarded as different, floral areas. It is, of course, highly desirable that a universal system of nomenclature should be

Table 3.
Geographical subdivision of Algal Vegetation

Börgesen and Jónsson	Nienburg	Smith Setchell, Hoyt	Okamura	Feldmann
Arctic	Arctic	Boreal		
Subarctic	Arctic-French			
Boreal-Arctic	North-European	Upper Boreal		
Cold Boreal	West-European-Baltic			
Warm Boreal	Meridional-Norwegian	N. Temperate	Temperate	Boreal Atlantic
		N. Subtropical	Subtropical	Tropical Atlantic
		Tropical	Tropical	Pantropical
		S. Subtropical		
		S. Temperate		
		Upper Australian		
		Australian		
	Cosmopolitan		Cosmopolitan	Cosmopolitan
			Indo-Pacific	Indo-Pacific
	Meridional		Japanese-Okhotsk	Mediterranean

used. Such a system, mainly based on the classic one by Börgesen and Jónsson, has been suggested by Chapman (1946, Table 4). Several authors deal with the subject endemism in various regions. The interesting problem of vicariating species is mentioned, amongst others, by Svedelius (1924).

Table 4.
Geographic Elements
(According to Chapman, 1946)

Arctic		Antarctic	
Subarctic		Subantarctic	
Boreal-Arctic		Austrial-Antarctic	
Cold-Boreal-Atlantic	Cold-Boreal-Pacific	Cold-Austrial-Atlantic	Cold-Austrial-Pacific
Boreal-Atlantic	Boreal-Pacific	Austrial-Atlantic	Austrial-Pacific
North-Subtropical-Atlantic	North-Subtropical-Pacific	South-Subtropical-Atlantic	South-Subtropical-Pacific
Tropical-Atlantic		Tropical-Pacific	
Pantropical		Cosmopolitan	
Caribbean		Arabican	
Indo-Pacific			

Vertical Distribution

A number of horizontal zones and fringes can be distinguished in the benthos vegetation. This zonation is caused by such factors as tide, degree of exposure, reduced and changed spectral composition of the submarine illumination etc. Therefore, one finds a l i t t o r a l z o n e, which is designated by an intermittent process of desiccation and which, above all, is limited by the high-tide and low-tide levels on all coasts subject to tidal influences. However, the upper and the lower limit of the littoral zone are not taken as two purely physical lines, but rather as two natural physiological boundaries which are influenced also by other factors. Areas where tidal activity is absent or negligible must be regarded as exceptional, as, for example the Mediterranean or the coasts of the

Kattegat, the Skagerrak, and the Baltic. More or less irregular changes of the water level due to winds etc. constitute there a "reduced" littoral zone corresponding to that of a normal tidal coast. Above the littoral zone begins the supralittoral zone which is never submerged but may be reached by spray; and below it expands the sublittoral zone which reaches down to the cessation of vegetation and is always submerged. There have been discussions about the nomenclature of the zones, and attempts to produce universal systems with zones of different ranks have also been made (cf. Levring 1969 a).

A diagram showing the algal zonations on a typical tidal coast in northwestern Europe is reproduced in Fig. 2. The lower limit

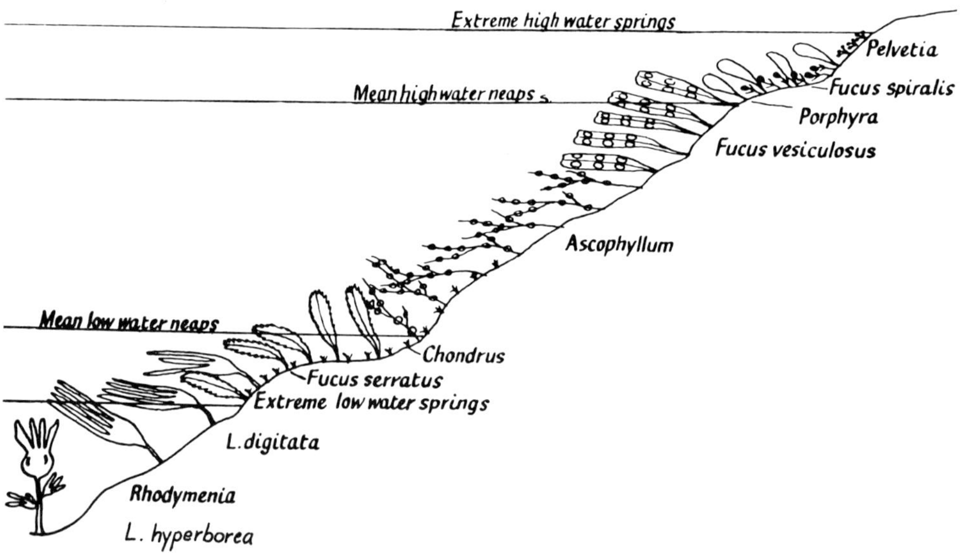

Fig. 3. Typical zonation on a European rocky shore (after Chapman).

of the vegetation is of course mainly determined by the reduction of the light intensity: in coastal waters of for instance North Europe about 30 metres depth, in tropical waters or in the Mediterranean with clearer water about 100 or perhaps occasionally 200 metres.

CLASSIFICATION

The algae do not constitute any uniform or systematic group. Instead they comprise various strains of primitive plants of low order which have undergone a more or less parallel evolution. Formerly a distinction was made between flagellates and algae. Flagellates were then so defined as to include coloured unicellular organisms moving by means of flagella but lacking cell walls. However, several colourless species, too, were classified among the flagellates. Several of them are closely related to the coloured species - they have only lost their photosynthetically active pigments. The coloured flagellates resemble several motile stages of certain algae (zoospores, gametes.)

As early as 1893 Klebshad drawn attention to the close relationship between coloured flagellates and algae. In this century, many scientists have clarified this relationship with greater precision. The flagellate-like species occur in primitive stages in various parallel strains of development.

In the older systematic classification of algae colour was not allowed for as a feature of distinction. Harvey (1836) was the first to distinguish between these various groups: Chlorospermeae (green algae), Melanospermeae (brown algae) and Rhodospermeae (red algae). These names were later more or less altered and several algal groups added (see below). The different groups contain special types of chlorophyll, carotene, xanthophyll and other pigments. Their metabolism, too, differs. This is also true of final products of photosynthesis: the Chlorophyceae produce starch, the Phaeophyceae laminarin, while the Cyanophyceae produce cyanophycean starch, and the Rhodophyceae rhodophycean starch etc. The flagellar mechanisms of the reproductive organs of the various groups are differently shaped. Other features indicate that not all groups are interrelated: rather they belong to different strains of development. Today the algae are , in general, classified into seven groups (strains or divisions):

I. C h l o r o p h y t a: 1st class Chlorophyceae, green algae. Large group of very differently developed organisms - unicellular or multicellular, filamentous, expanded etc. The Chlorophyta contain the same pigments as the higher plants.

II. E u g l e n o p h y t a: 1st class Euglenophyceae, green flagellates. No true starch is formed. Mainly in fresh water.

III. C h r y s o p h y t a: 1st class Xanthophyceae (Heterokontae), yellow-green algae, mainly in fresh water, some on soil or planktonic in the sea.

2nd class Chrysophyceae, yellow brown algae. Mainly in fresh water. The coccolithophorides play an important role as plankton in the sea.

3rd class Bacillariophyceae, diatoms. Characterized by their two strongly silicified cell-membranes. In fresh water as well as in the sea. The centric diatoms constitute an important part of the marine phytoplankton.

IV. P y r r o p h y t a: 1st class Desmophyceae (Desmokontae). Primitive, mainly marine phytoplankton.

2nd class Dinophyceae. Free-living, unicellular organisms. Typical in marine phytoplankton.

3rd class Cryptophyceae. Small, poorly known group of flagellates in fresh water and in the sea.

V. P h a e o p h y t a: 1st class Phaeophyceae, brown algae. With few exceptions marine. Multicellular, thallus showing very different degrees of differentiation and development from microscopic filaments to complicated build bodies, often of considerable size. The majority occur in the littoral and upper sublittoral zones, where they often form an important part of the vegetation. Many species are used commercially (*Ascophyllum*, *Fucus*, *Macrosystis*, *Laminaria* etc.).

VI. C y a n o p h y t a: 1st class Cyanophyceae, blue green algae. Mainly macroscopic layers and cushions on soil, rocks, wood etc. Others are microscopic. Some planktonic forms are occasionally produced in such numbers that the water is coloured. In fresh water and marine.

VII. R h o d o p h y t a: 1st class Rhodophyceae (Subclasses Bangioideae and Floriceae), red algae. A large group of algae which almost all belong to the sea. The thallus shows a very different development from microscopical cushions or filaments to complicated build cell bodies. The red algae grow mainly in the lower littoral and sublittoral zones. Many species are used commercially (*Chondrus*, *Eucheuma*, *Furcellaria*, *Gelidium*, *Gigartina*, *Gracilaria*, *Porphyra*, *Pterocladia* etc.).

REFERENCES

Biebl, R., Seaweeds. In Lewin, Physiol. and Biochem. of Algae, 799 - 815 (1962).

Börgesen, F., Jónsson, H., The distribution of the marine algae of the Arctic Sea and the northernmost part of the Atlantic. Botany of the Faeröes, 3 Appendix (1905).

Burkholder, P. R., Vitamin producing bacteria in the sea. Preprints Intern. Oceanogr. Congr. New York, 912 - 913 (1959).

Chapman, V. J., Marine algal ecology. Bot. Review, 12, 628 - 672 (1946).

Droop, M. R., Organic micronutrients. In Lewin, Physiol. and Biochem. of Algae, 141 - 159 (1962)

Dyrssen, D., Havsvattnets kemiska sammansättning. Svensk Naturvetensk., 278 - 288 (1966).

Harvey, W. H., Algae. In J. T. Mackay, Flora Hibernica. Dublin (1836).

Höffler, K., Hypotonietod und osmotische Resistenz einiger Rotalgen. Österr. Bot. Z., 80, 51 - 71 (1931).

Hoffmann, C., Die Atmung der Meeresalgen und ihre Beziehung zum Salzgehalt. Jahrb. wiss. Bot., 71, 214 - 268 (1929).

Jensen, A., Ascorbic acid in Ascophyllum nodosum, Fucus serratus and Fucus vesiculosus. Proc. IV Intern. Seaweed Symp., 319 - 325 (1964).

-, Carotenoids of Norwegian brown seaweeds and seaweed meals. Rep. Norw. Inst. Seaweed Res. 31, 138 pp (1966).

Jerlov, N. G., Marine optics. Elsevier Oceanography Ser., 14, 231 pp. (1976).

Kalle, K., Zum Probleme der Meerwasserfarbe. Ann. Hydr. u. mar. Meteorol., 66, 1 - 13 (1938).

-, Über die gelösten organischen Komponenten im Meerwasser. Kieler Meeresforsch., 18, 128 - 131 (1962).

Klebs, G., Flagellatenstudien II. Ztschr. Wiss. Zool., 55, 353 - 445 (1893).

Kylin, H., Studien über die Algenflora der schwedischen Westküste. Diss. Uppsala (1907).

-, Über den osmotischen Druck und die osmotische Resistenz einiger Meeresalgen. Svensk Bot. Tidskr., 32, 238 - 248 (1938).

Larsen, B., The influence of season, habitat and age of tissue on the niacin content of some brown algae. Norsk Inst. Tang of Tareforskn., 19 (1958).

Levring, T., Submarine daylight and photosynthesis of marine algae. K. Vet. o. Vitt. Samh. Handl. Göteborg, 4:B 5 (1947).

-, Some modern aspects of growth and reproduction in marine algae in different regions. Union Intern. Sciences Biol., B, 24, 57 - 65 (1958).

-, The vegetation in the sea. Classification of the Algae. In Levring, Hoppe, Schmid, Marine Algae. Hamburg (1969 a).

-, Light conditions, photosynthesis and growth of marine algae in coastal and clear oceanic water. Proc. VI Intern. Seaweed Symp., 235 - 244 (1969 b).

-, The marine algae of the archipelago of Madeira. Bol. Museu municipal Funchal, 28, 5 - 111 (1974).

Lund, S., The marine algae of East Greenland. Medd. om Grøn-land, 156:2 (1959).

Michanek, G., Seaweed resources of the ocean. FAO Fisheries techn. P., 138, 1 - 127 (1975).

Stemann Nielsen, E., Havets planteverden. Skr. Komm. Havunders. Köbenhavn, 13, 108 pp. (1944).

Svedelius, N., On the discontinuous geographical distribution of some tropical and subtropical algae. Ark. f. bot., 19, (1924).

Taylor, W. R., Marine algae of the Northeastern coasts of the Americas. Ann Arbor, 870 pp. (1960).

-, Distribution in depth of marine algae in the Caribbean and adjacent seas. Rec. adv. bot. Toronto, 193 - 197 (1961).

Wilce, R. T., Heterotrophy in arctic sublittoral seaweeds, an hypothesis. Bot. Mar., 10, 185 - 197 (1967).

Zaneveld, J. S., The occurrence of benthic marine algae under shore-fast ice in the western Ross Sea, Antarctica. Proc. V Intern. Seaweed Symp., 215 - 231 (1966).

MARINE ALGAE AND THEIR PRODUCTS AND CONSTITUENTS IN PHARMACY

Heinz A. Hoppe
Finkenstrasse 5
D-2150 Buxtehude (near Hamburg) - F. R. Germany

The utilization of algae for medicinal purposes is not new, but the discoveries of the last years have opened new, perhaps important possibilities in this field. The earliest particulars on seaweed utilization originate from the Chinese "Materia Medica" of Shên-nung 2700 B.C. Dioscorides (c. 50) mentions algae in medicines of earlier civilizations and also indicates their importance during those times. The role of macroscopic algae and also of plankton as medicine in ancient times has been described in detail by M. & D. Schwimmer (302). The Chinese, Japanese, Filipinos and other South and East Asiatic peoples in particular used algae not only as food but also in medicine. Zaneveld (362) reports on utilization of marine algae in tropical South and East Asia, especially in folk medicine.

During the 17th century soda and potash were obtained by ashing seaweed in kelp kilns. Seaweed was also an important raw material for extraction of iodine. Of the phycocolloids only agar was known until a few decades ago. From 1662 until World War I the production of this substance was a Japanese monopoly and although "Irish Moss" was in use for several centuries as thickener, it is during the last 50 years that the many industries making agar, alginates, carrageenans and other algal products have been established. These developments have not come to an end, and, furthermore, the possibilities of using marine algae for pharmaceutical purposes are steadily growing but only a small proportion of the thousands of marine algal species that are described have yet been examined for these

Marine Algae in Pharmaceutical Science
edited by H. A. Hoppe, T. Levring, Y. Tanaka

purposes.

Long before scientific research began, numerous algae were used as medicaments, especially in coastal countries. They were used in folk medicine against goitre, nephritic diseases, anthelmintic, catarrh and so on, and an investigation would certainly give interesting knowledge in this field. In different pharmacopeias and pharmaceutical handbooks a number of algae are mentioned as medicinal drugs. Many of their constituents have a considerable importance in pharmacy. They are used as medicaments or as auxiliaries. According to recent studies there are a number of species with substances of pharmaceutical interest, whose structure could be found out.

The increasing interest in research and utilization of marine algae, better methods of analysis, the discovery of pharmaceutical substances, the increasing uses of phycocolloids etc. have led to an important increase of the special literature. In addition to the Proceedings of the International Seaweed Symposia and the handbooks mentioned under "General Literature" there are a great many single publications of which only some can be reviewed here.

It is understandable that we are particularly well informed about distribution and utilization of algae on the Japanese coasts. Tokida, The Marine Algae of Southern Saghalien (335), Okazaki, Seaweed and their Uses in Japan (240), and Tokida and Hirose, Advance of Phycology in Japan (30), should be mentioned.

The valuable review of Hoffmann "Die praktische Bedeutung der Meeresalgen" (10) should not be omitted.

A study of the literature makes clear that research in the pharmaceutical uses of algae has increased considerably in recent years. The problems of modern research and utilization have been discussed not only at the International Seaweed Symposia, but also in 1969 in Kiel during the congress of the "Gesellschaft für Arzneipflanzenforschung" when "Planzliche Arzneistoffe aus dem Meer" (105) was reported.

The following papers are from the symposium on "Marine Biomedicinals", Corvallis, Oregon, 1969: (68), (88), (361), (365).

Symposia on "Drugs from the Sea" took place in 1967 and 1969 at the University of Rhode Island. In March, 1977 well-known scientists and practitioners in this special field met in Salishan Lodge, Oregon, in order to discuss the marine plant biomass of the northwest Pacific, an area of great economic potential (168). From this meeting Levring (177) reports:

"Generally speaking the seaweed resources of the world are still exploited only to a fairly low extent. The possibility of increasing harvesting depends to a great deal on economic factors - the demand for products and the cost of harvesting and processing. The global production of red algae could probably be increased by about 50%. The biomass of the brown algae would probably allow an increase of the utilization by 20 times.

"One way to satisfy the demand for algal raw material is probably by means of large scale cultures in the sea of selected species.

"Probably new, additional ways of employment for seaweeds will be found in future, which will raise the demand for raw material, maybe of very selected species for use in the pharmacy, agriculture etc. I am quite convinced that the importance of the seaweeds to mankind will increase."

In "Seaweed Resources of the Ocean" Michanek (207) reports:

"Quantitative studies of seaweeds are much rarer than those of fish stocks. The exceptions could be indicated by a few spots on a world map - the southwest coast of Norway, Scotland, Nova Scotia, the Californian kelp beds. The rest of the map is a blank space. This review is a first attempt to fill that space, but is primarily a challenge to field phycologists to produce the data from which a better evaluation of the world's seaweed resources can be made.

"Since this study concentrates on resource assessment, the potential harvest, and related quantitative studies, less emphasis has been given to studies on relative distribution, seasonal variation, mode of reproduction, etc. except where these are of direct interest to present or future utilization. However, for the areas for which suitable quantitative data are not yet available, ecological and similar information has been used to give an idea of the structure and composition of the flora, and thus provide a first approximation to the potential harvest.

"Any study of this kind must distinguish between biomass (or standing stocks) on the one hand and production or potential harvest on the other; further, a distinction needs to be made between the theoretically attainable harvest and that which is feasible in the light of practical and economic constituents."

MARINE ALGAE AS MEDICINAL DRUGS AND IN FOLK MEDICINE

Green Algae

Acetabularia major

The alga is used in South-East Asia in medicine as a means of treating gallstone and other stones. About Acetabularia, esp. A. mediterranea detailed papers are published (64, 66, 179).

Caulerpa

Some C.species are used for human consumption. They are cultivated in large containers.

C. lamourouxii is toxic for some individuals. Caulerpicin, caulerpin and sterols have been isolated. Caulerpicin, responsible for the peppery taste, causes a mild anesthesia when placed in the mouth. The neurotropic activity of caulerpicin is thought to be of clinical value (53, 57, 91, 292).

C. racemosa. Caulerpicin has been isolated (57).

Cladophora glomerata

The alga was used in folk medicine for burns.

Codium

Codium species are used in folk medicine as vermifuge, especially against Ascaris lumbricoides (106).

Cymopolia barbata

Purified extracts contain the sarganin-complex, a broad-spectrum antibiotic substance, isolated originally from Sargassum natans (190).

Pleurococcus naegelii (=Protococcus vulgaris)

Erithrit, a phytoncid effective against Staphylococcus and Streptococcus, was isolated and used against skin disease and parasitic illness in the form of ointments (342).

Rhizoclonium rivulare

The alga lives in fresh water and is used as vermifuge. Rh. rivulare and Rh. riparium are also remedies for wounds in folk medicine.

Trentepohlia iolithus

Aerophytic alga, used in folk medicine for skin diseases.

Ulva

In some regions U. lactuca is used in popular medicine for gout. Ulva lactuca is perhaps the BRYON THALASSION of Dioscorides. It is astringent and has an effect on irritations and podagra (28). Dragendorff also reports on the utilization of Ulva for scrofula and gout.

U. penniformis is used as a medical preparation in some parts of Siberia (337).

U. pertusa - "Sea Lettuce". The anthelmintic effects and the influence on the muscle contraction of worms has been investigated. In China for fever.

U. species are also used as vermifuge.

Brown Algae

Ascophyllum nodosum

The drug is a constituent of obesity preparations. Fluid extracts are to have an effect on rheumatism, strains etc. Stahl et al. (303) report on A. nodosum and Fucus vesiculosus as medicinal drugs in pharmacopeias.

Chordaria flagelliformis

The alga is eaten in East Asia. On Kamchatka it is known as "nebbpett". It is reported to have a minor anticoagulent effect.

Cutleria multifida

The mucilage is said to relieve gastric ulcer.

Cystoseira barbata

Coasts of the Adriatic and the Black Seas. Pharmacological studies have been carried out. One group had lipolytic activity, another one exhibited hypoglycemic effect (110, 113, 116, 118).

Dictyopteris polypodioides

This alga is used in popular medicine in the Mediterranean countries for lung diseases and scrofula.

Durvillea antarctica

Durvillea is used by the Maoris as a drug against scabies named "rimuroa" and also as vermifuge.

Eisenia bicyclis

The crude extract possesses anti-inflammatory and anticurare activities. The anti-inflammatory compound was found to be a laminarin-like polymer. A compound with anti-inflammatory and anticurare activity was found to be a complex polymer (363).

Fucus

F. esculentus. In Russia the alga is a constituent of sweets which are prescribed as prophylactic for sclerosis or for disturbance of the thyroid gland (178).

Sterols of F. gardneri in animal experiments reduced the plasma-cholesterin level (260).

F. vesiculosus has been used as a medicinal drug, mainly on account of its iodine content, for obesity defects, goitre, Basedo's disease etc. (303, 338, 341, 346).

Fucus vesiculosus tostus

(=Aethiops vegetabilis) was used in former times against scrofulosis and goitre.

Heterochordaria abietina

The alga contains the amino acid laminine (see Amino acids) with hypotensive and other pharmacological properties (57, 325). Johnston (153) gives a detailed chemical analysis.

Laminaria

L. species were employed mainly in Japanese folk medicine for lowering the blood pressure.

Laminaria powder (from L. saccharina) is applied in Russia for chronic constipation, goitre and as prophylactic for arteriosclerosis.

Laminaria was used as "Hai-tai", "Kwanpu" or "Kai-Wan" by Chinese for menstrual difficulties (5, 302).

L. angustata

Extracts have shown a benign effect on patients with arteriosclerosis and hypertension. The amino acid laminine is the effective principle (159).

L. hyperborea

Treated dried stipes, the so-called "Stipites Laminariae" or "Laminaria stalks" have played a role in surgery and gynaecology. The stipes swell up in water to a size five times as large as their original circumference. See Feochari's contribution in this monograph.

In some pharmacopeias the stipes of Laminaria digitata are also mentioned. An interesting paper "Laminaria digitata: A Checkered Career" was published by Barbara Rubin (262).

L. japonica

is used as a medicinal drug in Southern China for producing a cooling and blood cleaning effect, for treatment of glandular weakness and for normalising blood pressure (74). Cheng (72) reported that L. japonica in China is a remedy for hyperthyroidism.

L. saccharina

was used in India as "Galhdr-ka-palta" against goitre and in the Himalayas and in Turkestan against skin diseases and syphilis (5, 302).

Macrocystis pyrifera

is used for the production of pharmaceutical and dietary products (vitamins, inorganic constituents, trace elements). In the U.S.A. Macrocystis pyrifera is used in tablets for pregnancy anaemia (106).

Nereocystis luetkeana

The alga is used in folk medicine of the Indians in North America.

Sargassum

S. confusum: Sargaline, an extracted substance, has been reported to reduce the blood sugar level (265).

S. fusiforme: In China the alga is used in pharmacy for producing a cooling and blood-cleaning effect, for treatment of glandular weakness, apoplexy and for normalising blood pressure (74).

S. liniifolium: In India the alga is used for urinary troubles and calculous diseases.

S. muticum: Sterols of the alga in animal experiments significantly reduced the plasma-cholesterine level (260).

S. natans (=S. bacciferum)
In South America the alga is used in medicinal preparations to cure goitre and renal disorders.
The sarganin-complex (sarganin A and sarganin B) isolated from S. natans has been found to be present also in the red alga Chondria littoralis (chonalgin) and in the green alga Cymopolia barbata. Sarganin is a broad-spectrum antibiotic substance with low toxic effect (189, 190, 191).
The antibiotic substances in this alga are effective in reducing the mortality of chicken infected with Salmonella gallinarum (166).
S. thunbergii
In China the alga is used as a medicinal drug like S. fusiforme (74).
S. vulgare
Substances obtained from S. vulgare are effective in increasing serum lipolytic activity. The hypoglycemic substance was found to lower blood sugar by 17%. A certain percentage was shown to have antilipemic activity and a polypeptide structure (37, 114, 120).

Stilophora rhizoides
In folk medicine of Mediterranean countries for lung diseases and scrofula.

Undaria pinnatifida
contains a compound which cures nicotine poisoning. This substance is present in the unsaponifiable material from the ether extract (344/345). See also "Wakame".

Red Algae

Alsidium helminthochorton
The alga contains a floridoside, an alkaloid-like substance. The active principle is not known. "Corsican Moss" is an anthelmintic remedy.

Centroceras clavulatum

The alga is used in folk medicine in East Asia.

Ceramium loureirii

A gelatinous preparation of the alga is used for chest diseases in coastal areas on the Pacific.

C. rubrum

The pigments can be used as a dye in cosmetics

Chondria

Ch. species have been investigated several times during the last years. The active substances may be of great interest for the pharmacy.

Ch. armata

Coasts of southern Japan and South East Asia. The Japanese term for this alga is "domoi". A crystalline acid component was named domoic acid (derivative of proline).
Domoic acid has an anthelmintic effect to roundworm and pinworm and is also effective against flies. Other constituents of Ch. armata are l-citrulline, D-aspartic acid, L-glutamic acid, beta-amino-glutaric acid and hypoxantine (81, 316, 317, 318, 319, 323, 324, 358).

Ch. crassicaulis

Chondrine is an amino acid isolated from this red alga (170).

Ch. littoralis

Chonalgin is an antibiotic substance of this alga, specifically against Escherichia coli and Candida albicans (189, 191).

Ch. oppositiclada

The alga contains chondriol and cycloeudesmol. Cycloeudesmol is effective against Staphylococcus aureus and Candida albicans (97, 98).

Chondria sanguinea and Ch. vermicularis

are used in Brazil as anthelmintic agents and for indigestion.

Chondrus crispus

Ch. crispus is often found in a mixture with Gigartina stellata. The algae are known as "Carrageen" or "Irish Moss". Carrageen has been used for a long time as a medicinal drug as a mucilage against diarrhoea, dysentery, gastric ulcers and as a component of numerous health teas, for example for colds. The drug contains 50 - 70% slime (the carragan) with at least five polysaccharides, 7 - 10% proteins, numerous amino acids, vitamins, mineral salts and trace elements.

Corallina

Corallina species were used as a vermifuge towards the end of the 18th century as "Muscus corallinus". In former times external absorbent for burns and internal absorbent for chronic gastritis and diarrhoea.

C. rubens

The protein part has been extracted and fractionated. Pharmacological studies have shown that fraction I has lipolytic effect while fraction III has revealed hypoglycemic activity. Fraction II has exhibited fibrinolytic activity (117, 119).

Delesseria sanguinea

In spring this species contains a blood anticoagulant. This substance is similar to heparin in its chemical and physilogical properties (95).

Digenea simplex

The alga is of significance because of its anthelmintic character, esp. in the treatment of ascariasis. It contains the vermifugal components L-alpha-kainic acid and its stereoisomer L-alpha-allokainic acid (see Amino acids) (48, 315, 358).

In South China D. simplex forms the basic stock of "TSE Ko-Tsoi" taken by adults and children as an anthelmintic (74).

Eucheuma

In the Philippines a mixture of E. species and different drugs is prepared as a cough remedy.

Gelidium

G. cartilagineum is used in Japan for colds and scrofula in pediatric medical science.

G. latifolium
The antilipemic properties of the alga have been studied. It was concluded that heparin-induced lipase has the same characteristics as the lipase released by the substance obtained from G. latifolium (Greville) (38, 113).

Gracilaria confervoides (=G. verrucosa)
The antilipemic properties of this alga have been studied. It was concluded that heparin-induced lipase has the same characteristics as the lipase released by the substance obtained from G. confervoides (38, 113).

Grateloupia
Some G. species are used as anthelmintics (106).

Hypnea
The effects of H. japonica extract on bone metabolism in rat and the isolation of a collagenolytic enzyme by the implantation of algal polysaccharides under the rat skin have been reported. The result have shown that the intravenous administration of a sulfated galactan causes an increase in the calcium uptake of the bone (135, 136).
H. musciformis
Because of its anthelmintic properties this alga is used as a vermifuge, esp. in Greece and Turkey, but also in Indonesia.
H. nidifica
Hawaiian Islands. This species is used for stomach treatment.

Iridaea flaccida (=I. laminaroides)
An aqueous extract could gain significance in preventing the coagulation of blood.

Lithothamnion species
Calcaricus algae are used as a neutralizer and as a filter element to remove undesirable components from water, such as poisonous elements, radioactive substances etc.
Constituent of pharmaceutical and cosmetic preparations.

Phyllophora nervosa
The antilipemic and anticoagulant properties of Ph. nervosa

occurring on the Turkish coasts have been studied. One substance isolated from this alga shows antilipemic but no anticoagulant properties (109, 113).

Plocamium coccineum

The red pigment phycoerythrin was used in classical times for cosmetics.

Polysiphonia subulifera

Substances obtained from this alga are effective in increasing serum lipolytic activity (37, 113).

Porphyra

P. atropurpurea is used medically for poultices and cataplasms.
P. columbina: In New Zealand the fermented alga is used by the Maoris as an aperient (67).

Rhodymenia indica

In East Asia this alga is used as a vermifuge.

Sphaerococcus cartilagineus

This alga is used as a medicament in China.

Additional literature to "Marine Algae as Medicinal Drugs and in Folk Medicine" (3, 5, 6, 10, 12, 25, 28, 57, 64, 88, 138, 140, 302, 314, 362).

PRODUCTS FROM MARINE ALGAE

The most important industrial products of marine algae are the phycocolloids of Rhodophyta and Phaeophyta. They are, above all, necessary auxiliaries in numerous industries as emulsifying agents, gelling agents, stabilizers, suspension agents and thickeners.

In this connection a comment by Czetsch Lindenwald (80) from his paper on "Products of Algae in Pharmacy" is worth noting: " We consider many products as indifferent auxiliary substances, as their colloid-chemical behaviour influences the absorption of drugs, they therefore become active agents and should be judged as such."

The term "phycocolloid" has been defined by Tseng as the polysaccharide-complexes from Rhodophyta or Red Algae and Phaeophyta or Brown Algae, which form colloidal systems when dispersed in water. But this collective name has become more and more specific. Water-soluble polysaccharides form the main part of the carbohydrates of algae. The chemistry and enzymology of marine algal polysaccharides are described by Percival and Mc-Dowell (248).

A g a r and c a r r a g e e n a n are the most important polysaccharides of the Rhodophyta.

Agar

Raw materials

Only the most important Rhodophyta, used for the production of agar, are to be mentioned here:

Acanthopeltis japonica - East-Asian coastal areas,
Ahnfeltia plicata - Coasts of the White Sea, Sakhalin, Korea and the Japanese islands,
Gelidium amansii - Coasts of Japan, Korea, China, phycocolloid content 25 - 30% of the dry weight,
G. cartilagineum - Coasts of California, Mexico, South Africa, phycocolloid content 40 - 45%,
G. corneum - Atlantic coasts of Spain, Portugal, Morocco,
G. coulteri - utilized in the U.S.A.,
G. japonicum - Coasts of Japan and Korea,
G. lingulatum - Coasts of Chile, phycocolloid content 18 - 23%,
G. nudifrons - utilized in the U.S.A.,
G. pacificum - Coasts of Japan,
G. spinulosum - Coasts of Morocco, phycocolloid content appr. 33%,

Gelidium pusillum from the Saurashtra coast (India) can be considered as a source for agar production. The alga yields 24 %. Culture studies have been carried out (187).

Gűzmán-del Prőo (124) reported: "Gelidium robustum es una Rhodophycea de valor comercial en Baja California, Mexico, que ha venido siendo objeto de estudio desde 1967." G. robustum supplies the greatest quantity of raw material for agar production in U.S.A. A resource management study has been published by Barilotti and Silverthorne (50). Environmental factors have been studied (51). G. robustum is abundantly found from Central California to Central Baja California (283), see "Cultivation of Marine Algae".

Gelidium floridanum, Gracilaria species, esp. G. debilis, Pterocladia capillacea are found on Brazilian coasts (241). Gelidium species and other agarophytes are cultivated esp. on the coasts of Japan.

Gracilaria - The most important of more than a hundred species are G. confervoides (coasts of the Indian Ocean, East Asia and the American continent) and G. lichenoides (coasts of the Indian Ocean). On these and other G. species numerous studies have recently appeared. The amount of phycocolloid content apparently depends on the effects of environmental factors, the time of harvesting etc. The figures differ between 15 and 30 % of the dry weight. Some examples: Yield and quality of agar from cultivated Gracilaria edulis on Indian coasts were determined by Thomas and Krishnamurty (334). Matsuhashi and Hayashi (195) reported on agar producing a firm gel, from G. foliifera harvested along the Florida West Coast. Seasonal, chemical and reproductive changes in G. foliifera from New Hampshire have been studied (247).

Pterocladia lucida, Australia, Tasmania, New Zealand, Japan is also used as agar raw material. P. capillacea is similarly utilized.

Numerous authors have given Indian *Gracilaria* and *Gelidiella* species their particular attention, see Gopal, contribution in this monograph.

Up to now less known species, too, have been investigated to guarantee the agar raw material basis. These raw materials are often exploited in mixture with other red algae.

Countries of Production

Europe: Portugal, Spain,- Africa: Morocco, - Asia: Japan, India, Korea, USSR, - America: U.S.A., Chile are the most important countries of production.

Definition

Agar, in former times usually called "agar-agar", is a gel-forming extract of certain Rhodophyta, the agarophytes. Agaroids are the extracts of other Rhodophyta, the agaroidophytes, forming, in contrast to agar, softer gels. The name "agar-agar" is of Malayan origin and was originally the term of the *Eucheuma*.

Structure

Agar varies much in chemical and physical properties depending on the raw material. Selby and Selby (273) report in detail on raw materials, processing, grade and type specifications, structure, properties and uses. For further information there is a voluminous special literature. At the Vth International Seaweed Symposium 1965 in Halifax Araki (45) reported on studies on the polysaccharides of agarophytes. Examinations of agar have been shown to be mixtures of agarose and agaropectin in varying proportions depending upon their sources.

In 1971 at Sapporo Yaphe and Duckworth (356) referred to the studies by Araki on the structural analysis of agarose and agaropectin and the relationship between structures and biological properties of agars. Young, K. *et al*. (360) also reported there on enzymic hydrolysis of agar and properties of bacterial agarases. Nisizawa and Sasaki (233) present, in "Cell wall composition of algae", a synopsis in tabulated form about percentage composition of agarose and agaropectins from various red algae. The contents differ from species to species.

See also the contribution in this monograph: "Enzymic and N.M.R. Spectroscopic Analysis of Agar-type Polysaccharides" by Bhattacharjee and Yaphe.

Utilization

The most important property of agar is its high melting point. In pharmacy agar is used as a laxative for chronic constipation, often with the addition of anthraquinone drugs, as drug vehicle and as a substratum for bacterial cultures agar also plays an important role. Agar also serves as a stabilizer for emulsions, as a constituent of ointments, lotions etc. Hawkins and O'Neill (130) report that granuloma will arise after injection of agar. According to Gerber _et al._ (104) agar and also carrageenan protect chicken embryos against infection by influenza B and mumps-virus. An interesting field in the use of agar is the inoculation of corn in the cultivation of _Claviceps purpurea_. Agar is also used in dentistry. In laboratory practice agar is used as an auxiliary means for numerous investigations. For agar in food technology and in industry see Levring - Hoppe - Schmid (25).

Agar is the oldest and perhaps the best known of the algal products. It is therefore understandable that a number of detailed works about this product were published some years ago (1, 14, 27, 30, 248, 272, 273, 305).

At the International Seaweed Symposia the topic of agar has been dealt with in many lectures.

Agarose

The use of agarose in immunology is of great interest. Agarose gels have proved more useful than unfractionated agar gels because of the low sulphate content and because it gives a clear gel (52, 57, 248). Guiseley (123) reported on viscometric determination of agarose. Virologists and bacteriologists need products with a low gelling temperature. Further agarose research was reported in 1974 in Bangor (86, 87).

Carrageenan

Raw materials

The most important raw materials for the production of carrageenan, carrageenates and their derivates are *Chondrus crispus* and various *Gigartina* species, particularly *G. stellata* but also *Eucheuma* and *Hypnea* species. Moreover, *Chondrococcus hornemannii*, *Halymenia venusta*, *Laurencia papillosa*, *Sarconema filiforme*, and *Endocladia*-, certain *Gelidium*-, *Gymnogongrus*-, *Rhodoglossum*-, *Rissoella*-, *Yatabella*- species and other red algae are utilized as raw materials (see also 59).

Chondrus

Ch. crispus is often found in a mixture with *Gigartina stellata*. The two species are mostly harvested and processed jointly and are known in the trade as "Carrageen" or "Irish Moss". *Ch. crispus* is found in many forms.

At the beginning of the 19th century the raw material was imported into the U.S.A. from Europe. In 1835, however, abundant occurences were discovered on rocky coasts from Massachusetts to Newfoundland. Stoloff (306) has produced a comprehensive compilation entitled "Irish Moss - From an Art to an Industry" in which he discusses the raw materials, harvesting, drying and bleaching, extraction and utilization. MacFarlane (185) has published a synopsis of *Chondrus crispus* with many details about this important alga and Taylor (333) has written on the biology and ecology of *Ch. crispus*.

In the meantime, further knowledge about this alga was gained by working in particular with the origin and the properties of carrageenan. Besides kappa- and lambda-carrageenan some further carrageenan-fractions have been found.

Ecological studies in connexion with potential mariculture of *Chondrus crispus* have been discussed by Mathieson and Tveter (194). Differential gelling properties occur in the tetrasporic and gametophytic stages. The non-gelling fraction lambda-carrageenan occurs in tetrasporic plants. McCandless and Craigie (200) have investigated the carrageenan production and Simpson *et al*. (284) the effect of pH on the growth and production of this substance. A method for vegetative propagation in tanks of flowing seawater has been developed.

Other *Chondrus* species, such as *Ch. canaliculatus* from the coasts of Japan can also be regarded as raw materials for carrageenan.

Gigartina

Gigartina is a widespread genus (appr. 90 species) and a number of them can be used as raw materials for carrageenan, for instance *G. stellata*. The content of kappa-carrageenan is comparable to that of *Chondrus crispus*.

G. acicularis, *G. asperifolia*, *G. canaliculata*, *G. chamissoi*, *G. cristata*, *G. decipiens*, *G. pistillata*, *G. radula*, *G. serrata*, *G. skottsbergii*, *G. stiriata* and others also contain carrageenans (25, 308).

Eucheuma

Eucheuma species occur in South East Asian coastal areas, and on the East African coast. *E. muricatum* is referred to in the trade as "Singapore Weed", *E. serra* and *E. cottonii* are sold as "Zanzibar Weed".

"Eucheuman" is the extract from these two groups. The products are regarded as agaroids.

The raw materials are often used as an additional agar raw material or for the production of carrageenan (25, 291).

In range of utilization "eucheuman" falls between agar and carrageenan. The applications are mainly watery and milk-containing media in pharmacy, cosmetic industry and food technology.

Recently numerous Eucheuma species have been investigated in order to find new raw materials for the increasing demand of carrageenans. Cheyney and Dawes (71) reported on ecological studies of Eucheuma along the Florida coasts, esp. of E. nudum (82).

Five forms of carrageenan are known: kappa-, lambda-, my-, ypsilon- and jota-carrageenan. These forms differ in the levels of sulfation and the ratio of galactose to 3,6-anhydrolactose but also in their physical properties (83). The Pacific forms E. cottonii, E. procrusteanum, E. serra, E. speciosum, E. striatum contain pure kappa-carrageenan. E. arnoldii, E. serra, E. spinosum may contain pure jota-carrageenan. E. odontophorum contains a mixture of kappa- and jota-carrageenan. E. uncinatum contains a hybrid form of jota- and ypsilon-carrageenan. E. gelidium, E. isiforme, E. nudum of the Caribbean coasts contain a "deviant" form of jota-carrageenan.

Dawes et al. (84, 85) have reported on physiological and biochemical studies on the jota-carrageenan producing Eucheuma uncinatum from the Gulf of California. The "typical" jota-carrageenan of this alga is different from the "deviant" jota-carrageenan found in E. isiforme, E. nudum, E. gelidium, and E. acanthocladum from Florida where a lower sulfation level is apparent.

The carrageenan yields of Tanzanian Eucheuma species have been determined by Mshigeni and Semesi (220).

E. spinosum contains appr. 72.8% with absorption peaks of jota-carrageenan. *E. striatum* appr. 69.0% do. of kappa-carrageenan. *E. platycladum* appr. 65.0% do. akin to jota-carrageenan. *E. okamurai* appr. 58.0% do. of kappa-carrageenan and *E. speciosum* f. *mauritianum* 54.0% do. akin to jota-carrageenan.

Because of the increased demand for carrageenan, some *Eucheuma* species are produced by farming. In 1968 at the 13th Session of the Indo-Pacific Fisheries Council the problem of cultivation of *E. muricatum* (=*E. spinosum*) and *E. edule* was discussed. The first experiments have given encouraging results.

Doty and Alvarez (93) reported on the productivity of *Eucheuma*-farms. The clean anhydrous yield from *E. striatum* contained about 50.0% kappa-carrageenan.

In the Philippines about 700 farms existed in 1973. They exported more than 100 tons of dry *Eucheuma* per month.

Ricohermoso and Deveau (261) report that there are now over 1000 *Eucheuma*-farms in this region, producing more than 300 tons per month for the world market.

Doty and Santos (94) report on studies to obtain comparative morphological and gel-chemical information on some fourteen *Eucheuma* species.

Hypnea

Extracts of *Hypnea* species, esp. *H. musciformis* are known as "hypnean" which forms gels of a great firmness. It has often been considered to be a special kind of agar but very little is known about the chemical structure of it except that it is somewhat similar to carrageenan containing kappa- and lambda-fractions (12, 25, 291).

Recently *Hypnea* species have been more precisely investigated. Today they are important raw materials for the production of carrageenans.

A study has been made of the preparation and the properties of the phycocolloid from H. musciformis from the Gujarat Coast (India). Rama Rao and Krishnamurthy (256) reported at that time that it would seem that the alga could be used as a source of agar. But the processing of the seaweed should be modified. Studies on growth cycle and phycocolloid content in H. musciformis (appr. 24%) have been carried out. It would be advisable to harvest the alga from January to March (257). H. musciformis, West Coast of India, has been investigated by Rama Rao (258). The seasonal variation in phycocolloid content has been studied by Rama Rao and Krishnamurthy (259).

Mollion (211) published a preliminary survey of the species of Hypnea from Senegal: H. musciformis, H. cervicornis and H. ceramioides. These algae are a raw material for phycocolloids which bears resemblance to carrageenan of Chondrus crispus with kappa- and lambda-fractions. H. musciformis occurs in abundance on the Brazilian coastline. See Oliveira Filho (241) Mshigeni (214 - 218) reports on development studies in H. cervicornis and H. chordacea and on the possibilities of field cultivation of Hypnea. The author gives in his introduction an interesting review of the problems of the increased industrial demand for carrageenan as a gelling, stabilizing, thickening or emulsifying agent in the pharmaceutical, modern food and cosmetics industries.

Studies have been conducted by Mshigeni on seasonal changes in the standing crops of H. cervicornis, H. chordacea and H. nidifica in Hawaii (219, 221). An understanding of seasonal changes is an important aspect in the case of economic algae for deciding what time of the year commercial harvesting could be undertaken most profitably. Rama Rao (258) has studied in the South Indian islands the H. valentiae. The seasonal variation in phycocolloid content has been studied by Rama Rao and Krishnamurty (259). Cultural studies of H. valentiae have been realized by Mshigeni and Lorri (223). See also Mshigeni and Mshigeni and Mziray, contributions about Hypnea in this monograph. H. cervicornis, H. nidifica, Chondrus and Gymnogongrus species, occurring in Hawaii, have been investigated for their carrageenan content by Santos and Doty (268).

Recently a number of further algae have been investigated for their carrageenan content. These works are of particular interest.

The phycocolloid from Chondrococcus hornemannii, shores of Tanzania, has been investigated by Semesi and Mshigeni (275). A yield of appr. 45% carrageenan of the dry weight akin to lambda-carrageenan was obtained. Another paper by Mshigeni (222) reported on the fine structure of the cell wall of Ch. hornemannii.

The phycocolloid from Halymenia venusta, widespread species on the shores of East Africa, has been studied by Semesi and Mshigeni (274). The phycocolloid has been found to be a carrageenan which is more closely related to lambda- than kappa-carrageenan. A total phycocolloid content of appr. 60% on dry weight was obtained.

The genus Laurencia seems to be also of interest as a possible source of phycocolloids. The content and nature of the phycocolloid from L. papillosa of Tanzanian shores have been investigated. A yield of appr. 33 % with peaks akin to lambda-carrageenan was obtained on a dry weight basis (224).

The phycocolloid from Sarconema filiforme from Tanzania has been investigated by Semesi and Mshigeni (275). A yield of appr. 35% of the dry weight was obtained. The phycocolloid showed absorption peaks characteristic of jota-carrageenan.

Countries of Production of Carrageenan

The U.S.A. is the greatest producer. Canada has significant raw material resources for the production of this phycocolloid. France, Great Britain and Norway also produce and export carrageenan.

Definition and Structure

Carrageenan is the unchanged extract of carrageenophytes. Carrageenates are the defined salts of carrageenic acid. Carrageenan is a highly sulphated hydrocolloid. The chemical constitution, the fractination etc. have been investigated by many authors. Stoloff (307) gives a summarizing report in "Industrial Gums", which, of course, has to be completed by new knowledge. Carrageenan, esp. of Chondrus crispus and some other red algae, can be separated into two fractions, appr. 40% kappa-carrageenan and appr. 60% lambda-carrageenan. The sulphate content in kappa-carrageenan is 23 - 28 % and in lambda-carrageenan 24 - 33 %. According to the investigations of Springer and Middendorf (290) the kappa-carrageenan fraction corresponds to a hot water extract and the lambda-carrageenan fraction corresponds to a cold water extract.

Schmitt (298) discusses the molecular formulas and physiochemical properties of carrageenans and their importance to practical application. Gel strength and gelation temperature can be varied between wide limits to suit the purpose, for which the product is to be used. The products are dealt in various grades of viscosity, gelating and non-gelating qualities.

Studies of carrageenan synthesis and biochemical analysis are reported in McCandless and Richter (198), McCandless (199) and Craigie and Wong (79). Further investigations of carrageenan were discussed at the Symposium in Bangor, 1974; for instance Gordon-Mills and McCandless (108) reported on kappa- and lambda-carrageenans in cell walls on

Chondrus crispus, and Bowtle and Anderson (65) on detection and determination of carrageenans in biological media. Besides kappa- and lambda-carrageenan further fractions are to be found, for example jota- and ypsilon-carrageenan (see raw materials). The determination of jota-carrageenan was described by Anderson and Bowtle (40).
My-carrageenan is presumably the precursor of kappa-carrageenan.

Utilization
Carrageenans are frequently used in the pharmaceutical industries as emulsifiers, for example in cod liver oil emulsions, as granulation and binding agents for example in tablets, elexiers, cough drops, syrups etc. Interesting papers about the application of carrageenans in ulcer therapy have been published. Animal experiments and investigations of the effects of decomposed carrageenan for acute gastric ulceration with positive effects have been described by Anderson and Soman (39). The molecular weight distribution of carrageenans was investigated by Stanley and Renn (304). It has been found out, that highly depolymerized jota-carrageenan, used as a drug in gastric ulcer therapy, may have undersirable physiological side effects.

Hawkins and Leonard (131) reported on antithrombotic activity of carrageenan in human blood. Schneider (299) has suggested the importance of Chondrus as regards carrageenan for the therapy of disease of the blood vessels.

Studies on carrageenan have been reported by Tanaka et al. (330) with respect to the binding reactions with lead and other heavy metal ions. Lead was found to be one of the metals bound most effectively by carrageenan and fucoidan. Paskins-Hurlburt, Tanaka and Skoryna (244) also discuss carrageenan and the binding of lead.

Carrageenans are used in the cosmetic industry as a stabilizer, suspending and thickening agent. The products are often used in the manufacture of skin ointments, creams, lotions, toothpastes, shaving creams, hair tonics, foam stabilizers, sun protection oils etc.

There are numerous possibilities of application for carrageenan in food technology and a large number of studies have been published on them. Technical qualities moreover are used in quite a number of other industries, for example in the leather industry, paper industry, textile industry etc.

Further Products of Rhodophyta

Ceylon Moss

Definition: washed, bleached and dried *Gracilaria* species, as *G. lichenoides*, *G. confervoides* (=*G. verrucosa*.)
Utilization: as a pharmaceutical product used in the treatment of colds; in Sri Lanka, India, China gelating material for food.

Chinese Moss

Definition: washed, bleached and dried *Gracilaria confervoides*.
Utilization: gelation and thickening agent esp. in China.

Dulsan

Definition: polysaccharides without gel-forming properties, isolated from *Rhodymenia palmata* (291).
Utilization: it appears more expedient to use the whole alga with all its constituents as amino acids, vitamins etc.

Funoran

Raw materials: *Gloiopeltis furcata*, *G. tenax*, *G. complanata*, *Ahnfeltia concinna*, *Dumontia incrassata*, *Grateloupia* species, *Gymnogongrus pinnulata*, *Iridaea* species
Definition: in contrast to other phycocolloids funoran is not an extract, but rather the total alga which is made

soluble by special treatment. Funoran is a sulphated polysaccharide. Fractination and structural investigation were made by Hirose and Watanabe (132).
Utilization: in the pharmaceutical industry funoran has been proposed as a blood anticoagulant. Fraction I of Gloiopeltis furcata showed considerable activity in inhibiting the growth of Sarcoma 180 tumor - Hirase and Watanabe (132). Funoran is also used in cosmetic industries.

Furcellaran

Definition: the raw material basis of furcellaran, also known as "Danish Agar" is Furcellaria fastigiata, which is harvested in large quantities, above all, in the Kattegatt. Furcellaran is a galactoglycane sulphate. Detailed descriptions of the physical and chemical properties have been given by Christiansen (75) and Schachat and Glicksman (291).
Utilization: in dietetic preparations, in particular in baby and diabetic food, in the production of milk and ice cream. In the cosmetic industry furcellaran is used for example in the manufacture of tooth paste.

Ginnansô

Definition: the raw materials of ginnansô are Iridaea cornucopiae, Rhodoglossum pulchrum, Neodilsea yendoana, Turnerella mertensiana and some Gigartina species. Ginnansô belongs to the group of "algal-slimes", similar to funoran.
Utilization: Ginnansô is used in East Asia like funoran, on account of its adhesive properties.

Gracilaria Gum

Raw materials are Gracilaria species.
Definition: The products made from the various G. species differ in their gelling temperature and gel strength. Schachat and Glicksman reported on gelling properties of common G. species. It appears possible to define various products more precisely, so that special fields of application can be found (291).
Utilization: these softer gelling materials are used particu-

larly wherever harder gelling products, such as agar from Gelidium species, are unsuitable.

Gulaman and Gulaman-Dagat

Definition: Gulaman is the Philippine name for Agardhiella tenera and incidentally also for Gracilaria lichenoides. Gulaman-Dagat comes from Indonesia.

Utilization: the materials are used for making jelly products.

Iridophycan

The most important raw material for the production of iridophycan is Iridaea flaccida on the Pacific coast. Numerous further I. species can be also regarded as possible raw materials (291).

Definition: Iridophycan can be compared with carrageenan in its properties. According to Yaphe (355) appr. 36% of the mucilage are of the type of kappa-carrageenan. Matsuhiro and Zanlungo (196) report that the polysaccharide of I. laminarioides approaches the composition and the properties of a carrageenan.

Utilization: Iridophycan has increased in commercial significance, particularly in its use as cold-mix stabilizer. In the pharmaceutical industry it could gain significance as an aqueous extract to prevent coagulation of blood.

Maërl

Raw materials are Lithothamnion calcareum (Phymatolithon calcareum), L. corallioides and other L. species - Cabioch (69), Adey and McKibbin (35), Oliveira (241).

Definition: the algae are entirely calcified. The constituents of Maërl are appr. 80% calcium carbonate, 10 - 12% magnesium carbonate, many trace elements, variable quantities of sand and shells in the raw material - Blunden (60, 61). The calcium content of maërl varied from 25.5 to 33.3%, magnesium content from 1.7 to 3.3%, major component was found to be calcite with magnesium carbonate, 10 to 15% aragonite.

Utilization: cleaned material is utilized in pharmaceutical

and cosmetic products, as a neutralizer for the filtration of water to remove poisonous elements, radioactive substances etc.

Phyllophoran

The raw materials of phyllophoran are *Phyllophora rubens* and *Ph. nervosa*.

Definition: Phyllophoran is an agaroid with 13 to 15% mineral constituents. It is completely soluble in hot water.
Utilization: in general the range of application of phyllophoran is identical to that of agar and agaroids in food technology. Phyllophoran can be used in bacteriological media - Güven *et al.* (111, 115).

Porphyran

Definition: Porphyran is the substance extracted from species of *Porphyra* by means of hot water. Content and composition differ in the individual species. Appr. 30% of porphyran has been obtained from *P. capensis* - Batey *et al.* (54).
Utilization: in human nutrition (see "nori" - products). Schachat and Glicksman (291) consider that porphyran will be of importance because of its interesting structure.

Products of the Phaeophyta

Alginic Acid and Alginates

Raw materials

The most important raw materials for the production of alginic acid and alginates are *Laminaria* and *Macrocystis* species. Particularly the following *Laminaria* species play a role:

Laminaria hyperborea

Distribution, growth, density, possibilities of culture, seasonal fluctuations of the constituents, harvesting and drying are reported in detail in the special literature, see also Levring (177). The seasonal growth and cultivation research of *L. hyperborea* has been studied in the Biologische

Anstalt Helgoland (181, 182).

The content of constituents is partly subjected to enormous fluctuations. The content of alginic acid differs in the fronds and stipes according to the season and the descent. Fronds contain in spring appr. 30% of dry weight, in autumn appr. 20%. Stipes contain 30 - 37 %. The content of laminaran is practically zero, 18 - 36 % in September. The content of mannitol is appr. 5 % in May, appr. 18 % in July, depending on relative depth. At 3 - 4 m it can be appr. 18 %, at 6 m depths up to 36 %. With laminaran the opposite holds: content declines with increasing depth.

We shall not enter here into particulars of the other constituents of *Laminaria* such as proteins, amino acids, vitamins, mineral substances etc. A review of the biological data available on this species was prepared by Kain (156).

Larsen and Haug (172) have determined the content of alginic acid, iodine, protein and ash in the periferal tissue, in outer and inner cortex and in the medulla of *L. hyperborea* stipes. The increasing amounts of alginic acid towards the inner parts of the stipe is in agreement with the assumption that alginic acid is present in the primary membranes. On the French channel coast the growth of *L. hyperborea* and *L. ochroleuca* was studied by Sheppard *et al.* (277). The findings are discussed in terms of several environmental factors, particularly strong currents, wave action and grazing.

Laminaria digitata
contains on an average 15 - 40% of the dry weight alginic acid. Gayral and Cosson (103) have published an "Exposé synoptique des données biologiques sur la Laminaire digitée". A study by Cardinal and Breton-Provencher (70) deals with a three-year analysis of variations in alginic acid content in *L. digitata*,

L. longicruris, *L. saccharina* and *Alaria esculenta* in maritime regions of the St. Lawrence estuary.

Laminaria saccharina
contains 15 - 35% alginic acid.

Laminaria japonica
Coasts of Japan (esp. of Hokkaido) Korea, Siberia, China. The alga is of great economic importance. Hasegawa (127) reported on forced cultivation of Japanese *Laminarias*. *L. japonica* contains 17.1 - 32% alginic acid, 8.4 - 28.4% mannitol. The alga has a very high sugar content. Cheng (72) reported on the production of kelp-products in China, esp. alginates. Johnston (153) reported on chemical analysis of some Japanese *Laminaria* species, as *L. angustata*, *L. japonica*, and *L. religiosa*. Information is presented esp. on the nutritionally important inorganic element content.

The alginic acid content of the Fucaceae of the St. Lawrence estuary was measured by Bérard-Therriault and Cardinal (55) at different periods of the year with respect to their habitat and their level in the littoral zone. The alginic acid content varies greatly for each species in relation to the parameters considered.

Macrocystis
Extensive fields of the "Giant Kelp" along Southern California and North-Western Mexico have been investigated by several authors. The alga occurs on the coasts of South America (Peru, Chile, Argentina), Australia, New Zealand and also in other parts of the world.

M. pyrifera is regarded to be the largest marine alga and one of the most important raw materials for alginic acid and alginates, as well as for other brown alga products. Extensive fields, often exclusively *Macrocystis*, permit mechanical harvesting methods, thereby giving special significance to this alga. M. pyrifera contains 14 - 21 % alginic acid. The other constituents have also been investigated in detail, particularly those of M. angustifolia and M. integrifolia. Growing, harvesting and utilization of the "Giant Kelp" from the coast of California is reported in Commercial Fisheries Review (77).

The importance of Macrocystis in particular was demonstrated in Santa Barbara, August 1977, in a special symposium with new contributions concerning georgraphical distribution, growth and reproduction, ecology, harvesting, cultivation, hybridization and so on.

Further brown algae for alginate production are the alginophytes Alaria species, Cystoseira species, Durvillea species, Ecklonia maxima, Egregia species, Ishige okamurai, Lessonia species, Nereocystis luetkeana, Pelagophycus porra, Sargassum species, Turbinaria species and such "rockweeds" as Ascophyllum nodosum. Most of the Phaeophyta contain alginic acid. The brown algae of numerous coastal areas have been examined and new sources of raw material utilized. Here some examples:

Studies of the chemical composition of Cystoseira barbata by Güven and Bergişade (116) brought a yield of 27% alginic acid. In this paper studies are cited which indicate for C. species a content of 17 - 44% alginic acid.

Alginic acid concentrations in Durvillea antarctica and D. willana, New Zealand and D. potatorum, Australia, have been investigated (288).

The possibility of cultivating Hormophysa triquetra to augment the alginophyte resources has been studied by Bhanderi and Trivedi (58). H. triquetra was collected on Indian coasts. This alga is reported to yield about 18% algin.

The effects of drying on the alginate component of Nereocystis *luetkeana* and *Macrocystis integrifolia* have been researched by Whyte *et al*. (349).

An exploitable resource for alginate production is the *Sargassum* species on the southern coast of Brazil, esp. *S. cymosum*, *S. stenophyllum* and *S. vulgare*. The mean value of sodium alginate yield amounts to about 20% on a basis of dry weight. Oliveira Filho and de Paula (242). *Sargassum* from India represents several species, mainly *S. tenerrimum*, *S. wrightii*, *S. myreocystum*, *S. cinereum* var. *berberifolia*. Average yield 25% alginic acid on air dried weeds (250).

Alginic acid content and mannitol content in *Stoechospermum marginatum* from Indian coasts have been investigated by Kalimuthu *et al*. (158). The alginic acid content was at its maximum from September to December 17.73 - 23.76%. The highest mannitol content was in October with 2.74%.

Kaliaperumal and Kalimuthu (157) investigated the changes in growth, reproduction, alginic acid and mannitol contents of *Turbinaria decurrens*. The yield of alginic acid varies from 16.3 to 26.3% and the estimated mannitol content from 1.5 to 8.7%. The alga should be harvested between December and February for extraction of alginic acid.

Countries of Production

The most important are the U.S.A., Great Britain, France, Japan, Norway, the USSR. But some other countries also produce alginic acid and alginates.

Definition

Alginic acid is an acid membrane mucilage regarded as a typical constituent of brown algae. Alginic acid is the base product of numerous economically interesting alginates and alginic acid derivates. The term "algin" is used as a collective name for alginic acid and the salts of alginic acid, the alginates, but also as a trade designation for sodium alginate. Alginic acid has a good reactivity and can be mixed with alkali, ammonium, amines, and organic bases to form water soluble and water insoluble salts with heavy metals and earth alkalis. The most important soluble alginates are sodium alginates, potassium alginates and ammonium alginates. The best known insoluble alginate is calcium alginate. Alginic acid derivates are alginic acid ester, amino acid compounds, amides, complex compounds, depolymerisates, mixed products. Patents have been published for the production of numerous alginates and derivates among others by Maass (183).

Structure

Many workers have studied the structure and biosynthesis of alginic acid. Detailed descriptions of investigations of the constitution of this important material have been given in handbooks, for instance Humm (13), Smith and Montgomery (29), McNeely (203), Maass (183), Percival and McDowell (248) and in numerous papers presented at the different Seaweed Symposia.

According to Haug and Larsen (129) alginic acid is not a single compound but a complex of polysaccharides with a varying composition of guluronic and mannuronic acid. See also Fischer and Dörfel (99).

Nisizawa and co-workers presented a paper on the fine structure of alginic acid in Sapporo 1971 (230). Larsen and Haug reported on biosynthesis in the same symposium (174).

Studies of lyase activities have been carried out above all by Japanese scientists. Kashiwabara _et al._ (162), Nisizawa _et al._ (231).

The problem of the mechanisms of the translocation of inorganic substances has been studied on _Laminaria digitata_ by Floc'h and Penot (100). It depends on the protein metabolism.

A short but excellent survey of the ever increasing knowledge of cell wall composition of algae was published by Nisizawa and Sasaki 1975, in "Advance of Phycology in Japan" (233).

Utilization

Alginates in pharmaceutical industry: Alginates of a very high degree of purity have been used as an excipient in pharmacy and their application has been very wide and is increasing. They are used as emulsifiers in watery emulsions with fats, oils and waxes, as well as suspension agents or stabilizers. Alginates serve as film forming, as jelling and as protective colloids. They are used as fillers in the manufacturing of tablets, pills, and lozenges, as vehicles and desintegrating agents. Alginic acid is used as tablet swelling agent, as it has a very high swelling strength.

The base of many ointments is made with alginates. As they are resorbable, also form the base of various dressing materials which can be left in the wounds. An alginate gauze is used as a blood-stopping plaster.

Vogt (340) published a paper on the application of alginic acid and alginates in pharmaceutical technology. Schwandt (300) has reported on alginates in ointment bases. Lesser (175), Schwimmer and Schwimmer (302), Bolliger (63) and Czetsch Lindenwald (80) have published detailed results on the studies of properties of alginic acid and alginates and their use in pharmaceutical industry.

Sodium alginate is reported to be a useful adjuvant in immunization against two strains of inflenza virus. It is also found effective in diminishing hyper calciuria in urolithiasis and is found useful in the treatment of esophagitis. Bhakuni and Silva (57). Sodium alginate is also used in slimming agents. The alginate forms a jelly in the stomach, which produces a feeling of satiety. Several reports have appeared in the pharmaceutical literature on the use of sodium alginate as a plasma substituting colloid.

Calcium alginate is used for the production of capsules. Calcium alginate is also reputed to be a hemostatic agent which stimulates the clotting of blood and is subsequently absorbed in the tissue. Bhakuni and Silva (57). McDowell (201) reported on the reactions of propylen-glycol-alginate esp. for capsule-preparations, pellets and granulates.

Alginates in the cosmetic, soap and detergent industries: Alginates of different viscosity are an important auxiliary agent in cosmetic industry, in the soap and detergent sector. Alginates serve above all as thickening and dispersing agents in ointments, creams, jellies, liquid emulsions, lotions, tooth paste, compact powder, bathing preparations, as well as an additive in numerous hair cosmetics. Alginates have a very good compatibility with other products, such as starches, gums, pectin etc. Alginates are used as foam stabilizers in the soap and detergent industry. They influence the hardness of water favourably. The ability of alginates to form films on the skin is made use of in skin ointments and creams. Alcohol containing preparations make use of amino alginates, which, in contrast to the sodium alginates, are compatible with alcohol.

Alginates in dental technology: Alginates are used in dental technology for making denture mouldings, as well as in denture fixatives. See Ahing and Hébert "The Use of Alginates in Dentistry", contribution in this monograph.

A new field of application of marine algae in an important sphere of public health was described in Halifax at the Vth International Seaweed Symposium, 1965, by Skoryna *et al.* (285): the use of alginic acid to suppress the absorption of radioactive strontium. Since that time this problem has been pursued with success. Tanaka published (326) a paper on inhibition of intestinal absorption of radioactive strontium and the relationship of biological activity to chemical composition of alginates. 1969 the author reported on this problem in Kiel (327). 1969 Skoryna and Tanaka reported on the ion exchange properties of fractionation products of brown algae (286). Marine algae are able to concentrate certain metals from sea water. The strontium-binding capacities, the binding of iron of certain algal fractions and the utilization of inert neutral polysaccharides to produce metal binders were researched. The possibilities thus demonstrated have attracted the increasing interest of experts and official authorities, as has been reported in nomerous lectures and dissertations. Tanaka *et al.* (328), Skoryna *et al.* (287).

"The principle of this method lies in the in vivo ion exchange reactions between acidic polysaccharides and the metal ion in question resulting in the formation of insoluble salts of the macromolecules, which are unabsorbable and are thus excreted. The use of algal polysaccharides may be practical as a preventive measure against poisoning from metal pollutants, particularly in areas where high degree of pollution from the metals exists".

These studies are gaining more and more practical importance. See also Carrageenan and Fucoidan. See Tanaka and Stara, contribution in this monograph.

Alginates in food technology and alginates in industry: see special literature and Levring - Hoppe - Schmid (25) and Hoppe (12). A report in the "Commercial Fisheries Review" (77) referred to more than 200 uses for algin.

Fucoidan

Raw materials

Fucoidan probably occurs in nearly all brown algae. The most important raw materials are *Ascophyllum nodosum*, *Ecklonia* radiata, Fucus species, Laminaria species, Pelvetia species. Ascophyllum nodosum contains 6 - 8%, Fucus species 9 - 11%, Laminaria species 5 - 20%, Pelvetia canaliculata up to 20% of fucoidan, calculated on the basis of the dry raw material. Seasonal fluctuations in fucoidan content have been investigated, with a maximum content in autumn.

Definition

Fucoidan is a mucilaginous intercellular polysaccharide with a high viscosity, containing 30 - 70% L-fucose and other sugars. It occurs together with alginic acid and laminaran in brown algae. Isolation, purification and structure are described in the following papers (25), (204), (233), (235), (248).

The problem of fucoidan as a factor in ecological adaptation has been studied by de Lestang and Quillet (176).

A highly purified fucoidan was isolated from Pelvetia wrightii and was found to contain L-fucose and D-galactose in a ratio of appr. 10:1, and it is considered to be a galactofucan sulphate (41). To study its chemical structure, the fucoidan was partially hydrolyzed (42), and L-fucose 4-sulphate was isolated (43). An attempt to fractionate this fucoidan has been made, and fucan sulphate as well as galactofucan sulphate has been isolated (44).

Utilization

Experiments have been made to use fucoidan in pharmacy. In therapy it can be used as a blood anticoagulant. It probably has a stronger antithrombotic effect than heparin (56, 289).

Fucoidan is a base material for fucose production.

Tanaka et al. (330) reported on binding of heavy metals by fucoidan, and lead was found to be one of those bound most effectively.

Another investigation involves the isolation, purification and metal binding characteristics of this substance from *Ascophyllum nodosum*, both in vitro and in vivo (245).

This research has been continued by Paskins-Hurlburt *et al.* (246) who concluded that this naturally occurring non-toxic poly-electrolyte is a suitable binding agent for lead, and that the insignificant binding of calcium is an important criteria for biological and clinical application. The high biological activity of this compound depends on position and availability of functional groups for the ion-exchange process, and the viscosity of the solution.

Tanaka *et al.* (331) reported at the IXth International Seaweed Symposium on the inhibition of iron absorption by fucoidan. The results were discussed with respect to significance of fucoidan in cases of excessive iron intake and preferential binding of biologically important metals by fucoidan.

Ascophyllan

Ascophyllan is a polysaccharide extracted from *Ascophyllum nodosum* and *Fucus spiralis* together with alginate. It can be separated from alginic acid by precipitation with acid and from other accompanying polysaccharides by fractional precipitation with salt and alcohol. Ascophyllan is a substance similar to fucoidan. *Ascophyllum nodosum* contains appr. 6% of the dry weight (173).

Laminaran

Raw materials

Laminaran is found in the fronds of *Laminaria* species and other brown algae (see Alginic acid and Alginates). The maximum content was found to be present in autumn and in early winter. The content of laminaran in raw materials dried after harvesting immediately under controlled conditions amounted to, for example, 23 - 32% for *Laminaria hyperborea*, 13 - 25% for *L. digitata* and 18 - 33% for *L. saccharina*. No laminaran was found in the stipes of *Laminaria*. The seasonal and other fluctuations in laminaran content were investigated (253, 263).

Definition

Laminaran is a reserve carbohydrate of certain marine brown algae comparable to starch in higher plants. We differenciate between "soluble" laminaran which mostly occurs in *L. digitata* and "insoluble" laminaran from *L. hyperborea* and other species. The insoluble form is practically insoluble in cold water, but soluble in hot water. The soluble form is readily soluble in cold water and gives neutral solutions. The insoluble form contained 1.7 - 2.7% mannitol. Laminarose and laminaritol are components of laminaran which can be produced as a by-product for the production of sodium alginate. A number of international patents have been issued (156, 206, 352, 232).

Utilization

The application of sodium laminaran sulphate as a blood anticoagulant and antilipemic agent has been studied (12, 25, 57). Adams *et al*. (34) have reported on the effects of laminaran sulphate on bone growth. In animal experiments the cholesterol level in the blood could be reduced by laminaran sulphate (90). Also in animal experiments an inhibitive effect was of laminaran sulphate on sarkoma 180 was found (154).

Mannitol

Raw materials

Mannitol is found in brown algae and is constituent of the cell sap. (see Alginic acid and Alginates). It also occurs in some red algae, e.g. *Furcellaria fastigiata*, *Porphyra umbilicalis*, and in other plants, e.g. *Fraxinus ornus*, *Olea glandulifera*.

The content of mannitol depends on seasonal fluctuations and on the depths at which the algae grow. *Laminaria saccharina* for example contains about 5% in January and up to 22% in May. The stipes of *Laminaria* have only a very small mannitol content. The fronds have a maximum in summer. Fronds of L. hyperborea from a depth of appr. 7.5 meters gave a mannitol content of appr. 36 % of the dry weight. Mannitol is produced from marine algae in the U.S.A., Great Britain, France and Japan.

Definition

Mannitol is a sugar alcohol of the hexite series also occurring as mannitan (anhydride) (206, 248).

The pathway of mannitol formation during photosynthesis in brown algae was studied in Eisenia bicyclis (232, 353).

Utilization

Mannitol is used in the production of tablets, for diabetic food, for making chewing gum, as dusting powder etc.

Kelp products

The historical development and definition are described in Levring - Hoppe - Schmid (25) and in many other papers. The most important raw materials are some brown algae: Laminaria species, Macrocystis species, Nereocystis luetkeana, Pelagophycus porra and others.

Since the 17th century "kelp" was the term for a compound obtained from the incineration of brown algae, used for the production of iodine, potassium- and sodium-salts. In the U.S.A. "kelp" is the term used for the large brown algae in the fresh state, for example "Giant Kelp" (Macrocystis pyrifera). Some authors differentiate between the terms "kelp" for the large brown algae and "rockweeds" for the marine algae growing on the rocks between the low and higher water marks, for example Ascophyllum nodosum and Fucus species.

During the First World War there was a widespread "kelp"-industry on the American Pacific coast for production of acetone, ethylacetate, ethylpropionate, ethylbutyrate, iodine,

iodine salts, potassium chloride and "kelpchar", a kind of charcoal used as a de-colouring agent (267).

In 1951 experiments were made in the USSR for the production of alcohol, acetone, lactic acid, and other organic acids from marine algae (6). In the U.S.A., too, the kelp-industry is merely of historical interest.

Iodine and Iodine Salts

Numerous investigations have been made on this subject. Many patents are valid for the production of iodine from marine algae. According to some patents iodine is isolated electrolytically or by extraction methods (183). In the cold water region of the Nishintoro peninsula large quantities of marine algae with a high content of iodine and potassium salts occur, e.g. *Alaria fistulosa*, *Arthrothamnus kurilensis*, *Laminaria diabolica* (335). In general, brown algae are utilized for the production of iodine. Iodine is less widespread in red algae. An exception to this is the *Phyllophora* species which contain 0.27% to 0.45% and which are utilized in the USSR for iodine production.

Bromine

Ocean water contains 0.007% of bromine on an average. It is strongly enriched in some marine algae. The red alga *Rhodomela larix* contains appr. 3% of the dry weight. Investigations made in Hokkaido have proved that *Rhodomela macracanthua*, *Odonthalia corymbifera* and other *Odonthalia* species, *Polysiphonia morrowii* and others are also raw materials for the production of bromine. Thus the content of *Odonthalia corymbifera* amounts to almost 6% of the dry weight (335).

The bromine of the red alga *Lenormandia prolifera* is localized in the chloroplasts (134).

Potassium Salts

In Japan there was once a considerable production of

potassium from marine algae, which, however, was carried out by a not very profitable ashing process. A number of brown algae, inter alia, *Laminaria saccharina*, *Costaria costata*, Sargassum species were utilized for potassium salt production (335).

Products of the Chlorophyta

Water-soluble sulphated polysaccharides were also extracted from Chlorophyta (205). Recently numerous further papers on the constituents of Chlorophyta such as protein, amino acids, inorganic constituents, sterols etc. have been published. Within the scope of this survey they are to be mentioned only in general. Esp. on *Chlorella* and *Scenedesmus* there is a comprehensive special literature.

Planktonic Algae

The reviews in this monograph and some contributions about the constituents of algae present new knowledge. The nerves toxines saxitonin and gonyautoxin from *Gonyaulax* are only an example. But even in the Middle Ages for instance *Nostoc* was held in great respect. It was mentioned about 1500 A.D. by Brunswig and used for gout, cancer and fistula.

CONSTITUENTS OF MARINE ALGAE

Besides the polysaccharides there are in algae numerous constituents of which some groups have attracted increasing interest in pharmacy.

Acids

The high acidity and the content of free sulphuric acid are noteworthy in *Desmarestia aculeata* (L.) Lamour. Malic, citric, and succinic acids are present in smaller quantities. Fucinic acid was reported in appreciable amounts in Ascophyllum nodosum and appears to be confined to the larger brown algae.

Alkaloids

Alkaloids in marine algae were little known till recently 1969 Güven and his co-workers (112) analysed at first an unknown substance in Phyllophora nervosa (D. C.) Grev. as hordenin. A survey of the literature demonstrated that this was the first time that an alkaloid had been represented in marine algae. 1976 Güven and his co-workers found three alkaloids in Cystoseira corniculata (122). But the yields were insufficient for further investigations. In some mikro-algae, too, alkaloids were found.

Amines

In green, brown, and red algae the occurrence of volatile amines such as methylamine, dimethylamine, trimethylamine, ethylamine, propylamine, isobutylamine, isoamylamine was found (57). The distribution of trimethylamine and its oxide have been worked out colorimetrically for 20 species of green, brown, and red algae by Fujiwara-Arasaki and Mino (101). Simple monoamines are widely distributed in Rhodophyceae. Formation and function of monoamines have been analysed (125). In "Recent Progress in the Physiology of Chlorella" Hase reports in "Advance of Phycology" (126) that aliphatic amines, such as ethanolamine, putrescine and spermidine occur in relatively large quantities in algal cells. Kneifel (167) screened mikro-algae for amines. He analysed primary and secondary amines to search for new natural products of possible pharmacological activity and reported on the polyamines in algae. See Kneifel, contribution in this monograph.

Antibacterial, Antibiotic, Antifungal and Antiviral Substances

The search for new biomedicinals has occupied numerous work groups in recent years. It is therefore not astonishing that the algae were also studied for substances with these activities. Further discussion is unnecessary here, as various authors of this monograph report on this topic in detail: see Aubert *et al.*, Ehresmann *et al.*, Glombitza, McConnell and Fenical, Pesando *et al.*, Tsuchiya.

Cellulose

The cell walls of marine algae contain cellulose. It has been named algulose by several authors, because it differs from cellulose (eucellulose) of higher plants. Algulose is also the term used for the insoluble cellulose-like constituents of algae, which are produced for example as a residue from the production of phycocolloids. Cellulose-alginates are less purified commercial qualities, which often contain cellulose and which are utilized for industrial purposes.

Enzymes

The enzyme systems and the knowledge of their specific activity in marine algae was described by Jacobi (147). Since then the enzyme problem has been researched. The following publications should be noted: Kashiwabara *et al.* (162), Young *et al.* (360), Ikawa *et al.* (143), Okazaki (240). On the enzyme activities Nisizawa read a paper at the IXth International Seaweed Symposium 1977 in Santa Barbara (23).

Floridean Starch

Floridean starch is a reserve polysaccharide found to be present in various red algae, for example in *Dilsea edulis* and *Furcellaria fastigiata*.

Glycosides

Floridosid and isofloridosid have been isolated from various red algae. These glycosides have been regarded as reserve carbohydrates or metabolic products. The concentration of floridosid for instance in *Rhodymenia palmata* may reach 15%.

Inorganic Constituents

For the use of algae in pharmacy and as supplementary nutrition and food inorganic constituents play a sizeable role.

Marine algae are capable of concentrati and accumulating many elements. For example some Phaeophyta and Rhodophyta are rich in iodine. Its concentration may occasionally reach 30,000 times that in seawater, and can constitute 1 % or more of the dry weight. Iodine is chiefly found in such algae as NaI and KI. A small part is bound organically. The contents of elements and minerals depend on the season. For the algae themselves the inorganic components are required for their biosynthetic pathways.

A comprehensive survey of inorganic components of marine algae is reported in "The Elementary Chemical Composition of Marine Organisms" by Vinogradow (339). In the course of recent years numerous single works and reviews on the occurrence of inorganic elements in marine algae have been published.

The content of trace elements in seaweeds is also the object of a systematic study by Yamamoto and Ishibashi (354). Seaweeds grow in fixed positions in the sea and absorb nutrients from seawater. In general, the concentration factor of elements in the same seaweed varies inversely with the increase of residence times of elements in seawater.

The mineral content of some benthic algae of the Kenya Coast has been investigated. The results for five elements (nitrogen, phosphorus, potassium, calcium, and magnesium)are discussed by Imbamba (144).

Biodeposited trace metals and mineral content of some tropical marine algae have been studied by Sivalingam (282). Based on this investigation the author proposed that certain algal species within the tropical zone could be employed as indicators for pollution studies in the near future. In a preleminary report, the concentrations of the trace elements Co, Cu, Zn, and Mn were determined in some red and brown algae from the Icelandic coast. - Munda (227). Concentrations of Ag, Hg, Cr, Pb, Ca, Cd, Co, Zn, and Ni the shelf waters of Nova Scotia have been determined in relation to the primary production by Devi and Durvasula (89).

Whyte and Englar (348) reported on the seasonal changes in the inorganic constituents of Macrocystis integrifolia. The dry matter contained an average ash content of 39.4%. The principal cations are potassium, sodium, calcium, magnesium, and phosphorus. Minor elements are strontium, iron, boron, aluminium, barium, and zinc. Trace elements are chromium, manganese, copper, cadmium, lead, cobalt and mercury.

The principal anions are chloride, sulphate, nitrate and silicate. Minor anionic components are iode, carbonate and bromide.

Agadi et al. (36) have reported on the metal concentrations in some seaweeds. Seventeen species of marine algae of the coasts of Goa (India) were analysed for Co, Cu, Fe, Mn, Ni, Pb and Zn, using atomic absorption spectrophotometer. The seven metals showed considerable variations in their concentration. The role of different seaweeds as indicators of metal pollution is discussed in relation to the degree of accumulation of metals.

Yamamoto and co-workers report in this monograph on the distribution of chemical elements in algae.

Lipids

Chlorophyta generally contain a small quantity of fatty oils with myristic-, palmitic-, lauric-, stearic-acid and a number of unsaturated fatty acids. Phaeophyta contains 0.6-3% lipids with highly unsaturated fatty acids. Variations in the composition and content of polar lipids in Fucus serratus, Fucus vesiculosus and Pelvetia canaliculata during their annual cycle have been analysed. Phospho-, glyco-, and sulfo-lipids have been found by Pham Quang and Laur (249).

Another component, for example in Fucus vesiculosus, is the hydrocarbon n-hentriacontan (9).

Rhodophyta contains only a small quantity of lipids (appr. 3 %) They consist mainly of highly unsaturated fatty acids.

Kaneda and Ando (161) have reported that Porphyra tenera contains small amounts of lipids which consist mainly of highly unsaturated fatty acids. However, a rancid odor is not detectable even after long storage. This fact indicates the presence of some antioxygenic substances. The lipids are probably protected from oxidation by both cell and outer cell membranes. Another factor is presumably the antioxygenic effect of the phospholipids.

Sterols

Algae contain sterols in the unsaponifiable lipid fraction. Sterol glucosides have been found in a few species (208). Sterols as sitosterol, fucosterol, chondrillasterol, ergosterol were pointed out in the unsaponifiable lipidfraction of Chlorophyta. The occurrence of sterols might be as various as in higher plants.

Phaeophyta contain a characteristic sterol, fucosterol, which was first isolated from *Fucus vesiculosus*. Some Phaeophyta also contain sargasterol (=20-isofucosterol). Safe *et al.* (269) reported that *Laminaria saccharina* contains fucosterol, 24-methylcholesterol, cholesterol, 24-ketocholesterol, saringosterol and *Ascophyllum nodosum* moreover brassicasterol. Sterols of *Fucus gardneri* and of *Sargassum muticum* in animal experiments significantly reduced the plasma-cholesterine level. Reiner *et al.* (260). Fucosterol may grow in significance as a base material for the manufacture of sexual hormones and for steroid synthesis - Mautner (197). Tsuchiya (336) shows that fucosterol decreases plasma cholesterol in chickens but not in rats.

It has been demonstrated by Ito and Tsuchiya (145) that *Heterochordaria abietina* has a high potency for lowering the blood cholesterol level for rats. The authors fractionated the substances responsible for this effect and found them to be phaeophytin, phaeophorbide and iodoamino acids. The hypocholesterolemic activity of algal polysaccharides remains unsolved.

Rhodophyta contain sitosterol, fucosterol, cholesterol, chalinasterol. Cholesterol is the most commonly occurring sterol in red algae. It has the ability to reduce blood cholesterol level - Bhakuni and Silva (57). Cholesterol was also obtained of *Corallina granifera* (121).

A paper by Abe and Kaneda (33) concerns the effectiveness of seaweed on the lowering of the plasma cholesterol level in rats and also the isolation of the effective substance from seaweed. The feeding experiments indicated that basic compounds were most effective in reducing plasma cholesterol level. In the effective fraction betaines and an unidentified compound were detected. The results indicated that all the green algae used in the experiment and *Porphyra* markedly reduced plasma cholesterol level, while brown algae were not effective.

Chondrillasterol from *Scenedesmus obliquus* and *Navicula pelliculosa* could be employed as starting material for the synthesis of pharmacologically important steroid hormones such as cortisone (208). See Youngken and Soliman, contribution in this monograph.

Steroids

Very little is known about the occurrence of Δ^4-3-ketosteroids in plants. It is a well-known fact that keto-steroids occur widely in animals and play an important role as sexual and adrenocortical hormones. Kanazawa and Yoshioka (160) reported on the occurrence of Δ^4-3-keto-steroid in some red algae.

Fatty Acids

Analyses of the total fatty acids of green, brown, and red algae, the saturated and the unsaturated fatty acids have been published by Miller (208). The fatty acid composition and the ratio of the percentage of unsaturated to saturated acid of the total fatty acids in the oil from some Japanese marine algae were anlysed by Ito and Tsuchiya (146). The content of crude fat decreased with increasing depths of immersion. The fatty acids of different Rhodophyta, Phaeophyta, and Chlorophyta were analysed by Pohl *et al*. (251), Pohl (252). See Pohl and Zurheide, contribution in this monograph.

Phenolic Compounds

Tanning substances with phenolic character occur in marine algae in the physodes or fucosan bladders of the Phaeophyta, for example *Ascophyllum nodosum* and *Sargassum* species. These brown algae have a positive tannin reaction. According to Kylin, fucosan is a tanninlike substance. Aqueous extracts from *Sargassum ringgoldianum* are used for tanning in areas where the alga occurs in sufficient quantities.

Sargaline, a phenolic substance from *Sargassum confusum*, has been reported to depress the blood sugar level (265).

Glombitza and co-workers reported above all on the phenolic compounds of marine algae. Some of these substances have antibiotic properties. In many species of the Phaeophyceae phloroglucinol could be identified.

1974 Glombitza reported the VIIIth International Seaweed Symposium on screening "Phenols from Phaeophyceae": The higher molecular compounds possess tannin characters. The name "Phlorotannins" is suggested for this new class of tannins (107). See Glombitza, contribution in this monograph.

Studies on the contents and composition of the phenolic components of a number of common brown algae have been also carried out by Jensen and Zavodnik (150). Although physodes of brown algae are primarily composed of phloroglucinol derivatives, they may also contain other tannins and terpenes. The recent discoveries of homologous series of phloroglucinol polymers in brown algae extracts provide an attractive chemical rationalization for the observed heterogenety of physodes (255).

Phytohormones

Growth substances like auxins and gibberellins in algae are described by Conrad and Saltman (78) and by Mowat (212, 213). Auxins or auxin-like substances may be synthesized by many algae. The plant growth regulatory substances in seaweed extracts have been investigated by several authors. Augier (47) started the essential problems in his work "Les hormones des algues. Etat actuel des connaissances". His bibliography lists 834 publications on this subject. H.-W. Schmidt (297) reported that besides the vitamins phytohormones also play a part in the development of certain remedies and cosmetics.

Pigments

The pigments of algae, chlorophylls, carotenes, xanthophylls, phycobilins and others, esp. of green, brown and red algae and their importance were the subject of numerous recent investigations. Pigments are contained in algae in varying composition, specified in the following table:

Chlorophyta

- chlorophylls
  - chlorophyll a
  - chlorophyll b
- carotenes
  - α- carotene
  - β- carotene (haematochrom, the most important yellow pigments)
  - γ- carotene
- xanthophylls
  - lutein
  - astaxanthin
  - neoxanthin
  - siphonein
  - siphonoxanthin
  - taraxanthin
  - violaxanthin
  - zeaxanthin

Phaeophyta

- chlorophylls
  - chlorophyll a
  - chlorophyll c
- carotenes
  - α- carotene
  - β- carotene
- xanthophylls
  - fucoxanthin
  - diatoxanthin
  - flavoxanthin
  - lutein
  - violaxanthin
  - zeaxanthin

A little-known pigment is responsible for the lilac holdfasts of Macrocystis pyrifera. It can be extracted with alcohol. It is evidently not an anthocyanin, a phycobilin or a carotenoid.

Yellow-brown substances were found under certain conditions as excretion products from brown algae. Chemical damage to the plants was the probable cause. The production of the yellow substances is inversely proportional to the disappearence of chlorophyllis and carotenoids in ground algae (169, 357).

Rhodophyta

- chlorophylls
  - chlorophyll a
  - chlorophyll d
- carotenes
  - α- carotene
  - β- carotene
- xanthophylls
  - lutein
  - taraxanthin
  - zeaxanthin
- phycobilins (bilichromoproteins)
  - phycocyanin
  - phycoerythrin

Floridorubin (phycohaematin) has been discovered in some Rhodomelaceae. The water-soluble substance, which turns orange-yellow when acids are added, was used as a dye.

Pigments are also studied in other algal classes such as Cyanophyta, Chrysophyta, Bacillariophyta.
(62, 133, 229, 237, 238).
See Wettern and Weber, contribution in this monograph.

Proteins

The nitrogen-containing constituents of algae are important for their uses in nutrition and food. The proteins participate in the substances of marine algae, which include nitrogen, a part of 60 - 70%. The rest is distributed on peptides, amino acids and other N-compounds.

A comparison of the soluble proteins in fresh specimens of marine algae has been investigated by Young (359). Extractable soluble proteins from 11 genera of Rhodophyta were analyzed by absorption spectroscopy and by polyacrylamide disc gel electrophoresis by Mallery and Richardson (188).

Some reports concerning the total proteins obtained from various algae and their fractions published by Güven and coworkers showed that these fractions have different pharmacological activities. The total protein and fraction I have lipolytic, total protein and fraction II anticoagulant and fraction III hypoglycemic activities.

Mason (192) reported at the IXth International Seaweed Symposium, 1977, on further studies of marine algal proteins.

Peptides

Marine algae are characterized by specific organic compounds. Only a few peptides have been isolated purely, for example octaglutamic acid from *Pelvetia canaliculata*, pentaaspartic acid from *Corallina officinalis*, L-arginyl-L-glutamine from *Cladophora* species, eisenine from *Eisenia bicyclis* and *Ecklonia cava*, fastigiatin from *Pelvetia fastigiata*, analipine from *Analipus japonicus*, a Japanese brown alga. Analipine is a peptide consisting of glutamic and aspartic acids or their amides (312, 313, 314).

A novel UV-absorbing substance "334" was isolated in a crystalline form from the red alga *Porphyra yezoensis* Ueda. It was found to be a peptidal compound (279, 280, 281). The authors have investigated the physico-chemical properties of the substance and also the physiological roles. The results from all these investigations suggested that the substance "334" plays some important role as a metabolic regulator or a temporal energy transferring substance at some still unknown sites in the photosyntetic pathways of algae, particularly in relation to UV-light absorption.

The substance "334" may possibly occur in all algae.

Amino Acids

More than 150 amino acids are known. Many of them have been isolated from marine algae.

Composition of amino acids in the ethanolic extracts from various algae (mg% on dry basis).

Amino acid	*Ulva pertusa*	*Enteromorpha linza*	*Undaria pinnatifida*	*Sargassum confusum*	*Chondria crassicaulis*	*Neodilsea yendoana*	*Laurencia nipponica*
Alanine	17.7	23.8	617	278	71.4	25.8	61.4
Alloisoleucine	0	0	2.7	0	0	0	0
Arginine	2.5	1.9	36.5	1.6	trace	1.3	1.4
Aspartic acid	4.4	13.7	5.4	27.5	87.1	6.1	42.2
Chondrine	28.7	43.0	41.0	22.8	193	29.0	24.7
Citrulline	0	0	0	0	0	40.7	0
Cysteic acid	35.5	13.1	4.8	3.7	13.3	9.8	8.7
D-Cysteinolic acid	152	73.4	1.9	2.2	57.4	0	0
Glutamic acid	31.8	55.0	89.8	28.5	49.5	0.7	219
Glycine	9.1	5.2	455	3.7	29.1	5.6	45.9
Histidine	0	0	2.1	0	3.3	0.6	0
Isoleucine	3.7	5.8	11.2	9.7	6.4	0.4	37.9
Leucine	6.6	6.8	19.6	14.3	5.3	1.7	11.6
Lysine	0.9	0.8	34.6	0	3.6	0.5	0.3
Methionine	0	0	1.7	0	trace	0	1.7
Ornithine	0	0	0	4.0	0	1.6	1.5
Phenylalanine	4.1	3.8	9.2	11.3	0	0	1.9
Proline	40.0	51.1	156	24.2	50.8	8.8	32.7
Serine	11.9	33.7	64.4	51.6	12.9	2.6	33.3
S-Hydroxymethyl-L-homocysteine	0	0	0	0	0	4.6	2.7
Taurine	1.8	2.0	11.8	9.7	14.5	218	24.9
Threonine	5.7	3.8	90.3	12.1	12.0	2.5	48.6
Tryptophan	2.4	1.4	5.8	8.0	12.6	0	3.3
Tyrosine	2.1	2.1	10.1	7.5	0	0	1.1
Valine	3.5	3.9	11.1	14.0	7.6	1.7	87.6
Unknown compound A	+	–	–	++	++	++	+
〃 B	–	+	+	–	–	–	–
〃 C	+	+	+	+	+	+	–
〃 D	–	–	+	–	–	–	–
NH_3	3.7	3.4	16.6	4.0	10.5	4.2	5.3
Total	368.1	347.7	1698.6	538.4	640.3	366.2	697.7

Takagi <u>et</u> <u>al</u>. (310).

Among the new amino acids found in marine algae L-α-kainic acid, N-methyltaurine, N-dimethyltaurine, D-glyceryltaurine, D-cysteinolic acid, domoic acid, chondrine, rhodoic acid, S-hydroximethyl-L-homocysteine, laminine, gongrine and gigartine are some which have medicinal effects (309, 312, 314, 311).

L-α-kainic acid, a derivative of proline, from <u>Digenea</u>

simplex is used as anthelmintic, esp. against ascaris. A steric isomer L-α-allokainic acid has an inferior anthelmintic effect (315). Domoic acid, a proline derivative, from Chondria armata is also effective as an anthelmintic. Chondrine has been isolated from Chondria crassicaulis and other algae (170).

Laminine, a hypotensive constituent, is a basic amino acid from Laminaria angustata. Laminine has also been found in other Laminaria species and in Ecklonia cava, Eisenia bicyclis, and Undaria species but in small quantities. A large amount of laminine was detected in Heterochordaria abietina. This alga has been said to be effective against hypertension in folk medicine (320, 321, 322, 325).

Takagi (314) reported that laminine suppressed heart action in vivo and in vitro, and temporarily depressed blood pressure without change in the pulse rate. It caused relaxation in smooth muscle of the small intestine, blood vessels and trachea, and blocked the action of acethylcholine and histamine, drugs which cause the contraction of certain smooth muscle.

In view of the great importance of the amino acids numerous green, brown and red algae have been researched in different coastal areas. Here some indications:
Investigations were made by Silva et al. (278) on chilean algas. They reveal all the essential amino acids and confirm the work made by other investigators on other algae.

The free amino acid pool of fresh and dry samples from Lessonia nigrescens, Durvillea antarctica, and Macrocystis integrifolia show qualitative and quantitative differences, either in fresh collected or dried samples, due to the drying process (171, 301).

A study of the free, nonprotein and common amino acid pool was undertaken by Madgwick and Ralph (186) in 34 Australian algae.

In particular, the non-protein constituents were sought because of their potential anthelmintic and medicinal properties. Baikian, chondrine, 1-methylhistidine, 1,3-dimethylhistidine are rare amino acids, which have been isolated in a range of different algal species.

The amino acids either free or combined (in the peptide form) during the different seasons in Laurencia papillosa, Cystoseira barbata and Enteromorpha intestinalis from the Egyptian coast have been studied. Since the amino acid metabolism depends on the individual algae, it is supposed that algal classes may differ from each other in their amino acid patterns constituting the individual proteins (210).

The total amino acid pool of 10 Rhodophyceae, 10 Phaeophyceae, and 4 Chlorophyceae, common in Icelandic waters, was determined. Seventeen of the common amino acids have been determined. In the edible species, the essential amino acids are present in notable quantities (226).

The amino acid composition of the protein hydrolyzates of the whole Macrocystis pyrifera is discussed by Mateus and coworkers (193).

Sugar Alcohols

Sugar alcohols, abundant in some algae, are sorbitol, dulcitol, scyllitol, laminitol, mytilitol, volemitol, trehalose, glucose, fructose, sucrose, and others. See also Mannitol.

Vitamins

Marine algae are an excellent source of vitamins. It would be beyond the scope of this introduction to cite all the works dealing with the vitamin content of algae.

It was reported that ascorbic acid (vitamin C),biotin (vitamin H), carotene (vitamin A precursor), choline, cobalamin (vitamin B_{12}), folic acid, folinic acid, inositol, nicotinic acid, pantothenic acid, pyridoxine (vitamin B_6), riboflavin (vitamin B_2), thiamin (vitamin B_1), tocopherols (vitamin E) and vitamin K are constituents of marine algae.

Johnson and Levring (152) report that vitamin D discovered in plankton and plants in the sea is produced under the influence of irradiation.

In view of the importance of vitamins in marine algae for the indirect medicinal effects of food and for feeding purposes it seems to be necessary to give particular attention to this subject in the future (314).

Vitamins are also essential growth factors and in their absence no growth can occur or growth is limited by the rate of synthesis of a vitamin.

Volatile Constituents

Japanese scientists such as Katayama (163) and others researched several Chlorophyta, Phaeophyta and Rhodophyta. The amounts of the volatile constituents differ according to the species, the habitat, the season of collection, the kind of harvest and drying. The following compounds were found out: acetic-, acrylic-, butyric-, n-caproic-, caprylic-, formic-, isovaleric-, myristic-, palmitic-, propionic-acid and some sulfur compounds, aldehydes, alcohols, terpenes, phenols, hydrocarbons and other substances not yet identified.

Several green algae have a characteristic, sometimes even unpleasant smell. These smells often come from sulphureous compounds.

The contents of volatile constituents in marine algae are shown in some examples:

Dictyopteris divaricata	2.54 - 3.56%
Dictyota dichotoma	0.39%
Undaria pinnatifida	0.03%
Sargassum confusum	0.19%
Laurencia nipponica	0.82%
Porphyra tenera	0.037 - 0.85%

Some other species contain only traces of volatile constituents.

Toxic Substances

Some algae doubtless contain toxic substances. *Pelvetia fastigiata* contains fastigiatin, an acidic tripeptide, antimicrobial substance and a toxin of an as yet unknown structure. Watery extracts of *Egregia menziesii* and *Macrocystis pyrifera* have shown toxic effects on mice.

Caulerpicin, responsible for the peppery taste of some *C.* species, could be isolated from dry *C. racemosa* (91). It seems that caulerpicin and caulerpin, produced by *Caulerpa* are physiologically active and toxic to rats and mice, and are transferred along the food chains and concentrated in the process at least in some herbivores (92).

Hashimoto *et al.* (128) reported on a screening of toxic algae. Some specimens were found to contain toxins lethal to mice or hemolytic to rabbit blood cells. The chemical studies on toxic components of algae are to be continued.

For further constituents of marine algae see (25, 294, 295).

At institutes in several countries recently numerous algae have been studied with regard to their constituents and active substances in order to test their therapeutic potential.

On marine biotoxins as a new source of medicinals Halstead remarked (365): "The need for a new and broader spectrum of more effective pharmaceuticals is obvious." New biodynamic substances are continually discovered from marine algae and other marine organisms. Some of them can find their field of activity in diseases which until recently were difficult to prevent. The review of Bhakuni and Silva (57) deals with the chemistry and the biological activity of the biodynamic substances isolated from marine algae. The studies of *Laurencia*, *Sargassum* and various algae are examples of new detailed chemical investigations with interesting aspects for medicine.

Besides the so-called explorations by chance and the scientific research of old native medicine the screening programs play a great part in discovering active substances. But it is a troublesome way, too, until a substance comes to a clinical test. But, "Our modern therapeutics owe to empirical uses and native medicine of all countries and peoples more than is generally supposed", as Hänsel remarked (364) at the 37th "Ärztlichen Fortbildungslehrgang".

Indirect Medicinal Effects

Besides the specific medicinal use of algae and algal constituents and the importance of the indirect medicinal effects is steadily growing as our knowledge increases. With constituents such as amino acids, trace elements, vitamins and so on, marine algae, being rich in these substances, are not only valuable food in many countries but also preventive remedies and therefore have a great indirect medicinal effect.

As Dragendorff (5) wrote in 1898 in his famous book "Die Heilpflanzen der verschiedenen Völker und Zeiten", the Porphyras are a valuable restorative.

Takagi (314) wrote about the direct and indirect medicinal effects of algae in "Advance of Phycology in Japan".

In 1976 the well-known German physician R. F. Weiss (347), author of "Moderne Pflanzenheilkunde", stressed the importance of increasing the resistance of the human organism to the influences of exitants of disease, and of increasing the reactions of defence: "The increase of resistance is a general therapeutic principle with comprehensive applicability. It is therefore little wonder that pharmacy endeavors furthermore to find additional drugs to increase resistance".

In terrestrial drugs there are often no active substances of specific efficacy which produce the desired effect. It is the sum of the constituents in the drug. And such a sum of constituents is particularly existent in many marine algae.

The algae are also valuable constituents of a number of dietary foodstuffs.

About the indirect medicinal effect of some products of marine algae it is certain that "the characteristic of many old remedies with small acute toxicity consists of bringing about a change of the morbid disturbance of metabolism pretty much as would hormones and vitamins; thus they not only remove symptoms but become real remedies", Müller-Plettenberg, 1977 (225).

ALGAE AS FOODSTUFFS

The use of algae as nutriment must not be unmentioned in this survey. Beside the well-known Kombu- and Nori-products numerous further products of algae are used in nutrition. Regarding algae in human nutrition there is a comprehensive scientific and popular literature, ranging from detailed chemical analyses to cookbooks with attractive and delicious recipes. Optimum human nutrition requires the supply of numerous dietary factors including at least inorganic elements, a set of vitamins and essential amino acids. Johnston for instance elaborated this point in his lecture "A Detailed Chemical Analysis of Some Edible Japanese Seaweeds" at Sapporo, 1971 (153).

Algal Products for Human Nutrition

Aonori

The raw materials of the "aonori" products are the various *Monostroma* species from the Japanese coasts. The algae cultivated mainly for this purpose, are *M. latissimum* and *M. nitidum*. Interesting modern methods of the culture have been reported by Segi and Kida (270). Production is limited to Japan, although there are interesting possibilities of using *Monostroma* for edible algal products in other parts of the world, above all because of their vitamin content. Japanese authors (164, 165, 271, 332) report in detail on the life history, cultivation and so on of *Monostroma*. Johnston (153), reported on the analysis of *M. latissimum*.

Enteromorpha species are often also used for this purpose, called "awo-nori" in Japan.

Hijiki

The raw material is *Hizikia fusiforme*. "Hijiki" is the Japanese name for the product. According to "Standard Tables of Food

Composition", Japan, the alga contains 6 to 7% raw protein, appr. 35% sugar, provitamin A, vitamin B_1, B_2, niacin and arginin.

Kombu

The raw materials of "kombu" products are East Asian Laminaria species and some other brown algae. All important substances required for human nutrition are contained in the "kombu" products. In addition to their vitamin, trace element, and salt content the value of these products lies in their high sugar content. Amino acids are esp. glutaminic acid, asparaginic acid, prolin, alanin (239). Utilized in human nutrition. When used as food "kombu" also has many indirect medicinal effects (314).

Nori

"Nori" is a collective term used for the various edible products made from species of Porphyra, a tender genus of red algae. They are used for food, have been known for many centuries, particularly as a vegetable or salad, and enjoy increasing popularity in many parts of the world.

The value of "nori" is due to the high content of protein, vitamins and minerals. A large number of investigations have been undertaken of the constituents of different Porphyra species. Only some examples can be mentioned here. P. tenera - constituents: appr. 40% polysaccharides, appr. 29-35% protein, amino acids, responsible for the flavour esp. glutamic acid, glycine, alanin, volatile constituents responsible for the characteristic odour. The alga contains 0.037 to 0.085% of volatile oils. Nucleic acid. Pigments: chlorophylls, carotenoids, chromoproteins. Vitamins: provitamin A, α and β-caroten, vitamin B_1, B_2, C, folic acid, biotin, cobalamin, cholin, nicotinic acid. Inorganic constituents: Ca, Na, Ph, Fe. Lipids: 35% of the lipids were phospholipids and 25% were triglycerides (antioxygenic activities). A substance lowering cholesterol content of the blood (46, 161).

P. linearis - responsible for the desirable flavor are a high concentration of isofloridoside and several free amino acids. Free amino acids: lysine, histidine, arginine, aspartic acid, threonine, serine, glutamic acid, proline, glycine, alanine, methionine,isoleucine, leucine, tyrosine, phenylalanine (202).

P. nereocystis contains 50% more vitamin C than oranges and considerable quantities of B-vitamins.

Miura (209) reported on _Porphyra_ cultivation in Japan. The _Porphyra_ culture is one of the largest parts of the fishing industry in Japan. It is obvious that physiological studies are indispensable in constructing a fundamental basis for this important industry (236). It has also become evident, said Chihara (73) in the same book, that the characteristics of certain algae, esp. in several species of _Porphyra_, are not always fixed but rather flexible depending upon environmental factors.

American "nori" of _Porphyra nereocystis_ is deemed to be similar in quality to foreign "nori" products being marketed in the U.S. (350, 351). Coll and Cox (76) reported on new species of _Porphyra_ from North Carolina. The new species are _P. carolinensis_ and _P. rosengurttii_. A new species, _P. thulaea_, from East Iceland and West Greenland is described by Munda and Pedersen (228). The amino acids have been analysed. A number of _Porphyra_ species also form the basic raw material of the "laver-bread" industry in Europe (coast of Wales) and other countries.

Sea Lettuce

Ulva lactuca and other _U._ species are eaten as a salad in many areas. The amino acids, vitamins, inorganic components and so on have been analysed. Cultivation experiments were made. Oza and Sreenivasa Rao (243) reported on the results of their studies.

U. fasciata is one of the potential sources of protein for food and fodder. It contains appr. 26 % protein in air dried material.

Wakame

The raw materials of "wakame" are *Undaria* species. According to the "Standard Table of Food Composition", in Japan, the alga *Undaria pinnatifida* contains about 15 % raw protein, appr. 57 % sugar, provitamin A, vitamin B_1, B_2, niacin and vitamin C. Saito reported (266) that the demands for *Undaria* have increased because of their vitamin, mineral and special protein content. *U. pinnatifida* in particular has been recognized as an important marine resource worthy of artificial propagation.

Professor K. Nisizawa, the president of the Japanese Society of Phycology, and some specialists in medical, pharmacological, and nutritional research recommend the *Undaria* products for repressing of curing various diseases such as high blood pressure, arterial sclerosis, liver complaints and so on.

Regarding the subject "Algal Products for Human Nutrition" the following aspect deserves special attention. The supply of food for the increasing population of the world is a very serious problem.

Considering the problem of animal and vegetable protein, the possibility of utilizing algae and the many advantages of their use for human nutrition should be discussed, not only for East-Asia, but for other countries as well. Some important points of view are enumerated by Loose (180) which ought to be observed in the cultivation, harvesting and preparation of algae, in order to obtain optimum results.

Schlichting, Jr. reported in Economic Botany (293) that in the future it may be feasible to extract amino acids from marine algae for food supplementation. Mixtures of *Laminaria*, *Caulerpa*, *Rhodymenia* and *Grateloupia* at present look the most promising as a high-quality algal protein source.

Hoppe (141) and Schmid (296) also reported on marine algae, which are important for the human nutrition.

Seaweed Meal

In connection with "Algal Products for Human Nutrition" the "Seaweed Meal" or "Kelp Meal" should also be mentioned. Raw materials are *Ascophyllum nodosum*, *Alaria esculenta*, *Laminaria* species, *Macrocystis* and other *Phaeophyta*.

Esp. Ascophyllum nodosum is used to a great extent. Baardseth (49) published a synopsis of biological data on *Ascophyllum*. Jensen (149) gave a detailed analysis of the average composition of a Norwegian Seaweed Meal. "Seaweed Meal" is a valuable feed additive with a high content of minerals, trace elements and vitamins, esp. used in breeding of calves, poultry etc. Various authors such as Jensen (148), Seterlund *et al.* (276), Jones *et al.* (155) report on experiments and experiences.

The indirect medicinal effects of "Seaweed Meal" also play a role.

Cultivation of Marine Algae

The increasing demand for marine algae has a raw material has already led to the cultivation of important species, with *Porphyra*-cultures predominating. Providing raw material for the production of phycocolloids gave rise to problems recently, but after intensive preparation the cultivation of these algae was successful. There is a comprehensive literature on this subject, esp. in the Proceedings of the International Seaweed

Symposia. A summarizing report is published by FAO Techn. Paper No. 130, 1974 (142) in the catalogue of cultivated aquatic organisms, ed. by Ihingran and Gopalakrishnan.

On the status and future of seaweed culture and algae culture techniques competent specialists reported in 1976, in Kyoto, Japan, see Report of the FAO Technical Conference on Aquaculture (96).

An interesting paper "Optimal Production from a Seaweed Resource" was published in 1977 by Silverthorne (283). A dynamic programming model of a harvesting industry, including an initial population model of the resource, is developed and used to assess the potential contributions to economic efficiency of information which could be expected to result from alternative additional research projects.

The commercially valuable agar raw material Gelidium robustum, which occurs abundantly from Central California to Central Baja California was taken by the author as an example.

"The Industrial Utilization of Seaweeds -- Past, Present and Future" was described by Jensen at the IXth International Seaweed Symposium, 1977, at Santa Barbara, California (151), where he explained the current situation: "The industrial utilization of seaweeds has changed from production of soda and iodine to organic extractives. Only a small part of the world resources was used. In some areas extensive cultivation and foodstuff production is based on seaweeds. Large scale cultivation of some red and brown algae for technical purposes has recently been introduced and this together with genetic improvements and systematic research may create a new era of seaweed utilization for production of chemicals, food and energy."

The papers of the present monograph give us not only a survey of the state of present research but also indicate possible uses of marine algae in the pharmaceutical sciences, in curing diseases and in strengthening resistance to pathological influences which are becoming more and more noticeable in all countries.

It cannot be disputed that terrestrial and marine plants are the main source of raw materials for remedies. For many plants used in therapeutics the principle of the biological activity is still unknown.

Here a great field of activity lies before us. But I would like to cite E. Pringsheim (254): "It is extraordinarily advantageous to push forward in new fields and to discover qualities unknown till now. The multiplicity of nature exceeds our imagination."

The research of algae and their substances is being carried on in many countries and we can be sure that it will still be producing interesting knowledge and new points of view in the near future. To accomplish this work a cooperation of all branches of knowledge is necessary and also energic support from official authorities. These discoveries can yield valuable contributions to the prevention or cure of some diseases.

General Literature

1. Araki, Ch.: Seaweed Polysaccharides. In: Carbohydrate Chemistry of Substances of Biological Interest. Wolfrom, M. L., ed., Pergamon Press Ltd., London (1959).

2. Black, W. A. P.: The Algae. In: Processed Plant Proteins Foodstuffs. Altschul, ed., Academic Press Inc., New York (1958).

3. Chapman, V. J.: Seaweeds and their Uses. Methuen & Co. Ltd., London. First Edition (1950), second Edition (1970).

4. Colloques Internationaux du Centre National de la Recherche Scientifique Dinard (1957): Écologie des algues marines. In: Chimie et physico-chimie des principes immédiates tirés des algues (1961).

5. Dragendorff, G.: Die Heilpflanzen der verschiedenen Völker und Zeiten. Neudruck der Ausgabe, Stuttgart (1898). Werner Fritsch, München (1967).

6. Fott, B.: Algenkunde. Gustav Fischer Verlag, Jena (1959) (1971).

7. Fritsch, F. E.: The Structure and Reproduction of the Algae. University Press, Cambridge (II, 1952) (I, 1956).

8. Hagers: Handbuch der Pharmazeutischen Praxis. 4. Ausg., Band I-VI, Springer-Verlag, Berlin-Heidelberg-New York (1967-1977).

9. Hegnauer, R.: Chemotaxonomie der Pflanzen, Band I. Birkhäuser Verlag, Basel und Stuttgart (1962).

10. Hoffmann, C.: Die praktische Bedeutung der Meeresalgen. Kieler Meeresforschungen, III (1), 165-232 (1938).

11. Hoppe, H. A., Schmid, O. J.: Phykokolloide als moderne Industriehilfsmittel und ihre Einsatzmöglichkeiten in der kosmetischen und verwandten Industrie, Seifen - Öle - Fette - Wachse. 88 (21) (1962).

12. Hoppe, H. A.: Drogenkunde. 8. Aufl., Band 2. Walter de Gruyter Berlin-New York (1977).

13. Humm, H. J.: The Brown Algae - Algin from Kelps and Fucoids.

14. Humm, H. J.: The Red Algae of Economic Importance - Agar and Related Phycocolloids. In: Marine Products of Commerce. Tressler, D. K. and Lemon, J. Mc. W., ed., Reinhold Publishing Corporation, New York (1951).

15. Ist International Seaweed Symposium. Edinburgh, 1952.

16. IIne International Seaweed Symposium. Trondheim, 1955. Pergamon Press, London-New York (1956).

17. IIIrd International Seaweed Symposium. Galway, 1958 (Abstracts).

18. IVth International Seaweed Symposium. Biarritz, 1961 (Proceedings). Pergamon Press, Oxford-London-New York-Paris (1964).

19. Vth International Seaweed Symposium. Halifax, 1965 (Proceedings). Pergamon Press, Oxford-London-New York-Paris (1966).

20. VIth International Seaweed Symposium. Santiago de Compostela, 1968. Dirección General de Pesca Marítima, Madrid (1969).

21. VIIth International Seaweed Symposium. Sapporo, Japan, 1971 (Proceedings). University of Tokyo Press (1972).

22. VIIIth International Seaweed Symposium. Bangor, 1974 (Abstracts). Marine Science Laboratories, Anglesey, U. K. (1974).

23. IXth International Seaweed Symposium. Santa Barbara, Cal., 1977 (Abstracts). Journal of Phycology 13 (Suppl.) (1977).

24. Kylin, H.: Die Gattungen der Rhodophyceen. C. W. K. Gleerups Förlag, Lund (1956).

25. Levring-Hoppe-Schmid: Marine Algae - A Survey of Research and Utilization. Cram, de Gruyter & Co. Hamburg (1969).

26. Neushul, M.: Botany. Hamilton Publishing Company, Santa Barbara, Cal. (1974).

27. Okazaki, A.: Seaweeds and their Uses in Japan. Tokai University Press (1971).

28. Schneider, W.: Lexikon der Arzneimittelgeschichte. Pflanzliche Drogen V/1-3. Govi-Verlag GmbH. - Pharmazeutischer Verlag, Frankfurt a. M. (1974).

29. Smith, F., Montgomery, R.: The Chemistry of Plant Gums and Mucilages. Reinhold Publishing Corporation, New York (1959).

30. Tokida, J., Hirose, H.: Advance of Phycology in Japan. VEB Gustav Fischer Verlag, Jena (1975).

31. Whistler, R. L., BeMiller, J. N.: Industrial Gums. Academic Press Inc., New York and London (1959).

32. Whistler, R. L, Smart, Ch. L.: Polysaccharide Chemistry. Academic Press Inc., New York (1953).

References

33. Abe, S., Kaneda, T.: The Effects of Edible Seaweeds on Cholesterol Metabolism in Rats. In: Proceedings of the VIIth International Seaweed Symposium. Sapporo, 1971. University of Tokyo Press (1972).

34. Adams, S. S. et al.: Lancet, 618 (1958).

35. Adey, W. H., McKibbin, D. L.: Studies on the Maerl Species Phymatolithon calcareum (Pallas) nov. comb. and Lithothamnium coralloides Crouan in the Ria de Vigo, Botanica Marina XIII (2): 100-106 (1970).

36. Agadi, V. V. et al.: Metal Concentration in Some Seaweeds of Goa (India). Botanica Marina XXI (4): 242-250 (1978).

37. Aktin, E., Güven, K. C.: Studies on Antilipemic Properties of the Algae Collected from Turkish Coasts. Qualitas Plantarum et Materiae Vegetabiles XII (2): 210-215 (1965).

38.: Aktin, E., Güven, K. C.: Studies on the Lipase Released by the Substances Obtained from Gracilaria verrucosa (Huds.) and Gelidium latifolium (Greville). Arzneim.-Forsch (Drug Res.) 19: 190-193 (1969).

39. Anderson, W., Soman, P. D.: Nature. 199: 389 (1963).

40. Anderson, W., Bowtle, W.: Determination of jota-carrageenan with 2-Thiobarbituric Acid, Analyst, Cambridge. 99 (1176): 178-183 (1974).

41. Anno, K. et al.: Isolation and Purification of Fucoidan from Brown Seaweed Pelvetia wrightii. Agr. Biol. Chem. 30, (5): 495-499 (1966).

42. Anno, K. et al.: Structural Studies on Fucoidan from Pelvetia wrightii. In: Proceedings of the VIth International Seaweed Symposium, Santiago de Compostela, 1969. Direccion General de Pesca Marítima, Madrid (1969).

43. Anno, K. et al.: Isolation of L-Fucose 4-sulfate from Fucoidan. Carbohydrate Research 13: 167-169 (1970).

44. Anno, K., Uemura, K.: Heterogeneity of Fucoidan Obtained from Pelvetia wrightii. In: Proceedings of the VIIth International Seaweed Symposium, Sapporo 1971. University of Tokyo Press (1972).

45. Araki, Ch.: Some Recent Studies on the Polysaccharides of Agarophytes. In: Proceedings of the Vth International Seaweed Symposium, Halifax, 1965. Pergamon Press, Toronto (1966).

46. Aruga, Y., Miura, A.: In Vivo Absorption Spectra and Pigment Content of the two Types of Color Mutants of Porphyra. In: Abstracts of the IXth International Seaweed Symposium, Santa Barbara, Cal., 1977. Journal of Phycology 13, Suppl. (1977).

47. Augier, H.: Les hormones des algues. État actuel de connaissances. Botanica Marina XIX, (3): 127-143 (1976). XIX, (4): 245-254 (1976). XIX, (6): 351-377 (1976). XX, (1): 1-11 (1977). XX, (3): 187-203 (1977). XX (6): 363-379 (1977). XX, (7): 440-470 (1977). XXI, (3): 175-197 (1978).

48. Auterhoff, H.: Die Japanische Pharmakopoe von 1961. Deutsche Apothekerzeitung 104 (16), (1964).

49. Baardseth, E.: Synopsis of Biolgical Data on Ascophyllum nodosum, (Linnaeus) Le Jolie, FAO Fisheries Synopsis No. 38 (1968).

50. Barilotti, C. C., Silverthorne, W.: A Resource Management Study of Gelidium robustum. In: Proceedings of the VIIth International Seaweed Symposium, Sapporo, 1971. University of Tokyo Press (1972).

51. Barilotti, D. C.: Environmental Factors that effect the Distribution and Abundance of Gelidium robustum. In: Abstracts of the VIIIth International Seaweed Symposium, Bangor, 1974. Marine Science Laboratories, Anglesey, U. K. (1974).

52. Barteling, S. J.: Clinische Chemie 15, 1002-1005 (1969).

53. Baslow, M. H.: Marine Pharmacology. The Williams and Wilkins Co. Baltimore (1969).

54. Batey, J. et al.: Variation in Structure among Galactan Sulphates Related to Agar. In: Abstracts of the VIIIth International Symposium, Bangor, 1974. Marine Science Laboratories, Anglesey, U. K. (1974).

55. Bérard-Therriault, L., Cardinal, A.: Variations de la teneur en acide alginique des Fucacées de l'estuaire du Saint-Laurent (Québec) en fonction de certains facteurs écologiques. Botanica Marina XVI, (2) 96-102 (1973).

56. Bernardi, G., Springer, G. F.: J. biol. Chem. 237, 75 (1962).

57. Bhakuni, D. S., Silva, M.: Biodynamic Substances from Marine Flora. Botanica Marina XVII, (1): 40-51 (1974).

58. Bhanderi, P. P., Trivedi, Y. A.: Rope Culture of Algin Yielding Seaweed Hormophysa triquetra (Linnaeus) Kutzing. Botanica Marina XX, (3): 183-185 (1977).

59. Black, W. A. P. et al.: The Evaluation of some Red Marine Algae as a Source of Carrageenan. Journal Sci. Food Agr. 16, (10): 543-585 (1965).

60. Blunden, G. et al.: Commercial Collection and Utilization of Maerl. Economic Botany, 29, (2): 140-145 (1975).

61. Blunden, G. et al.: The Composition of Maerl from the Glenan Islands of Southern Brittany. Botanica Marina, XX, (2): 121-125 (1977).

62. Bogorad, L.: Chlorophylls. In: Physiology and Biochemistry of Algae. Lewin, R. A. ed., Academic Press, New York and London (1962).

63. Bolliger, R.: Schweiz. Apotheker Zeitung, 96, 77-81 and 95-99 (1958).

64. Bonotto, S. et al.: Recent Advances in Research on the Marine Alga Acetabularia. Adv. mar. Biol., 14, 123-250 (1976).

65. Bowtle, W. J., Anderson, W.: Detection and Determination of Carrageenans in Biological Media. In: Abstracts of the VIIIth International Seaweed Symposium, Bangor, 1974. Marine Science Laboratories, Anglesey, U. K. (1974).

66. Brachet, J., Bonotto, S. (ed): Biology of Acetabularia. Academic Press, New York and London (1970).

67. Brooker, S. G., Cooper, R. C.: New Zealand Medicinal Plants. Economic Botany, 15, (1): 1-10 (1961).

68. Burkholder, P. R., Sharma, G. M.: Antimikrobial agents from the sea. Lloydia 32, (4): 466 (1969).

69. Cabioch, J.: Contribution à l'Étude Morphologique, Anatomique et Systématique de deux Mélobésiées: Lithothamnium calcareum (Pallas) Areschoug et Lithothamnium corallioides Crouan. Botanica Marina IX, (1/2): 33-53 (1966).

70. Cardinal, A., Breton-Provencher, M.: Variations de la teneur en acide alginique des Laminariales de l'estuaire maritime du Saint-Laurent. Botanica Marina, XX, (4): 243-251 (1977).

71. Cheney, D. P., Dawes, C. J.: Ecological Studies of Florida Eucheuma. In: Abstracts of the VIIIth International Seaweed Symposium, Bangor, 1974. Marine Science Laboratories Anglesey (1974).

72. Cheng, Tien-Hsi: Production of Kelp - A Major Aspect of China's Exploration of the Sea. Economic Botany, 23, (3): 215-236.

73. Chihara, M.: Rhodophyta, their Lif-histories. In: Advance of Phycology in Japan, Tokida, Hirose ed. VEB Gustav Fischer Verlag, Jena (1975).

74. Chiu, B.: Seaweeds of Economic Importance in South China. In: International Seaweed Symposium, Trondheim, 1955. Pergamon Press London-New York (1956).

75. Christiansen, E.: Danish Agar. In: Industrial Gums. Whistler, R. L., Be Miller, J. N., ed.,Academic Press Inc., New York (1959).

76. Coll, J., Cox, J.: The Genus Porphyra C. Ag. (Rhodophyta, Bangiales) in the American North Atlantic. I. New Species from North Carolina. Botanica Marina, XX, (3): 155-159 (1977).

77. Commercial Fisheries Review. California's Giant Kelp, 33, (3): 30-32 (1971). WFA 22, (3): 47 (1971).

78. Conrad, H. M., Saltman, P.: Growth Substances. In: Physiology and Biochemistry of Algae. Lewin, R. A. ed.,Academic Press, New-York and London (1962).

79. Craigie, J. S., Wong, K.: Carrageenan Biosynthesis. In: Abstracts of the IXth International Seaweed Symposium, Santa Barbara, 1977. Journal of Phycology 13, Suppl. (1977).

80. Czetsch Lindenwald, H. v.: Algenprodukte in der Pharmazie. Botanica Marina III, (1): 22-29 (1961).

81. Daigo, K.: Studies on the Constituents of Chondria armata L. Pharmaceutical Faculty, University of Osaka. 79, (3), Nr. 76-80 (1959).

82. Dawes, C. J.: Ecological Studies of Florida Eucheuma. In: Abstracts of the VIIth International Seaweed Symposium, Bangor, 1974. Marine Science Laboratories, Anglesey (1974).

83. Dawes, C. J.: Seasonal and Reproductive Aspects of Plant Chemistry and ι-Carrageenan from Floridian Eucheuma. Botanica Marina, XX, (3): 137-147 (1977).

84. Dawes, C. J.: A Photosynthetic and Biochemical Comparison of Eucheuma from the Gulfs of Mexico and California. In: Abstracts of the IXth International Seaweed Symposium, Santa Barbara, Cal., 1977. Journal of Phycology, 13, Suppl. (1977).

85. Dawes, C. J. et al.: Physiological and Biochemical Studies on the ι-Carrageenan Producing Red Alga Eucheuma uncinatum Setchell and Gardner from the Gulf of California. Botanica Marina, XX, (7): 437-442 (1977).

86. Day, D. F., Yaphe, W.: Enzymatic Hydrolysis of Agarose. In: Abstracts of the VIIIth International Seaweed Symposium, Bangor, 1974. Marine Science Laboratories, Anglesey, U. K. (1974).

87. Dea, J. C. M., Scott, W. E.: The Agarose Double Helix and its Function in Agarose Gel Structures. In: Abstracts of the VIIIth International Seaweed Symposium, Bangor, 1974. Marine Science Laboratories, Anglesey, U. K. (1974).

88. Der Marderosian, A. H.: Current status of marine biomedicinals. Lloydia 32, (4): 438 (1969).

89. Devi, J. S., Durvasula, S. V.: Primary Production in Relation to Tracemetal Variations. In: Abstracts of the IXth International Seaweed Symposium, Santa Barbara, Cal., 1977. Journal of Phycology, 13, Suppl. (1977).

90. Diluzio, N. R.: U. S.-Patent 3081226, March 12, 1963.

91. Doty, M. S., Aguilar-Santos, G.: Nature, 211, 990 (1966).

92. Doty, M. S., Aguilar-Santos, G.: Transfer of Toxic Algal Substances in Marine Food Chains. Pacif. Sci. 24, (3): 351-355 (1970).

93. Doty, M. S., Alvarez, V. B.: The Productivity of Eucheuma Farms. In: Abstracts of the VIIIth International Seaweed Symposium, Bangor, 1974. Marine Science Laboratories, Anglesey (1974).

94. Doty, M. S., Santos, G. A.: Eucheuma and its Gels. In: Abstracts of the IXth International Seaweed Symposium, Santa Barbara, Cal., 1977. Journal of Phycology, 13, Suppl. (1977).

95. Elsner, H. et al.: Tierversuche mit einem gerinnungshemmenden Algenstoff aus Delesseria sanguinea (L.) Lam. Naunym-Schmiedebergs Archiv f. exp. Pathologie und Pharmakologie, 190, (5): 510-514 (1938).

96. FAO Fisheries Report No. 188, (1976). Culture of Algae and Seaweeds. In: Report of the FAO Technical Conference of Aquaculture, Kyoto, Japan.

97. Fenical, W., Sims, J. J.: Chondriol. Tetrahedron Lett. (1973): 313. Chemiker Zeitung 97, (7): - (1973).

98. Fenical, W., Sims, J. J.: Cycloeudesmol. Tetrahedron Lett. (1974): 1137. Chemiker Zeitung 98, (6): - (1974).

99. Fischer, F. G., Dörfel, H.: Die Uronsäuren der Braunalgen. Hoppe-Seyler's Zschft. physiol. Chem. 302, 186-203 (1955).

100. Floch, J.-Y., Penot, M.: Influence de la Cycloheximide sur le transport a longue distance du ^{32}P chez Laminaria digitata (Linné) Lamouroux. Botanica Marina, XXI, (1): 7-13 (1978).

101. Fujiwara-Arasaki, T., Mino, N.: The Distribution of Trimethylamine and Trimethylamine Oxide in Marine Algae. In: Proceedings of the VIIth International Seaweed Symposium, Sapporo, 1971. University of Tokyo Press (1972).

103. Gayral, P., Cosson, J.: Exposé synoptique des données biologiques sur la Laminaire digitée- Laminaria digitata. Synopsis FAO sur les pêches, No. 89, Rom (1973).

104. Gerber, P. et al.: Proc. Soc. Exptl. Biol. Med. 99, 590 (1958).

105. Gesellschaft für Arzneipflanzenforschung: 17. Vortragstagung, Kiel, 1969. Deutsche Apotheker-Zeitung, 109, (25): 955 (1969).

106. Gessner, F.: Vortrag. Gesellschaft für Arzneipflanzenforschung, Kiel (1969).

107. Glombitza, K.-W.: Highly Hydroxylated Phenols from Phaeophyceae. In: Abstracts of the VIIth International Seaweed Symposium, Bangor, 1974. Marine Science Laboratories, Menai Bridge, Anglesey, U. K. (1974).

108. Gordon-Mills, E., McCandless, E. L.: Studies on Carrageenan in the Cell Walls of Chondrus crispus. In: Abstracts of the VII th International Seaweed Symposium, Bangor, 1974. Marine Science Laboratories, Anglesey, U. K. (1974).

109. Güven, K. C., Aktin, E.: A Study on the Antilipemic and Anticoagulant Activities of Phyllophora nervosa (D. C.) Grev. Eczacilik, Bülteni 4, (12): 199-206 (1962).

110. Güven, K. C., Aktin, E.: Studies on Antilipemic and Anticoagulant Properties of the Algae Collected from Turkish Coasts. Botanica Marina, VII, (1/4): 1-3 (1964).

111. Güven, K. C. et al.: Investigations on the Properties of the Agar Obtained from Phyllophora nervosa (D. C.) Grev. Eczacilik Bülteni, VIII, (1): 10-14 (1966).

112. Güven, K. C. et al.: About Alkaloid Content of Marine Algae. Eczacilik Bülteni, XI, 177-184 (1969).

113. Güven, K. C. et al.: Untersuchungen von Meeresalgen. Deutsche Apotheker Zeitung 110, (30) (1970).

114. Güven, K. C., Tekinalp, B.: About Sargassum vulgare Agardh. Botanica Marina, XIV, (1): 63-64 (1971).

115. Güven, K. C. et al.: Investigations on Phyllophora nervosa (D. C.) Grev. Botanica Marina, XV, (1): 46-48 (1972).

116. Güven, K. C., Bergiadi, N.: Studies on Cystoseira barbata J. Ag. Botanica Marina, XVI, (1): 49-51 (1973).

117. Güven, K. C. et al.: Anticoagulant-Antithrombin and Fibrinolytic Actions of Extract of Marine Alga Corallina rubens L. Haemostasis, 2: 260-268 (1973/4).

118. Güven, K. C. et al.: Lipolytic and Hypoglycemic Compounds Obtained from Cystoseira barbata J. Ag. Arzneim.-Forsch.(Drug Res.) 24, (2): 144-147 (1974).

119. Güven, K. C. et al.: Chemical and Pharmacological Investigations on Corallina rubens. Planta medica, 27, (1): 5-13 (1975).

120. Güven, K. C., Güven, N.: Analysis of Protein Fractions Obtained from Sargassum vulgare Agardh. Eczacilik Bülteni, XVII, 13-16 (1975).

121. Güven, K. C., Bergiadi, N.: Sterol of Corallina granifera Ell. et Sol. Eczacilik Bülteni, XVII, 30-32 (1975).

122. Güven, K. C. et al.: Biochemical Investigations of Cystoseira corniculata Hauck. Botanica Marina, XIX, (5): 323-325 (1976).

123. Guiseley, K. B.: Viscosimetric Determination of Agarose Gelling Temperature. In: Proceedings of the VIIth International Seaweed Symposium, Sapporo, 1971. University of Tokyo Press (1972).

124. Guzmán-del Próo, S. A. et al.: Shedding Rhythm and Germination of Spores in Gelidium robustum. In: Proceedings of the VIIth International Seaweed Symposium, Sapporo, 1971. University of Tokyo Press (1972).

125. Hartmann, T., Aufermann, B.: Formation and Function of Simple Monoamines in Red Algae. In: Abstracts of the VIIIth International Seaweed Symposium, Bangor, 1974. Marine Science Laboratories, Anglesey, U. K. (1974).

126. Hase, E.: Recent progress in the physiology of Chlorella. In: Advance of Phycology in Japan. Tokida,Hirose, ed., VEB Gustav Fischer Verlag, Jena (1975).

127. Hasegawa, Y.: Forced Cultivation of Laminaria. In: Proceedings of the VII th International Seaweed Symposium, Sapporo, Japan, 1971. University of Tokyo Press (1972).

128. Hashimoto, Y. et al.: Screening of the Toxic Algae on Coral Reefs. In: Proceedings of the VIIth International Seaweed Symposium, Sapporo, 1971. University of Tokyo Press (1972).

129. Haug, A., Larsen, B.: Studies on the Composition and Properties of Alginates. In: Proceedings of the IVth International Seaweed Symposium, Biarritz, 1961. Pergamon Press, Oxford-London-New York-Paris (1964).

130. Hawkins, W. W., O'Neill, A. N.: Can. J. Biochem. Phys., 33, 545 (1955).

131. Hawkins, W. W., Leonard, V. C.: Antithrombic Activity of Carrageenan in Human Blood. Can. J. Biochem. Physiol., 41, (5): 1325-1327 (1963).

132. Hirase, S., Watanabe, K.: Fractionation and Structural Investigation of Funoran. In: Proceedings of the VIIth International Seaweed Symposium, Sapporo, 1971. University of Tokyo Press, 451-454 (1972).

133. Hirose, H.: Photosynthetic pigments of algae and algal phylogeny. In: Advance of Phycology in Japan. Tokida, Hirose, ed., VEB Gustav Fischer Verlag, Jena (1975).

134. Hofsten, A. v. et al.: Localization of Bromine in the Chloroplasts of the Red Alga Lenormandia prolifera. Botanica Marina, XX, (4): 267-270 (1977).

135. Hong, K. C. et al.: The Effects of Hypnea japonica Extract on Bone Metabolism in the Rat. In: Proceedings of the VIIth International Seaweed Symposium, Sapporo, Japan, 1971. University of Tokyo Press (1972).

136. Hong, K. C., Cruess, R. L.: Isolation of a Collagenolytic Enzyme by the Implantation of Algal Polysaccharides under Rat Skin. In: Abstracts of the VIIth International Seaweed Symposium, Bangor, 1974. Marine Science Laboratories, Anglesey, U. K. (1974).

137. Hoppe, H. A.: Geschichtliche Entwicklung der Auswertung von Meeresalgen. Botanica Marina, III, Suppl., 12-15 (1962).

138. Hoppe, H. A., Schmid, O. J.: Meeresalgen als moderne Industrieprodukte. Botanica Marina, III, Suppl., 16-66 (1962).

139. Hoppe, H. A.: Beiträge zur Terminologie der Phykokolloide aus Rhodophyceen. Botanica Marina, IX, (1/2): 5 (1966).

140. Hoppe, H. A.: Meeresalgen und Meeresalgenprodukte in der Pharmazie. Deutsche Apotheker-Zeitung, 106, (7): 197-200 (1966).

141. Hoppe, H. A.: Nahrungsmittel aus Meeresalgen. Botanica Marina, IX, Suppl., 19-40 (1966).

142. Ihingran, V. G., Gopalakrishnan, V.: A catalogue of cultivated aquatic organisms. FAO Fisheries Technical Paper No. 130 (1974).

143. Ikawa, T. et al.: Comperative Studies on Fructose Diphosphate Aldolases mainly in Marine Algae. In: Proceedings of the VIIth International Seaweed Symposium, Sapporo, 1971. University of Tokyo Press (1972).

144. Imbamba, S. K.: Mineral Element Content of some Benthic Marine Algae of the Kenya Coast. Botanica Marina, XV, (2): 113-115 115 (1972).

145. Ito, K., Tsuchiya, Y.: The Effect of Algal Polysaccharides on the Depressing of Plasma Cholesterol Levels in Rats. In: Proceedings of the VIIth International Seaweed Symposium, Sapporo, 1971. University of Tokyo Press (1972).

146. Ito, K., Tsuchiya, Y.: Differential Fatty Acid Composition of Some Marine Algae Associated with Their Habitat Depths. In: Abstracts of the VIIIth International Seaweed Symposium, Bangor, 1974. Marine Science Laboratories, Menai Bridge, Anglesey, U. K. (1974).

147. Jacobi, G.: Enzyme Systems. In: Physiology and Biochemistry of Algae. Lewin, R. A., ed. Academic Press, New York and London (1962).

148. Jensen, A.: Journal of the Science of Food and Agriculture 20, (10): 622-626 (1969). WLD Fish. Abstr. 21, (2): 23 (1970).

149. Jensen, A.: The Nutritive Value of Seaweed Meal for Domestic Animals. In: Proceedings of the VIIth International Seaweed Symposium, Sapporo, 1971. University of Tokyo Press (1972).

150. Jensen, A., Zavodnik, N.: Studies on the Phenolic Content of Some Brown Algae. In: Abstracts of the VIIIth International Seaweed Symposium, Bangor, 1974. Marine Science Laboratories, Menai Bridge, Anglesey, U. K. (1974).

151. Jensen, A.: The Industrial Utilization of Seaweeds - Past - Present and Future. In: Abstracts of the IXth International Seaweed Symposium, Santa Barbara, Cal., 1977. Journal of Phycology 13, Suppl. (1977).

152. Johnson, N. G., Levring, T.: On the Production of Vitamin D in the Sea. Svenska Hydrografisk-Biologiska Kommissionens Skrifter: Hydrografi 1, (3): 3-7 (1947).

153. Johnston, H. W.: A Detailed Chemical Analysis of Some Edible Japanese Seaweeds. In: Proceedings of the VIIth International Seaweed Symposium, Sapporo, 1971. University of Tokyo Press (1972).

154. Jolles, B. et al.: Brit. J. Cancer 17, 169 (1963).

155. Jones, R. T. et al.: Effects of Dietary Ascophyllum nodosum on Blood Parameters of Rabbits, Rats and Pigs. In: Abstracts of the VIIIth International Seaweed Symposium, Bangor, 1974. Marine Science Laboratories, Anglesey, U. K. (1974).

156. Kain, J. M.: Synopsis of Biological Data on Laminaria hyperborea. FAO Fisheries Synopsis, No. 87, Rom (1971).

157. Kaliaperumal, N., Kalimuthu, S.: Changes in Growth, Reproduction, Alginic Acid and Mannitol Contents of Turbinaria decurrens Bory. Botanica Marina, XIX, (3): 161-178 (1976).

158. Kalimuthu, S. et al.: Alginic Acid and Mannitol Contents in Relation to Growth in Stoechospermum marginatum (C. Agardh) Kuetzing. Abstracts of IXth International Seaweed Symposium, Santa Barbara, Cal., 1977. Journal of Phycology 13, Suppl. (1977).

159. Kameda: Japan. Medical Gazette 3, (5): 5 (1966).

160. Kanazawa, A., Yoshioka: The Occurrence of Cholest-4-en-3-one in the Red Alga Gracilaria textorii. In: Proceedings of the VIIth International Seaweed Symposium, Sapporo, 1971. University of Tokyo Press (1972).

161. Kaneda, T., Ando, H.: Component Lipids of Purple Laver and their Antioxygenic Activity. In: Proceedings of the VIIth International Seaweed Symposium, Sapporo, 1971. University of Tokyo Press (1972).

162. Kashiwabara, Y. et al.: Alginate Lyases of Pseudomonads. The Journal of Biochemistry 66, (4): 503-512 (1969).

163. Katayama, T.: a) Nippon Suisan Gakkaishi 24, 346, 925 (1959). CA 53, 11532 c (1959), CA 57, 15512 f (1962). b) Nippon Suisan Gakkaishi 26, 818 (1960). CA 55, 15630 h (1961). c) Nippon Suisan Gakkaishi 27, 75 (1961).

164. Kida, W.: Studies on the Morphology and Ecology of Monostroma in Japan. Journal of Faculty of Fisheries. Prefectural University of Mie. 7, (1): 81-164 (1967).

165. Kida, W.: The Present State and Future of Green-laver Cultivation in Japan. Bull. Jap. Soc. Phycol. XV, (3): 1-3 (1967).

166. Kim, Ch. S.: The Effects of Dietary Sargassum natans and Ascophyllum nodosum on Salmonella gallinarum Infection in chicks. In: Proceedings of the VIIth International Seaweed Symposium, Sapporo, Japan, 1971. University of Tokyo Press (1972).

167. Kneifel, H.: Untersuchungen über Inhaltsstoffe von Algen. Chemiker-Zeitung 101, (4): 165-168 (1977).

168. Krauss, R. W. ed.: The Marine Plant Biomass of the Pacific Northwest Coast. Oregon State University Press, Corvallis (1978).

169. Kroes, H. W.: Excretion of Mucilage and Yellow-brown Substances by some Brown Algae from the Intertidal Zone. Botanica Marina XIII, (2): 107-110 (1970).

170. Kuriyama, M. et al.: On a New Amino Acid "Chondrine", isolated from the Red Alga Chondria crassicaulis. Bulletin of the Faculty of Fisheries, Hokkaido University 11, (2): 58-66 (1960).

171. Landsberger, W. et al.: Determinacion cromatografica de amino acidos libres en algas pardas Chilenas. Botanica Marina XII, (1-4): 218-227 (1969).

172. Larsen, B., Haug, A.: The Distribution of Iodine and Other Constituents in Stipe of Laminaria hyperborea (Gunn.) Foslie. Botanica Marina II, (3/4): 250-254 (1961).

173. Larsen, B. et al.: Preparation and Structural Investigation of Ascophyllan. In: Proceedings of the Vth International Seaweed Symposium, Halifax, 1965. Pergamon Press, Toronto (1966).

174. Larsen, B., Haug, A.: Biosynthesis of Alginate. In: Proceedings of the VIIth International Seaweed Symposium, Sapporo, 1971. University of Tokyo Press (1972).

175. Lesser, M. A.: Alginates in Drugs and Cosmetics. Economic Botany 4, (4): 317-321 (1950).

176. Lestang, G. de, Quillet, M.: The Sulphated Fucoidan of Phaeophyceae as a Factor of Ecological Adaptation. In: Abstracts of the VIIIth International Seaweed Symposium, Bangor, 1974. Marine Science Laboratories, Anglesey (1974).

177. Levring, T.: Potential Yields of Marine Algae - with Special Reference to European Species. In: The Marine Plant Biomass of the Pacific Northwest Coast ed. by R. W. Krauss. Oregon State University Press, Corvallis (1978).

178. Litvinov, V. S.: Ref. ex Nahrung 6, (4): Beiheft 50 (1962).

179. Loni, M. C. et al.: Cap Morphology Statistics on a Natural Population of Acetabularia mediterranea. Archives de Biologie 86, (4): 413-421 (1975).

180. Loose, G.: Die Bedeutung mariner Algen für die Welternährung. Botanica Marina IX, Suppl., 7-18 (1966).

181. Lüning, K.: Cultivation of Laminaria hyperborea in situ and in continuous darkness under laboratory conditions. Helgoländer wiss. Meeresunters. 20, 79-88 (1970).

182. Lüning, K.: Seasonal Growth of Laminaria hyperborea under Recorded Underwater Light Conditions Near Helgoland. IVth European Marine Biology Symposium, ed. by D. J. Crisp, Cambridge University Press (1971).

183. Maass, H.: Alginsäure und Alginate. Strassenbau, Chemie und Technik, Verlagsgesellschaft mbH., Heidelberg (1959).

184. Maass, H.: Über die Verwendungsmöglichkeiten von Alginaten. Botanica Marina III, Suppl., 86-100 (1962).

185. MacFarlane, C. J.: Chondrus crispus Stackhouse - A Synopsis. Seaweed Division, Nova Scotia Research Foundation, Halifax, Canada (1968).

186. Madgwick, J. C., Ralph, B. J.: Free Amino Acids in Australian Marine Algae. Botanica Marina XV, (4): 205-209 (1972).

187. Mairh, O. P., Sreenivasa, Rao, P.: Culture Studies on Gelidium pusillum (Stackh.) Le Jolis. Botanica Marina XXI, (3): 169-174 (1978).

188. Mallery, Ch. H., Richardson, N.: Disc Gel Electrophoresis of Proteins of Rhodophyta. In: Proceedings of the VIIth International Seaweed Symposium, Sapporo, 1971. University of Tokyo Press (1972).

189. Martínez Nadal et al.: Sarganin and Chonalgin, New Antibiotic Substances from Marine Algae from Puerto Rico. Antimicrobial Agents and Chemotherapy, 68-72 (1963).

190. Martínez Nadal et al.: Antibiotic Properties of Marine Algae. Botanica Marina IX, (1/2): 21-26 (1966).

191. Martínez Nadal et al.: Low Toxic Effect of Antimicrobial Substances from Marine Algae. Botanica Marina IX, (1/2): 62-63 (1966).

192. Mason, C. P.: Studies of Marine Algal Proteins Using Acrylamide Gel Electrophoresis. Abstracts of the IXth International Seaweed Symposium, Santa Barbara, Cal., 1977. Journal of Phycology 13 Suppl. (1977).

193. Mateus, H. et al.: The Amino Acid Composition of the Marine Brown Alga Macrocystis pyrifera from Baja California, Mexico. Botanica Marina XIX, (3): 155-159 (1976).

194. Mathieson, A. C., Tveter, E.: Carrageenan Ecology of Chondrus crispus and Gigartina stellata. In: Abstracts of the VIIIth International Seaweed Symposium, Bangor, 1974. Marine Science Laboratories, Anglesey, U. K. (1974).

195. Matsuhashi, T., Hayashi, K.: Rheological Behavior of Agar Gel Processed from Gracilaria foliifera of Florida. In: Proceedings of the VIIth International Seaweed Symposium, Sapporo, 1971. University of Tokyo Press (1972).

196. Matsuhiro, B., Zanlungo, A. B.: Polysaccharides from Chilean Seaweeds, II. Studies on Iridaea laminarioides. Botanica Marina XIX, (5): 317-321 (1976).

197. Mautner, H. G.: The Chemistry of Brown Algae. Economic Botany 8, (2): 174-192 (1954).

198. McCandless, E. L., Richer, S.: ^{14}C Studies of Carrageenan Synthesis. In: Proceedings of the VIIth International Seaweed Symposium, Sapporo, 1971. University of Tokyo Press (1972).

199. McCandless, E. L.: Biological Regulation of Carrageenan Structure. In: Abstracts of the VIIIth International Seaweed Symposium, Bangor, 1974. Marine Science Laboratories, Anglesey, U. K. (1974).

200. McCandless, E. L., Craigie, J. S.: Reevaluation of Seasonal Factors Involved in Carrageenan Production by Chondrus crispus: Carrageenans of Carposporic Plants. Botanica Marina XVII, (3): 125-129 (1974).

201. McDowell, R. H.: J. Soc. Cosmet. Chem. 21, (7): 441-457 (1970).

202. McLachlan, J. et al.: Porphyra linearis Grev. - An Edible Species of Nori from Nova Scotia. In: Proceedings of the VIIth International Symposium, Sapporo, 1971. University of Toky Press (1972).

203. McNeely, W. H.: Algin. In: Industrial Gums. Whistler, R. L. - J. N. Be Miller, ed. Academic Press Inc., New York (1959).

204. McNeely, W. H.: Fucoidan. In: Industrial Gums. Whistler, R. L. - J. N. Be Miller, ed. Academic Press Inc., New York (1959).

205. Medcalf, D. G. et al.: Chemical Characterization of Mucilaginous Polysaccharides from Ulvaceae Species. In: Proceedings of the VIIth International Seaweed Symposium, Sapporo, 1971. University of Tokyo Press (1971).

206. Meeuse, B. J. D.: Storage Products. In: Physiology and Biochemistry of Algae Lewin, R. A., ed. Academic Press, New York and London (1962).

207. Michanek, G.: Seaweed Resources of the Ocean. FAO Fisheries Technical Paper No. 138 (1975).

208. Miller, J. D. A.: Fats and Steroids. In: Physiology and Biochemistry of Algae Lewin, R. A., ed. Academic Press, New York and London (1962).

209. Miura, A.: Porphyra Cultivation in Japan. In: Advance of Phycology in Japan. Tokida, Hirose ed. VEB Gustav Fischer Verlag, Jena (1975).

210. Mohsen, A. F. et al.: Amino Acid Pattern and Seasonal Variation in Some Marine Algae from Alexandria. Botanica Marina XVIII, (3): 167-178 (1975).

211. Mollion, M. J.: Étude preliminaire des Hypnea au Senegal comme Source de phycocolloides. Botanica Marina XVI, (4): 221-225 (1973).

212. Mowat, J. A.: Auxins and Gibberellins in Marine Algae. In: Proceedings of the IVth International Seaweed Symposium, Biarritz, 1961. Pergamon Press, Oxford - London - New York - Paris (1964).

213. Mowat, J. A.: A Survey of Results on the Occurrence of Auxins and Gibberellins in Algae. Botanica Marina VIII, (1): 149-155 (1965).

214. Mshigeni, K. C.: Development Studies in Hypnea cervicornis J. Agardh and Hypnea chordacea Kützing: Spore Germination. Botanica XIX, (4): 217-221 (1976).

215. Mshigeni, K. C.: Holdfast and Shoot Differentiation in the Sporelings of Hypnea cervicornis J. Agardh. XIX, (4): 223-226 (1976).

216. Mshigeni, K. C.: Field Cultivation of Hypnea Spores for Carrageenan: Prospects and Problems. Botanica Marina XIX, (4): 227-230 (1976).

217. Mshigeni, K. E.: On the Development of Hyaline Hairs in Hypnea Lamouroux. Botanica Marina XIX, (5): 304-311 (1976).

218. Mshigeni, K. E.: A Note on the Fate of Tetra-sporangial Stichidia in Hypnea Lamouroux after Spore Release. Botanica Marina XIX, (5): 313-315 (1976).

219. Mshigeni, K. E.: Studies on the Reproduction of Selected Species of Hypnea from Hawaii. Botanica Marina XIX, (6): 341-346 (1976).

220. Mshigeni, K. E., Semesi, A. K.: Studies on Carrageenans from the Economic Red Algal Genus Eucheuma in Tanzania. Botanica Marina XX, (4): 239-242 (1977).

221. Mshigeni, K. E.: Seasonal Changes in the Standing Crops of Three Hypnea Species in Hawaii. Botanica Marina XX, (5): 303-306 (1977).

222. Mshigeni, K. E.: On the Ultrastructure of the Cell Wall and Pit Plugs of Chondrococcus hornemannii (Rhodophyta, Cryptonemiales). Botanica Marina XX, (6): 345-346 (1977).

223. Mshigeni, K. E., Lorri, W. S. M.: Spore Germination and Early Stages of Development in Hypnea valentiae (Turner) Montagne (Rhodophyta, Gigartinales). Botanica Marina XX, (6): 381-383 (1977).

224. Mshigeni, K. E., Nzalalila, E. V.: Contributions on the Content and Nature of the Phycocolloid from Laurencia papillosa (Forssk) Greville (Rhodophyta, Ceramiales). Botanica Marina XX, (7): 443-447 (1977).

225. Müller-Plettenberg, D.: Bericht über den 52. Fortbildungskongreß für Naturheilverfahren, 1977. Deutsche Apotheker Zeitung 117, (31): (1977).

226. Munda, J. M., Gubensek, F.: The Amino Acid Composition of Some Common Marine Algae from Iceland. Botanica Marina XIX (2): 85-92 (1976).

227. Munda, J. M.: Trace Metal Concentrations in Some Icelandic Seaweeds. Botanica Marina XXI, (4): 261-263 (1978).

228. Munda, J., Pedersen, P. M.: Porphyra thulaea sp. nov. from East Iceland and West Greenland. Botanica Marina XXI, (5): 283-288 (1978).

229. Nakayama, T. O. M.: Carotenoids. In: Physiology and Biochemistry of Algae. Lewin, R. A., ed. Acad. Press, New York and London (1962).

230. Nizizawa, K. et al.: Further Studies on the Fine Structure of Alginic Acid. In: Proceedings of the VIIth International Seaweed Symposium, Sapporo, 1971. University of Tokyo Press (1972).

231. Nisizawa, K. et al.: Alginate Lyases in Brown Algae. In: Abstracts of the VIIIth International Seaweed Symposium, Bangor, 1974. Marine Science Laboratories,Anglesey (1974).

232. Nisizawa, K., Ikawa, T.: Photosynthesis in Algae other than Chlorella. In: Advance of Phycology in Japan, Tokida, Hirose ed. VEB Gustav Fischer Verlag, Jena (1975).

233. Nisizawa, K., Sasaki, S.: Cell wall composition of algae from a phylogenetic point of view. In: Advance of Phycology in Japan. Tokida, Hirose, ed. VEB Gustav Fischer Verlag, Jena (1975).

234. Nisizawa, K. et al.: Circadian Rhythm of Several Enzyme Activities Related to the Photoassimilation of CO_2 in Brown Algae. In: Abstracts of the IXth International Seaweed Symposium, Santa Barbara, Cal., 1977. Journal of Phycology 13, Suppl. (1977).

235. O'Colla, P. S. Mucilages. In: Physiology and Biochemistry of Algae. R. A. Lewin, ed. Acad. Press, New York and London (1962).

236. Ogata, E.: Physiology of Porphyra. In: Advance of Phycology in Japan, Tokida, Hirose ed. VEB Gustav Fischer Verlag, Jena (1975).

237. Ogino, C.: Tannins and Vacuolar Pigments. In: Physiology and Biochemistry of Algae, Lewin, R. A., ed. Acad. Press, New York and London (1962).

238. Óh Eocha, C.: Phycobilins. In: Physiology and Biochemistry of Algae, Lewin, R. A., ed. Acad. Press, New York and London (1962).

239. Oishi, K. et al.: Bulletin of the Japanese Society of Scientific Fisheries 33, (1): 41-44 (1967).

240. Okazaki, M.: Some Enzymatic Properties of Ca^{2+}- dependent Adenosine Triphosphatase from a Calcareous Red Alga, Serraticardia maxima and Its Distribution in Marine Algae. Botanica Marina XX, (6): 347-354 (1977).

241. Oliveira Filho, E. C.: The Occurrence and Distribution of Seaweeds of Economical Value in Brazil. In: Abstracts of the VIIIth International Seaweed Symposium, Bangor, 1974. Marine Science Laboratories, Anglesey, U. K. (1974).

242. Oliveira Filho, E. C., Paula, E. J. de: Potentiality for Alginate Production in S. Paulo Littoral (Brazil). In: Abstracts of the IXth International Seaweed Symposium, Santa Barbara, Cal., 1977. Journal of Phycology 13, Suppl. (1977).

243. Oza, R. M., Sreenivasa Rao, P.: Effects of Different Media on Growth and Sporulation of Laboratory Raised Germlings of Ulva fasciata Delile. Botanica Marina XX, (7): 427-431 (1977).

244. Paskins- Hurlburt, A. J. et al.: Carrageenan and the Binding of Lead. Botanica Marina XIX, (1): 59-60 (1976).

245. Paskins-Hurlburt, A. J. et al.: Isolation and Metal Binding Properties of Fucoidan. Botanica Marina XIX, (5): 327-328 (1976).

246. Paskins-Hurlburt, A. J. et al.: Fucoidan: Its Binding of Lead and Other Metals. Botanica Marina XXI, (1): 13-22 (1978).

247. Penniman, C.: Seasonal, Chemical and Reproductive Changes in Gracilaria foliifera (Forssk) Børg. from Great Bay, New Hampshire (USA). In: Abstracts of the IXth International Seaweed Symposium, Santa Barbara, 1977. Journal of Phycology 13, Suppl. (1977).

248. Percival, E., McDowell, R. H.: Chemistry and Enzymology of Marine Algal Polysaccharides. Acad. Press Inc. Ltd., London and New York (1967).

249. Pham Quang, L., Laur, M.-H.: Variations de la composition et de la teneur en lipides polaires des trois Fucacées communes des côtes bretonnes: Pelvetia canaliculata (L.) Decn. et Thur., Fucus vesiculosus L., Fucus serratus L. au cours de leur cycle annuel. Botanica Marina XVIII, (2): 105-113 (1975).

250. Pillay, K. S.: Colorless High Polymer Alginic Acid from Mixed Indian Sargassum. In: Abstracts of IXth International Seaweed Symposium, Santa Barbara, Cal., 1977. Journal of Phycology 13, Suppl. (1977).

251. Pohl, P. et al.: Inhaltsstoffe von Algen über die unterschiedliche Fettsäurezusammensetzung von Salz- und Süsswasseralgen. Phytochemistry 7, 1565-1572 (1968).

252. Pohl, P.: Stickstoff im Stoffwechsel von Algen. Deutsche Apotheker-Zeitung 113, (42): (1973).

253. Powell, J. H., Meeuse, B. J. T.: Laminarin in Some Phaeophyta of the Pacific Coast. Economic Botany 18, (2): 164-166 (1964).

254. Pringsheim, E. G.: Vergangenheit und Zukunft der Algenforschung. Naturw. Rdsch. 23, (12): 502-504 (1970).

255. Ragan, M. A.: Physodés and the Phenolic Compounds of Brown Algae. Composition and Significance of Physodes in vivo. Botanica Marina XIX, (3): 145-154 (1976).

256. Rama Rao, K., Krishnamurthy, V.: Study of the Preparation from Hypnea musciformis (Wulf) Lamour from Veraval, Gujarat Coast. Botanica Marina XI, (1/4): 129-133 (1968).

257. Rama Rao, K.: Studies on Growth Cycle and Phycocolloid Content in Hypnea musciformis (Wulf) Lamour. Botanica Marina XIII, (2): 163-165 (1970).

258. Rama Rao, K.: Studies on Indian Hypnaceae. Botanica Marina XX, (1): 33-39 (1977).

259. Rama Rao, K., Krishnamurthy, V.: Seasonal Variation in Phycocolloid Content in Two Species of Hypnea. Botanica Marina XXI, (4): 257-259 (1978).

260. Reiner, E. et al.: Can. J. Bioch. Physiol. 40, 1401 (1962).

262. Ricohermoso, M. A., Deveau, L. E.: Commercial Propagation of Eucheuma spp. Clones in the South China Sea. IXth International Seaweed Symposium, Santa Barabara, Cal., 1977.

262. Rubin, B.: Laminaria digitata: A checkered Career. Economic Bot. 31, (1): 66-71 (1977).

263. Rusanowski, P. C., Vadas, R. L.: Localization of Laminaran and a Model for Cell Metabolite Utilization. In: Abstracts of the VIIIth International Seaweed Symposium, Bangor, 1974. Marine Science Laboratories, Anglesey (1974).

264. Safe, L. M. et al.: Sterols in marine algae. J. Pharmaceut. Sci. 63, (3): 464-466 (1974).

265. Saito, K., Nakamura, J.: J. Chem. Soc. Japan, Pure Chem. Sect. 72, 992, (1950).

266. Saito, Y.: Undaria. In: Advance of Phycology in Japan, Tokida, Hirose, ed. VEB Gustav Fischer Verlag, Jena (1975).

267. Sanford, F. B.: Fishery Leaflet 4 69 (1958). US Dep. of the Interior, Wash., D. C.

268. Santos, G. A., Doty, M. S.: Carrageenan from Hawaiian Red Algae. In: Abstracts of the IXth International Seaweed Symposium, Santa Barbara, Cal., 1977. Journal of Phycology 13, Suppl. (1977).

269. Segawa, S.: The Seaweeds of Japan. Hoi kusha, Osaka, Japan (1957).

270. Segi, T., Kida, W.: On the Relation between Distribution of Early Germlings of Monostroma and Tidal Current in the Culture Ground. Botanica Marina II, (3/4): 223-228 (1961).

271. Segi, T., Kida, W.: Synopsis of Biological Data on Monostroma latissimum Wittrock in Japanese Cultivation. FAO Fisheries Synopsis No. 39 (1968).

272. Selby, H. H.: Natural Plant Hydrocolloids. Advances in Chemistry Series 11 (1954).

273. Selby, H. H., Selby, Th. A.: Agar. In: Industrial Gums. Whistler, R. L., BeMiller, J. N., ed. Academic Press Inc., New York (1959).

274. Semesi, A. K., Mshigeni, K. E.: Contributions on the Content and Nature of the Phycocolloid from Halymenia venusta Boergesen. Botanica Marina XX, (4): 233-237 (1977).

275. Semesi, A. K., Mshigeni, K. E.: Studies on the Yield and Infrared Spectra of Phycocolloids from Chondrococcus hornemannii (Lyngbye) Schmitz and Sarconema filiforme (Sonder) Kylin from Tanzania. Botanica Marina XX, (5): 271-275 (1977).

276. Seterlund, W. et al.: The Agricultural College of Norway, Institute of Poultry and Fur Animals. Report No. 22, WLD Fish. Abstr. 21, (3): 47 (1970).

277. Sheppard, C. R. C. et al.: Studies on the Growth of Laminaria hyperborea (Gunn.) Fosl. and Laminaria ochroleuca de la Pylaie on the French Channel Coast. Botanica Marina XXI, (2): 109-116 (1978).

278. Silva, G. F. et al.: Determinacion Cromotografica de Aminoacidos en Algas Marinas. Botanica Marina VIII, (2/4): 244-251 (1965).

279. Sivalingam, P. M. et al.: Biochemistry and Physiology of the Characteristic 334 nm UV-Absorbing Substance in Algae. In: Abstracts of the VIIIth International Seaweed Symposium, Bangor, 1974. Marine Science Laboratories, Menai Bridge, Anglesey, U. K. (1974).

280. Sivalingam, P. M. et al.: Isolation and Physico-Chemical Properties of a Substance 334 from the Red Alga Porphyra yezoensis Ueda. Botanica Marina XIX, (1): 1-7 (1976).

281. Sivalingam, P. M. et al.: Physiological Roles of a Substance 334 in Algae. Botanica Marina XIX, (1): 9-21 (1976).

282. Sivalingam, P. M.: Biodeposited Trace Metals and Mineral Content Studies of Some Tropical Marine Algae. Botanica Marina XXI, (5): 327-330 (1978).

283. Silverthorne, W.: Optimal Production from a Seaweed Resource. Botanica Marina XX, (2): 75-98 (1977).

284. Simpson, F. J. et al.: The Cultivation of Chondrus crispus. Effects of pH on Growth and Production of Carrageenan. Botanica Marina XXI, (4): 229-235 (1978).

285. Skoryna, St. C. et al.: Suppression of Intestinal Absorption of Radiostrontium by Substances occurring in Phaeophyceae. In: Proceedings of the Vth International Seaweed Symposium, Halifax, 1965. Pergamon Press, Toronto (1966).

286. Skoryna, St. C., Tanaka, Y.: Biological Activity of Fractionation Products of Brown Marine Algae. In: Proceedings of the VIth Internationals Seaweed Symposium, Santiago de Compostela, 1968. Dirección Genéral de Pesca Marítima, Madrid (1969).

287. Skoryna, St. C. et al.: The Effects of Enzymic Degradation Products of Alginates on Intestinal Absorption of Radiostrontium. In: Proceedings of the VIIth International Seaweed Symposium, Sapporo, 1971. University of Tokyo Press (1972).

288. South, G. R.: Alginate Levels in New Zealand Durvillea with Particular Reference to Age Variations. In: Abstracts of IXth International Seaweed Symposium, Santa Barbara, Cal., 1977. Journal of Phycology 13, Suppl. (1977).

289. Springer, G. F. et al.: Proc. Soc. exp. Biol. Med. N. Y. 94, 404 (1957).

290. Springer, R., Middendorf, L.: Arch. Pharmazie 294/66, 693 (1961).

291. Schachat, R. E., Glicksman, M.: Some Lesser Known Seaweed Extracts. In: Industrial Gums. Whistler, R. L., BeMiller, J. N. ed. Academic Press Inc., New York (1959).

292. Schimmer, O.: Algen als Arzneimittel. Deutsche Apotheker Zeitung 112, (47): 1879-1880 (1972).

293. Schlichting, Jr., H. E.: Protein Quality of Some Fresh-Water Algae. Economic Botany 25, (3): 317-319 (1971).

294. Schmid, O. J.: Die Inhaltsstoffe der Meeresalgen. Botanica Marina III, Suppl., 67-74 (1962).

295. Schmid, O. J., Hoppe, H. A.: Neuere Erkenntnisse über Inhaltsstoffe der Meeresalgen. Chemiker-Zeitung 89, (16): 549-553 (1965).

296. Schmid, O. J.: Für die menschliche Ernährung wichtige Inhaltsstoffe mariner Algen. Botanica Marina IX, Suppl.: 41-46 (1966).

297. Schmidt, H.-W.: Die Phytohormone, Seifen - Öle - Fette - Wachse. 97, (23): 876 (1971).

298. Schmitt, M.: Les Carraghenans. Botanica Marina III, (1): 17-21 (1961).

299. Schneider, G.: Arzneipflanzen von aktueller medizinischer Bedeutung. Deutsche Apotheker Zeitung 102 (47): 1529-1533 (1962).

300. Schwandt: Deutsche Apotheker Zeitung 97 (32): 706 (1957).

301. Schwencke, J., Landsberger, W.: Polychromatic Detection of Amino Acids on Thin Layer Chromatograms. Botanica Marina XII, (1-4): 228-232 (1969).

302. Schwimmer, M., Schwimmer, D.: The Role of Algae and Plankton in Medicine. Grune and Stratton, New York and London (1955).

303. Stahl, E. et al.: Knoten- und Blasentang als Arzneibuchdroge. Deutsche Apotheker Zeitung 115, (49): 1893-1896 (1975); 116, (3): 51-54 (1976).

304. Stanley, N. F., Renn, D. W.: Estimation of Molecular Weight Distribution of Carrageenans by Electrophoresis in Rehydrated Agarose Gel. In: Abstracts of the VIIIth International Seaweed Symposium, Bangor, 1974. Marine Science Laboratories, Anglesey, U. K., (1974).

305. Stoloff, L., Lee, Ch. F.: Fishery Leaflet 173 (1946). U. S. Dep. of the Interior, Washington D. C.

306. Stoloff, L.: Irish Moss - From an Art to an Industry. Economic Botany 3, (4): 428-435 (1949).

307. Stoloff, L.: Carrageenan. In: Industrial Gums. Whistler, R. L., BeMiller, J. N., ed. Academic Press Inc., New York (1959).

308. Stoloff, L.: Algal Classification - An Aid to Improved Industrial Utilization. Economic Botany 16, (2): 86-94 (1962).

309. Takagi, M. et al.: Free Amino Acid Components of Rhodomela larix and Chondrus ocellatus. Bull. of the Japanese Soc. of Sci. Fish. 30, (3): 284-290 (1964).

310. Takagi, M. et al.: Free Amino Acid Composition of Some Species of Marine Algae. Bull. of the Japanes Soc. of Sci. Fish. 33, (7): 669-673 (1967).

311. Takagi, M. et al.: Seasonal Variation of Free Amino Acid Composition of Chondrus ocellatus. Bull. of the Japanese Soc. of Sci. Fish. 33, (9): 834-838 (1967).

312. Takagi, M.: Low Molecular Nitrogen Compounds of Marine Algae. Bull. of the Fac. of Fish., Hokkaido University 21 (3): 227-233 (1970).

313. Takagi, M. et al.: Isolation and Some Chemical Properties of a New Peptide Analipine from a Brown Alga, Analipus japonicus. Bull. of the Japanese Soc. of Sci. Fish. 39 (9): 961-967 (1973).

314. Takagi, M.: Seaweeds as medicine. In: Advance of Phycology in Japan, Tokida, Hirose, ed. VEB Gustav Fischer Verlag, Jena (1975).

315. Takemoto, T.: Kainic Acid. Pharmaceutical Faculty, University of Osaka, May 1 (1957).

316. Takemoto, T., Daigo, K.: Chem. and Pharm. Bull. 6 (5): 578-580 (1958).

317. Takemoto, T., Daigo, K.: Bull. Jap. Soc. Phyc. 7, (3): 91-93 (1959).

318. Takemoto, T.: Pharmazeutische Studien über die Bestandteile von Chondria armata. Japanese Journal of Pharmacy and Chemistry XXXI, (8): 30-42.

319. Takemoto, T. et al.: Der fliegentötende Bestandteil von Chondria armata. Japanese Journal of Pharmacy and Chemistry XXXIV, (10): 21-22.

320. Takemoto, T. et al.: Studies on the Hypotensive Constituents of Marine Algae. I. A New Basic Amino Acid "Laminine" and the Other Basic Constituents isolated from Laminaria angustata. Yakugaku Zasshi 84, (12): 1176-1179 (1964).

321. II. Synthesis of Lamine and Related Compounds. Yakugaku Zasshi 84, (12): 1180-1182 (1964).

322. III. Determination of Laminine in Laminariacean. Yakugaku Zasshi 85, (1): 37-40 (1965).

323. Takemoto, T., Sai, T.: Studies on the Constituents of Chondria armata. Yakugaku Zasshi 85, (1): 33-37 (1965).

324. Takemoto, T. et al.: Studies on the Constituents of Chondria armata. Yakugaku Zasshi 85, (1): 83-85 (1965).

325. Takemoto, T.: Studies on Biologically Active Amino Acids. Japan Medical Gazette 3, (5): 5 (1966).

326. Tanaka, Y. et al.: Studies on Inhibition of Intestinal Absorption of Radioactive Strontium. The Canadian Medical Assoc. Journal 99, 169-175 (1968).

327. Tanaka, Y.: Biologische Aktivität von Alginaten. 17. Vortragstagung der Deutschen Gesellschaft für Arzneipflanzenforschung, Kiel, 1969. Deutsche Apotheker Zeitung 109, (25): 955 (1969).

328. Tanaka, Y. et al.: The Preventive and Therapeutic Use of Alginate in Strontium-90 Poisoning. American Association for the Advancement of Science, Symposium, Chicago (1970).

329. Tanaka, Y. et al.: Application of Algal Polysaccharides as in vivo Binders of Metal Pollutants. In: Proceedings of the VIIth International Seaweed Symposium, Sapporo, 1971. University of Tokyo Press (1972).

330. Tanaka, Y. et al.: Binding of Heavy Metals by Sulfated Algal Polysaccharides as Compared to Effects of Polyuronates. In: Abstracts of the VIIIth International Seaweed Symposium, Bangor, 1974. Marine Science Laboratories, Anglesey (1974).

331. Tanaka, Y. et al.: Inhibition of Iron Absorption by Fucoidan. In: Abstracts of the IXth International Seaweed Symposium, Santa Barbara, Cal., 1977. Journal of Phycology 13, Suppl. (1977).

332. Tatewaki, M.: Culture Studies on the Life History of some Species of the Genus Monostroma. Scientific Papers of the Institute of Algological Research, Fac. of. Sci., Hokkaido Univ. VI, (1): 1-56 (1969).

333. Taylor, A. R. A.: Studies of the Biology and Ecology of Chondrus crispus Stack. (Irish Moss) on Prince Edward Island. Industrial Development Branch Fisheris Service, Project Report No. 35 (1969).

334. Thomas, P. C., Krishnamurthy, V.: Agar from cultured Gracilaria edulis (Gmel.) Silva. Botanica Marina XIX, (2): 115-117 (1976).

335. Tokida, J.: The Marine Algae of Southern Saghalien. Fac. of Fish., Hokkaido University 2, (1), (1954).

336. Tsuchiya, Y.: Comparative Hypocholesterolemic Activities of Marine Algae. In: Proceedings of the VIth International Seaweed Symposium, Santiago de Compostela, 1968. Dirección General de Pesca Marítima, Madrid (1969).

337. Uphof, J. C. Th.: Dictionary of Economic Plants. H. R. Engelmann (J. Cramer), Weinheim (1959).

338. Vevera, E.: Fucus vesiculosus. Deutsche Apotheker Zeitung 99, (10), (1959).

339. Vinogradov, A. P.: The Elementary Chemical Composition of Marine Organisms. Sears Foundation for Marine Research, Yale University, New Haven (1953).

340. Vogt: Pharmaz. Zentralhalle 93, (9): 323 (1954).

341. Volk: Deutsche Apotheker Zeitung 97, 421 (1957).

342. Vonderbank: Pharmazie 5, 213 (1950).

343. Watanabe, A.: Nitrogen fixation by algae. In: Advance of Phycology in Japan, Tokida, Hirose ed. VEB Gustav Fischer Verlag, Jena (1975).

344. Watanabe, Y.: Harmful effects of nicotine can be prevented by intake of Undaria. General Meeting of Nippon Dietetical and Food-technological Society (1968). In: Advance of Phycology in Japan. Tokida, Hirose, ed. VEB Gustav Fischer Verlag, Jena (1975).

346. Weber, U., Gerhard, H.: Deutsche Apotheker Zeitung 53, 1350 and 1373 (1938).

347. Weiss, R. F.: Moderne Pflanzenheilkunde. Kneipp-Verlag G.m.b.H., Bad Wörishofen (1976).

348. Whyte, J. N. C., Englar, J. R.: Seasonal Changes in the Inorganic Constituents of Macrocystis integrifolia. In: Abstracts of the IXth International Seaweed Symposium, Santa Barbara, Cal., 1977. Journal of Phycology 13, Suppl. (1977).

349. Whyte, J. N. C. et al.: Effects of Drying on the Alginate Component of Nereocystis luetkeana and Macrocystis integrifolia. In: Abstracts of the IXth International Seaweed Symposium, Santa Barbara, Cal., 1977. Journal of Phycology 13, Suppl. (1977).

350. Woessner, J. W.: The Measure and Harvest of the California Marine Crop Plant Porphyra. In: Abstracts of the VIIIth International Seaweed Symposium, Bangor, 1974. Marine Science Laboratories, Anglesey, U. K. (1974).

351. Woessner, J. W. et al.: The Economic Potential of Harvesting and Marketing an American Porphyra. In: Abstracts of the IXth International Seaweed Symposium, Santa Barbara, Cal., 1977. Journal of Phycology 13, Suppl. (1977).

352. Woodward, F. N.: Laminaran. In: Industrial Gums. Whistler, R. L., BeMiller, J. N., ed. Academic Press Inc., New York (1959).

353. Yamaguchi, T. et al.: Pathway of Mannitol Formation during Photosynthesis in Brown Algae. Plant and Cell Physiol. 10, 425-440 (1969).

354. Yamamoto, T., Ishibashi, M.: The Content of Trace Elements in Seaweeds. In: Proceedings of the VIIth International Seaweed Symposium, Sapporo, 1971. University of Tokyo Press (1972).

355. Yaphe, W.: Can. J. Botany 37, 751 (1959).

356. Yaphe, W., Duckworth, M.: The Relationship Between Structures and Biological Properties of Agars. In: Proceedings of the VIIth International Seaweed Symposium, Sapporo, 1971. University of Tokyo Press (1972).

357. Yentsch, Ch. S., Reichert, C. A.: The Interrelationship between Water-Soluble Yellow Substances and Chloroplastic Pigments in Marine Algae. Botanica Marina III, (3/4): 65-74 (1962).

358. Yosisige, K.: Present Status and Development Trends of the Plant Drugs in Japan. herba polonica XVI, (1): 96-103 (1970).

359. Young, E. G.: A Comparison of the Soluble Proteins in Various Species of Algae by Disc Electrophoresis in Polyacrylamide Gels. Phytochemistry 9, 2167-2174 (1970).

360. Young, K. et al.: Enzymic Hydrolysis of Agar and Properties of Bacterial Agarases. In: Proceedings of the VIIth International Seaweed Symposium, Sapporo, 1971. University of Tokyo Press (1972).

361. Youngken, H. W.: The biological potential of the oceans to provide biomedical material. Lloydia 32, (4): 407 (1969).

362. Zaneveld, J. S.: The Utilization of Marine Algae in Tropical South and East Asia. Economic Botany 13, (2): 89-131 (1959).

363. Zelenski, St. G., Worthen, L. R.: Two Biologically Active Polymers from Eisenia bicyclis (Kjellman) Setchell. Botanica Marina XVII, (4): 191-195 (1974).

Addendum

364. Hänsel, R.: Medizinische Empirie und moderne Arzneimittelforschung. Deutsche Apotheker Zeitung 117, (41): 1672-1679 (1977).

365. Halstead, B. W.: Marine biotoxins: a new source of medicinals. Lloydia 32, 484 (1969).

LIST OF MULTICELLULAR ALGAE OF COMMERCIAL USE.

S. Bonotto
Department of Radiobiology, C.E.N.-S.C.K., 2400 Mol, Belgium.

1. Introduction.

Marine multicellular algae are useful for man in many ways. They are used as human and animal food, as manure in terrestrial farming projects, for various medical purposes, and provide raw materials for several industries (2, 3, 10). Seaweeds have been used as food since ancient times : species of *Laminaria* and *Gracilaria* were used by the Chinese several thousand years ago and the red seaweeds *Porphyra* have been eaten for more than 2000 years in Japan, as well as in China, America, Hawaii, Philippines, Malaysia and Indonesia (5). In European countries (mainly in Iceland, Ireland and Scotland) species of *Chondrus*, *Rhodymenia* and *Ulva* are used as food to a limited extent (5, 6).

The importance of seaweeds as food is expected to increase in the future, since terrestrial ecosystems alone cannot meet the increasing nutritional demands of the growing human population (14). Some unicellular green algae (*Chlorella*) and blue-green algae (*Spirulina*) could also prove to be useful for man in the future (13).

Seaweed utilization for many different purposes led to the industrialization of their production, and in recent years far-

Marine Algae in Pharmaceutical Science
edited by H. A. Hoppe, T. Levring, Y. Tanaka

ming of seaweeds has reached considerable proportions, involving millions of people (5). The practical applications of a number of algal derivatives in many fields of modern industry have received detailed attention from various authors (see ref. 10 for the literature). In the field of energy, several species of algae seem to be valuable. The large brown alga *Macrocystis pyrifera* is being studied as the source of fermentable material for the production of commercial quantities of methane gas (4, 7). Moreover, the phosynthetic formation of hydrogen in freshwater (*Anabaena cylindrica*) as well as in marine filamentous blue-green algae (*Calothrix membranacea, C. scopulorum, Oscillatoria brevis*) is under investigation as a potential source of fuel (1, 9).

The lists reported in this paper concern only the macroscopic multicellular marine algae of commercial importance.

2. List of green, brown and red algae of commercial use.

The lists are based on papers published by various authors (see ref. 3) and on papers presented at the IXth International Seaweed Symposium in Santa Barbara (8, 12, 15, 16). Predominant use is indicated by : hf : human food, af : animal food, mp : medical purposes, ma : manure, iu : industrial uses, en : possible energy source.

2.1. Green Algae.

Codium sp.	hf, mp
C. dichotomum	hf
C. divaricatum	hf
C. fragile	hf
C. intricatum	hf
C. lindenbergii	hf
C. muelleri	hf
C. tenue	hf
C. tomentosum	hf, iu
Enteromorpha compressa	hf, af
E. flexuosa	hf
E. intestinalis	hf, af
E. linza	hf, af
E. prolifera	hf
Monostroma crassissimum	hf, af
M. grevillei	hf, af
M. latissimum	hf, af
M. nitidum	hf, af
M. tubuliforme	hf, af
Prasiola japonica	hf
Ulva fasciata	hf
U. lactuca	hf, af, mp, ma
U. latissima	hf, af
U. nematoida	hf
U. penniformis	mp
U. pertusa	hf, af, mp
U. reticulata	mp

2.2. Brown Algae

Agarum cribrosum	iu
A. fimbriatum	iu
Alaria esculenta	hf, af, ma
A. fistulosa	hf
A. marginata	ma
A. ochotensis	hf
A. yezoensis	hf
Arthrotamnus bifidus	hf
A. kurilensis	hf
Ascophyllum nodosum	af, ma, iu
Chnoospora pacifica	hf
Chorda filum	hf
Chordaria flagelliformis	hf, mp
Costaria costata	hf
Cystophora sp.	af
Cystophyllum sp.	hf
Dictyopteris polypodioides	hf, mp
(Syn : *D. membranacea*)	
D. plagiogramma	hf
Dictyota acutiloba	hf
D. apiculata	hf
Durvillea antarctica	hf, af, mp, ma, iu
Ecklonia bicyclis	iu
E. cava	iu
E. kurome	hf
E. latifolia	hf
E. maxima	af, ma, iu
Eckloniopsis sp.	hf, iu
Egregia menziesii	ma, iu
Eisenia bicyclis	hf, ma, iu

Endarachne sp.	hf
Fucus serratus	af, ma, iu
F. vesiculosus	af, mp, ma, iu
Heterochordaria abietina	hf
Himanthalia elongata	ma, iu
(Syn: *H. lorea*)	
Hizikia fusiforme	hf
Hormosira banksii	af
Hydroclathrus clathratus	hf, ma
(Syn: *H. cancellatus*)	
Kjellmaniella gyrata	hf
(Syn: *Laminaria gyrata*)	
Laminaria angustata	hf, iu
L. cichorioides	hf
L. diabolica	hf, iu
L. digitata	af, ma, iu
L. fragilis	iu
L. hyperborea	af, ma, iu
(Syn: *L. cloustonii*)	
L. japonica	hf, mp, iu
L. longipedalis	hf
L. longissima	hf, iu
L. religiosa	hf
L. saccharina	af, ma, iu
Lessonia sp.	iu
L. variegata	ma
Macrocystis integrifolia	af, ma, iu
M. pyrifera	af, ma, iu, en
Mesogloia crassa	hf
M. decipiens	hf
Nemacystis decipiens	hf
Nereocystis luetkeana	hf, af, ma, iu
Padina australis	af, ma

Pelagophycus porra	ma, iu
Pelvetia canaliculata	af, ma
Petalonia fascia	hf
(Syn: *Phyllitis fascia*	
Ilea fascia)	
Sargassum aquifolium	hf
S. echinocarpum	hf
S. enerve	hf, iu
S. fusiforme	hf, mp, ma, iu
S. granuliferum	hf
S. hemiphyllum	ma
S. horneri	iu
S. linifolium	iu
S. micracanthum	iu
S. natans	mp
(Syn: *S. bacciferum*)	
S. nigrifolium	iu
S. polycystum	hf
S. serratifolium	hf, iu
S. siliquosum	hf
S. sp.	hf, af, ma
S. tennerrimum	ma
S. thumbergii	af, mp, ma
S. vulgare	hf
S. wrightii	ma
Scytosiphon sp.	hf
Splachnidium rugosum	hf
Tinocladia sp.	hf
Turbinaria sp.	ma
T. ornata	hf
Undaria peterseniana	hf
U. pinnatifida	hf, mp
U. undarioides	hf

2.3. Red Algae

Acanthopeltis japonica	hf, iu
Acanthophora spicifera	hf
Aeodes nitidissima	iu
Agardhiella tenera	iu
Ahnfeltia concinna	hf, iu
A. plicata	iu
Alsidium helminthochorton	mp
Asparagopsis sanfordiana	hf
Beckerella sp.	iu
Bostrychia radicans	hf
Caloglossa adnata	hf
C. leprieurii	hf
Campylaephora crassa	iu
C. hypnaeoides (Syn: *Ceramium hypnaeoides*)	hf
Carpopeltis affinis	iu
C. flabellata	iu
Catenella impudica	hf
C. nipae	hf
Ceramium aduncum	iu
C. boydenii	iu
C. cimbricatum	iu
C. cliatum	iu
C. codii	iu
C. crassum	iu
C. fastigiatum	iu
C. fimbriatum	iu
C. gracillimum	iu
C. kondoi	iu
C. nakamurai	iu

C. paniculatum	iu
C. rubrum	iu
C. tenerrimum	iu
C. tenuissimum	iu
Chondria armata	mp
Chondrus crispus	hf, af, mp, iu
C. elatus	hf
C. ocellatus	iu
C. pinnulatus	iu
C. platinus	hf
C. yendoi	iu
Corallina officinalis	mp
Cryptonemia decumbens	hf
Cystoclonium armatum	hf
Digenea simplex	hf, mp
Dilsea edulis	hf
Dumontia incrassata	hf, iu
Eucheuma amakusaensis	iu
E. cottonii	iu
E. crustaeforme	iu
E. cupressoideum	hf, iu
E. edule	hf
E. gelatinae	hf, iu
(Syn: *E. gelatinosa*)	
E. horridum	hf, iu
(Syn: *Gigartina horrida*)	
E. isiforme	iu
E. muricatum	hf, iu, mp
(Syn: *E. spinosum*,	
E. denticulatum,	
Fucus muricatum)	
E. okamurai	hf, iu

E. serra	hf, iu
E. striatum	hf, iu
Furcellaria fastigiata	iu
Gelidiella acerosa	hf, iu
(Syn: *Gelidium rigidum*)	
Gelidiopsis rigida	hf, iu
Gelidium amansii	iu
G. arborescens	iu
G. arbuscula	iu
G. attenuatum	hf
G. capillaceum	iu
G. cartilagineum	iu
G. corneum	iu
G. crinale	iu
G. decumbensum	iu
G. divaricatum	iu
G. elegans	iu
G. incurvatum	iu
G. japonicum	iu
G. johnstonii	iu
G. kintaroi	iu
G. latifolium	hf
G. liatulum	iu
G. lingulatum	iu
G. linoides	iu
G. micropterum	iu
G. nudifrons	hf, iu
G. pacificum	hf, iu
G. planiusculum	iu
G. polycladum	iu
G. polysticum	iu
G. pristoides	iu

G. pulchellum	iu
G. pulchrum	iu
G. pulvinatum	iu
G. purpurascens	iu
G. pusillum	iu
G. pyramidale	iu
G. rigens	hf
G. sp.	hf, mp, iu
G. subcostatum	hf
G. subfastigiatum	iu
G. tenue	iu
G. vagum	iu
Gigartina chamissoi	iu
G. exasperata	iu
G. clavifera	iu
G. decipiens	iu
G. intermedia	iu
G. ockotensis	iu
G. pacifica	iu
G. pistillata	hf, iu
G. stellata	iu
(Syn: *G. mamillosa*)	
G. teedii	iu
G. tenella	iu
Gloiopeltis complanata	iu
G. furcata	iu
G. tenax	iu
Gracilaria arcuata	iu
G. blodgettii	iu
G. bursa-pastoris	iu
G. caudata	iu
G. chorda	iu

G. compressa	hf, iu
G. confervoides	hf, ma, iu
G. cornea	hf
G. coronipifolia	hf, iu
G. crassa	iu
G. denticulata	iu
G. dura	iu
G. edulis	hf, iu
G. eucheumoides	hf
G. gigas	iu
G. incurvata	iu
G. multipartita	iu
(Syn: *G. lacinulata*,	
G. foliifera)	
G. punctata	iu
G. purpurascens	iu
G. salicornia	iu
G. sublittoralis	iu
G. taenioides	hf, iu
G. textorii	iu
G. verrucosa	iu
Gracilariopsis chorda	iu
G. rhodetricha	iu
G. vermiculophylla	iu
Grateloupia divaricata	hf, iu
G. filicina	hf, iu
G. imbricata	hf, iu
G. okamurai	hf, iu
Gymnogongrus flabelliformis	hf, iu
G. griffithsiae	iu
G. javanicus	hf
G. pinnulata	hf, iu

Halosaccion glandiforme	hf
Halymenia durvilliae	hf
Hypnea boergensii	iu
H. cenomyce	hf, iu
H. cervicornis	hf, iu
H. charoides	iu
H. chordacea	iu
H. cornuta	iu
H. divaricata	hf
H. esperi	iu
H. flabelliformis	iu
H. hamulosa	iu
H. japonica	iu
H. musciformis	iu
H. nidifica	hf, af, mp
H. nidulans	iu
H. pannosa	iu
H. saidana	iu
H. spicifera	iu
H. valentiae	iu
H. variabilis	iu
Iridea cornucopiae	iu
I. edulis	hf
I. flaccida	iu
(Syn: *I. laminaroides*)	
Laurencia botryoides	hf
L. obtusa	hf
L. papillosa	hf
L. perforata	hf
L. pinnatifida	hf
Liagora farinosa	hf
L. decussata	hf

Lithothamnion calcareum	ma
L. crassum	ma
Mastocarpus klenzianus	hf
Meristotheca papulosa	hf, iu
Nemalion multifidum	hf
Neodilsea yendoana	iu
Odonthalia corymbifera	iu
Pachymenia himantophora	ma
Pachymenia sp.	iu
Pachymeniopsis elliptica	hf, iu
Phyllophora brodiaei	iu
P. membranifolia	iu
P. nervosa	iu
P. rubens	iu
Plumaria plumosa	ma
Porphyra angusta	hf
P. atropurpurea	hf, mp
P. columbina	hf
P. dentata	hf
P. kuniedai	hf
P. kunthiana	hf
P. laciniata	hf
P. leucosticta	hf
P. nereocystis	hf
P. ochotensis	hf
P. okamurai	hf
P. onoi	hf
P. perforata	hf
P. pseudo-linearis	hf
P. purpurea (Syn: *P. umbilicalis f. laciniata*)	hf

P. suborbiculata	hf
P. tenera	hf
P. umbilicalis	hf
P. vulgaris	hf
P. yezoensis	hf
Pterocladia capillacea	iu
(Syn: *P. pinnata*)	
P. densa	iu
P. lucida	hf
P. nana	iu
P. tenuis	iu
Rhodoglossum hemisphaericum	iu
R. japonicum	iu
Rhodymenia indica	mp
R. palmata	hf, af, mp
Sarcodia montagneana	hf
Soliera chordalis	ma
Sphaerococcus cartilagineus	mp
Suhria vittata	hf, iu
Turnerella mertensiana	iu

3. Concluding remarks.

The seaweed industry has known ups and downs during the last 100 years (10). However, at present seaweed utilization continues in an economically satisfactory way and it is expected to increase in the future as numerous kelp beds in different partsof the world are still unexploited (11). Many results obtained in basic research have helped to build the fundament for large-scale commercial utilization of marine algae.

More recently, a rapid progress in phycological research has been obtained by the application of several modern biological techniques : tissue culture, protoplasts technique, fusion of vegetative cells, inter-specific and inter-generic crossing experiments and selection of fast growing strains. Moreover, genetical studies on marine algae are in progress in several laboratories. It is to be noted, however, that marine pollution may progressively interfere with life in the oceans and coastal waters and seriously impede the culture of marine algae, unless regulations are issued and effectively controlled.

4. Acknowledgements.

This work has been supported in part by the "Fonds de la Recherche Scientifique Fondamentale Collective" and by NATO Research Grant n° 1027. I thank Mrs. Romeyer-Luyten J. for typing the text.

5. References.

1. Benemann, J.R., Weare, N.M. : Hydrogen evolution by nitrogen-fixing *Anabaena cylindrica* cultures. Science 184, 174-175 (1974).
2. Boney, A.D. : Aspects of the biology of the seaweeds of economic importance. Adv. mar. Biol. 3, 105-253 (1965).
3. Bonotto, S. : Cultivation of plants. Multicellular plants. In : Marine Ecology, O. Kinne Ed., John Wiley & Sons, London, 3, 467-501 (1976).
4. Bryce, A.J. : A research & development program to assess the technical and economic feasibility of methan production on a commercial scale from giant brown kelp. J. Phycol., Suppl. 13, 10 (1977).
5. Dawson, E.Y. : Marine Botany, an Introduction. Holt, Rinehart and Winston, New York (1966).
6. Hallsson, S.V. : The uses of seaweed in Iceland. Int. Seaweed Symp. 4, 398-405 (1964).
7. Hart, M.R., de Fremery, D., Lyon, C.K., Kohler, G.O. : Processing of *Macrocystis pyrifera* for fermentation to methane. J. Phycol., Suppl. 13, 28 (1977).
8. Hong, K.C., Cruess, R.L. : The effects of *Undaria* extracts on calcium and bone metabolism. J. Phycol., Suppl. 13, 31 (1977).
9. Lambert, G.R., Smith, G.D. : Hydrogen formation by marine blue-green algae. FEBS Letters 83, 159-162 (1977).
10. Levring, T., Hoppe, H.A., Schmid, O.J. : Marine algae. A survey of Research and Utilization. Cram, De Gruyter & Co. Hamburg (1969).
11. Michanek, G. : A preliminary Appraisal of World Seaweeds Resources. FIRM-C 128, Food and Agriculture Organization of the United Nations, Rome (1971).

12. Ricohermosa, M., Devean, L.E. : Commercial propagation of *Eucheuma* spp. clones in the South China Sea. A discussion of trends in cultivation technology and commercial production patterns. J. Phycol. Suppl. 13, 58 (1977).

13. Santillán Sánchez, C., Durand-Chastel, H. : The utilization of *Spirulina* algae for industrial photosynthesis. J. Phycol. Suppl. 13, 60 (1977).

14. Stirn, J. : Ecological consequences of marine pollution . Rev. Int. Océanogr. Méd., 24, 13-46 (1971).

15. Waaland, J.R. : Growth and strain selection in the marine alga *Gigartina exasperata*. J. Phycol., Suppl. 13, 71 (1977).

16. Woessner, J.W. : The economic potential of harvesting and marketing an American *Porphyra*. J. Phycol. 13, 74 (1977).

SEAWEEDS IN PHARMACEUTICALS AND MEDICINE: A REVIEW

V. J. CHAPMAN
University of Auckland, New Zealand

SUMMARY

After a brief survey of the state of the art in the use of seaweeds in pharmaceuticals and medicine as around 1969, a review is provided of subsequent published material in these two fields. Some compounds, eg: chondriol, zonarol, are listed together with their potential uses. Particular attention is paid to the carrageenans and to products that show promise to common ailments of mankind. Finally an appeal is made for much more work in this promising field of algal utilization.

INTRODUCTION

Seaweeds have been used as drugs or drug sources over a large number of years going back into folk medicine. In the case of some of the earlier uses, eg. as vermifuges, the active principles have still not been extracted. Such is the case with *Alsidium helminthocorton* a red seaweed. Another red seaweed, *Digenea simplex* is used for the same purpose in Asia and in this case the active principle is a mixture of kainic and allokainic acids (Chapman 1970). Kainic acid with santonin added to it is sold in Japan under the name of Digesan as a remedy against tapeworms.

Agar and Carrageenan

The best known and most widely used products of seaweeds in medicine and pharmaceuticals are the agars, carrageenans and alginates, and at present there do not appear to be any artificially prepared alternatives. Agar is an important product prepared from a range of red algae known as agarophytes, mainly belonging to the genus *Gelidium*. The best known

Marine Algae in Pharmaceutical Science
edited by H. A. Hoppe, T. Levring, Y. Tanaka

use of agar is in the production of bacteriological plates on which pathogenic bacteria and fungi can be grown for identification purposes. Whilst this is a very important use of the product it does not, in fact, represent the major use of agar, which is outside the pharmaceutical industry. In the U.S.A. 10,000 lbs of agar were used for pharmaceutical purposes in 1956, and this had doubled to 20,000 lbs in 1966, so that if the trend has continued the amount used last year may well be close to 40,000 lbs. Apart from the bacteriological use agar is also employed as a binding for tablets, as a laxative, and it can also be a useful anticoagulant for blood. In dentistry it is used under the name of 'Dentocoll' for making impressions of gums. In this use it has, however, very largely been replaced by the alginates.

Another major phyco-colloid obtained from red seaweeds is carrageenan, which comprises a group of closely allied compounds. The original source of the material was Irish Moss or _Chondrus crispus_, but it is now obtained from a number of other genera, among which _Eucheuma_ has assumed considerable importance in the Pacific region. A refined extractive, known as _decoctum chondri_ is used in emulsions, as a cough syrup and in tooth pastes. In the presence of potassium it is used as a binding agent for tablets, in shaving soaps and haircreams and as a blood anti-coagulant. It may also have a future in providing growth of connective tissue. More recently (Bhakuni & Silva 1974) it is reported, especially in France and Great Britain, as being valuable in control of ulcers through action with the mucous lining of the stomach. An allied compound in the same group. Furcelleran, obtained from _Furcellaria_ sp., is used for the same purpose pharmaceutically, though to a lesser extent. Iridophycan, obtained from the genus _Iridophycus_, and chemically in the same group, has anti-coagulant properties and could probably be used for the same purposes as carrageenan from _Chondrus_. The principle problem here would be difficulty of harvesting adequate supplies.

The capacity of algae to produce products that can be used as blood anti-coagulants appears quite widespread. Fucoidan, which is probably the calcium salt of carbohydrate ethereal sulphate, is obtainable from a number of the larger brown algae, mainly within the Fucaceae, and is believed to be an anti-coagulant. Laminaran sulphate obtained from members of the Laminariaceae could also be used for the same purpose.

Algin

The alginate industry is extensive in the northern hemisphere and manufacturers in Japan and the United Kingdom have started to use supplies of raw weed from New Zealand and Tasmania respectively. Sodium alginate is an excellent suspensory agent for a wide range of drugs and alginates generally have replaced gums in the manufacture of greaseless lubricating jellies. Since alginates are not broken down by stomach juices but are by intestinal ones, algin capsules can be used to enclose substances that must be liberated intestinally. There are a number of dental uses, eg. moulds, control of bleeding, as well as a binding material for a wide range of pills. At least 17 different general pharmaceutical uses have been listed (Chapman 1970).

A recently reported use is for the removal of Sr^{90} from the body. With increasing involvement of radioactivity for the generation of electricity and potential liberation or radioactive compounds, including Sr^{90}, this could be of great significance (Bhakuni & Silva 1974).

Antibiotics

A wide range of algae, mostly Rhodophyceae, have been tested for antibiotic activity, but so far none appear to have been used in medicine. In most of these cases the active agent is probably acrylic acid. This acid is produced by the planktonic alga *Phaeocystis pouchetii* which is consumed by penguins and is said to make them naturally sterile. Kelp pills or kelp infusions, because of the iodine they contain, can be effective goitre remedies, and the low incidence of goitre in Japan is probably due to the considerable amount of seaweed consumed in foods.

Successive International Seaweed Symposia have always resulted in the presentation of one or more papers dealing with pharmaceutical or potentially pharmaceutical products, and I expect the present Symposium will be no exception. At the sixth Symposium Anderson (1969) discussed the pharmacology of the carrageenans and pointed out that these sulphated polysaccharides differed from those of other sulphated polysaccharides, and that there were differences also depending upon the use of the κ or λ- forms as well as between extracts from different seaweeds. In species of Gigartina total carrageenan levels show little variation as between male, female and tetrasporic plants. The proportions of κ to λ do however, vary and there is practically no κ form in tetrasporophytes (Stuart, Parsons & Bailey 1973). Varying molecular size appeared to exert different pharmacological effects.

A study by Johnston & McCandless (1969) at the same symposium, reported that specific precipitating anti-bodies had been obtained from both κ - and λ - carrageenans of Chondrus crispus, and they raised the questions of whether tertiary structure of the two polysaccharides was involved in their immunochemical specificity. This does not appear to have been prosecuted further.

Varied Uses

A compound, Algasol - T_{13}, has been prepared and been available in Italy for a number of years. It appears to exert a useful oncologic effect and has also been tried in association with normal treatment for rheumatoid arthritus (Schiavetti & Tuzi 1969). At the 8th International Symposium Cavi & Guiseppe (1974) reported that the product had had a beneficial effect on 100 patients suffering from bronchitis or emphysema. The use of this product would seem to offer considerable possibilities.

In 1969 Tsuchiya reported on the hypocholesterolemic activities of a number of marine algae. The responsible compounds are sterols, eg. ergosterol, chondriosterol, poriferasterol etc. The following species among others, were more or less responsible for the depression of blood pressure in human atheroclerosis or the lowering of blood cholesterol levels in rats and rabbits: - Enteromorpha compressa, E. prolifera, Ulva pertusa, Monostroma nitidum, six species of Laminaria, Costaria costata, Sargassum muticum, Fucus gardneri, two species of Kjellmaniella, Arthrothamnus bifidus, Undaria (two species), Heterochordaria abietina, Chondrus and Porphyra tenera. From the laminarians the compound laminine was found to be the effective agent. Laminine monocitrate is a derivative that could be a potentially useful pharmaceutical agent (Bhakuni & Silva 1974). Baker and Murphy (1976) report that this compound also possesses antihelminthic as well as antihypertensive properties. Two chemical formulae are quoted for the compound, $C_{13}H_{24}N_2O_{10}$ (Bhakuni & Silva 1974), and $C_9H_{22}N_2O_3$ (Baker and Murphy 1976). It would seem that further work is necessary in order to establish which is correct. Domoic acid, which is obtained from the red alga Chondria armata, has also been studied in recent years as an antidote to Ascaris worms. Phaeophytin and Phaeophorbide from Heterochordaria lowered blood cholesterol in rats. More recently (Abe and Kaneda 1975) state that a new hypocholesterol agent has been extracted from Monostroma nitidum and named ulvaline. This compound, however, does not have such a depressant effect as homobetaine which is obtained from the same alga.

One of mankind's more tiresome maladies is caused by the family of viruses responsible for herpes sores. Because of the known anti-bacterial properties of some red seaweeds, two University of Califonia scientists, Deig and Ehresmann (reported in Time magazine) have examined some 29 varieties of Pacific Rhodophyceae to determine if extracts would be effective against the herpes virus group. The extract was found to be 99% effective in stopping viral multiplication and 50% effective in reducing spread from an existing infection. Studies showed that the compound, a polysaccharide, acted by blocking the viral absorption point in the cell membrane. The extract could be Chondriol ($C_{15}H_{18}BrClO_2$), which is obtained from Chondria oppositicladia and which has anti-viral properties (Baker and Murphy 1976).

Active Compounds in Seaweeds

Chemical studies have been carried out recently on a number of seaweeds in order to determine the nature of compounds that could be of medical or pharmaceutical benefit. Thus Bhakuni and Silva listed quite a number of compounds with varying properties, some useful, some not. Brominated oxygen compounds as follows:-

Laurencin	$C_{17}H_{23}O_3Br$	from *Laurencia glandulifera*
Laureatin	$C_{15}H_{20}O_2Br_2$	from *Laurencia nipponica*
Laurinterol	$C_{15}H_{19}Br$	from *Laurencia intermedia*
Laurene	$C_{15}H_{20}$	from *Laurencia glandulifera*

were all found to be toxic and therefore of no clinical value. Domoic acid, $C_{15}H_{21}O_6N \cdot 2H_2O$, derived from *Chondria armata* acted as an anti-helminthic as also did Kainic acid (see above for *Digenea*) obtained from *Corallina officinalis*. If allied species of *Corallina* behave similarly there should be no lack of raw material.

Species of *Caulerpa* yielded caulerpin ($C_{24}H_{18}N_2O_4$) and caulerpicin ($C_{48}H_{87}O_2N$) both of which functioned as mild anaesthetics and were suggested as having a possible clinical value. If this were established, beds of *Caulerpa* can be extensive and actual cultivation of such algae should not be impossible. Baker and Murphy (1976) also report several other compounds that possess antifungal and antibacterial properties. These include Pachydictyol A($C_{20}H_{32}O_2$), from *Pachydictyon coriaceum*, Isozonarol ($C_{21}H_{30}O_2$), zonarol ($C_{21}H_{30}O_2$), both from *Dictyopteris zonaroides*, and squalene ($C_{30}H_{50}$) from *Fucus vesiculosus*. Only the last named could be obtained in sufficient quantity from raw material.

Cultivation

The importance of the agar and alginate industries has, in recent years, compelled workers to turn to the possibilities of cultivating significant species either in the field or under artificial conditions. Workers at Woods Hole have been studying the artificial growth of *Chondrus crispus* and in New Zealand Luxton (PhD thesis) has made preliminary studies on the cultivation of *Pterocladia lucida*, and has pointed out the possibilities for commercial growth of *Gracilaria secundata* in association with sewage waste water disposal. In Japan Hasegawa (1976) has initiated the forced growth of *Lammaria japonica* on cremona strings, and, starting in 1969 when 30 tons wet weight was obtained, the output for 1974 was 7500 tons with a potentiality of 150,000 tons. In the case of *Gelidium* and agar a somewhat different approach, involving optimal production under natural conditions, has been developed by Silverthorne (1977).

CONCLUSIONS

The significance of these studies to the pharmaceutical industry lies in the fact that should a really valuable product be found, that cannot be manufactured artifically, there is the prospect that under a suitable cultivation technique adequate supplies to meet demand could be grown. The present rarity or low supply of raw material is therefore not necessarily a bar to its use. For this reason many more marine algae could perhaps be investigated, including phytoplanktonic forms. So far it appears that only one species, *Gonyaulax catenella*, has been examined in detail and that produces saxitoxin ($C_{10}H_{16}N_2O_4$) which acts as a block to nerve conduction. The alga, *Prymnesium parvum*, is said to produce two toxins, one affecting the central nervous system, and the other functioning as a hemolytic agent. Whether these could be of value pharmaceutically has yet to be shown.

Finally, whilst a number of fresh-water Cyanophyceae may possess carcinogenic or central nervous system toxins, only one marine species, *Lyngbya maiuscula*, has been reported as possibly possessing an antibacterial compound. The use of this alga, however, is not likely to be promoted since it may also produce a dermatitis (Yamazoto 1976). One may suggest that other blue-green algae that occur abundantly on salt marshes and in mangrove swamps should be investigated.

In conclusion it may be stated that there is evidence of potentially valuable drugs from marine algae, sufficient to warrant much more detailed study and the interest of more investigators in this specialised field.

REFERENCES

Abe, S., Kaneda, T.: Studies on the effect of marine products on Cholesterol metabolism in rats. XI. Isolation of new Betaine, Ulvaline from a green laver, Monostroma nitidum, and its depressing effect on plasma cholesterol levels. Bull. Jap. Soc. Sci. Fish. 41(15), 567-71. (1975).

Abel, R.B.: Drug pushing in the Ocean in Food Drugs from the Sea Symposium. Mar. Tech. Soc. Wash. D.C. (1969).

Baker, J.T., Murphy, V.: Handbook of Marine Science Compounds from Marine Organisms. C.R.C. Press. Vol. 1. (1976).

Bhakuni, D.S., Silva, M.: Biodynamic Substances from Marine Flora. Bot. Mar. 17(1), 40-51. (1974).

Cavi, G. de, Guiseppe, S.: Bronchopathies, complications and outcome under the action of 'Algasol T-331'. Abst. 8th Intern. Seaw. Symp. Bangor. B. 54. (1974).

Chapman, V.J.: Seaweeds and their uses. 2nd Ed. Methuens. (1970).

de Marderosan, A.: Marine Pharmaceuticals. J. Pharm. Soc. Jan. 1969. (1969)

Food Drugs from the Sea. Mar. Techn. Soc. Wash. D.C. (1970).

Hasegawa, Y.: Progress of Laminaria culture in Japan. J. Fish. Res. Bd. Canada 33(4), 1002-06. (1976).

Johnston, K.H., McCandless, E.L.: Immunochemistry of Carrageenans. Abst. 6th Intern. Seaw. Symp. Spain No. 76. (1969).

Schiavetti, L., Tuzi, T.: Les Algues Marines en rheumatologia. Resultate d'une etude clinique et experimentale avec 'l'Algasol T-331'. Abst. Intern. Seaw. Symp. Spain No. 79. (1969).

Silverthorne, W. Optimal Production from a Seaweed. Resource Bot. Mar. 20(21), 75-98.

Stuart, E.P., Parsons, M.J., Bailey, R.W.: Composition of Gigartina Carrageenan in relation to sporophyte and gametophyte stages of the life cycle. Phytochem. 12, 2441-44. (1973).

Trans. Drugs from the Sea Symposium 1967. Ed. H.D. Freudenthal. Publ. Ma . Techn. Soc. Wash. D.C. (1968).

Tsuchiya, Y. Comparative hypocholesterolemic activities of marine algae. Abst. 6th Intern. Seaw. Symp. Spain No. 80. (1969).

Yamazoto, K., Y. Some aspects of biology of a dermatitis producing alga, *Lyngbya maiuscula* Gomont. Abst. Intern. Symp. Ecol. & Manag. trop. shallow water communities, p. 24 Indones. Inst. Sciences. (1976).

Contributions and Potentialities of Caribbean Marine Algae in Pharmacology

M. Diaz-Piferrer

Department of Marine Sciences
University of Puerto Rico, Mayaguez, Puerto Rico, 00708

From the ancient Chinese culture even before the first millennium B. C. mentions are found of the "hai tsao" (algae). The Pentsao recommends all of the medicinal algae in the treatment of several kinds of maladies afflicting mankind. Practically all the medicinal properties of plants are attributed to the semimythical Shen Nung, known as the Father of Husbandry and Medicine, who purportedly lived around 3000 B. C.

Since the dawn of Western civilization there have been indications that algae were used as remedies to kill pain and heal wounds. Nevertheless, Virgil, the prince of Latin poets, who lived in 70-19 B. C., used the phrase "vilor algae" meaning "more vile or worthless than algae". 65-8 B. C. also Horace shared Virgil's poor estimation of algae when in his Satires he writes ... "viler than seaweed ..." or "useless seaweed". Gradually such concepts changed so that Pliny the Elder (23-79 A. D.), the Roman naturalist, wrote of the excellences and uses of seaweeds (5).

During the centuries that followed, seaweeds increasingly drew the attention of western man, and here and there reports are found of scholars looking to algae as a possible source of useful products to help to cure diseases of mankind. The XVII and XVIII centuries witnessed European students gradually directing their attention to medical aspects of marine plants. Within the scope of the Linnean development of modern taxonomy, the scene became set for biochemical research on marine algae during the XIX century. The search for drugs from the sea stirred the interest of Europeans, a fact which greatly inflecnced the progress of understanding and set the fundaments of the role of marine algae in Pharmacology.

Marine Algae in Pharmaceutical Science
edited by H. A. Hoppe, T. Levring, Y. Tanaka

Saint-Yves (20) in De L'Utilité des Algues Marines gave a resumé of the uses of algae in medicine at that time. Stanford (1837-1899) made a great contribution with his studies of phycocolloids from the Phaeophyta or brown algae, bringing alginic acid within pharmaceuticals. Koch and Hesse introduced the use of agar-an outstanding Japanese contribution-into the micro-biological laboratory. British, Germans, French and other Europeans, as well as American and Japanese, made the XIX century the historical starting point for the scientific role and uses of seaweeds in Pharmacology (12, 19, 29).

During the XX century, particularly after World War II, research in the biochemistry of algae has been steadly progressing in Europe, America, and Japan, and the prospects are at present most promising for obtaining pharmaceuticals from marine plants (13, 21, 25, 27).

Interest in Economic Phycology in the Caribbean began to attract the attention of universities in the early 1950's. Research began in Cuba under the sponsorship of the Cuban Institute for Technological Research, (I. C. I. T.), in Havana, searching for sources of industrial phycocolloids such as agar, carrageenan and alginic acid, and also for nutrimental values of seaweeds such as vitamins, minerals and other products derived from marine plants. Research continued in Puerto Rico in the early 1960's under the sponsorship of the Industrial Research Department of the Puerto Rico Development Administration (Fomento) and the Department of Marine Sciences of the University of Puerto Rico, in Mayaguez. At present several institutions are actively conducting research in Economic Phycology in the Caribbean. A report of the results related to Pharmacology follows.

AGAR, CARRAGEENAN AND AGAROIDS

The term agarophytes for the agar producing algae has been commonly used. The term "agaroidophytes" was introduced in 1964 for the red seaweeds that yield phycocolloids (agaroids) which are chemically similar to agar, but with different physical properties (10). The term "carragenophytes" could be used for the species of red algae yielding carrageenan, and the term "alginophytes" for the species of brown algae which are good producers of alginates.

As a practical laboratory routine test, the criteria used for the classification of gels as agaroids was the low syneresis, i. e. the gel obtained reabsorbs the water during the process of dehydration by the freezing method, the water ist not exuded but reabsorbed. Agar will exude most of the water by the same process. Differences were found in the yield of gels between the same species treated in Cuba or in Puerto Rico (9, 10). These differences were sound to be associated with the methods of extraction, the nature of the habitats, and the time of the year when the plants were collected. Table V presents the results of gel extraction from agarophytes from Puerto Rico.

After about 20 years of research, 10 genera with 31 tested species habe been recorded as agarophytes, and listed in Table I; 2 genera with 8 tested species as carragenophytes in Table II; 10 genera with 24 tested species as agaroidophytes producers of a "carragenaan-type" of gel in Table III; and 10 genera with 23 species of agaroidophytes producers of mucilages in Table IV. It makes a total of 23 genera with 86 species of Rhodophyta as a source of industrial phycocolloids in the Caribbean (1, 9, 10, 11, 22). Species have been listed alphabetically for convenience, and for taxonomy Taylor (26) has been followed.

TABLE I. Caribbean Rhodophyta producers of Agar (Agarophytes)

Brongniartella mucronata (Harvey) Schmitz
Bryothamnion seaforthii (Turner) Kützing
Bryothamnion triquetrum (Gmelin) Howe
Digenia simplex (Wulfen) C. Agardh
Enantiocladia duperreyi (C. Agardh) Falkenberg
Gelidiella acerosa (Forsskal) Feldmann & Hamel
Gelidium crinale (Turner) Lamouroux
Gelidium corneum (Hudson) Lamouroux
Gelidium serrulatum J. Agardh
Gracilaria blodgettii Harvey
Gracilaria cervicornis (Turner) J. Agardh
Gracilaria compressa (C. Agardh) Greville
Gracilaria crassissima Crouan ex J. Agardh
Gracilaria cuneata Areschoug
Gracilaria curtissiae J. Agardh
Gracilaria cylindrica Børgesen
Gracilaria damaecornis J. Agardh
Gracilaria debilis (Forsskal) Børgesen
Gracilaria domingensis Sonder
Gracilaria ferox J. Agardh
Gracilaria foliifera (Forsskal) Børgesen
Gracilaria mammillaris (Montagne) Howe
Gracilaria sjoestedtii Kylin
Gracilaria venezuelensis Taylor
Gracilaria verrucosa (Hudson) Papenfuss
Polysiphonia echinata Harvey
Polysiphonia ferulacea Suhr
Pterocladia americana Taylor
Pterocladia bartlettii Taylor
Pterocladia pinnata (Hudson) Papenfuss
Vidalia obtusiloba (Mertens) J. Agardh

TABLE II. Caribbean Rhodophyta producers of Carragenaan (Carragenophytes)

Eucheuma acanthocladum (Harvey) J. Agardh
Eucheuma echinocarpum Areschoug
Eucheuma gelidium (J. Agardh) J. Agardh
Eucheuma isiforme (C. Agardh) J. Agardh
Hypnea cervicornis J. Agardh
Hypnea cornuta (Lamouroux) J. Agardh
Hypnea musciformis (wulfen) Lamouroux
Hypnea spinella (C. Agardh) Kützing

TABLE III. Caribbean Rhodophyta producers of Carragenaan-type of gels (Agaroidophytes)

Acanthophora muscoides (Linnaeus) Bory
Acanthophora spicifera (Vahl) Børgesen
Agardhiella ramosissima (Harvey) Kylin
Agardhiella tenera (J. Agardh) Schmitz
Amansia multifida Lamouroux
Bryocladia cuspidata (J. Agardh) De Toni
Bryocladia thyrsigera (J. Agardh) Schmitz
Coelothrix irregularis (Harvey) Børgesen
Chondria atropurpurea Harvey
Chondria littoralis Harvey
Chondria tenuissima (Goodenough & Woodward) J. Agardh
Gigartina acicularis (Wulfen) Lamouroux
Gigartina teedii (Roth) Lamouroux
Gymnogongrus tenuis (J. Agardh) J. Agardh
Laurencia corallopsis (Montagne) Howe
Laurencia gemmifera Harvey
Laurencia microcladia Kützing
Laurencia obtusa (Hudson) Lamouroux
Laurencia papillosa (Forsskal) Greville
Laurencia poitei (Lamouroux) Howe
Laurencia scoparia J. Agardh
Spyridia aculeata (Schimper) Kützing
Spyridia clavata Kutzing
Spyridia filamentosa (Wulfen) Harvey

TABLE IV. Caribbean Rhodophyta producers of Mucilages

Ceramium nitens (C. Agardh) J. Agardh
Cryptonemia bengryi Taylor
Cryptonemia crenulata J. Agardh
Cryptonemia luxurians (Mertens) J. Agardh
Dasya corymbifera J. Agardh
Dasya pedicellata (C. Agardh) C. Agardh
Dasya ramosissima Harvey
Dictyurus occidentalis J. Agardh
Grateloupia cuneifolia J. Agardh
Grateloupia filicina (Wulfen) C. Agardh
Grateloupia gibbesii Harvey
Halymenia floresia (Clemente) C. Agardh
Halymenia floridana J. Agardh
Halymenia gelinaria Collins & Howe
Halymenia pseudofloresia Collins & Howe
Heterosiphonia gibbesii (Harvey) Falkenberg
Liagora ceranoides Lamouroux
Liagora farinosa Lamouroux

TABLE V. PHYSICAL CHARACTERISTICS OF GELS OBTAINED FROM 10 SPECIES OF AGAROPHYTES

Algae	Yielding %	Gel hardness 1.5%	Point gelification 1.5%	Point fusion 1.5% °C
Bryothamnion triquetrum	16.9	132	41.5	77.5
Gelidiella acerosa	44.5	214	41	81
Digenia simplex	26.7	247	35	75
Hypnea musciformis	36.4	402	49	68
Gracilaria crassissima	28.2	132	42	78.5
Gracilaria compressa	27.1	90	37	80
Gracilaria curtissiae	33.2	176	48	91
Gracilaria debilis	47.2	67	36	77.5
Gracilaria domingensis	44.0	313	43.5	74
Gracilaria mammillaris	18.4	20	41	74
Bacto Agar Difco (Control)		235	35	84

Liagora pinnata Harvey
Liagora valida Harvey
Ochtodes secundiramea (Montagne) Howe
Polysiphonia denudata (Dillwyn) Kützing
Polysiphonia subtilissima Montagne

ALGINIC ACID

Nothing in the Caribbean resembles the giant source of alginic acid found in the Pacific Kelps; neither the impressive large eulittoral populations of Laminaria, Ascophyllum or Fucus found in the higher lattitudes. The largest Caribbean brown algae (Phaeophyta) as source of alginic acid are Turbinaria, Sargassum, Spatoglossum, Stypopodium, and two species of Dictyopteris as presented in Table VI. Nevertheless, the pelagic sargaza, Sargassum natans and Sargassum fluitans, should be taken into consideration. Since the days of Columbus the large amount of floating sargaza drifting about the so-called Sargasso Sea has attracted the attention of scientists. Pelagic sargaza make up about 99% of the floating masses drifting toward tropical America most of which arrive at Caribbean coasts. The yearly production of pelagic sargaza has been estimated from 4 to 11 millions tons (18).

Several species of brown algae were tested in Cuba. Again, the nature of the habitats exerted strong influence in the percentage of alginates obtained, e. g. T. turbinata and S. polyceratium from a small bay influenced by a river, yielded, respectively, 13.25% and 13.9% of sodium alginate, while the same species from exposed, open coasts gave 20.20% of the same product (7). Research with alginophytes is being carried in Puerto Rico at present related to the yield of alginates from pelagic as well as several species of sessile Sargassum. Table VII presents yields of sodium alginate obtained from alginophytes collected at different times of the year in Puerto Rico.

TABLE VI. Caribbean producers of Alginic Acid: Alginophytes

Dictyopteris hoytii Taylor
Dictyopteris justii Lamouroux
Sargassum acinarium (Linnaeus) C. Agardh
Sargassum bermudense Grunow
Sargassum cymosum C. Agardh
Sargassum filipendula C. Agardh
Sargassum fluitans Børgesen
Sargassum furcatum Kützing
Sargassum hystrix J. Agardh
Sargassum natans (Lannaeus) J. Meyen
Sargassum platycarpum Montagne
Sargassum polyceratium Montagne
Sargassum pteropleuron Grunow
Sargassum ramifolium Kützing
Sargassum rigidulum Kützing

Sargassum vulgare C. Agardh
Spatoglossum schroederi (Mertens) Kutzing
Stypopodium zonale (Lamouroux) Papenfuss
Turbinaria tricostata Barton
Turbinaria turbinata (Linnaeus) Kuntze

TABLE VII. Yield of Sodium Alginate from 7 species of Phaeophyta

Species of Algae	Time of Collecting	% Yield
Sargassum fluitans	Summer (July)	18.5
Sargassum fluitans	Winter (February)	16.8
Sargassum natans	Summer (July)	17.6
Sargassum natans	Winter (February)	15.3
Sargassum polyceratium	Summer (July)	21.0
Sargassum polyceratium	Winter (February)	17.9
Turbinaria turbinata	Summer (July)	22.5
Turbinaria turbinata	Winter (February)	19.7
Stypopodium zonale	Summer (July)	15.2
Stypopodium zonale	Winter (February)	12.6
Dictyopteris justii	Summer (July)	14.5
Dictyopteris justii	Winter (February)	16.7
Spatoglossum schroederi	Summer (July)	11.5

VITAMINS, PROTEINS AND MINERALS

The preparation of tablets of high nutritional values with concentration of vitamins, proteins and minerals obtained from seaweeds has been a steadly growing pharmaceutical industry involving millions of dollars at present. Marine algae as source of such nutrimental values have been also the subject of research in the Caribbean resulting in 24 genera with 39 tested species recorded. Table VIII presents the nutrimental values of marine algae from Cuba. If taken into consideration that the samples were sundried according to the method used (6) the vitamin content of the algae tested is remarkable as is the case of the green alga Enteromorpha lingulata. Similarly, the content of proteins and amino acids were high as found in the red alga Agardhiella tenera with 882 mg % in lysine, and 340 mg % in methionine (6).

Seaweeds is a useful and valuable source of minerals particularly of all known trace elements. Tables IX and X present the mineral values of seaweeds from Puerto Rico. The figures obtained illustrate the valuable source of minerals available from marine algae in the Caribbean; such nutrimental values are highly suggestive for the manufacturing of concentrates and tablets for dietary and medical purpose (3, 4, 8).

TABLE VIII. NUTRITIONAL VALUES OF 11 SPECIES OF CARIBBEAN MARINE ALGAE (CUBA)
Results from 100g of dried samples

	ALGAE	g E	g A	g F	mg Ca	mg P	mg Fe	g N	Cal	g CH	g P	mg T	mg M	mg L	mg Car.	mg AA	mg Th	mg R	mg Nc
CHLOROPHYTA	Anadyomene stellata	2.01	39.99	16.60	2.08	18.14	855.0	2.75	250	40.92	17.08	181	187	710	0.13	Trace	0.05	0.38	1.62
	Caulerpa racemosa	1.84	12.70	7.76	3.69	178.7	258.0	1.76	239	44.45	11.01	146	185	397	0.10	1.66	0.05	0.55	1.61
	Codium isthmocladum	2.55	36.70	1.92	169.0	22.70	200.5	2.26	267	46.61	14.14	233	213	802	0.18	1.32	0.07	0.45	1.41
	Enteromorpha lingulata	3.68	21.55	4.39	2.37	29.20	249.8	3.15	320	52.07	19.70	298	193	754	12.80	95.20	0.72	2.25	2.63
	Ulva fasciata	1.00	38.70	3.88	3.59	200.0	223.5	3.52	250	38.28	22.02	229	166	918	2.82	15.85	0.44	0.84	5.17
PHAEOPHYTA	Sargassum natans	1.23	19.75	14.25	2.89	57.2	57.4	0.77	326	74.21	4.81	33	44	174	0.01	3.70	0.02	0.31	1.46
	Sargassum polyceratium	1.96	26.70	8.70	53.40	88.2	197.0	1.91	303	59.44	11.90	115	141	445	2.46	4.36	0.02	0.35	1.48
	Turbinaria turbinata	1.23	33.20	14.25	25.30	82.6	127.0	0.77	326	74.21	4.81	33	44	174	0.87	19.57	0.01	0.27	2.88
RHODOPHYTA	Agardhiella tenera	0.94	6.93	4.47	1.35	8.42	207.2	3.88	272	41.73	24.88	224	340	882	2.94	20.82	0.42	1.65	1.52
	Digenia simplex	0.85	30.60	11.16	78.20	156	200.5	1.64	282	58.28	10.27	110	149	634	0.44	1.80	0.18	0.48	1.35
	Gracilaria mammillaris	4.76	34.85	5.48	128	23.23	87.5	3.02	285	41.49	18.90	127	254	760	0.56	8.97	0.36	1.35	2.06

TABLE IX MINERAL ANALYSIS OF 10 SPECIES OF CHLOROPHYTA FROM PUERTO RICO.

	mg/g							ng/g		
ALGAE	N	P	K	Na	Ca	Mg	Fe	Mn	Zn	Cu
Enteromorpha flexuosa	10.6	0.71	5.52	0.69	19.33	30.13	2.85	13.54	48.57	41.58
Caulerpa prolifera	27.5	1.74	0.07	0.23	8.29	3.84	1.20	72.05	84.15	30.65
Caulerpa sertularioides	39.0	2.33	0.76	6.15	11.17	2.09	0.43	10.24	30.00	15.44
Caulerpa taxifolia	37.2	2.70	0.18	0.66	21.39	4.17	0.51	52.19	32.83	15.24
Cladophoropsis membranacea	13.9	0.70	1.48	10.01	20.69	5.27	0.90	58.29	29.71	15.13
Codium decorticatum	15.8	0.91	5.55	9.69	11.97	14.51	0.25	7.20	72.51	10.48
Cymopolia barbata	9.6	1.10	22.78	6.45	171.86	3.65	1.18	24.72	9.42	7.26
Halimeda discoidea	23.9	1.53	0.15	0.87	131.31	10.21	0.25	32.60	21.36	10.00
Udotea flabellum	24.5	1.15	0.46	1.50	108.59	1.83	0.23	33.28	13.09	7.78
Ulva lactuca	29.9	1.24	0.63	0.37	2.94	19.35	0.46	35.31	25.93	14.95

TABLE X. MINERAL VALUES OF 6 SPP. OF RHODOPHYTA AND 3 SPP. OF PHAEOPHYTA FROM PUERTO RICO.

	mg/g							ng/g		
ALGAE	N	P	K	Na	Ca	Mg	Fe	Mn	Zn	Cu
Agardhiella tenera	20.5	1.13	16.51	0.9	19.02	8.39	1.03	26.38	16.71	7.75
Acanthophora muscoides	22.7	1.02	16.03	5.98	66.59	9.65	0.50	10.83	14.17	7.72
Chondria littoralis	24.6	0.81	7.17	1.73	19.96	3.34	0.20	27.48	27.38	14.59
Gracilaria debilis	15.2	1.06	53.94	7.15	1.22	2.15	0.10	61.29	12.28	7.32
Gracilaria domingensis	16.6	4.00	85.96	4.57	4.20	4.80	0.81	21.39	11.61	7.25
Hypnea musciformis	20.5	1.50	5.50	9.43	19.57	7.76	0.74	15.02	26.59	22.11
Dictyopteris justii	18.8	1.20	3.69	5.65	7.44	4.03	2.37	33.43	14.63	14.03
Sargassum platycarpum	8.5	1.39	46.51	5.65	23.86	9.95	1.04	38.42	15.50	3.88
Sargassum polyceratium	11.4	0.70	2.16	0.92	40.06	10.85	0.57	29.32	70.02	29.10

KEY TO THE SYMBOLS

A= Ashes	M= Methionine
AA= Ascorbic Acid	mg= miligrams
Ca= Calcium	N= Nitrogen
Cal= Calories	Nc= Niacin
CH= Carbohydrates	ng= nanograms
Car= Carotene	P= Phosphorous
E= Ether extract	Pt= Protein
F= Fiber	R= Riboflavin
Fe= Iron	Th= Thiamine
L= Lysine	Trp= Tryptophane

ANTIBIOTICS

The search for antibiotic substances extracted from Caribbean marine algae has been active for the last 20 years with promising results. Martínez Nadal and associates conducting research in Puerto Rico found new antibiotic substances tested against 11 species of bacteria and 9 species of fungi: SARGANTIN, originally isolated from *Sargassum natans*, and CHONALGIN extracted from the red alga *Chondria littoralis*. According to Martínez Nadal both substances were related but not identical, and both were volatile acids (pH 4.5). Their fumes inhibited total growth of pathogenic microorganisms. In vitro demonstrated their potent antibacterial properties agains gram-positive and gram-negative, acid-fast bacteria and fungi such as *Staphylococcus aureus, Bacillus subtilis, Escherichia coli, Proteus vulgaris, Pseudomonas aeruginosa, Mycobacterium smegmatis, Candida albicans*.

Sarganin showed stronger action that did penicillin controls against *B. subtilis*; also stronger than streptomycin and chlortetracycline controls against *M. smegmatis*, and than nystatin controls against *C. albicans*. Chonalgin was more specific against *E. coli* and *C. albicans* (14, 15). Antibiotic properties similar to Sarganin and Chonalgin were also found in purified extracts of the green alga *Cymopolia barbata* (16, 17).

Burkholder and associates made a rough screening of species of marine algae from Puerto Rico against *Staphylococcus aureus, Escherichia coli, Mycobacterium smegmatis* and *Candida albicans*. Among the more active species of algae were *Chondria littoralis, Dictyopteris justii, Falkenbergia hillebrandii, Laurencia obtusa, Murrayella periclados* and *Wrangelia argus* (2).

Welch (28) conducted a survey of fungistatic properties of marine algae in Puerto Rico. Homogenized preparations of 35 species of seaweeds were tested for inhibitory activity against 6 pathogenic or opportunistically pathogenic fungi: *Aspergillus niger, Penicillum sp., Mucor racemosus, Rhizopus cryzae, Candida albicans* and *Cryptococcus neoformans*. Although almost all the species of marine algae showed at least a trace of activity against one or more tested organisms, the most potent inhibitors were *Laurencia obtusa, Laurencia papillosa, Wrangelia argus* and *Lynbya majuscula*.

Caulerpa racemosa was particularly active against *Candida* and *Cryptococcus*.

Starr and associates reported activity of seaweed extracts with virus. The maturation of mouse meningopneumonitis virus in a tissue culture line was visualized after acridine orange staining as a sequence of cytochemical changes involving nucleic acid synthesis. Of 45 extracts of marine algae assayed by this system, 5 showed some degree of inhibition of this predictable maturation sequence. *Lyngbya majuscula* and one species of *Cladophora* were most active in this respect. In addition, extracts of both algae were active against strains of *Pseudomonas fluorescens*, *Micrococcus pyogenes* var. *aureus* and *Mycobacterium smegmatis* (23).

Conducting research with 40 species of marine algae from Puerto Rico, Starr and associates also reported biological active substances which induced nuclear and mitotic anomalies in the McCoy tissue-cultured cell line. The cytological abnormalities were associated with the phenomena of amitosis, multiple cytokinesis, and micronucleation without cell division. Amitosis or direct constriction of the interphase nucleus into a binucleate cell was associated with bundles of filaments located near the region of constriction. Multiple cytokinesis was associated with the formation of miniature cell fragments each of which usually contained some of the nuclear complement of the mother cell. Micronucleation occurred in the absence of cell division and the process appeared similar to that observed after treatment of tissue-cultured cells with colchicine. From all the species of marine algae tested, the more active substances were extracted from *Acanthophora muscoides*, *Acanthophora spicifera*, *Asparagopsis taxiformis*, *Hypnea musciformis*, *Spyridia filamentosa* and *Padina sanctae-crucis* (24).

According with the literature cited above, from about 150 species tested, 25 species of marine algae have been recorded as producers of extracts of strong antibiotic properties as presented in Table XI. As results of the years of research in the various fields of Economic Phycology, the Caribbean marine flora offers at present an attractive field of potentialities to Pharmacology with a contribution of 53 genera with 130 tested species of marine algae.

TABLE XI. *Caribbean marine algae with active antibiotic properties*

CYANOPHYTA:
- *Lyngbya majuscula* (Dillwyn) Harvey

CHLOROPHYTA:
- *Caulerpa racemosa* (Forsskål) J. Agardh
- *Cladophora fascicularis* (Mertens) Kützing
- *Cymopolia barbata* (Linnaeus) Lamouroux
- *Enteromorpha prolifera* (Muller) J. Agardh

PHAEOPHYTA:
- *Dictyopteris justii* Lamouroux
- *Sargassum natans* (Linnaeus) J. Meyen
- *Padina sanctas-crucis* Børgesen

RHODOPHYTA:

Acanthophora muscoides (Linnaeus) Bory
Acanthophora spicifera (Vahl) Børgesen
Asparagopsis taxiformis (Delile) Collins & Hervey
Chondria littoralis Harvey
Digenia simplex (Wulfen) C. Agardh
Falkenbergia hillebrandii (Bornet) Falkenberg
Gelidiella acerosa (Forsskål) Feldmann & Hamel
Halymenia floresia (Clemente) C. Agardh
Hypnea cervicornis J. Agardh
Hypnea musciformis (Wulfen) Lamouroux
Laurencia obtusa (Hudson) Lamouroux
Laurencia papillosa (Forsskål) Greville
Murrayella periclados (C. Agardh) Schmitz
Spyridia filamentosa (Wulfen) Harvey
Wrangelia argus Montagne

REFERENCES:

1. Anderson, W. Pharmacological aspects of carragenaans. 6th Internat. Seaweed Symp., Spain, 627-635 (1968).

2. Burkholder, P. R. et al. Antibiotic activity of some marine algae from Puerto Rico. BOTANICA MARINA 2 (1-2):149-156 (1960).

3. Caudales, R. & M. Díaz-Piferrer. Micronutrients from Caribbean Marine algae. Manuscript.

4. Céspedes, G. Distribución de vitamina B12 (Cobalaminas) en materiales cubanos. I.C.I.T., Mem. 21:1-36 (1962).

5. Chase, F. M. Useful algae. Smithsonian Inst., Washington. Publ. 3667: 401-452, 9 pls. (1942).

6. Díaz-Piferrer, M. & H. López. Taxonomía, ecología y valor nutrimental de algas marinas cubanas, I.C.I.T., Habana, Mem. 6: 1-51 (1959).

7. Díaz-Piferrer, M. Acido algínico de algunas especies de algas pardas cubanas. Bol. Inform. I.C.I.T. 5 (1): 3-7 (1961).

8. Díaz-Piferrer, M. et al. taxonomía, ecología y valor nutrimental de algas marinas cubanas, II. Utilización de algas en alimentación de aves. I.C.I.T., Habana, Mem. 16: 1-87 (1961).

9. Díaz Piferrer, M. Taxonomía, ecología y valor nutrimental de algas marinas cubanas, III. Algas productoras de Agar. IBID. 17: 1-84 (1961).

10. Díaz-Piferrer, M. & C. Caballer de Pérez. Taxonomía, ecología y valor nutrimental de algas marinas de Puerto Rico: Algas productoras de Agar. Adm. Fomento Econ. P. R. 1-145 (1964).

11. Díaz Piferrer, M. Los recursos marinos de Venezuela: Algas de importancia económica. FAROL, Creole Corp. 222: 18-22 (1967).

12. Dillon, T. E. C. C. Stanford and the discovery of alginic acid. 3rd. Internat. Seaweed Symp., Ireland, 41-42 (1958).

13. Lewin, R. A. Physiology and biochemistry of algae. Academic Press, N. Y. (1962).

14. Martínez Nadal, N. et al. Sarganin and Chonalgin, new antibiotic substances from marine algae from Puerto Rico. Antimicrobial Agents and Chemoterapy. Amer. Soc. Microbiol. 68 -72 (1964).

15. Martínez Nadal, N. et al. Isolation and characterization of Sarganin complex, a new broad-spectrum antibiotic isolated from marine algae. IBID 131-134 (1965).

16. Martínez Nadal, N. et al. Low toxic effect of antimicrobial substances from marine algae. BOTANICA MARINA 9 (1-2): 62-63 (1966).

17. Martínez Nadal, N. et al. Antibiotic properties of marine algae III. *Cymopolia barbata*. IBID 9 (1-2): 21-26 (1966).

18. Parr, A. E. Quantitative observations on the pelagic *Sargassum* vegetation of the Western North Atlantic. Bull. Bingham Oceang. Coll. 6 (7): 1-94 (1939).

19. Perrot, E. & C. L. Gatin. Les algues marines utiles et en particular les algues alimentaires d'Extreme-Orient. Ann. L'Inst. Oceanóg. 3 (1): 1-101 (1911), Mónaco.

20. Saint-Yves, A. De L'utilité des algues marines. Libraire Médicale de Louis Leclerc, 1-55 (1879). Paris.

21. Smith, F. & R. Montgomery. The chemistry of plant gums and mucilages. Reinhold Publish. Corp., N. Y. (1959).

22. Soloni Toural, F. G. Ficocoloides de algas marinas cubanas. Centro Invest. Pesqueras, Habana. Cont. 4: 1-20, 2 pt. (1954).

23. Starr, Th. J. et al. Antibacterial and activiral activities of algal extracts studies by acridine orange staining. Texas . Rep. Biol. & Medic. 20 (2): 271-278 (1962).

24. Starr, Th. J. et al. Mitotic anomalies in tissue-cultured cells treated with extracts derived from marine algae. IBID 24 (2):208–221 (1966).

25. Stewart, W. D. P. Algal physiology and biochemistry. Blackwell Scientific Publicat., London (1974).

26. Taylor, W.R. Marine algae of the Eastern tropical and subtropical coasts of the Americas, Univ. Michigan Press, Ann Arbor (1972).

27. Tressler, D. K. & J. McW. Lemon. Marine products of commerce. Reinhold Publish. Corp. N. Y. (1960).

28. Welch, A. M. Preliminary survey of fungistatic properties of marine algae. J. Bacteriol. 83 (L): 97–99 (1961).

29. Whistler, R. L. & J. N. BeMiller. Industrial gums. Academic Press, N. Y. (1959).

MARINE PHARMACOLOGY

Focus on Algae and Microorganisms

Ara Der Marderosian, Ph. D.

Since the publication of my last review of marine pharmacology in 1974, numerous research papers have appeared in the literature as well as reviews and texts. The simple, prosaic marine drugs of our older pharmacopoeias have been catalogued and include such examples as agar from Gelidium, Gracilaria, Hypnea and other seaweed genera; cod liver oil and sodium morrhuate from codfish, Gadus morrhua; protamine sulfate from the sperm or mature testes of salmon; the alginates from the species of Fucus and Macrocystis; carrageenin from Chondrus; spermaceti from the sperm whale; ichthammol from bituminous schists containing fossil fish and numerous others. Many of these represent a substantial industry, e. g. the processing of alginates from the giant kelp which flourishes along the Pacific coasts of the U. S. and Mexico and which provides the starting material for a line of over 70 algin products. The stabilizing, suspending, emulsifying, film forming and thickening properties make algin derivatives useful in the processing and manufacture of foods, pharmaceuticals, beverages, paints, cosmetics and paper. Sales in 1972 totaled more than $21 million in one plant alone.

Consistently, one finds in the literature more and more exotic and unique biodynamic molecules being isolated from marine plants and animals. The marine environment has allowed for eons of evolutionary "trials and errors" in marine organisms and their protective mechanisms in the form of physiological forms and physiologically active chemical agents. We

Marine Algae in Pharmaceutical Science
edited by H. A. Hoppe, T. Levring, Y. Tanaka

have progressed to the state where we can now catalog certain types of these agents from the various classes of marine organisms. The literature is replete with examples of antibiotic, antifungal and antiyeast agents present in the monerans; anthelmintic, antiviral, anticoagulant, antispasmodic and ichthyoticix substances from the protistans; antibiotics, aggregation factors and growth regulators from sponges; neurohumoral and anticoagulant agents from coelenterates; sperm immobilizing, neuromuscular blocking and neuroactive substances from echinoderms; antiviral, neuromuscular, growth inhibiting nad hemolytic agents from molluscs; insecticides from marine annelids; cardioactive agents from arthropods and numerous compounds in the various classes of fish, amphibians and marine reptiles. But the surface has only been scratched.

Of these, several examples of drugs which have been developed and marketed may be given. These include the cephalosporin family of antibiotics originally discovered in the marine fungus, Cephalosporium acremonium; the anthelmintic kainic acid from the red alga, Digenea simplex; the anticoagulant, hemostatic, radioproectant and pharmaceutical adjuvant agent alginic acid from Fucus and Macrocystis algal species; the antiviral and antitumor nucleic acid derivatives from the marine sponge Cryptotethya crypta, which served as models for the ultimate synthesis of D-arabinofuranosylcytosine (Cytarabine) used in the treatment of leukemia; the powerful insecticide nereistoxin from the marine annelid Lumbriconercis heteropoda which served as a prototype for the synthesis of 1,3 bid-(carbamoylthio)-2-N,N-(dimethylamino) propane, now marked as Padan or Cartep in Japan; the powerful neuroactive drug tetrodatoxin from the puffer fish which has been used clinically to relax muscle spasms and as a palliative in terminal cancer; and the prostaglandin derivative, 15-epi-PGA_2, isolated from the gorgonian Plexaura homomalla and converted to active isomers (PGE_2 and PGF_2-alpha) which have great potential as agents useful for birth control, prevention of peptic ulcers, treatment of asthma, etc. (Baslow, 1969; Der Marderosian 1969; Martin and Padilla, 1973).

One of the areas of great significance has been the use of numerous biologically active agents from marine sources as "tools" or "molecular probes" in basic pharmacology research. These will no doubt have as great an impact as did muscarine from the terrestrial fungus Amanita muscaria which brought attention to a "muscaric" effect pharmacologically, which later helped uncover and explain the general parasympathetic effect of other drugs, even though it never had drug itself. Current analogous marine drugs include the use of tetrodotoxin for explaining certain effects on nerve cells by other drugs and the use of eledoisin from the octopus, as one of the most potent hypotensive drugs known.

Marine animals have unique physiological systems which lend themselves to basic pharmacological studies. Cafruny et al. (1967) have pointed out that in many cases an appropriate experimental animal may be found that has an anatomy favoring the exploration of a pharmacological system common to many animals. Comparative studies may reveal unique pharmacologically significant differences between lower and higher organisms. Frequently, the mechanisms for disposition of drugs and other foreign compounds are surprisingly similar in the evolutionarily more primitive water-dwelling species. This means that it may be possible to do more direct "structure-activity" type studies in marine species to determine the locus and mode of action of a particular drug. Already, marine animals have proven directly useful to man in developing new drugs or for screening purposes. For example, the use of the electric eel helped to find the cholinesterase reactivator, pralidoxime, which is now used as an antidote in posticide poisoning. The sea urchin is used in screening studies for water pollution. The shells of crustaceans like the Alaskan king crab are being exploited to yield chitin and chitosan, marine polymers being evaluated for use in the encapsulation of pharmaceuticals, as additives in antacid formulations, and in the treatment of wounds. The highly sensitive Limulus amebocyte lysate test which uses the blood cells of the horsehoe crab, Limulus polyphemus, for the detection of endotoxins (pyrogens) in various parenteral pharmaceuticals is currently being evaluated. It is a whole order of magnitude more sensitive than current tests. Numerous other examples may also be given (Der Marderosian, 1975; Robinson, 1974; Thomson, 1975; Lewis, 1976).

With respect to focusing attention to marine research and its potential for yielding drugs we can look to Emerson and Taft (1945) who as far back as 1945 presented an excellent review of the thenknown pharmacologically active agents from the sea. They presented 358 references to studies and publications around the world on marine bioactive substances. However, as so often happens with pioneers, they were so far ahead in time that their efforts were unappreciated. Then too, they were caught up in the emerging era of synthetic drugs in medicine which led to a further dampening of enthusiasm for marine pharmacognostical research. Nevertheless, we cannot help but appreciate what they said then, because it still holds true today, some thirty years later, and it will probably hold true for some many years hence:

"Use of marine products in medicine shared the fate of the general materia medica when therapeutic nihilism reduced the number of all drugs in use to a few of demonstrable value. Whereas considerable progress has since been made in developing rational therapeutic agents through chemical synthesis and collaborative botanical and chemical studies, the science of marine biology has been virtually neglected as a possible contributor to better therapeutic means."

Since this early report several other pioneers in the field like Nigrelli (1958), Halstead (1957), Burkholder (1963), Baslow (1969) and others have continued writing many articles and books on their extensive studies as well as the work of numerous others. However, it was not until the early sixties that "drugs from the sea" became an idea whose time had come. Perhaps it rode on the crest of the "back-to-nature" wave which itself emerged from a rising concern for ecology, pollution, chemical carcinagens in the environment and the like. In addition, tremendous advances in natural product chemical separation, and analysis now make it possible to carry out extractions and identifications of molecules at a rate which was not possible in previous decades.

A national awareness of marine pharmacological research perhaps had its origin in the first symposium on the biochemistry and pharmacology of compounds from marine sources which was held in New York under the direction of Nigrelli (1960) and under the auspices of the New York Academy of Sciences in 1960. A total of 32 papers were read and published.

The next meeting followed in a few years by the first formal "Drugs from the Sea" Conference, organized by the Marine Biology Committee of the Marine Technology Society and held in recognition of the substantial progress attained in the first hald of the sixties. As Freudenthal (1968) pointed out at this meeting, the concept of utilizing substances of marine origin in medicine was suddenly simultaneously both very new and very old. The several disciplines of botany, zoology, chemistry, oceanography, pharmacology, pharmacognosy, and others found common ground at this symposium and a total of 25 papers were presented and published. These meetings were so successful that two biennial symposia followed, with the publication of 35 papers from the first symposium in 1969 (Youngken, H. W., 1969) and 42 papers from the second in 1972 (Worthen, L. R., 1972). The 1974 meetings were hold at the University of Puerto Rico in Mayaguez with a total of 52 papers (Ruggieri, G. and Webber, H., 1976). These symposia included, in addition to drugs from the sea, papers on the impinging and related areas of food from the sea, governmental support of biological aspects of oceanography, mariculture, public health related to the ingestion of various marine toxins, specimen collection problems, and radiological transvectoring and marine biopollution among other topics.

The national meetings, common problems and a flood of review articles and texts have helped prompt the first marine biomedical workshop. This was held at the Smithsonian Institution in Washington on May 2-3, 1974, under the auspices of the Committee on Objectives and Procedures (COOP) of the U. S. Marine Technology Society, with the support of several agencies including the Office of the Sea Grant, National Science Foundation and National Institutes of Health, and several industrial sponsors. The workshop participants came from five governmental agencies, five pharmaceutical companies and sixteen academic institutions.

The primary purpose of this meeting was to assess the current status of the national research effort in harnessing biologically significant compounds from marine life forms, to recognize relevant developments to date, to focus upon those areas of research which show most promise for basic scientific discovery or for discovery and development of useful new drugs, and to develop guidelines which may further catalyze the research effort in the marine biomedical field. To these ends, the invited participants presented synopses of their ongoing research efforts, the problems they encounter, and the direction they invision. The complete report of this meeting was given at the Food-Drugs from the Sea Conference held at the University of Puerto Rico in Mayaguez in 1974. Specifically, the report took the form of a "Position Paper on Marine Biomedical Research" and was prepared by P. Kaul (1974) as chairman of the COOP Task Force, Biological Resources Committee, U. S. Marine Technology Society, at the University of Oklahoma, Norman, Oklahoma. In it were given definitions of various aspects of marine biomedical research, the accomplishments and current status of research on physiologically active substances, substances of commercial value, basic and ecologic research, significant potential pharmacological tools, major obstacles and possible specimens. Several methodological problems including collection of marine specimens, identification, and isolation of constituents were also covered. The major problem of communication and funding as well as the role which federal agencies and pharmaceutical industry could play were among the final topics.

On February 13-14, 1975 another marine biomedical research meeting was held at the Smithsonian Institute Museum of Natural History in Washington, D. C., this time sponsored by the National Institute of Environmental Health Sciences (NIEHS), National Institutes of Health (NIH) and the Smithsonian Institute. It was attended by approximately 125 scientists, directors and representatives from many of the county's marine biological laboratories, and government agencies whose interests and responsibilities are in the marine biology and health areas. The purpose of the meeting was to uncover the areas of undeveloped research in marine biology which may have a direct application to human health problems as well as to provide information on current research in various marine biology laboratories. Four general areas were covered during the first day of the meeting which included: 1. Marine Species as Models for Human Disease; 2. Environmental Carcinogenesis and Mutagenesis; 3. Human Health and the Marine Evironment- Infectious Agents and Naturally Occurring and Foreign Toxins; and 4. Drugs from the Sea. The second day was devoted to discussions by representatives from twelve of the country's approximately 40 marine laboratories on their organizational structure, developmental history, scientific programs, facilities and present status of support. A report on this NIEHS sponsored meeting was prepared on June 26, 1975 by Dr. David Rall, director of NIEHS, Research Triangle Park, North Carolina.

The publication of a workshop on Aquatic Sciences appeared in January of 1975 (Baiardi, J. and Ruggieri, G., 1975) in the Annals of the New York Academy of Sciences. In it were papers which dealt with fish health, the marine dinoflagellate, _Crypthecodinium_, aquacultural ecosystems, aquatic animals in biomedical research, poisonous and venomous marine animals and their toxins and the evolution of the nation's ocean program.

Continuing evidence of interest in marine research was seen at the Gordon Research Conference on Marine Natural Products held from Juni 30-July 4, 1975 at Santa Barbara, California. These meetings were chaired by Dr. Paul Scheuer, Chairman (University of Hawaii) and Dr. Yuzuru Shimizu, Vice Chairman (University of Rhode Island, College of Pharmacy). Although unpublished as such, presentations on continuing research with sponges, coelenterates, sea hares, Hawaiian seaweeds, tetrodotoxin, halogenated marine natural products, was given, as well as applying newer chemical methodologies to solving marine natural product research problems.

Among the reviews which have dealt with the general subject of drugs and bioactive substances from the sea, the following may be cited. They include, Marine Biomedicinals (Der Marderosian, 1969); Continuing Research on Biomedicinals from the Sea (Hales and Der Marderosian, 1974); Drugs from the Deep (Bhate and Seshadrinathan, 1974); Pharmacology of Toxins of Marine Organisms (Russell, 1971); Pharmaceuticals from the Sea (Halstead, 1972); Potential Biomedical Substances in the Extracts of Marine Organisms (Yasumoto, 1971); Marine Toxins (Baslow, 1971); Progress in Marine Poison Research (Asano, 1971); Chemistry of some Toxins Isolated from Marine Organisms (Scheuer, 1969); Toxins from Fish and other Marine Organisms (Scheuer, 1970); Toxins from Marine Invertebrates (Scheuer, 1971); Pharmaceutical Research Potential of the Sea (Giloli, 1970); Bluegreen and Green Algal Toxins (Gentile, 1971); Annotated Bibliography of Drugs from the Sea (Auffhammer and Deichmann, 1971); chemistry of Marine Toxins (Scheuer, P., 1972); Chemistry of Natural Marine Poisons (Scheuer, 1975); Marine Biotoxicology (Halstead, 1974); Analysis of Organic Halogen Compounds in Marine Organisms (White, 1974); Marine Biochemistry (Kittridge, 1974); Marine Natural Products other than Pigments (Chang, 1973); Marine Natural Product Chemistry and the Interim (Bland, 1974); Biochemical composition of marine zooplankton (Raymont, 1972); Development of sample preparation methods for analysis of marine organisms (McKee, 1974); Halogenated compounds in marine algae (Keiichi, 1975); Marine biotoxins (Alam, 1975); Poisonous and Venomous Marine Animals and their Toxins (Russell, 1975); Recent developments in the chemistry of marine toxins (Scheuer, 1975).

Schizophyta and Eumycophyta (Bacteria and Fungi)

Gauthier (1969) has studied the antibiotic activity of twenty-six marine bacterial stock cultures, all of which were gram-negative, mobile, asporagenous and strongly proteolytic. These marine microfloral representatives included species of the genera Pseudomonas, Achromobacter, Marobacterium, Xanthomonas, Chromobacterium, and Vibrio. They act antagonistically to both gram-negative and gram-positive bacteria, but exert a stronger effect against the latter, a phenomenon which is apparently fairly common in the marine benthic environment.

Using microbiological assays, Kurata (1970) investigated the production of Vitamin B_{12} by a variety of marine bacteria and its subsequent transudation from marine bottom muds in Maizuru Bay, in the Sea of Hinchi as well as in Wakasa Bay around Japan. Vitamin B_{12} from these bacteria was suggested as an important source of this vitamin for other marine organisms in the area.

Merkel and Sipos (1971), in an article on marine bacterial proteases have characterized an endopeptidase produced by Vibrio B-30. It has a molecular weight of 49,000 and was isolated by ammonium sulfate precipitation, ion-exchange chromatography, and gel filtration. The endopeptidase was shown to be a serine protease which could be irreversibly inhibited by certain metal-complexing agents.

Tunstall et al. have studied the susceptibility to antibiotics of marine strains of Pseudomonas. These marine pseudomonads, which cause large scale spoilage of fish and fish products, were identified and classified by antibiotic sensitivity tests (Tunstall, A. and Gowland, G., 1974).

A good coverage of antimicrobial agents from the sea is provided in a review by Burkholder and Sharma (Burkholder, P. and Sharma, G., 1969) in which these authors summarize data on this activity in algae, sponges, protochordates, and the bacteria Pseudomonas bromoutilis. This organism was found to contain a highly brominated pyrrole compound, specifically 2-(3,5-dibromophenol) 3,45-tribromopyrrole. Through the numerous studies of these authors it is suggested that the open sea is probably not as fertile an area for the isolation of antibiotic bacteria as coastal waters and sediments. They believe that estuaries, lagoons and reefs, where organic substances are abundant, are more favorable habitats for antibiotic producing bacteria.

Although not directly related to marine drug research, it is interesting to note in studies by Desai and Betrabet (1971) dealing with cotton research, that three marine fungi (two Aspergillus species and one Paecilomyces species) possessed enzymes with cellulase activity. These marine fungi were isolated from canvas cloth exposed to seawater in Bombay, India.

Other interesting studies on steroid-type compounds from marine ascomycetes and Fungi Imperfecti have been reported by Kirk and Catalfomo (1970). The compounds include choline and ergosterol. Teshima (1971) has also reported on the sterol composition of marine yeasts and identified both ergosterol and campesterol.

Horgen and Griffin (1971) have used the aquatic fungus Blastocladiella emersonii as an experimental test organism in studying the action of several drugs on three RNA polymerases in the organism.

Cyanophyta (Blue-Green Algae)

The blue-green algae are unicellular, filamentous or colonial algae. Because their nuclear material and plastids are not delimited by membranes as in higher plants, they are usually segregated and treated separately. Some authors classify them together with bacteria and fungi into the monera group.

As far as biological activity is concerned, numerous blue-green algae have been studied for toxic properties, particularly the fresh-water species responsible for algal blooms in drinking water reservoirs and lakes (Der Marderosian, 1969). More recent studies include focus on characterizing these toxins. Murthy (1970) has reported a new isolation technique and structure for the endotoxin from Microcystis aeruginosa NRC-1. His studies indicate that the "fast-death" factor is a cyclic polypeptide containing seven amino acids: aspartic acid, glutamic acid, serine, voline, ornithine, alanine and leucine. All except serine are L-amino acids. The unique CNS activity of this toxin begs further study.

Alam et al. (1973) have studied the toxin of Aphanizomenon flos-aquae and have found it to be a strongly basic substance, which gives a positive ninhydrin reaction and appears to be a guanidine derivative. It proved to be quite toxic to mice. A dose of 2.7 μg injected i. p. into 20 g mice caused death in 4.9 minutes. The purified toxin can reversibly block action potentials and tension development in amphibian nerve-muscle preparations without alteration of the resting potential. These authors compared this toxin with saxitoxin and found that it differed in certain chemical and physical properties as well as certain pharmacological properties on the neuromuscular systems studied.

Numerous studies have been carried out on various constituents of blue-green algae and include toxopherol and phytoquinone content (Dasilva, 1971; Whistance, 1970; Botham, 1971); biosynthesis and content of biotin, pyridoxine, thiamine, riboflavin, vitamin B_{12}, nicotinic and pantothenic acids (Kopteva, 1970; Kopteva, 1970) and glycogen (Chao, 1971).

In the area of chemotaxonomy, comparative analysis of lipids of blue-green algae have shown some relationship and evolutionary value (Kirilova, 1969; Pelloquin, 1970; Kenyon, 1970).

The use of the larvae of marine organisms for determining the toxicity of blue-green algae has been reported by some investigators (Ilyaletdinova, 1972).

In continuing studies on the mode of action of Lyngbya majuscula, it was learned that the toxic factor of this dermatitis-producing alga fails to induce either a protective or sensitizing effect with pre-exposure, acting as a general cell toxin (Chu, 1969; Moikeha, 1971).

Several authors have hinted at the possibility of controlling the growth of blue-green algae in water blooms by the use of viruses (Goryushin, 1974).

Protista

This group includes the Pyrrophyta (dinoflagellates), Chrysophyta (Diatoms, yellow-green and golden-brown algae), Chlorophyta (green algae), Phaeophyta (brown algae) and Rhodophyta (red algae).

A recent review of planktonic nerve toxins has been produced by Glombitza, 1970. In it may be found the toxins produced by dinoflagellates, and other flagellated unicellular organisms which are widely distributed throughout the oceans.

More concentrated studies by Martin et al., 1970, have resulted in the isolation and partial characterization of two toxins from the red-tide organism, Gymnodinium breve. To one of these structures, the formula

$C_{90}H_{162}O_7P$ has been assigned. In addition, it has been found to lack acetylcholinesterase inhibitor activity. The organism was isolated from Florida "red-tide" waters. Similarities may be noted in the toxic substances isolated from the organism in the natural environmental and from laboratory cultures of the organism (Alam, 1973) studied later. "Red tides" continue to be associated with extensive fish mortality along the west coast of Florida and elsewhere. In most cases, Gymnodinium breve is strongly incriminated by toxicological evidence as the primary source of the two types of toxic components detected in oysters, clams, coquinas and even seawater at times (Cummins, 1971). It has also been reported (Paster, 1969) that G. breve toxin hemolyzes rabbit erythrocytes. The toxicity to annelids, crustaceans, molluscs and fish has been compared with a related organism, Gonyaulax monilata (Sievers, 1969). Species of Gonyaulax and Saxidomus giganteus have been revealted as cources of saxitoxin (with a reported empirical formula of $C_{10}H_{17}N_7O_4 \cdot 2HCl$), which has also been reported in clams and mussels. Saxitoxin is a well-known neuromuscular paralyzer (Cheymol, 1969).

Several articles have appeared recently on the pharmacology and toxicology of saxitoxin which include comparative biological effects (Evans, 1971); its blockade and binding of sodium channels of excitable membranes (Smythies et al., 1974) and (Henderson et al., 1973); its general pharmacology and relationships with tetrodotoxin (Kao, 1972); and its cardiovascular properties (Nagasawa, 1970). The chemistry of saxitoxin has been pretty well worked out (Wong et al., 1971) as has the production of saxitoxin by cultures of Gonyaulax catenella (Proctor et al., 1975). Reports continue on the sporadic occurrence of paralytic shellfish poisonings around the world (Prakash, 1971; Steidinger et al., 1973; Murakami, 1973).

In summary, saxitoxin may be characterized as a potent marine biotoxin produced by dinoflagellate algae of various species of the genus Gonyaulax which include G. catenella, G. tamarensis and G. polyhedra. It is apparently the major toxin responsible for "red-tide" poisoning, caused by the vast multiplication of these organisms. These dinoflagellates release saxitoxin into the water. The toxin is capable of accumulating in various shellfish, notably the Alaskan butter clam (Saxidomas giganteus - whence the name of the toxin is derived) and the California mussel. When these are consumed by humans, paralytic shellfish poisoning occurs.

The symptoms of shellfish poisoning appear in about thirty minutes following ingestion. The typical symptoms include a burning sensation about the mouth, gums and face (paresthesia) which may progress to the limbs. This is usually followed by gastrointestinal distress, viz., nausea, vomiting, diarrhea and abdominal pain. Muscular incoordination and respiratory distress follow. Death can occur within 12-24 hours due to respiratory failure. The FDA has established the maximum human tolerance of paralytic shellfish poison to be 1200 Sommer mouse units (a mouse unit is that amount of toxin which can kill a 20 gram mouse in 15 minutes). Quarentine is the best prevention against human shellfish intoxication since visual detection in poisoned shellfish is impossible. Chemical analysis will reveal the poison and it is quite stable. Cooking, frying and baking do not destroy it.

Chemically saxitoxin is a non-crystalline, highly polar and non-volatile compound. The structure has been elucidated from a single crystal x-ray diffraction study of p-bromobenzene-sulfonate saxitoxin (Schantz, 1975).

Saxitoxin has a relatively small molecular weight of 299 and is a diacidic (pKa 8.1, 11.5) base of composition $C_{10}H_{17}N_7O_4$, which readily forms a double salt with HCl. The molecule is tricyclic and is described as 3,4,6-trialkyltetrahydropurine. The biosynthesis of saxitoxin has been studied by Proctor, 1975.

A chemical assay for saxitoxin has been recently developed (Bates et al., 1975). Normally, saxitoxin is detected in shellfish by a mouse bioassay procedure which is sensitive to as little as 0.3 mcg of saxitoxin per gram of shellfish. However, this lower limit cannot be attained because the presence of sodium ions in the shellfish counteracts the effects of the toxin. In addition, there are several problems with this assay in that several dilutions must be prepared and assayed before the concentration can be measured with any accuracy, many mice of the proper strain must be used for significant results, and most importantly, the mouse assay is not specific for saxitoxin.

Saxitoxin has similar pharmacological properties to tetrodotoxin, also a well known extremely lethal marine toxin but from the puffer fish rather than dinoflagellates. These two toxins rank among the deadliest poisons known to man. The minimum lethal dose, as determined in the mouse assay for both of these toxins is 8 mcg/kg. Although comparable in activity, their chemical structures are different and distinct. Tetrodotoxin has only one guanidinium moiety, while saxitoxin has two. Also, the two toxins can be differentiated chemically by comparing their relative solubilities when heated in acid solution. Other differences are present.

Both tetrodotoxin and saxitoxin act as neuromuscular blocking agents and interfere with the early, transiently open ionic channel through which sodium ions pass in most common excitable membranes. They inhibit the conductance of sodium ions. Thus, both saxitoxin and tetrodotoxin, as well as others with similar pharmacological action, have been the subject of intense investigations designed to determine the molecular structure of the sodium channel. The specifics of these studies is beyond the scope of this coverage.

In different pharmacological studies, both saxitoxin and tetrodotoxin were found to lower the set point for physiological thermoregulatory responses (Clark et al., 1974).

Both of the toxins also were found to possess severe hypotensive activity. They have been shown to cause a fall of 2/3 the normal arterial pressure in a dose of 2-3 mcg/kg. In both dog and cat experiments it was determined that the hypotension was due to a direct relaxant effect of the toxins on the vascular muscle at doses less than 1.5 mcg/kg and also blockade of vasomotor nerves at doses greater than 1.5 mcg/kg. In comparison with tetrodotoxin, saxitoxin has a lower hypotensive effect and tends to produce a late pressor effect that may be attributed to catecholamine secretion. This pharmacological observation may account for the apparent absence of hypotension in clinical cases of paralytic shellfish poisoning. However, it is not yet known whether saxitoxin can directly stimulate catecholamine release or whether the release is a reflex response to the lowered systemic pressure (Nagasawa, 1971).

In conclusion, it may be stated that saxitoxin has been the focus of many interesting studies in recent years. There still remains, however, need for developing an antidote for the shellfish poisoning and some means of inactivating it when high concentrations appear in the periodic "red-tides". Its continued use as a "molecular probe" or tool for further elucidation of the process of nerve conduction is apparent. Lastly, there is little doubt that saxitoxin may serve as a model for the synthesis of useful neuroactive agents.

Other interesting and related organisms and secretory products which have been implicated as toxins include glenodinine, an ichthyotoxic substance produced by Peridinium polonium (Hashimoto, 1968); a "red-tide" caused by Noctiluca scintillans (Morton, 1971); paralytic shellfish poisoning in eastern Canada caused by Gonyaulax tamarensis (Prakash, 1971); effect of Gymnodinium breve toxin on hemolysis induced by Prymnesium parvum toxin (Martin, 1972); anticoagulant properties of a red tide toxin (Doig, 1973); biological activity of a cell extract from the dinoflagellate, Amphidinium carteri (Thurberg, 1973); purification and biological activity of Gymnodinium breve toxins (Spiegelstein et al., 1973); the physiological characteristics of neritic red-tide flagellates (Iwasaki, 1973); the present state and problems on the red tide research in Japan (Irie, 1973); studies on the red tide dinoflagellate, Entomosigma spp. (Iwasaki, 1968) and parasitism of Gonyaulax catenella by the endoparasitic dinoflagellate Amoebophrya ceratii (Taylor, 1968).

The phytoflagellate, Prymnesium parvum yields toxins which are quite different from those produced by dinoflagellates (Ulitzur, 1970). The organism is eurhaline and the growth and toxigenesis are both a function of salinity (Padilla, 1970). The toxins here have been reviewed and found to possess cytotoxic, ichthytoxic, hemolytic, anti-spasmodic and antibiotic properties (Der Marderosian, 1969). Martin (1971) has investigated the kinetics and binding in hemolysis induced by the toxins in this organism and Ulitzer (1970) has compared the effect of Prymnesium toxins with that of certain surfactants on bacteria.

In 1973, two excellent reviews appeared on the pharmacology and structure-activity relationships of the toxin, prymnesin from the chrysomonad, Prymnesium parvum (Paster, 1973 and Padilla et al., 1973). Because of the variations in methods of extraction as well as complexity of activity, prymnesin is considered to be a mixture. These authors report studies which show separation of at least two hemolysins and an ichthyotoxic component. It was surprising to find that even though at least two different chemical compounds were purified from P. parvum, i. e. a glycolipid and a lipoprotein, both retained all the activities of the original extract.

In addition, they continued to share similar physical properties, viz., solubility, foaming activity and detergent-like properties, which apparently reside in their nonpolar lipid moiety and their polar, carbohydrate or proteinaceous moiety. One study showed a fraction of the toxin to be a mixture of six separate toxic compounds. The mixture contained 22 % protein, 10-12 % hexoses, 0.47 % inorganic phosphate, and twelve unidentified fatty acids. A variety of enzyme tests showed that only papain was capable of destroying the hemolytic activity without affecting the ichthyotoxic property.

Studies by Paster (1973) himself show that prymnesin is a high molecular weight glycolipid with uniformly high biological activities. Analytical ultracentrifugation shows a single peak and a single component is also seen by electrophretic methods. The molecular weight as calculated from ultracentrifugation is 23,000 $\pm$ 1800. The skeleton structure is a polysaccharide containing close to 100 hexase units made up of glucose, mannose, and galactose in a 2:1:1 ratio. This polysaccharide has 26 hydroxyl groups esterified by four long chain fatty acids, viz. myristic, stearic, palmitic and oleic acid in a 2:8:2:1 ratio as determined by gas chromatography. An elemental analysis showed C_9H_7 and O only without any P, N, or S.

As far as the biological activity of prymnesin is concerned, it probably results primarily from attachment or interaction with the cell membrane. It is a powerful hemolysin which very rapidly disrupts membrane to such an extent that there is rapid loss of not only ions, but amino acids and macromolecules from the cells as hemolysis is initiated.

As far as prymnesins overall activity is concerned, it is primarily toxic to gill-breathing species such as fish, mollusks, arthropods, and also to the gill-breathing stage of amphibians. Again, the broad spectrum of this toxin apparently can be related primarily to its detergent-like structure and its disruptive effect on biological membranes. While the precise mode of action remains unclear, the assumption that the "blocking action" of prymnesin on excitable membranes may not involve sodium conductance, but rather calcium movement, can make prymnesin an outstanding "molecular tool" in neurophysiological research. Perhaps the ultimate medical value of the toxin lies in its specific cytotoxic and neuromuscular blocking activities which may yet yield model compounds for further study.

Several toxic algal substances have been known which produce severe acute inflammatory reactions. One of these studied is the toxic factor of the dermatitis-producing alga, Lyngbya majusca. In studies by Chu, 1969 and Moikeha, 1971, it was shown that the toxic factor of the alga fails to induce either a protective or sensitizing effect with pre-exposure, acting mainly as a general cell toxin. Of particular interest is the fact that the toxic principle is capable of lysing rabbit erythrocytes and also possesses marked antibacterial activity against Mycobacterium species and slight to moderate activity against Bacillus cereus, Gaffkya tetragena and Sarcina lutea.

Acrylic acid, an antimicrobial elaborated by various sea algae, and first discovered in Phaeocystis pouchetii, has recently been screened for in several different algae by Glombitza, 1970. The studies were quantitative and concentrations of both acrylic acid and dimethyl-ß-propiothetin (its precursor) were determined.

It is now known that algae have potential medical uses beyond their antimicrobial properties. For example, the nitrogenous compound, laminine (trimethyl-(5 amino-5-carboxypentyl)-ammonium chloride) found in several members of the Laminariaceae, has hypotensive effects. More recently Takagi, 1970, has isolated a neutral amino acid, petalonine (N-(2-hydroxyl ethyl) glycine), in addition to laminine and pipecolic acid from the Formosan seaweed, Petalonia fascia which has hypotensive activity.

Another pharmacologically interesting property of certain marine algae is their hypocholesterolemic effect. Recently, Tsuchiya 1968, reviewed these cholesterol-lowering activities of several algae in a symposium publication. Of the various interesting findings it was noted that the unsaturated fatty acids isolated from Heterochordaria abietina were capable of depressing the blood cholesterol levels in rats at a dose of 27 mg per animal per day. Species differences were noted with some of the compounds studied, e. g. fucosterol decreases plasma cholesterol levels in chicks but not in rats. Further, it was found that carrageenan, agar and alginic acid all possess hypocholesterolemic properties.

Chunakova, 1967, has suggested the daily addition of sea food (e. g. sea kale, sea grass, etc.) to an antiatherosclerotic diet for the treatment and prophylaxis of coronary atherosclerosis. Another investigation (Besterman, 1970) focuses on the effects of laminarin sulfate on experimental atherosclerosis and on serum lipids in rabbits during long-term intermittent cholesterol feeding.

A number of algae have long been economically exploited for their nutritive value alone. Burkholder et al., 1971, has analyzed twenty-six species of marine algae for their ash, carbohydrate and protein contents. Many may find food use in years to come in the U. S. and European countries, as they have been throughout parts of Asia. In general, most species are high in carbohydrate and relatively low in protein. They currently appear to be popular based largely on their "fishy" or "marine flavor" and are found in soup products and in zesty "nibble" food items. Ohoka, 1971, has found that culturing sea weeds in the presence of triazine and its derivatives yields a higher protein content than usual. Imata et al., 1971, found that seaweed growth is enhanced in media containing carbon dioxide and sulfur dioxide while that of microorganisms is inhibited. It seems, therefore, that like their terrestrial counterparts, seaweeds can be growth enchanced and fortified through cultivation and selective fertilization techniques. As a final aside, seaweeds continue to be used in various cosmetic formulations, e. g. hair conditioners and cosmetic creams have incorporated seaweeds to provide amino acids, steroids and oils (Florent, 1969; Simon, 1970).

A number of interesting uses of algae may be cited including the use of Scendesmus obliquus in the postoperative care of the coagulation surface (Fabik, A., 1970); the use of Laminaria species as a "tent" for safe cervical dilation (Manabe, Y., 1971); a study of plasma-glucose and plasma-insulin levels in rats treated with Chlorella (Rodriguez-Lopez, M. et. al., 1971); the use of Chlorella protein hydrolysate to label typhoid bacteria and E. coli with radioactive carbon (Tumanian, M. et al., 1970); continuing studies on the pharmacological activity of carrageenan, known already for its antiulcer, anticoagulant and antithrombic properties (Di Rosa, M. et al., 1970; Maillet, M. et al., 1970; Marcus, R. et al., 1970; Salvaggio, J. et al., 1970; Schwartz, H. J. et al., 1969; Sharratt, M. et al., 1970; Sharratt, M. et al., 1971; Uruchurfu, M. A. et al., 1970); the use of marine diatoms to diagnose drowning (Timperman, J., 1969); diatomite treatment of late ulcerous dermal x-ray lesions (Kalantyrskaia, M. A. et al., 1968) and the antiviral, antifungal, antiyeast activity of a cellular extract of the diatom Asterionella notata (Gauthier, M., 1969).

Caulerpin is a new red pigment from green algae of the genus Caulerpa (Aguilar-Santos, G., 1970). It is of toxicological interest in that it produces dizziness, numbness of the tip of the tongue, weakening of the extremities and difficulty in breathing in some individuals. Besides caulerpin, the compounds, caulerpicin, palmitic acid, ß-sitosterol and the triterpene taraxerol, have also been isolated and quantified in Caulerpa lamourouxii (Santos, G., 1971). Both caulerpin and caulerpicin are apparently transferred along marine food chains and are concentrated in the process, at least in some herbivores (Doty, M. et al., 1970). These authors have suggested that these algae may in some way be implicated in the ciguatera type of fish intoxication because of the similarities in the toxicity syndromes.

Throughout the seventies numerous articles were published which dealt with the chemistry and biochemistry of numerous algae including ion exchanging properties of alginates; alginates as dental impression material; alginates as adjuvants and keratolytic agents; alginates useful in the treatment of secondary glaucomas; the variety of serols in algae; and the role of marine plants in radiation protection and the role of algae in assessing the extent of marine pollution (Hales, B., et al., 1974).

Another interesting compound isolated from algae is bifuhalol, a polyhydroxyphenyl ether, isolated from Bifurcaria bifurcata. It was found to possess antibiotic properties and is thought to be a precursor for the phaeophyte tannins. Chemically it is 2',3,4,4',5,6'-hexahydroxydiphenyl ether (Glombitza, K., 1974).

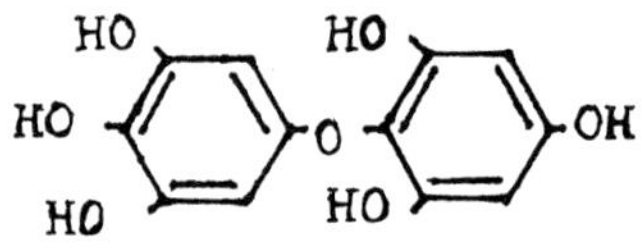

In a related study (Glombitza, K. et al., 1974) nine bromophenols, eight with antibiotic activity were isolated from 13 members of the Rhodomelaceae including Polysiphonia species. A typical structure is given below.

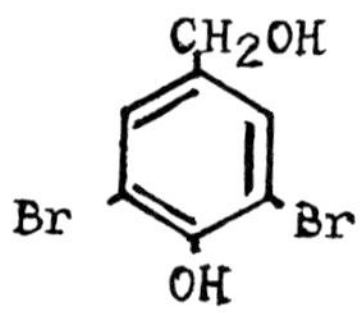

An interesting cytochemical study of the wound-healing protein in the alga Bryopsis hypnoides was carried out by Evert, 1972. The proteinaceous substance is not complexed with a polysaccharide, i. e. it does not seem to be a glycoprotein or a mucopolysaccharide. In addition the structures which were interpreted earlier as representing stages in the development of the wound-healing protein showed similarity in their reactivity in the various tests. This, according to the author, lends further support to the proposal that they are ontogenetically related.

A study by Takagi, N., 1970, on hypotensives in algae has revealed the amino acid composition of the widely spread (Formosan) seaweed, Petalonia fascia (Scytosinphonaceae). A neutral amino acid, petalonine was isolated, in addition to laminine and pipecolic acid. The chemical structure of petalonine was found to be N-(2-hydroxyethyl) glycine from its elemental analysis and synthesis.

The polysaccharide, L-fucose, derived from species of Fucus was found to cause reduction of tumor size by 20 % in rats with no toxicity to the animals except diuresis (Rosato, F., 1972). Another isomer, D-Fucose, showed both antibiotic and genetic effects in bacteria (Breverin, S. et al., 1971).

Marine Pollution

While it may not seem to bear a direct relationship to the topic of marine pharmacology, there are several interrelationships which exist between the topic of marine pollution and the health of man. First there is the direct deleterious effect of pollutants on the various marine organisms which need further study to reveal future drugs or to serve as animal models. Second, the various marine organisms can serve as monitors of the quality of the marine environment. Third, pollutants at one end of the food-chain may be transvectored and bioconcentrated through several marine organisms to ultimately poison man. And lastly, pollution may be in the form of nutrients from fertilizers which run-off into the oceans which can provide the rapid growth of undesirbale species, e. g. red tide.

Again, the number of references here is staggering, but a review of a few will be instructive. Several recent articles deal directly with the topic of marine pollution (Abelson, P., 1971; Paoletti, A., 1970). Goldberg, E. and Gross, M. (1970) have grouped marine pollution wastes into three categories to simplify the problem: particles from fuel combustion, elements from industrial activities, and synthetic organic chemicals.

Several investigators have studied the effects of oil and oil spill dispersants on the marine environment (LaRoche, G. et al., 1970; Shelton, R., 1971; Beyon, L., 1971; Button, D., 1971). Ahearn, D. et al. (1970) proposed a potential role for marine yeasts in the decomposition of these oils. Surface slicks can also act as concentrators of pesticides in the sea (Seba, D., 1969; Koeman, J., 1972). Here is a good example of a deleterious interaction of two different pollutants, one making the other more deadly than usual.

The DDT, DDD, DDE and PCB contents of marine animals along coastal California (Burnett, R., 1971) and from Swedish waters (Jensen, S. et al., 1969) have been reported. Studies like these led to the recent banning of these pesticides in the U. S. and other countries. Other studies have included the ecological effects of sewage discharge on the marine environment (Muchmore, D., 1973; Koerting, W., 1972; O'Sullivan, A., 1971) and the changes in intertidal sand communities due to thermal pollution (Barnett, P., 1971).

Large amounts of radioactive products, long-lived radionuclides have been introduced into the ocean from the atmosphere as a result of nuclear weapon tests (Miyake, Y., 1971). The possibility of applying the distribution of these radionuclides to the analysis of water dynamics has been considered (Khanaichenko, N., 1970), as has their potential use in isotopic marine ecology studies (Saiki, M., 1970). Marine fauna collect and frequently serve to concentrate these radionuclides (Seki, H., 1970;

Fowler, S., 1969; Williams, P. et al., 1970; Mascia, C. and Capone, W., 1969; Dutton, J., 1970; Ancellin, J. and Vilquin, A., 1968; Risik, N. et al., 1970; Polikarpov, G. et al., 1970; Lowman, F. et al., 1971). Elements that may be significantly concentrated by marine organisms include structural, catalytic and heavy divalent elements, heavy halogens and elements easily hydrolyzed at seawater pH.

Exclusive of fish eggs and invertebrate eggs and larvae, most marine organisms for which data exist are themselves relatively radio-resistant (Templeton, W. et al., 1971; Woodhead, D., 1971; Bryan, G., 1971); however, if man wishes to use the lower tropic levels (greater concentrations of radioactivity) as a food or drug source, then the capacity of the seas to accept waste radioactivity will decrease (Bowen, V. et al., 1971). Radionuclides and trace elements have recently been measured in marine protein concentrates and seafood (Beasley, T. et al., 1969 and 1970). The ^{210}Pb-^{210}Po pair were the most significant radionuclides present. If these proteins are used as human food Pb and Po may become the principal source of the skeletal radiation dose. Lowenstein, J. (1976) reviewed nuclear pollution of the seas and also pleads for control.

Organic discharges such as the acetaldehyde industry and agricultural chemicals, and also inorganic discharges such as mercury mines, the chlorine-alkali industry and miscellaneous other uses are responsible for the present increase in mercury pollution on a nationwide scale. Accumulation of methyl mercury in the water biomass through the food chain was found to be very high in various sea and fresh waters (Langley, D., 1973; Ui, J. and Kitamura, S., 1971). Of foremost interest has been the occurrence of mercury in edible fish (Skerfving, S., 1970; Block, J., 1971; Aoki, H., 1970; Wobeser, G., 1970; Berglund, F., 1969; Johansson, B. et al., 1970; Bache, C. and Liski, D., 1971; Gatti, G., 1971), but total mercury content has been determined in other marine plants and organisms, sediment and seawater (Yamane, Y. et al., 1971; Burton, J. et al., 1971; Smith, J. et al., 1971; Omang, S., 1971; April, R. et al., 1970).

It has been possible to provide only a glimpse on the tremendous potentialities of the sea as a source of marine biodynamic agents in this article. It is obvious that scientists are only at the threshold as far as research on compounds derived from marine sources are concerned. The importance of safeguarding these resources from wanton pollution cannot be overstated.

References

1. Abelson, P. H.: Marine pollution. Science 171, 21, (1971).
2. Aguilar-Santos, G.: Caulerpin, a new red pigment from the genus Caulerpa. J. Chem. Soc. (Org.) 6, 842-843 (1970).
3. Ahearn, D. G., Meyers, S., Standard, P.: Role of yeasts in the decomposition of oils in marine environments. Develop. Ind. Microbiol. 12, 126-134 (1970).
4. Alam, B. et al.: Purification of Aphanizomenon flos-aquae toxin and its chemical and physiological properties. Toxicon 11 (1), 65-72 (1970).
5. Alam, M., J. Sasner Jr., Ikawa, M.: Isolation of Gymnodinium breve toxin from Florida red tide water. Toxicon 11 (2), 201-202 (1973).
6. Alam, M.: Marine biotoxins. Tex. Rep. Biol. Med. 33 (1), 183-199 (1975).
7. Ancellin, J., Vilquin, A.: Experimental contamination of marine species with cesium-137, ruthenium-106 and cerium-144. Radioprotection 3 (3), 185-213 (1968).
8. Aoki, H.: Evironmental contamination by mercury (Hg series No. 14) 3. Inorganic and organic mercury in human hair and marine fisch. Jap. J. Hyg. 24, 556-562 (1970).
9. April, R. W., Hume, D.: Environmental mercury: rapid determination in water at nanogram levels. Science 170, 849-850 (1970).
10. Asano, M.: Progress in marine poison research. Kagaku No Ryoiki 25 (3), 214-223 (1971).

11. Auffhammer, I. W., Deichmann, W. B.: Sea Grant Tech. Bull. No 16.

12. Bache, C. A., Liski, D.: Gas chromatographic determination of organic mercury compounds by emission spectrometry in a helium plasma. Application to the analysis of methylmercuric salts in fish. Anal. Chem. 43, 950-952 (1971).

13. Baiardi, J., Ruggieri, G.: Aquatic Sciences, Annal. N. Y.Acad. Sci. 245, 3-69 (1975).

14. Barnett, P. R.: Some changes in intertidal sand communities due to thermal pollution. Proc. R. Soc. Lond. (Biol.) 177, 353-364 (1971).

15. Baslow, M. H.: Marine pharmacology, Williams & Wilkins, Baltimore, Md. (1969).

16. Baslow, M. H.: Marine Toxins. Ann. Rev. Pharmacol. 11, 447-454 (1971).

17. Bates, H. A., Rapoport, H.: Journal of Agriculture and Food Chem. 23 (2), 237-239 (1975).

18. Beasley, T. M., Jokela, T., Eagle, R.: Radionuclides and selected trace elements in marine protein concentrates. U. S. At. Energy Commun. RLo-2225-T-14-2, 15 pp. Avail. Dept. NTIS from Nucl. Sci. Abstr. 1971. 25 (16), 816-620 (1970).

19. Beasley, T. M., Osterberg, C., Jones, Y: Natural and artificial radionuclides in seafoods and marine protein concentrates. Nature (London) 221 (5187), 1207-1209 (1969).

20. Berglund, F.: Evaluation of the occurrence of mercury in fish from the point of view of food hygiene. Nord. Hyg. T. 50, 29-31 (1969).

21. Besterman, A. M.: Effects of laminarin sulfate on experimental atherosclerosis and on serum lipids in rabbits during longterm intermittent cholesterol feeding. Atherosclerosis 12, 85-96 (1970).

22. Beverin, S. et al.: D-Fucose as a gratuitous inducer of the L-arabinose operon in strains of Escherichia coli B/r mutant in gene ara C. J. Bacteriol. 107 (1), 79-86 (1971).

23. Beynon, L. R.: Oil spill dispersants. J. Inst. Petrol. London 57 (553), 1-7 (1971).

24. Bhate, B., Seshadrinathan, A.: Drugs from the deep. Bombay Technol. 24, 60-64 (1974).

25. Bland, J., Medcalf, D.: Marine natural product chemistry and the interim. J. Chem. Ed. 51 (10), 672 (1974).

26. Block, J. B.: Mercury pollution - a fish story. J. Ky. Med. Assoc. 69, 861-863 (1971).

27. Botham, K. M. et al.: The biosynthesis of toxopherols and related compounds in the blue-green alga Anabaena variabilis. Biochem. J. 122, 127-128 (1971).

28. Bowen, V. T., Olsen, J., Osterberg, C., Ravera, J.: Ecological interactions of marine radioactivity. Radioactiv. Mar. Environ. 200-222 (1971).

29. Breverin, S., Sheppard, D. E., Park, S. S.: D-fucose as a gratuitous inducer of the L-arabinose oepron in strains of Escherichia coli B/r mutant in gene ara C. J. Bacteriol. 107 (1), 79-86 (1971).

30. Bryan, G. W.: The effects of heavy metals (other than mercury) on marine and estuarine organisms. Proc. R. Soc. Lond. (Biol.) 177, 389-410 (1971).

31. Burkholder, P.: Drugs from the sea. Armed Forces Chem. J. 27, 1-8 (1963).

32. Burkholder, Pl, Sharma, G.: Antimicrobial agents from the sea, Lloydia 32, (1969).

33. Burkholder, P. R. et al.: Nutritive constituents of some Caribbean marine algae. Bot. Mar. 14 (2), 132-135 (1971).

34. Burnett, R.: DDT Residues. Distribution of concentration in Emerita analoga along coastal California. Science 174 (4009), 606-608 (1971).

35. Burton, J. D., Leatherland, T.: Mercury in a coastal marine environment. Nature (London) 231, 440-442 (1971).

36. Button, D. K.: Petroleum. Biological effects in the marine evironment. Impingement Man Oceans. 421-429. Ed. Hood, D. W. Interscience, New York, N. Y. (1971).

37. Cafruny, E., Cosmides, G. J., Rall, D. P., Schroeder, C. R., Weiner, I. M., Eds.: Proceedings of an international symposium on comparative pharmacology. Fed. Proc. 26, 964-1262 (1967).

38. Chang, C. W. J.: Marine natural products other than pigments. J. Chem. Educ. 50, 260-262 (1973).

39. Chao, L. et al.: Purification and properties of glycogen isolated from a blue-green alga, Nostoc muscorum. J. Bacteriol. 105, 331-338 (1971).

40. Cheymol, J., Toan, T.: Neuromuscular paralyzing action of saxitoxin on the rat. Therapie 24 (1), 191-198 (1969).

41. Chu, G. W.: A unique toxic lipid from a Hawaiian dermatitis producing marine alga. J. Formosan Med. Ass. 68, 570 (1969).

42. Chunakova, E. P. et al.: Changes in the permeability of blood capillaries in patients with coronary atherosclerosis under the effect of an antiatherosclerotic diet. Sov. Med. 30 (8), 30-35 (1967).

43. Clark, W. G., Lipton, J. M.: Journal of Physiology (London) 238 (1), 181-191 (1974).

44. Cummins, J. M. et al.: Occurrence of toxic bivaloe molluscs during a Gymnodinium breve "Red Tide". Trana. Amer. Fish Soc. 100 (1), 112-116 (1971).

45. Dasilva, E. J. et al.: Content of tocopherol in some blue-green algae. Biochim. Biophys. Acta 239, 345-347 (1971).

46. Der Marderosian, A. H.: Marine pharmaceuticals. J. Pharm. Sci. 58, 1-33 (1969).

47. Der Marderosian, A. H.: Sources of information concerning drugs from the sea. Drug Information Journal, May/Sept., 101-106 (1975).

48. Desai, A. J., Betrabet, S. M.: Cellulolytic activity of marine fungi. Curr. Sci. 40 (6), 423-426 (1971).

49. DiRosa, M. et al.: Some pharmacodynamic properties of carrageenan in the rat. Brit. J. Pharmacol. 38, (1970).

50. Doig, M. T., Martin, D. F.: Anticoagulant properties of a red tide toxin. Toxicon 11 (4), 351-355 (1973).

51. Doty, M. S., Aguilar-Santos, G.: Transfer of toxic algal substances in marine food chains. Pac. Sci. 24 (3), 351-355 (1970).

52. Dutton, J. W. R.: Determination of radiocesium in sea and fresh waters. Minst. Agric. Fish Food, (Rep.) FRL-6, 9 pp. Avail. Dep. NTIS from Nucl. Sci. Abstr. 1871, 25 (14), 31830.

53. Emerson, A. G., Taft, C. H.: Pharmacologically active agents from the sea. Tex. Rep. Biol. Med. 3, 302-338 (1945).

54. Evans, M. H.: A comparison of the biological effects of paralytic shellfisch poisons from clam, mussel and dinoflagellate. Toxicon 9, 139-144.

55. Evert, R. F.: Cytochemical study of the wound-healing protein in Bryopsis hypnoides. Cytobios 6, (24), 199-215 (1972).

56. Fabik, A.: Postoperative care of the coagulation surface following diathermocoagulation using the weed Scendesmus obliquus. Cesk. Gynek. 35, 4-6 (1970).

57. Florent, R. J.: Skin and scalp-treating preparation based on marine plankton extracts. Fr. 2,053,399 (Cl. A61K), 3 pp.

58. Fowler, Scott, W.: Distribution and transfer of zinc-65 accumulated from food and seawater by three marine crustaceans. Diss. Abstr. Int. B. 30 (3), 1233 (1969).

59. Freudenthal, H. D., Ed.: Drugs from the sea, Symposium Marine Technology Soc., Washington, D. C. (1968).

60. Gatti, G. L.: Presence of mercury in the environment and in fish. Sci. Aliment. 17 (5), 105-112 (1971).

61. Gauthier, M.: Antibacterial activity of a marine diatom: Asterionella notata. Rev. Int. Oceanogr. Med. 15/16, 103-171 (1969).

62. Gentile, J. H.: Bluegreen and green algal toxins. Microbial Toxins 7, 27-66.

63. Giloli, M., Giacobazzi, C.: Pharmaceutical research potential of the sea. Boll. Chim. Farm. 109 (6), 358-362 (1970).

64. Glombitza, K. W.: Antimicrobial contents of algae 2. Presence of acrylic in different sea algae. Planta Med. 18, 210-221 (1970).

65. Glombitzka, K. W.: Antimicrobial contents of algae 3. Quantitative determination of acrylic acid in sea algae. Planta Med. 18, 281-284 (1970).

66. Glombitza, K. W.: Nerve toxins from plankton. Deut. Apoth. Ztg. 110 (20), 713-718 (1970).

67. Glombitza, K. W. et al.: Antibiotics from algae IX Bromophenols from Rhodomelaceae. Planta Med. 25 (2), 105-114 (1974).

68. Glombitza, K. W. et al.: Antibiotics from algae XI Bifuhalol. Diphenylether from Bifurcaria bifurcata. Phytochemistry 13 (7), 1245-1247 (1974).

69. Goldberg, E. D., Gross, M. G.: Identification of globally distributed wastes in the marine environment. Chlorinated Hydrocarbons Mar. Environ., Rep. Panel Monit. Persistent Pestic. Mar. Environ. 37-42 (1970).

70. Goryushin, V. A., Chaplinskaya, S. M.: Viruses of blue-green algae. Aktual. Probl. Biol. Sinezelenykh Vodoroslei. 9-17. C. A. 13505s, May 26, 1975, Vol. 82, page 244, 1975 (1974).

71. Hales, B. F., Der Marderosian, A. H.: Continuing research on biomedicinals from the sea. Proceedings 10th Annual Conference, Marine Technology Society, Sept. 23-25, 1974, Washington, D. C., pp. 575-610 (1974).

72. Halstead, B.: Pharmaceuticals from the sea. New Sci. 56, 564-567 (1972).

73. Halstead, B.: Marine biotoxicology, Environ. Qual. Saf. 3, 212-239 (1974).

74. Halstead, B. W.: Unexploited ocean resources. Med. Arts Sci. 11, 72-75 (1957).

75. Hashimoto, Y. et al.: Glenodinine, an ichthytoxic substance produced by a dinoflagellate, Peridinium polonicum. Bull. Jap. Soc. Fish. 34 (6), 528 (1968).

76. Henderson, R., Ritchie, J. M., Strichartz, G. R.: The binding of labeled saxitoxin to the sodium channels in nerve membranes. J. of Physiology 235 (3), 783 (1973).

77. Horgen, P., Griffin, D.: Specific inhibitors of the three RNA polymerases from the aquatic fungus Blastocladiella emersonii. Proc. Nat. Acad. Sci. 68, 338-341 (1971).

78. Ilyaletdinova, S. G. et al.: Method of determining the toxicity of blue-green algae in relation to culicidae larvae. Izv. Akad. Nauk. Kaz. SSR Ser. Biol. 2, 39-42 (1972).

79. Imata, K., Saito, K.: Culture of seaweeds. Japan 71 02, 376 (Cl. A 01K, C12b) 3 pp. (1971).

80. Irie, H.: The present state and problems on the red-tide research in Japan. Bull. Plankton Soc. Jap. 19 (2), 115-124 (1973).

81. Iwasaki, H., Toraya, F., Eiji, Y.: Studies on the red tide dinoflagellates: I. On Entomosigma sp. in coastal area of Fukuyama. J. Fac. Fish Anim. Husb. Hiroshima Univ. 7 (2), 259-267 (1968).

82. Iwasaki, H.: The physiological characteristics of neritic red tide flagellates. Bull. Plankton Soc. Jap 19 (2), 104-114 (1973).

83. Jensen, S., Johnels, A., Olsson, M., Otterlind, G.: DDT and PCB in marine animals from Swedish waters. Nature (London) 224, 247-250 (1969).

84. Kalantyrshaia, M. A. et al.: Diatomite treatment of late ulcerous dermal X-ray lesions. Sov. Med. 31, 112-115 (1968).

85. Kao, C. Y.: Pharmacology of tetrodotoxin and saxitoxin. Federation Proceedings 31, No. 3, May-June (1972).

86. Kaul, P. N.: A Position Paper on Marine Biomedical Research. The University of Oklahoma, Norman, Oklahoma (1974).

87. Keiichi, O.: Halogenated compounds in marine algae. Kagaku To Seibutsu 13 (4), 252-259 (1975).

88. Kenyon, C. N. et al.: Possible evolutionary significance of polyunsaturated fatty acids in blue-green algae. Nature (London) 227, 1164-1166 (1970).

89. Khanaichenko, N. K.: Distribution of long-living radionuclides and water circulation in the tropical zone of the Atlantic Ocean.

90. Kirilova, V. S. et al.: Lipids of some strains of nitrogen-fixing blue-green algae. Mikrobiol. Zh. 31, 479-482 (1969).

91. Kirk, P. W. Jr., Catalfomo, P.: Marine fungi: Occurrence of ergosterol and choline. Phytochemistry 9 (3), 595-597 (1970).

92. Kittredge, J. S.: Comparative biochemistry, Marine biochemistry. Exp. Mar. Biol. 225-267. Ed. by Mariscal, R., Academic Press, N. Y.

93. Koeman, J.: Toxicological implications of pesticides in surface waters. Fold. Eur. Gewaesserschutz. Informationsbl. 18, 40-44 (1972)

94. Koerting, W.: Sewage damage to fish. Fisch. Flussbiol. 21, 82-87 (1972).

95. Kopteva, Zh. P., Biosynthesis of thiamine, riboflavin, and vitamin B12 by some blue-green algae. Mikrobiol. Zh. 32, 429-433 (1970).

96. Kopteva, Zh. P.: Biosynthesis of biotin, pyridoxine, nicotinic and pantothenic acids by some blue-green algae. Mikrobiol. Zh. 32, 555-560 (1970).

97. Kurata, A.: Vitamin B12 transudation from marine bottom muds. J. Oceanogr. Soc. Jap. 26 (2), 81-86 (1970).

98. Langley, D.: Mercury methylation in a aquatic environment. J. Water Pollut. Contr. Fed. 45 (1) 44-51 (1973).

99. LaRoche, G., Eisler, R., Tarzwell, C.: Bioassay procedures for oil and oil dispersant toxicity evaluation. J. Water Pollut. Control. Fed. 42, 1982-1989 (1970).

100. Lewis, Thomas: Marine models in modern medicine. Oceanus 19, 2-5 (1976).

101. Lowman, F. G., Rice, T., Richards, F.: Accumulation and redistribution of radionuclides by marine organisms. Radioactiv. Mar. Environ. 161-199 (1971).

102. Lowenstein, J.: Nuclear pollution of the seas. Oceans 9 (3), 60-65 (1976).

103. Maillet, M. et al.: Carrageenan: effects in animals. Lancet 2, 414-415 (1970).

104. Manabe, Y.: Laminaria tent for gradual and safe cervical dilatation. Am. J. Obstet. Gynecol. 110, 743-745 (1971).

105. Marcus, R. et al.: Carrageenan and the human gut. Lancet 2, 414-415 (1970).

106. Martin, D. F., Chatterjee, A. B.: Chemical and physical properties of two toxins from the red-tide organism Gymnodinium breve. Nat. Oceanic Atmos. Admin. (U. S.) (1970).

107. Martin, D. F. et al.: Hemolysis induced by Prymnesium parvum toxin. Kinetics and binding. Biochim. Biophys. Acta 241, 213-225 (1971).

108. Martin, D. F., Padilla, G. M., Heyl, M. G., Brown, A.: Effect of Gymnodinium breve toxin on hemolysis induced by Prymnesium parvum toxin. Toxicon 10 (3) (1972).

109. Martin, D., Padilla, G.: Marine phyrmacognosy. Academic Press, N. Y., pp. 1-295 (1973).

110. Mascia, C., Capone, W.: Presence and distribution of fallout strontium-90 in marine fauna. Rend. Semin. Fac. Sci. Univ. Cagliari 39 (3-4), 445-449 (1969).

111. McKee, H. C., Rarazi, D. S.: Development of sample preparation methods for analysis of marine organisms. U.S.N.T.I.S. PB Rep. No. 234201/2GA, 64 pp. (Eng.). Avail. NTIS. 3.75 dollars. From Govt. Rep. Announce (U. S.) 74 (20), 80 (1974).

112. Miyake, Y.: Radioactive models. Impingement Man Oceans 565-588 (1971).

113. Moikeha, S. N., Chu, G. W.: Dermatitis producing alga. Lyngbya majusca, Hawaii II. Biological properties of the toxic factor. J. Phycol. 7 (1), 8-13 (1971).

114. Morton, B., Twentyman, P. R.: The occurence and toxicity of a red tide caused by Noctiluca scintillans (Macartney) Shrenb. in the coastal waters of Hong Kong. Environ. Res. 4 (6), 544-557 (1971).

115. Muchmore, D.: The effects of chlorination of waste water on fertilizations in some marine invertebrates. Mar. Biol. 19 (2), 93-95 (1973).

116. Murakami, A.: Red tide in the Seto Inland Sea. Bull. Plankton Soc. Jap. 19 (2), 131-132 (1973).

117. Murthy, J. R. et al.: A new isolation and structure for the endotoxin from Microcystis aeruginosa NRC-1. Canad. J. Biochem. 48, 508-510 (1970).

118. Nagasawa, J., Spiegelstein, M., Kao, C. Y.: J. Pharmacol. Expt'l. Therap. 178 (1), 103-109 (1971).

119. Nagasawa, J., Spiegelstein, N.: Cardiovascular actions of saxotoxin. Federation Proceedings 29, 274 (1970).

120. Nigrelli, R. F.: Dutchman's baccy juice or growth promoting and growth inhibiting substances of marine origin. Trans. N. Y. Acad. Sci. Ser. II. 20, 248-262 (1958).

121. Nigrelli, R. F.: Biochemistry and pharmacology of compounds derived from marine organisms. Ann. N. Y. Acad. Sci. 90, 615-950 (1960).

122. Ohoka, T. et al.: Seaweeds with high protein contents. Japan 71 02, 377 (Cl. AO1K, C12b), 3 pp. (1971).

123. Omang, S. H.: Determination of mercury in natural waters and effluents by flameless atomic absorption spectrophotometry. Anal. Chim. Acta 53, 415-419 (1971).

124. O'Sullivan, A. J.: Ecological effects of sewage discharge in the marine environment. Proc. Roy. Soc. Lond. (Biol.) 177, 331-351 (1971).

125. Padilla, G. M.: Growth and toxigenesis of the chrysomonal Prymnesium parvum as a function of salinity. J. Protozool. 17 (3), 456-462 (1970).

126. Padilla, G., Martin, D.: Interactions of Prymnesin with erythrocyte membranes. In: Marine Pharmacology. Eds. Martin, D., Padilla, G. Academic Press, N. Y., pp. 265-295 (1973).

127. Paoletti, A.: Chemical pollution of the sea. Rev. Sanid. Hig Publica (Madr.) 44, 765-794 (1970).

128. Paster, Z. et al.: Hemolysis of rabbit erythrocytes by Gymnodinium breve toxin. Toxicon 7, 245 (1969).

129. Paster, Z.: Pharmacology and mode of action of prymnesin. In: Marine Pharmacology. Eds. Martin, D., Padilla, G. Academic Press, N. Y., pp. 241-263 (1973).

130. Pelloquin, A. et al.: Comparative study of the lipids of Spirulina platensis (Gom.) Geitler and Spirulina geitleri J. de Tonj. C. R. Acad. Sci. (D) (Paris) 271, 932-935 (1970).

131. Polikarpov, G. G., Zaitsev, Y., Strogonov, A., Tsytsugina, V., Zesenko, A.: Accumulation of radionuclides by Mediterranean Sea aquatic animals. Radioekologicheskie Issled. Stredizemnogo Morya. 111-119 (1970).

132. Prakash, A., Medcaf, J. C., Tennant, A. D.: Paralytic shellfish poisoning in eastern Canada. Fish Res. Bd. Can. Bull. 177, 2-87 (1971).

133. Proctor, N. H., Chan, S. L., Trevor, A. J.: Production of saxitoxin by cultures of Gonyaulax catenella. Toxicon 13, 1-9 (1975).

134. Raymont, J. E. G.: Biochemical composition of marine zooplankton. In: Essays Hydrobiol., Ed. by Clark, R. B., Wootton, R. J. Univ. Exeter, Exeter, Engl., pp. 83-91 (1972).

135. Risik, N. S., Strogonov, A., Zelezinskaya, L.: Content of uranium-238 in Mediterrenean Sea hydrobionts. Radioekologicheskie Issled. Sredizemnogo Morya 182-187 (1970).

136. Robinson, P.: Versatile chitin. Sea Frontiers. Jan./Feb.42-48 (1974).

137. Rodriguez-Lopez, M. et al.: Plasma-glucose and plasma-insulin in normal and alloxanized rats treated with Chlorella. Life Sci. (11) 10, 57-60 (1971).

138. Rosato, F. E. et al.: Continuous intravenous fucose treatment of rat mammary tumor. J. Surg. Oncol. 4 (2), 94-101 (1972).

139. Ruggieri, G., Webber, H., Eds.: Food, Drugs from the Sea, Proceedings The fourth Drugs from the sea conference, 1974. Marine Technology Society, 1730 M St., NW, Suite 410, Washington, DC, 20036 (1976).

140. Russell, F. E.: Pharmacology of toxins of marine organisms. Sec. 71, Vol. II, pp. 3-114. In: International encyclopedia of pharmacology and therapeutics, pharmacology and toxicology of naturally occuring toxins, Pergamon Press, N. Y. (1971).

141. Russell, F. E.: Poisonous and venomous marine animals and their toxins. Ann. N. Y. Acad. Sci. 245, 57-64 (1975).

142. Saiki, M.: Marine ecology and atomic energy, especially on marine-ecological studies by isotope technique. Radioisotopes 19 (7), 328-336 (1970).

143. Salvaggio, J. et al.: The use of carrageenan as a granuloma-producing agent and Freund's adjuvent. Proc. Soc. Exp. Biol. Med. 134, 1116-1119 (1970).

144. Santos, G. A., Doty, M. S.: Constituents of the green alga Caulerpa lamourouxii. Lloydia 34 (1), 88-90 (1971).

145. Schantz, E. J. et al.: Journal Am. Chem. Soc. 97 (5), 1238-1239 (1975).

146. Scheuer, P. J.: Chemistry of some toxins isolated from marine organisms. Fortschr. Chem. Org. Naturst. 27, 322-339 (1969).

147. Scheuer, P. J.: Toxins from fish and other marine organisms. Advan. Food Res. 18, 141-161 (1970).

148. Scheuer, P. J.: Toxins from marine invertebrates. Naturwissenschaften 58 (11), 549-554 (1971).

149. Scheuer, P. J.: Chemistry of marine toxins. Toxins Anim. Plant Origin, (Proc. Int. Symp. Anim. Plant Toxins), 2nd 1970 (Pub. 1972) 2, 545-556 (1972). Ed. by DeVries, Andre-Kochva, E., Gordon and Breach, New York, N. Y.

150. Scheuer, P. J.: Chemistry of natural marine poisons. Kaguku (Kyoto), 30 (8), 595-601 (1975).

151. Scheuer, P. J.: Recent developments in the chemistry of marine toxins. Lloydia 38 (1), 1-7 (1975).

152. Schwartz, H. J. et al.: Carrageenan and delayed hypersensitivity. II. Activation of Hageman factor by carrageenan and its possible significance. Proc. Soc. Exp. Biol. Med. 132, 1021-1024 (1969).

153. Seba, D. B.: Surface slicks as concentrators of pesticides in the marine environment. Pestic. Monit. J. 3, 190-193 (1969).

154. Seki, H.: Bioaccumulation of radioisotopes by microorganisms in the sea. Nippon Kaiyo Gakkaishi 26 (6), 367-372 (1970).

155. Sharratt, M. et al.: Carrageenan ulceration as a model for human ulcerative colitis. Lancet 2, 932 (1970).

156. Sharratt, M. et al.: Carrageenan ulceration and ulcerative colitis. J. Pathol. 103, 18-19 (1971).

157. Shelton, R. G.: Effects of oil dispersants on the marine environment. Proc. R. Soc. Lond. (Biol.) 177, 411-422 (1971).

158. Sievers, A. M.: Comparative toxicity of Gonyaulax monilata and Gymnodinium breve to annelids, crustaceans, molluscs and a fish. J. Protozool. 16, 401-404 (1969).

159. Simon, R.: Composition for the treatment of washing of hair. Fr. 1, 601,008 (Cl. A61K) 3pp.

160. Sherfving, S.: Methyl mercury in fish: an important health problem Lakartidningen 67, 5235-5243 (1970).

161. Smith, J. D., Nicholson, R.: Mercury in water of the tidal Thames. Nature (London) 232, 393-394 (1971).

162. Smythies, J., Benington, F., Bradley, R. J., Bridges, W. F., Morin, R. D.: The molecular structure of the sodium channel. J. Theoretical Biology 43, 29-42 (1974).

163. Speigelstein, M. Y., Paster, Z., Abbott, C. C.: Purification and biological activity of Gymnodinium breve toxins. Toxicon 11 (1), 85-93 (1973).

164. Steidinger, K., Joyce, E., Jr.: Florida red tides. Fla. Dep. Nat. Resour. Educ. Ser. 17, 1-26 (1973).

165. Takagi, N. et al.: Studies on the hypotensive constituents of marine algae. V. Amino acid constituents of Petalonia fascia. Yakugaku Zasshi 90 (7), 899-902 (1970).

166. Taylor, F. J. R.: Parasitism of the toxin-producing dinoflagellate Gonyaulax catenella by the endoparasitic dinoflagellate Amoebophrya ceratii. J. Fish Res. Board Can. 25 (10), 2241-2245 (1968).

167. Templeton, W. L., Jaske, R., Watson, D.: Radiation effects. Radioactiv. Mar. Environ. 223-239 (1971).

168. Thomson, P.: Value is extracted from a nuisance. Smithsonian 6, 40-45 (1975).

169. Thurberg, F. P., Sasner, J. J.: Biological activity of a cell extract from the dinoflagellate Amphidinium carteri. Chesapeake Sci. 14, (1), 48-51 (1973).

170. Timperman, J.: Medico-legal problems in death by drowning. Its diagnosis by the diatom method. J. Forensic Med. 16, 45-75 (1969).

171. Tsuchiya, Y.: Comparative hypocholesterolemic activities of marine algae. Proc. Int. Seaweed Symp., 6th, pp. 747-757. Ed. Margalef, R. (1968).

172. Tumanian, M. A. et al.: Use of Chlorella protein hydrolysate for labelling typhoid bacteria and E. coli with radioactive carbon. Zh. Mikrobiol. Epidermiol. Immunobiol. 47, 61-63 (1970).

173. Tunstall, A., Gowland, G.: Susceptibility to antibiotics of marine strains of Pseudomonas. J. Appl. Bacteriol. 37, 455-457 (1974).

174. Ui, J., Kitamura, S.: Mercury pollution of sea and fresh water. Its accumulation into water biomass. J. Fac. Eng., Univ. Tokyo, Ser B 31 (1), 271-306 (1971).

175. Ulitzur, S. et al.: Procedure for purification and separation of Prymnesium parvum toxins. Biochim. Biophys. Acta 201, 350-363 (1970).

176. Ulitzur, S. et al.: Effect of Prymnesium parvum toxin, cetytrimethylammonium bromide and sodium dolecyl sulfate on bacteria. J. Gen. Microbiol. 62, 363-370 (1970).

177. Uruchurfu, M. A. et al.: Carrageenin induced arthritis in the specific pathogen free pig. J. Comp. Pathol. 80, 607-611 (1970).

178. Westoo, G., Johannsson, B., Ryhage, R., Westoo, G.: Identification and determination of methyl mercury compounds in fish using combination gas chromatograph-mass spectrometer. Acta. Chem. Scand. 24, 2349-2354 (1970).

179. Whistance, G. R. et al.: Biosynthesis of phytoquinones. Homogentisic acid. a precursor of plastoquinones, tocopherols and alpha-tocopherolquinones, Tocopherols and alpha-tocopherolquinones in higher plants, green algae, and blue-green algae. Biochem. J. 117, 593-600 (1970).

180. White, R. H.: Analysis of organic halogen compounds in marine organisms. Ph. D. Thesis. Univ. Illinois, Wibana, Will., 1974. 233 pp. Avail. Xerox, Univ. Microfilms, Ann Arbor, Mich. Order #75-468 from: Diss. Abstr. Int. B., 1975 35 (7), 3196-7.(1974).

181. Williams, P. M., McGowan, J., Stuiver, M.: Bomb carbon-14 in deep sea organisms. Nature (London) 227 (5256), 375-376 (1970).

182. Wobeser, G.: Mercury poisoning from fish. Canad. Med. Ass. J. 102, 1209 (1970).

183. Wong, J. L., Oesterlein, R., Rapoport, H.: The structure of saxitoxin. J. Am. Chem. Soc. 93, 7344-5 (1971).

184. Woodhead, D. S.: The biological effects of radioactive waste. Proc. R. Soc. Lond. (Biol.) 177, 423-437 (1971).

185. Worthen, L. R. (Ed.): Food-drugs from the sea proceedings. Marine Technology Soc., Washington, D. C.

186. Yamane, Y., Motoichi, M., Kazui, S., Noriko, M., Yoshitaka, F., Hiroyuki, K., Eiichi, I., Chizoko, S., Rinko, N., Kazue, K., Takahiro, K., Sumiko, K.: Determination of total mercury in sediment. Eisei Kagaku 17 (4), 274-279 (1971).

187. Yasumoto, T.: Potential biomedical substances in the extracts of marine organisms. Nippon Suisan Gakkaishi 37, 800-805 (1971).

188. Youngken, H. W. (Ed.): Food-drugs from the sea proceedings. Marine Technology Soc., Washington, D. C. (1969).

SEAWEED RESOURCES FOR PHARMACEUTICAL USES

G. Michanek
Dept of Marine Botany, University of Gothenburg,
Carl Skottsbergs Gata 22, S-413 19 Gothenburg, Sweden

NEEDS AND DEMANDS

Which seaweed species have pharmaceutical properties? Without any doubt, the answer is: all. As an example, carrageenan, agar and alginic acid all exhibit hypocholesterolemic activity, especially the first one is an excellent agent (Tsuchiya 1969). At least one of these phycocolloids or some of their close relatives are present in all red and brown algae and also green algae show a significant effect on lowering plasma cholesterol level. Considering the fact that, in the well-fed part of the world, more people will die from circulatory disturbance than from any other disease group, there are good reasons for a substantial increase in pharmaceutical research on seaweed substances, and for increased production of ant-atherosclerotic drugs. This reason alone, the lowering effect on blood cholesterol level, is sufficient for an increase in human consumption of seaweed.

But there are other and stronger motives. Endemic goitre is still a torment in many areas where rain, flooding or glaciation have washed out the soil iodine (Fig. 1.) An extrapolation from available statistics yields at a world figure in the vicinity of 300 million incidents. Actually the quantities of seaweed harvested at present are so modest, 2 million tons a year, and their iodine content so high, 100 to 40 000 times that of sea water, that all of it should be distributed as

Marine Algae in Pharmaceutical Science
edited by H. A. Hoppe, T. Levring, Y. Tanaka

food within the populations of these areas, if we want to live up to the basic ideas upon which the United Nations are founded, among which the right to freedom from disease should not be forgotten. As long as hundreds of millions of people suffer from goitre, simply due to our failure to organize a fair distribution of the seaweed resources, we have not fulfilled our commitment. The main problem is that these inland and highland populations are often poor, with very weak purchasing power, and with little insight into the goitre problem and possible remedies . The coastal people who should collect and prepare the seaweed are better nourished, but nevertheless there are production and transportation costs. Governments should engage themselves in the initial phase; the organizing of collection transportation and distribution could advantageously be supported by internationa organizations.

Malnutrition is the gateway to many infections and other problems in the tropics. Seaweeds are rich in proteins, vitmins and trace elements. Their amino acids are present partly in different proportions from those of cereals and other land plants, and are not so favourably composed as those in meat and egg. The total quantities are much too small to "fill the protein gap". As a possible remedy for malnutrition, seaweed should be selected primarily for their content of vitamins and trace elements. Many species rank among the most vitmain-rich plants of the world. Their antiberberic activity has been known for centuries among the peoples of the Far East. In the long run tropical people would profit more from eating their algae than from exporting them. Certainly, phycocolloids have many blessings, such as use for bacteriological agar, and many advantages for luxury consumption, such as for ice-cream softening, at present a big commodity; but the main bulk of world deaweed resources should be reserved for human food and in particular for the groups mentioned who require iodine and other protectional substances.

In addition to these benefits of broad importance which may be attributed to most species of algae - iodine, vitamins and phycocolloids are typical of most of them - there are also a number of seaweeds with particular pharmaceutical properties.

A few species contain vermifugal substances. Most of them have been used in folk medicine for generations. In our days the active substances have been extracted, analyzed and introduced into modern pharmacopeias. Among people living in coastal regions these algae are still used in the traditional way, fresh or dried. With increasing understanding for the exhausting effect on populations sucked out by parasites, various campaigns have been started among which those that distribute indigenous seaweed products to inland populations should be supported in particular, as they could most easily form the basis for lasting improvements.

Among possible pharmaceutical uses there is one which may change the whole pattern of seaweed harvesting and trade: tumour inhibiting substances have recently been isolated from algae. If these preliminary findings finally result in remedies for cancer, this would not only give new hope to countless individuals, but would also create a new aquaculture and a new industry for the cultivation and processing of seaweed species containing such drugs. It is not unlikely that the concept of seaweed's general utility as a food with preventive properties will be brought home to the public. This speculation is mentioned because of the possibility that recent advances might cause a price development which could direct still more tropical resources to industrial countries.

ANTHELMINTICS

The most beneficial use of algae would be in the fight against some wide-spread diseases which are physically power-reducing, a health hazard to infants and a draw-back to society.

If we agree with this view, it will be of primary interest to compare the distribution areas of some vermifuges with the distribution of intestinal worm diseases. In a way the latter distribution coincides with that of man, as no area on the earth is known to be free from worms as human parasites. Among principal helminth parasites of man the one affecting the highest number of individuals is the relatively harmless *Ascaris* which is estimated to infect a third of mankind.

A break-down on parts of the world is found in the somewhat outdated World-Atlas of epidemic diseases (Rodenwaldt and Jusatz 1956):

		per cent	Incidence of ascariasis millions
North America		1,7	3
Europe		9,1	32
North, Central, Spain		1-7	
South-East including Italy		13-46	
Sovjet	Eur	7,1	13
	Asian	21,6	6,9
Central and S. Am.		33,9	42
Africa		35,1	59
Asia		39,3	488

A 90 per cent infection rate was found in certain groups e. g. of children and soldiers in S. Am., or in certain provinces of China, the Philippines, Bangladesh, Paulu, Caroline and Marshal Islands.

Worms put a heavy load on people in affected areas not only because so much organic matter in their meagre food intake to the building up of worm organisms, but these also interfere with the absorption of nutrients. The name "ascarase" has been assigned to an enzyme isolated from *Ascaris lumbricoides* that has an inhibitory effect on pepsin and trypsin and a significant capacity to decrease food absorption. Other worms cause loss of blood to feces and loss of serum albumin from blood.

Digenea simplex, a red alga of the Rhodomelaceae family, has been known as a vermifuge for more than a thousand years. It has the widest use at present and may be the most efficient among seaweeds in the treatment of *Ascaris* and *Oxyuris* as well as whipworm, *Trichuris*, and tape worm, *Taenia*.

Information on the geographical distribution of *Digenea* is often misleading and sometimes false. Usually it is summarized as "found in most warm seas" or "reported from the Atlantic, the Mediterranean, the Indian and the Pacific Oceans". These indications may give the impression that it covers the area. However, a check in some of the more comprehensive and complete local investigations from these warm seas shows that it is by no means a common alga.

Actually *Digenea* is n o t f o u n d in the Canary Islands (Boergesen 1930), nor in Morocco (Gayral 1958), Madeira (Levring 1974), Portugal (Ardré 1970), Côte des Albères, the Mediterranean end of the Pyrenees (Feldmann 1938), Jabuka Island in the centre of the Adriatic (Ercegovic 1957), nor along the coasts of Karachi and Kathiawar (Boergesen 1934), Bombay (Boergesen 1935), South India (Boergesen 1938), Sri Lanka (Durairatnam 1961), Bangladesh (Islam 1976), Vietnam (Phạm Hoàng Hộ 1962), Bogor, Java (Wildeman 1900), Kai Islands in eastern Banda Sea (Weber-van Bosse 1926), Caroline Island (Trono 1969), Bikini (Taylor 1960), or Tahiti (Setchell 1926). In this particular case a negative list is necessary to correct various misinterpretations. The positive list is likewise most interesting:

<u>Digenea simplex</u> is found in the entire Caribbean from Tampa Bay in the central part of Florida's west coast, through Cuba, Jamaica, Puerto Rico, Venezuela, to Brazil, and is common and often abundant in the intertidal belt and down to 20 m. On the African side it is noted only in lower latitudes: Sao Tomé, Ghana, Senegal and Cape Verde Isl. From Corsica two findings (in 1873 and 1975) are recorded. In Naples and the surrounding area it was found in calm bays just beneath the surface or at 3-10 m depth, flourishing 2.5 m. In the Aegean Sea it is reported from most places investigated and in the eastern Mediterranean it is known from Israel.

In the western part of the Indian Ocean it is reported from Mauritius, Réunion, Tanzania, Kenya, Sinai and the Iranean Gulf. In the Pacific it is reported from N and W Australia, from New Caledonia, and on the American coast from the Gulf of California to Costa Rica.

The centre for the collection of <u>Digenea</u>, however, is an amazingly small area in the south China Sea. Off the Chinese island of Pratas in an open lagoon of 16 km diameter, 300 tons a year were produced by some forty divers according to Tseng, who visited the island 45 years ago. At that time, in addition to the Chinese production, a large number of Japanes also visited Pratas each year from March to September, collecting the greater part of the production and exporting it to Taiwan and Ryukyu (Tseng 1935).

In Japan <u>Digenea</u> is collected only in subtropical waters along Kyushu in the three southernmost prefectures: Kumamoto 18 t, Miyazaki 38 t, and Kagoshima 225 t (Okazaki 1971). The product is used as folk medicine as well as for pharmaceutical preparations. Both Swiss, U. S. and Japanese companies are marketing products, the active agent of which is extracted from <u>Digenea</u>.

The genus *Chondria* also belongs to the family Rhodomelaceae, and like *Digenea* has found use as highly efficient anthelmintic working on *Oxyris* as well as *Ascaris* and on threadworm. The algal genus is represented in all warmer seas and at least four species are reported as vermifuges: *C. armata*, *C. littoralis*, *C. sanguinea*, and *C. vermicularis*. *Chondria armata*, like *Digenea simplex* growing in the intertidal belt along the coasts of Kyushu, is in Japanese called 'Domoi', for which the effective principle domoic acid has been named.

The discussion on how to control helminthic infections is mostly concerned with the relative effectivity of certain drugs. Of the two active agents of *Digenea*, α-kainic acid (Digenea acid) is said to be about ten times stronger in action than is Santonin, derived from *Artemisia maritima* while the α-allokainic acid is faintly anthelmintic. A single dose of 5-10 mg of kainic acid is capable of driving out *Ascaris* in adults and no unpleasant side effects have been observed. Concurrent administration of kainic acid and Santonin was found to cause synergetic action and to result in marked increase in ascaricidal effect (Takemoto 1957). The WHO expert committee (1967) refers to a mass chemotherapy performed in Japan, where piperazine maleate was compared to a mixture of 50 mg santonin and 5 mg kainic acid, both given for two days. The cure rates were 86% and 78% respectively.

Hypnea musciformis, a turf-forming red alga from the lower part of the tidal belt in warmer seas, is used as an anthelmintic especially in Greece and Turkey (Diannelidis 1949). According to Kylin (1954) this species is restricted to the warmer parts of the northern Atlantic and to the Mediterranean. There are, however, numerous records of *Hypnea musciformis* from other parts of the world: Brazil, Ghana, Cameroun, Angola, Tanzania, Oman (2 500 t standing crop), Pakistan, India (in abundance on the Gujarat coast), Bangladesh, Rottnest Island (off Western Australia), and Pacific Mexico. There may be identification problems between *Hypnea musciformis* and the 25 other *Hypnea* species but no investigations have been e to compare various local races and other species of the genus with regard to vermifugal activity.

There exists no doubt, however, that *Hypnea musciformis* - the most well-known among *Hypnea* vermifuge producers - is widely distributed from Virginia in the north through the Gulf and Caribbean where it often dominates the lower part of the intertidal to Brazil in the south. It is recorded not as a vermifuge raw material but as a possible source of hypnean, an agar-resembling phycocolloid of high quality. The gel-strength far exceeds that of other known phycocolloids and the melting point can be influenced during production. The range of uses includes bacterial cultures and various pharmaceutical products. *Hypnea* species constitute the most important source of carrageenan found in the Caribbean. Especially rich in Venezuela are *H. cervicornis* and *H. musciformis* (Díaz-Pifferer 1967). The problem is that it usually forms dense but rather low mats, and is hard to harvest in quantities. On Cuba it reaches exceptional sizes. Baardseth (1968) gives an example of yield per unit effort: In Rio Banes a man collected 510 kg in four hours. Nevertheless the total in this area was not sufficient to sustain an agar industry. No doubt here, as in elsewhere, vermifuge production would require as much as as a hypnean production.

During an assessment of seaweed cast ashore in Pakistan an almost monospecific occurrence of *Hypnea musciformis* was found in one of the points investigated. The wrack belt was two meters wide, the seaweed amounted to 11 kg/m^2, of which 92 per cent was made up of *Hypnea*.

In Europe, on the Atlantic coast as well as in the Mediterranean, the vermifuge from a calcareous alga, *Corallina officinalis*, was common , as already indicated in the name given by

Linneus. Along the same coasts another calcareous alga, *Jania rubens* was also used in the same way. *Rhodymenia palmata*, dulse, one of the algae most frequently eaten in Europe, was also considered to have an anthelminthic effect. From 1775, however, these were all abandoned for a more efficient remedy called Corsican moss, which is believed to be the same alga that was used as a vermifuge by the ancient Greeks (Chase 1942). The algae with this trade name were mainly collected in the vicinity of Ajaccio and were dried and marketed as Muscus corsicanus. Actually the commercial article is made up by a mixture of a number of small mossy red algae of which inly one, *Alsidium helminthocorton*, is for certain a vermifuge, though a very effective one. There are a few other records of this species from Naples, from the Adriatic and from Syria.

Also the bull kelp of the southern hemisphere, *Durvillea antarctica*, is known to be a vermifuge, used as such by the maoris of New Zealand. Other algae mentioned for this purpose are *Sargassum vulgare*, common in the Atlantic, the Mediterranean and the Pacific, and the green alga *Ulva lactuca* which has a more world-wide distribution than most other species.

IRON DEFICIENCY REMEDIES

Iron deficiency is, next to joule and protein deficiencies, the most common among malnutrition diseases, from an international point of view. It is often but not always linked to protein or other deficiencies and is often more common within population groups which are economically and socially less well off (Hallberg 1975).

There is a significant relationship between anaemia and physical work, especially maximum performances. This has been established *inter alia* during an investigation in Guatemala carried out by INCAP (Institute of Nutrition of Central America and Panama) (Viteri and Torun 1974).

It is believed that iron deficiency in man is not primarily due to low iron content in human food, as this content is actually often ten times as high as the need. The reason lies rather in a low ability of the body to take up iron from food. Ascorbic acid stimulates this activity, 25 mg increases iron uptake from maize by three times; bran is retarding, 10% bran in wheat meal decreases iron uptake by a half (Björn-Rasmussen 1974). Iron uptake from fish is very good, from fish meal none. Where does the active substance come from? Is it already available in the primary producer of organic substances in the sea, the algae? Iron deficiencies are not pronounced in Japan, where people eat more seaweed than in other parts of the world - but also more fish.

It is likely that this problem complex, including the many interactions of nutrition and infection, may contain a hidden key to public health in tropical countries and thus a key to development.

The question of the possible role of seaweed for an adequate diet is still open. We can draw attention to some data of interest: iron content in seaweed is strongly varying. In 133 samples of 59 species minimum was 52, maximum 3 410, and mean value 591 mg/kg of the dry substance (Yamamoto and Ishibashi 1972). The daily need is just 1-2 mg Fe, which quantity is available in as little as 10-20 g fresh weight of an average seaweed. However, due to uptake difficulties the recommended intake is higher: 5 mg for men and 14 for woman when over 25 per cent of the energy value of the diet comes from animal food, or 9 and 28 mg respectively when animal food represents less than 10 per cent of the energy value of food. The uptake stimulating ascorbic acid ("vitamin C") belongs to the group of vitamins which occur richly in seaweeds. In Norwegian seaweed meal it is 500 - 2 000 mg/kg (Jensen 1972). Recommended intake per day for adults is 30 mg (Passmore *et al.* 1974). This quantity is normally available in 25-35 g meal, which means that even small additions of seaweed are useful; every portion of 3 g contains a tenth of the daily need.

For tropical, subtropical and some warm temperate areas the richest source of vitamin C from algae would be *Sargassum* species. Among Indian seaweeds, *S. myriocystum* had an ascorbic acid content of 666 mg/kg or ten times the average for other Indian algae (Subba Rao 1965).

Sargassum species are traditionally consumed in Malaysia, Burma, Indonesia, Vietnam, China, Japan and Hawaii. Korea and Japan report harvests of about 2.2 thousand tons each, most of which is, however, used for the production of sodium alginate, which is likewise manufactured in India. *Sargassum* is a potential resource not utilized in Tanzania, Egypt and Saudi Arabia; in Oman the standing crop is estimated at 28 000 tons. In many countries with relatively modest marine floras *Sargassum* species are found **to predominate, often in fairly large** amounts, e.g. Pakistan and Sri Lanka, Yugoslavia and Greece, Senegal, Sierra Leone, Ghana and Cameroon. 'Gulf weed' is common and often dominating in many parts of the Caribbean, and is most abundant among the seaweeds washed up in quantities on the beaches of the Atlantic coasts of Florida.

In addition to all these national resources the Sargasso Sea has estimated quantities of 0.8 - 2.15 t/km^2 over an area of 5 million km^2, which would correspond to a total of 5 to 10 million tons. Other free-floating resources are some 90 000 tons in the Gulf of Mexico and large but not quantified accumulations along the Asian coast of the Pacific (Michanek 1975).

The important point, however, as to whether or not seaweed in addition to iron and vitamin C contains the particular iron uptake stimulating agent present higher up in the food pyramid, still remains to be investigated.

ANTIBIOTICS

One hundred and fifty marine algae of Puerto Rico were tested for antibiotic activity against four laboratory indicator bacteria. Sixty-six of them showed antimicrobial activity against *Staphylococcus aureus* or other common laboratory microbes. Also bacteria isolated from the sea were strongly inhibited by a few algae (Burkholder *et al.* 1960). Well over one hundred seaweed species are known to contain antibiotic substances, even when those with faint effect are first sorted out.

Do these figures entitle us to hope for new substitutes for penicillin from algae? At present the answer has to be in the negative or at least hesitant; actually none of the many antibiotic substances from seaweeds is pharmaceutically utilized. On the other hand investigations into antibiotic activity of marine plants are not usually initiated in search of new drugs but from an interest in the nature of competition. In an aqueous surrounding, the organisms must have an active defence against bacteria and fungi, and against each other, to keep their surfaces as free from epiphytes as necessary and to keep their living space free from competitors. Each species has its own means in the chemical warfare on rocks and in crevices. Consequently the distribution of antibiotic activity is universal in the world of algae.

Antibacterial activity is as common in algae from tropical seas as in those from warm or cold temperate waters. Within genera known for abundance and wide distribution antibacterial species have been found among the green algae

Acrosiphonia, Cladophora, Codium, Enteromorpha, Halimeda, Monostroma and Ulva. Among the brown algae we find records of activity in Cystoseira, Dictyota, Sargassum, and all the large kelp and rockweed species in cold waters; among red algae in Ceramium, Chondria, Chondrus, Corallina, Delesseria, Gigartina, Iridea and Polysiphonia. Also bluegreen algae, e.g. Spirulina, are recorded as showing antibacterial activity.

Antifungal activity has not been investigated to the same extent, but is found in Caulerpa, a genus of green algae frequent in all warm waters, in the brown Cystoseira, Halopteris and Scytosiphon, and in the red Laurencia and Wrangelia. Among phytoplankton the diatom Chaetoceros proved significantly antifungal against pathogenic fungi.

Antiviral activity is more sensational. A particular interest is directed towards the Californian red alga Constantinea, active against herpes, a virus infection which can go down into the bronchae and be a starting point for cancer. Antiviral effect was also found in Chondrus, one of the important commercial species, active against influenza B and mumps, and in Cladophora, which among other things inhibits maturation of mouse meningopneumonitis virus.

Two genera are notable for showing activity within all these three fields of antibiotism: against bacteria, viri and fungi. They are the agar raw material red alga Gelidium, represented by several commercial species, and the blue-green Lyngbya, a genus with world-wide distribution and a large number of species small in size and difficult to harvest.

It seems likely that a possible utilization might start either with a species that is already available on the world market or with one which has shown a strong and particular activity, like Constantinea. The latter sort would scarcely be sought out in natural surroundings but rather cultivated under controlled conditions.

ANTITUMOUR-ACTIVE ALGAE

In folk medicine seaweeds have been used for a variety of purposes, from anti-curare, cooling agent for fever, eczema, gallstone, gout (as early as Pliny), menstrual troubles, renal troubles, scabies, scrofula etc., to water purification. It appears that most of these uses have been regarded as superstition by serious therapists in particular the idea that certain algae could be used as a remedy for tumours. Corsican moss, bull kelp, dulse and _Hypnea_ have been used against "squirrhe" tumours (Erhardt 1973). Nor did the report at the third international seaweed symposium that "algasol", an Italian preparation, had proved to be active against Walcher carcinoma, Erlich carcinoma, sarcoma 180 and Galliera sarcoma (and also therapeutic for arthritis and rheumatism) (Rositto 1958) create any great sensation. This decade, however, has brought a series of Japanese reports showing that _Codium_, _Dictyopteris_ and _Sargassum_ species possess substances which inhibit, to different degrees, the growth of transplanted carcinoma and sarcoma in mice. Among substances which have been identified, is funoran from _Gloiopeltis._

For obvious reasons, the species at present known to be curative are found in Japanese waters. The four genera mentioned represent green, brown, and red algae. A variety of species were investigated, although the first reports did not specify exactly how many, which leaves the chance open that antitumour substances may be present in a considerable number of species from all regions. Evidently the resource will not be a problem for pharmaceutical quantities. _Codium_ and _Sargassum_ are abundantly available in all oceans and are not much in demand for other purposes, the widespread genus _Dictyopteris_ is consumed in many areas and the East Asian _Gloiopeltis_, 'funori', is used for silk sizing. The active substances are polysacharids found in crude water extracts. Therefore another open question is whether a regular intake of seaweed may be prophylactic. As already mentioned such an assumption might increase the demand for seaweed in human consumption.

METAL BINDING AND RADIATION PROTECTION PROPERTIES OF ALGINATE

Marine algae have a high ability to concentrate metal ions from sea water. For this reason they are an outstanding source for trace elements, and health-giving as food.

Among the ions concentrated are also radionuclides, with low concentration factors of 10 to 100, thus normally only at harmless levels (Unni 1967). It has been suggested that hazardous or inappropriate local contamination of sea areas could be detected by assaying radioactivity in marine algae.

In brown algae the ability to take up radionuclides is based on the ability of alginic acid to form insoluble complexes with strontium. In cases when dairy products, cereals, fish or meat may be contaminated, the body uptake can be prevented by addition of alginate to the food. Alginate samples in which the ratio mannuronic acid/guluronic acid is less than 1.0 are remarkably active, inhibition reaches 89%. The strontium isotope is bound to the alginate and passes through the intestines without being absorbed into the body. Highest inhibition values were found in Macrocycystis, Hedophyllum, Egregia, Laminaria, Costaria, Ascophyllum and Fucus (Waldron-Edward 1968).

If the body uptake is not prevented by alginates the strontium is absorbed by the blood and deposited in the skeleton, where it may cause bone tumours or leukaemia. Alginates can even be used to reduce strontium already deposited in the bone. It is supposed that some of this is released slowly, re-secreted, possibly through the bile, and then bound in the intestines by alginate and excreted, if alginate treatment is given; otherwise it is re-absorbed by the blood. Thus alginate can serve both as a preventive and a therapeutic measure. The phycocolloids of red algae do not possess this property and within brown algae there is great variation. In addition to a high level of guluronate units of the alginate a certain chain length (ca 150 units polyguluronate) is essential for the highest biological activity. Suitable combinations could be looked for within the natural resources, or obtained through partial hydrolysis of the alginates.

Another possible application of alginate is for the inhibition of absorption of lead, the concentration of which is steadily increasing in our environment due to the anti-knock agents in gasoline, from where it contaminates food products and finally raises blood lead-level. These findings can be extended to barium, cadmium and zink, and alginates are recommended as a preventive measure in areas such as the vicinity of mines and chemical plants, where a high degree of pollution from metals exists (Tanaka 1970, Tanaka _et al_. 1972).

If we include the populations of all industrial areas and all polluted large cities into the part of mankind which could need a chemical protection from heavy metals, there are still supplies enough. A good three quarter of the world's unexploited seaweed resources are made up by an estimated 13 million tons of brown algae, all in cold waters (Michanek, unpublished). This figure does not refer to the standing crop but to the potential of harvestable seaweed available with reasonably increased efforts. Largest in bulk may be the supplies of kelp and rock-weed belonging to the genera _Laminaria_, _Ascophyllum_ and _Fucus_ in the northern Atlantic, _Macrocystis_, _Nereocystis_, and various Laminariales in the northern Pacific, and _Macrocystis_, _Durvillea_ and _Lessonia_ in the southern hemisphere. The most vigorous growths of these brown algae are often found in regions with a very rough climate, such as the Straits of Magellan, where storms may obstruct harvesting and rain may complicate drying.

While it has previously been regarded as more or less insignificant whether brown or red algae were given as a source of supplementary trace elements, vitamins, and proteins, it is now evident there are important differences. The alginates of brown algae are protective against many kinds of metal poisoning, due to their ability to bind certain atoms. For the same reason it is possible that certain trace metals actually present in the seaweed ash will never be released from the alginates. These may have a low "natural" content of strontium - they are still capable of taking up all strontium ions present in other parts of our food. Ascophyllum has a not negligible content of copper. But ewes fed dominantly on seaweed gave birth to lambs with copper deficiency. Evidently their rock-weed fodder had not only unreleasably retained all its copper content, but had also absorbed all copper from other sources of food simultaneously present in the sheep intestine. It is most important in this connection to make sure that iron is not prevented from being taken up by the body in presence of alginates. That iodine is not thus prevented, must be regarded as having been satisfactorily established through generations of experience.

HELP FOR ENDEMIC GOITRE

Thyroid enlargement caused by low iodine content in food will often be accompanied by mental retardation if aquired early in life, and even by physical retardation or, in most severe cases, cretinism. Public information, social welfare and iodized salt or seaweed food can theoretically eradicate this unnecessary disease.

The U. S. serves as an example, however, that fast progress should not necessarily be expected just because resources are sufficient. The number of men rejected for military service on account of goitre was the same during the last great war as it was twenty-five years before - five in every thousand examined.

Figures forty times higher were found in eight Mexican states investigated in 1952. Over two million people were goitrous out of a total population of about ten million (Kelly and Snedden 1960).

Central America. A great goitre zone extends through Mexico and the Central American republics. Most of the goitre centres are situated in the mountains, but there are stretches along the Pacific Coast where the disease is also to be found, among these all Mexican Pacific states except for Lower California.

The distribution of seaweed is the opposite. Lower California is extremely rich, the giant kelp *Macrocystis* can sustain a yield of 150 000 t a year and some 25 000 t red algae are harvested for agar production. Much of this is exported. A draw-back for the local industry is the lack of fresh water, necessary for algin and agar production, but not required if the seaweed is merely to be dried for food. Among the species found in abundance but not yet utilized are many which are eaten in other parts of the world: *Eucheuma uncinatum*, *E. isiforme*, *Porphyra perforata*, *Gracilaria verrucosa* and *Hypnea musciformis*.

The strictly tropical waters of the Pacific are almost devoid of algae in spite of the rocky indented coastline. The Gulf and Caribbean coasts are made up of monotonous sand shores which are not very hospitable for seaweed. It seems that the use of algae for food is now almost non-existant along these coas
In ancient Mexico however, 'tecuítlatl' from fresh-water blue-

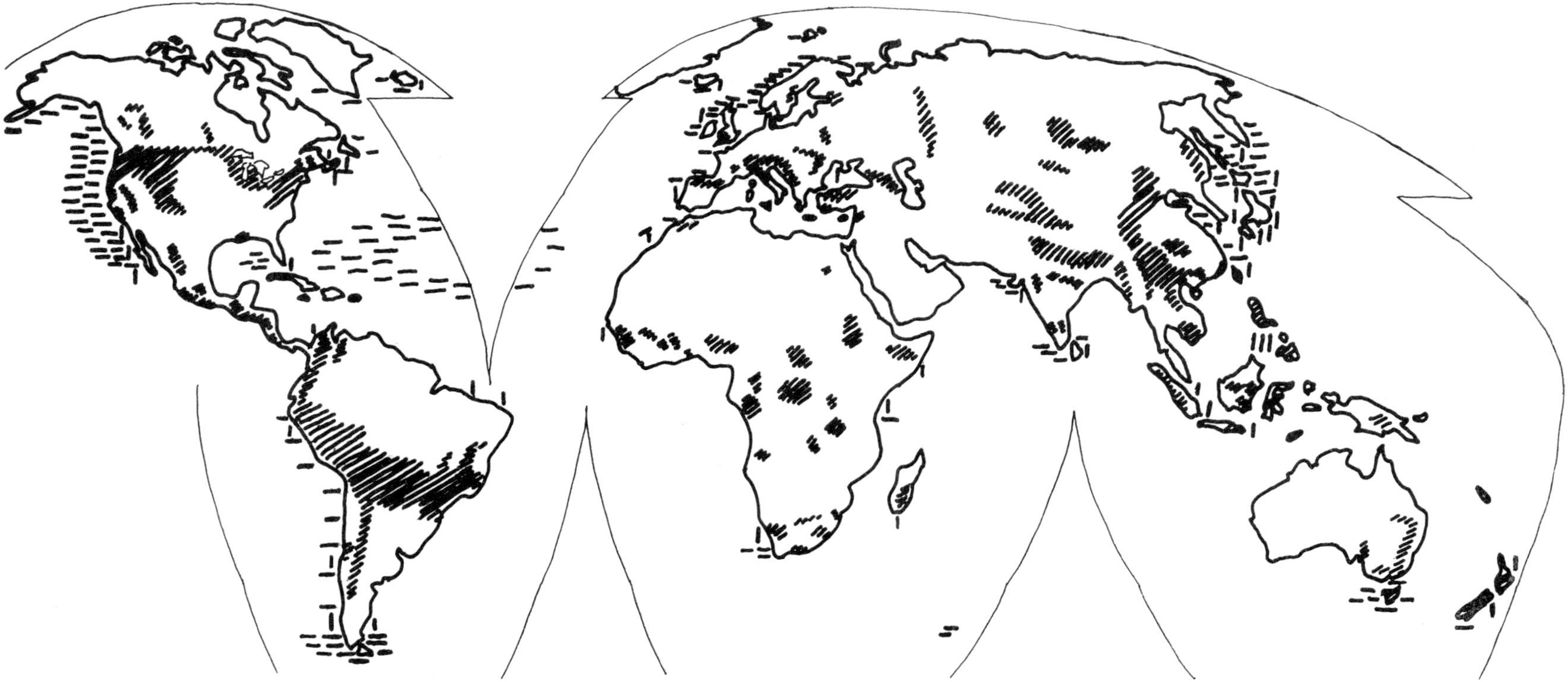

Fig.1. Distribution of endemic goitre and of seaweed resources.

//// Areas where endemic goitre has been found (after Kelly and Snedden 1960)
= Areas rich in resources of brown algae
|| Areas rich in red algae
Goode Homolosine Equal-Area Projection by permission of the University of Chicago

green algae was an important protein-rich food supplement. Tláhuac, a city of 40 000 inhabitants, may be translated by "the place which has tecuítlatl". A similar product is still sold under the name 'cocol de agua' (Ortega 1972).

South America. At a first glance at maps of goitre areas from any part of the world it is striking that the distribution of this disease coincides very much with high mountainous areas. Altitude by itself is, however, no reason for iodine deficiency. The correlated factor is how much the soil has been leached of its salt. Data on prevalence related to altitude are given from Peru (Burga 1956). Moving inwards from the coast one first encounters cases of goitre at 250 - 850 metres above sea-level. At an elevation of 1 000 - 3 000 metres the frequency increases, attaining average rates of 15% to 25% and above. At this level cretinism, myxoedema and hypometabolism are met. At even greater height, over 3 000 metres, the goitre rate diminishes again until it is almost insignificant in places of highest altitude. In the department of Amazonas the goitre rate in low-lying areas east of the Andes is 90%, which is much greater than in higher parts of the same department, and cretinism, mental deficiency and deaf-mutism are correspondingly more serious.

The goitre zone of the Rocky Montains and Central America extends in South America into Colombia and Venezuela and proceeds along the highlands and eastern slopes of the Cordillera through Equador and Peru to central Chile and western Argentina. In addition to this alpine zone, comparable to those of the European Alps and the Himalayas, there is a broad belt eastwards through Paraguay and Southern Brazil.

Along the coasts adjacent to the goitre areas there are rich resources of seaweeds, suitable for food. From the Caribbean coast of Colombia Gracilaria spp, Agardhiella, the popular 'gulaman' of the Filipinos, Sargassum and Caulerpa are all recorded as abundant. Quantities are not indicated, as they are not commercially utilized. Pacific Colombia and Equador on the other hand are not rich in algae while Peru, northern and central Chile are very rich, among others in Macrocystis, Lessonia, Gracilaria, Gelidium, Chondrus, Rhodymenia, Ahnfeltia and a few species which are actually eaten: Durvillea, Gigartina, Porphyra, Ulva. In Brazil the rich resources in the states of Rio Grande do Norte, Ceará and Paraíba are located far from the main goitre area in the Brazilian highlands. Southern Chile and Argentina enjoy still larger quantities - the Straits of Magellan may have the richest in the world - but in this area there is no endemic goitre.

A need, a resource, a distance. Would it be possible to bridge the difficulties, including the introduction of a strange foodstuff? The last point would probably not present the most difficult problem. Pilot projects have already been cariied through. Durvillea seaweed meal has been added to children's food in Chile in account of its high iodine content (Etcheverry 1953). There are also remnants of a competent folk medicine. Some Indian tribes in the Andean region used Phyllogigas as 'goitre sticks' to counter high incidence of thyroid enlargement. Sargassum bacciferum was also used by South American Indians to cure goitre (and renal disorders) (Schwimmer and Schwimmer 1955). In Peru the marine algae have formed part of the human diet since the pre-Inca and Inca periods. Inhabitants of the Andean region could obtain dried Ulva and Porphyra from the coast through exchange or more quickly by messengers. These algae are still brought to the market places of the interior in considerable quantities (Acleto 1971). It seems reasonable to suggest that an attack on the goitre problem in South America should start with an ethno-botanical study of lost habits and surviving uses to explore the possibillity of breathing new life into salutary old customs.

Africa. While in Europe the goitre areas are covered with

legally regulated distribution of iodized salts, in Africa there are some districts where seaweed could advantageously be included in prevention efforts. In Morocco endemic centres are scattered along the Rif mountain chain, while the rest of the Mediterranean and West African coasts have none. Morocco alone is likewise rich in seaweed while the North and West African coasts have remarkably little.

Among the Canary Islands, Santa Cruz de la Palma is an example that areas close to the sea are not necessarily free from goitre. Among 1 100 schoolchildren 46% of the girls and 36% of the boys were found to have goitre. The island is mountainou , with basalt and other volcanic rock ,and the iodine content of the drinking water is exceedingly low. The marine flora is dominated by Gelidium, very high in iodine and much appreciated in Japan, China and Indonesia. Certain bottoms are characterized by carpets of the green alga Caulerpa.

In southern West Africa, Senegal alone has quantities large enough for utilization. They are dominated by Hypnea, which is an export article, but also include Gracilaria, Gigartina and Ulva. Heavy rains prepare the grounds for goitre, common around the banks of the low-lying Songrougrou and Casamance rivers and the upper waters of the Gambia river. Other goitre centres in West Africa cannot be connected with any indigenous seaweed resources although Sierra Leone has some Sargassum accompanied by vigorous growth of grape-like masses of Caulerpa racemosa.

South Africa, washed by colder waters, has seaweed along its coasts and has goitre in certain mountain areas. On the east coast of the continent the best known resources and probably the only ones worth considering in this context are those of Tanzania, in particular south of Dar. The goitre region on the other hand is in the southern highlands 500 to 1 000 km west of the coast.

Asia. It is well known that centres for goitre and cretinism are found on the northern and southern slopes of the Himalaya mountains. They extend eastwards through Burma into China, where they are widely distributed in highlands distant from the sea. Adjacent endemic areas are found in southern Asiatic Russia.

The most goitrous province in all China is Yunnan on the borders of Burma. Extreme environmental iodine deficiency is aggravated by excessively hard water. In other areas goitrogenous effects may be caused by certain agents, the most well-known of which is present in cabbage. Also around Peking entire villages are reported to be almost 100% goitrous (Kelly and Snedden 1960).

The Chinese government has made stronger efforts than known in any other country to meet its desperate demand for iodine through seaweed cultivation. The legendary Pen-Ts'ao Tsing treatise on herbs and roots(2 700 B. C) is said to mention *Sargassum* as an efficacious remedy against goitre, and *Laminaria japonica* has been widely used by the Chinese as food and medicine for over 1 500 years. Previously most was imported, but now China has trained hundreds of scientists and tens of thousands of fishermen in kelp cultivation. The production rose from 328 t in 1952 to 145 680 t wet weight in 1959. Recent figures are not available, but 600 000 t were aimed at in the sixties and ten times as much is mentioned as required to meet the nation's demand for kelp as food and industrial raw material. New methods have been applied and it appears that every metre of China's coasts is now cultivated with seaweed. Of particular interest is the government's ambition to distribute the product also to inland populations, where it is especially popular during the season when green vegetables are not readily available (Cheng 1969). It could be worth recalling that the famous anthelmintic *Digenea* is also noted for a high level of iodine.

Of goitre centres in Siberia only one, the Primorsk Region from Vladivostok northwards to the mouth of the Amur, is situated near the sea. On the other hand this coastline as well as Sakhalin and Kamchatka is abundantly rich in seaweed, in particular *Laminaria japonica* but also *Alaria*, *Gelidium*, *Ahnfeltia*, *Porphyra* and other edible genera (Michanek 1975). They could easily provide all the raw material necessary for a campaign based on, say, an addition of a few per cent of seaweed meal to all bread.

The northern frontiers of the Indian peninsula along the southern valleys and foothills of the Himalayas have a goitre reputation more formidable perhaps than any other part of the world. In comparison to this 'hyper-endemic' centre other areas are of secondary importance: the Pakistan seaboard of Kutch, the central plateau from Pakistan to Bangladesh and minor occurrencies in the Dekkan and extreme south.

Along the entire Indian coasts there are only two centres for seaweed production: the Kathiawar peninsula in the northwest and the Mandapam district in the southeast, extending southwards along the Mannar Bay to the southernmost point.

With a population of 600 million, even if seaweed harvest were raised to 6 000 t a year there would be 100 Indians for every kg of fresh algae - a minute amount if used for food, and entirely insufficient for industry. As goitrous populations are relatively small and restricted to the areas mentioned, goitre prevention within these areas should be given high priority in relation to the limited resources.

Sri Lanka is mountainous, wet, and goitrous in the southwest, where population density is greatest. Heavy rainfalls have led to intensive chemical leaching, giving rise to laterite soils with low iodine content.

Prevalence of goitre among Sri Lanka children and adolescents

	Iodine content of water µg/l	Percentage with goitre boys	girls
Wet region	1.4 - 5.3	6 - 18	22 - 56
Dry region	6.6 - 61.0	nil	5 - 12

After Kelley and Snedden 1960 from Wilson 1950.

The main seaweed resources are found in the northern part of the country, but the south is rich in <u>Sargassum</u>, suitable for seaweed meal. Fishermen use seaweed for soups and salads, puddings and sweets.

Burma has two goitrous regions, one where a vitamin deficiency may be responsible, another where eating has aggravated the effects of the low iodine content of food and water.

Dietary deficiency of vitamin A is notorious and is regarded the most important single goitrogenic factor operating among the Chin Hills people in the west of the country (Kelley and Snedden 1960). Vitamin content is recorded for some Japanese algae. Of these only a few are mentioned in particular being rich in provitamin A. Among these, however, we find <u>Gracilaria gigas</u> and <u>Enteromorpha</u> spp. (Miller 1927, Kanazawa 1961, Takagi 1975). Marine algae from eight genera are reported to be eaten in Burma, among these <u>Gracilaria</u> and <u>Enteromorpha</u>, only along the coasts however, although in the hills they could work miracles.

"An exceptionally large number of cretins and deaf-mutes are seen in the Burmese goitre areas. Among the Kachins who inhabit the valleys and steep hill-sides along the north and north-east frontiers the enormous rate of 10 per 1 000 is recorded ... The Kachin people drink water from hill streams impregnated with lime and customarily eat large quantities of lime in powdered form. Stott and his colleagues believe that this peculiar habit is in large measure responsible for causing goitre among these northern Burmese peoples." (Kelley and Snedden 1960).

In Thailand as in Burma and in Vietnam goitre is prevalent in the highlands of the north and the north-east, and as in Burma and in Vietnam some regions show overall rates well over 50 % while some lowland regions are completely free. One obstacle to a programme of salt-iodization is that some communities do not use much salt. A suggestion is that iodate might be incorporated into the vitaminized premix already being added to polished rice in beriberi areas.

As beriberi is an avitaminose B_1, another suggestion is to make a double contribution by choosing the necessary source of iodine among seaweeds which are at the same time rich in thiamine. In Japan high values are found in the genera _Porphyra_, _Gracilaria_ and _Rhodymenia_, all of which are represented in Thailand even if by other species. In Songkhla _Porphyra_ is harvested and sold locally. Another species harvested on the east coast is known as _Gracilaria verrucosa_ in the literature (P. Edwards, AIT Bangkok, personal communication). Generally speaking the proportions of provitamin A, thiamine (B_1), and a number of other vitamins in seaweed equal or surpass their proportions in other kinds of food (Takagi 1975).

In Indonesia, centres of endemic goitre are found throughout the whole length of the archipelago, from Sumatra to Timor. In Java, the prevalence of goitre among school children in some villages has been known to reach even 100%. This sensational percentage for the prevalence of a disease is encountered also in some Malayan villages, where it is said to figure not only among school children and women in childbearing ages, both being vulnerable groups, but among the entire population

Both in Malaysia and in Indonesia seaweeds are traditionally used in cooking. Subba Rao (1965) mentions more than twenty species and their uses. It appears, however, that these traditional customs are now declining. In the olden days species of Laminaria were imported into Malaya as a cure for goitre.

Exporting seaweed is also a rather old custom. Half a century ago Malaysia exported a few hundred tons to China and Hong Kong (Kirby 1953), and a quarter of a century ago dried seaweed was shipped from all parts of the Indonesian archipelago. On the east coast of Sumatra, Gracilaria, Gelidium, Eucheuma muricatum and Caulerpa were trade seaweeds, on Java Gracilaria, Gelidiopsis and Eucheuma muricatum, in Lesser Sunda Islands Eucheuma speciosum, Hypnea cervicornis, Corallopsis and Caulerpa, in Bongok and Kangean also Acanthophora spicifera, in Celebes Eucheuma muricatum, Hypnea musciformis and Gracilaria spp. and in Ceram and surrounding islands Eucheuma and Gracilaria (Eisses 1952). These are all among the species reported by Subba Rao as eaten in Indonesia, but as trade seaweeds they will, except for Caulerpa, most likely end up as industrial phycocolloids.

At present there is a boom in seaweed production. Previously **not taken into account in internationa statistics, Indonesia pro**duced 3 000 t along the Pacific (Java Sea) coasts in 1973 and 74, the two following years this figure almost tripled to 8 400 and 8 800 t respectively. At the same time the western and southern coasts towards the Indian Ocean have started production, which rose from zero in 1973 to 89 t in 1976. No doubt the potential resources are still much higher in particular in the eastern part of the archipelago. A possible conflict with "economical interests" can be foreseen if these potential export products should be claimed for health campaigns.

Oceania. In the Pacific islands soils are much richer in iodine than the rocks. As precipitation is modest, rain adds more airborn iodine to the soils that it leaches out of them. Normally only a few mild cases of thyroid enlargement are found, this is also attributed to the large consumption of sea-foods. Exceptions are found within some populations isolated from the sea in larger islands like New Guinea, Bougainville, Fiji and Hawaii. In Australia and New Zealand the frequency distribution of goitre may be correlated with the extent of quaternary glaciation, and practically all inland parts of New Zealand and Tasmania can be said to be goitrous. The two islands last mentioned are extremely rich in seaweeds, the largest quantities represented by giant kelp, *Macrocystis*, the most edible, according to local customs, by *Porphyra*, *Gracilaria*, *Gigartina*, and green algae.

Conclusion. "Simple goitre is the easiest of all known diseases to prevent ... It may be excluded from the list of human diseases as soon as society determines to make the effort" was stated by a pioneer in goitre prophylaxis in 1920. The statement is still valid. Goitre is still a curse. One problem is that while smallpox was eradicated forever the day the last patient was cured, goitre can only be kept under control. Background iodine deficiency will always remain, the remedy must be applied perpetually. In our days many ambitious campaigns have posthumous reputations as: "there is little sign that the preventive programme continues to be pursued with vigour and determination". For injection of iodized oil or distribution of tablets, every single person has to be called on every six months, for iodized salt authorities must have continous control that no other kind is ever distributed within the vulnerable area, and it is necessary that people really use salt for cooking.

Wherever seaweed is introduced in the right way there is at least some hope that a custom will develop, a demand will be created, which could possibly survive periods when governments have other priorities. Other advantages are the multiple health benefit, the economic gain that seaweed does not have to be imported into any countries bordering the sea, and the positive impact on employment, as mariculture of seaweed creates lots of job opportunities. As seaweed has been known as an antigoitre prophylax and remedy for centuries or even millennia, in most countries the introduction of seaweed for food would not mean an innovation but a revival of old customs. The only new feature would be the possibility of offering seaside resources for inland needs.

REFERENCES

Acleto Osorio, C., Algas marinas del Perú de importancia económica. Univ. Mayor de San Marcos, Perú, Museo Hist. Nat. "Janvier Prado" Ser. Divulg., 5: 85 p (In Spanish) (1971).

Ardré, F., Contribution à l'étude des algues marines du Portugal. 1. La Flore. Portugaliae Acta Biol. (B)10: 1 - 423 (In French) (1970).

Baardseth, E., Cuba. Algología marina. Dec. 1966 - Oct. 1977. UNESCO BMS - RD/AVS, (533): 13 p (In Spanish) (1968).

Baslow, M. H., Marine Pharmacology. A study of toxins and other biologically active substances of marine origin. Baltimore, Williams and Wilkins: 55 - 85 (1969).

Björn-Rasmussen, E., Studies on food iron absorption in man. Thesis, 25 p (1974).

Boergesen, F., Marine algae from the Canary Islands, especially from Teneriffe and Gran Canaria. 3. Rhodophyceae 3 Ceramiales. K. Danske Vidensk. Selsk. Biol. Medd. 9(1): 1 - 159 (1930).

—, Some marine algae from the northern part of the Arabian Sea with remarks on their geographical distribution. Ibid. 9(6): 1 - 72 (1934).

—, A list of marine algae from Bombay. Ibid. 12(2): 1 - 64 (1935).

—, Contributions to a South Indian marine algal flora, 3. J. Indian Bot. Soc. 17(4): 205 - 242 (1938).

Burga, H. B., Rev. peru. Salud Públ.: 341 (1956).

Burkholder, P. H., Burkholder, L. M., Almodóvar, L. R., Antibiotic activity of some marine algae from Puerto Rico. Bot. Mar. 2: 149 - 156 (1960).

Chase, F. M., Useful algae. Smithsonian report for 1941: 401 - 452 (1942).

Cheng, T. H., Production of kelp - a major aspect of China's exploitation of the sea. Econ Bot. 23(3): 215 - 236 (1969).

Diannelidis, T., Greek marine flora and its utilization. Prakt. Hellen Hydrobiol. Inst. 3(2): 71 - 84 (In Greek, English summary) (1949).

Díaz-Pifferer, M., Los recursos marinos de Venezuela. Algas de importancia económica. El. Farol, 29(222): 18 - 22 (In Spanish) (1967).

Durairatnam, M., Contribution to the study of the marine algae of Ceylon. Bull. Fish. Res. Stat. Dept. of Fish., Ceylon 10: 1 - 181 (1961).

Eisses, J., The research of gelatinous substances in Indonesian seaweeds at the laboratory for chemical research, Bogor. J. Sci. Res., 1(2): 44 - 49 (1952).

Ercegovic, A., La flore sous-marine de l'îlot de Jabuka. Acta Adriatica 8(8): 1 - 130 (In French) (1957).

Erhardt, J. P., Contribution des algues a la médecine et a la biologie. Rev. Intern. Océanogr. Med. 31/32: 194 - 219 (In French with short English abstract) (1973).

Etcheverry, D. H., Seaweed resources of Chile. Proc. Int. Seaweed Symp., 1: 108 (1953).

Feldmann, J., Recherches sur la végétation marine de la Méditerranée. La Côte des Albères. Rev. Alg. 10: 1 - 339 (In French) (1938).

Gayral, P., Algues de la Côte Atlantique Marocaine. La Nature au Maroc 2.: 1 - 523 (In French) (1958).

Hallberg, L., Iron deficiency - a paradox in an affluent society. Näringsforskning 19: 4 - 11 (In Swedish with short English summary) (1975).

Islam, A. K. M. N., The marine algae of Bangladesh. Beiheft Nova Hedwigia (1976).

Jensen, A., The nutritive value of seaweed meal for domestic animals. Proc. Int. Seaweed Symp., 7: 7 - 14 (1972).

Kanazawa, A., Studies on the vitamin B-complex in marine algae. 1. On vitamin content. Mem. Fac. Fish. Kagoshima Univ. 10: 31 - 69 (1962).

Kelley, F. C., Snedden, W. W., Prevalence and geographical distribution of endemic goitre. In Endemic goitre. World Health Organization Monograph Series (44) pp. 27 - 233 (1960).

Kirby, R. H., Seaweeds in commerce. London, Her Majesty's Stationary Office; 157 p (1953).

Kylin, H., Die Gattungen der Rhodophyceen. Lund, Gleerups, 673 p (1956).

Levring, T., The marine algae of the archipelago of Madeira. Bol. Mus. Municipal do Funchal 28: 1 - 111 (1974).

Michanek, G., Seaweed resources of the ocean. FAO Fisheries Technical Paper (138): 1 - 127 (1975).

Miller, C. D., Food values of Poi, Taro and Limu. Bernice P. Bishop Mus. Bull. 37: 1 - 25 (1927).

Okazaki, A., Seaweeds and their uses in Japan. Tokay, Univ. Press, 165 p (1971).

Ortega, M. M., Estudio de las algas commestibles del Valle de Mexico. Rev. lat-amer. Microbiol. 14(2): 85 - 97 (In Spanish. Abbreviated English translation in Bot. Mar. 15: 162 - 166) (1972).

Passmore, R., Nicol, B. M., Narayana Rao, M., Handbook on human nutritional requirements. FAO Nutritional Studies (28), 70 p (1974).

Phạm-Hoàng Hộ, Contribution a l'étude du peuplement du littoral rocheux du Vietnam (sud). Ann. Fac. Sc. Saigon 1962: 249 - 350 (In French) (1962).

Rodenwaldt, E., Jusatz, H. J., World Atlas of Epidemic diseases 1 (1952).

Rositto, G., Pharmacological and clinical study of a new seaweed extract registered in the italian pharmacopeia as "Algasol T 331". Milano, 15 p (1958).

Schwimmer, M., Schwimmer, D., The role of algae and plankton in medicine. New York, Grune and Stratton, 85 p (1955).

Subba Rao, G. N., Use of seaweeds directly as human food. Reg. Stud. IPFC, (2): 32 p (1965).

Setchell, W. A., Tahitian algae. Univ. of Calif. Publ. Bot. 12(5): 61 - 142 (1926).

Takagi, M., Seaweeds as Medicine. In Advance of Phycology in Japan. Edited by Jun Tokida and Hiroyuki Hirose. Jena, G. Fisher, pp. 321 - 325 (1975).

Tanaka, Y., Application of metal-binding properties of marine algae in medicine. In Proc. Food-Drugs from the Sea 1969. Edited by Heber W. Youngken Jr. Washington D. C., Mar. Technol. Soc. pp. 351 - 357 (1970).

Tanaka, Y., Hurlburt, A. J., Angeloff, L., Skoryna, S. C., Application of algal polysaccharides as in vivo binders of metal pollutants. Proc. Int. Seaweed Symp. 7: 602 - 604 (1972).

Taylor, W. R., Plants of Bikini and Northern Marshall Islands. Ann Arbour. Univ. of Michigan Press: 227 p (1960).

Trono, Jr. G. C., The marine benthic algae of the Caroline Islands, 2. Phaeophyta and Rhodophyta. Micronesia 5(1): 25 - 119 (1969).

Tseng, C. K., Economic seaweeds of Kwangtung province, S. China. Lignan Sc. J. 14(1): 93 - 104 (1935).

Tsuchiya, Y., Comparative hypocholesterolemic activities of marine algae. Proc. Int. Seaweed Symp. 6: 747 - 754 (1969).

Unni, C. K., Natural radioactivity of marine algae. In Proceedings of the Seminar on Sea, Salt and Plants, edited by V. Krishnamurthy. Bhavnagar, Central Salt and Marine Chemicals Research Institute, pp 265 - 273 (1967).

Viteri, Torun, Anaemia and physical work capacity. Clinics in Haematol. 3: 609 (1914).

Waldron-Edward, D., The use of alginate in the prevention and treatment of radio-strontium toxicity. In Drugs from the Sea, Edited by Hugo D. Freudenthal. J. Ocean Technol. pp. 267 - 275 (1968).

Weber-van Bosse, A., Papers from Dr. Th. Mortensen's Pacific Expedition 1914-16. (33) Algues de l'Expédition danoise aux iles Kei. Vidensk. Medd. Dansk. Naturh. Foren. 81: 57 - 155 (1926).

WHO expert committee, Control of ascariasis. Wld. Hlth Org. techn. Rep. Ser. (379): 47 p (1967)

Wildeman, E. de, Les algues de la flore de Buitenzorg. Flore de Buitenzorg 3: 1 - 457 (In French) (1900).

Yamamoto, T., Ishibashi, M., The content of trace elements in seaweeds. *Proc. Int. Seaweed Symp.* 7: 511 - 514 (1972).

ALGAE AS DRUG PLANTS IN INDIA

Arunkumar Misra and Ramkumar Sinha

Botany Department, Bithila University, Darbhanga
864 004 India

The Indian System of Medicine called Ayurveda has been gaining ground in recent times, not only because of the aphrodisiacs but also on the overall merit of the system. It is a treasurehouse of the use of plants for medicinal purposes.

Jivak, a famous specialist of Ayurveda in ancient India, on completion of his training, was asked by his guru to go round the city of Taksila (now Lahore) and list the plants of medicinal use. After several days of survey, Jival returned to his guru only to say that each and every plant possessed one or another medicinal property, and none was devoid of them. Jivak got his degree, and served Lord Buddha as his personal physician. Ayurveda, meaning science of health, is several thousand years old and has its roots in Ethnobotany and folk-medicine. The science developed during the vedic age. The term ayus means duration or life-span, and veda, knowledge. Ayurveda is, thus, the prolongation of healthy life and prevention and curing of diseases (Sharma 1976). This term signifies a vast body of knowledge and literature from the vedic to the modern times. The beginnings of Ayurveda may be traced in Rigveda, and then in Atharveda, both of the second millennium BC. It is regarded as Upanga (part) of Atharveda and also as an Upaveda (secondary Veda), forming part of Rigveda. The ancient ayurvedic medical classics are the various Samhitas (treaties) of Carak, Susurta and others (of the 4th Century BC), Bhavaprakas (16th Century AD) etc. A specialist or practioner of the science is called Vaidya.

Marine Algae in Pharmaceutical Science
edited by H. A. Hoppe, T. Levring, Y. Tanaka

Algae, both marine and fresh-water,are known as Saivale in Ayurveda, and have many times been referred to as possessing medicinal properties. The word Saivala has been derived from the Sanskrit root see, which means "to lie down". By the addition of affix "balach", the term Saivala has been developed, meaning the green plant mass lying on the surface of water (Diksita 1962, Böhtlingk and Roth 1966, and Radhakantdeva 1961).

Let us now see to what medicinal uses the algae were put to in ancient India, and whether the previous knowledge may be corroborated by modern scientific investigations.

Srinivasan (1973) has given an account of the uses of the marine algae; however, the use of the sea-weeds for medicinal purposes are very poorly understood.

Digenia sinplex, the Rhodophycean alga,is made into a drug, and is important as antihelminthic agent. Experimental work has been done with sodium-laminarin-sulphate, on the blood anti-coagulant properties of the crude extracts of Digenia plants. Compounds similar to heparin, fucoidin etc. have been discovered (Chopra et al. 1969).

Extracts of Chondrus crispus and Gigartina spp. have agarlike substances and is said to be effective against coughing. The material is also used as a pharmaceutical emulsifier (Chopra et al. 1969).

The antibiotic chlorellin has been isolated from Chlorella. Extracts of Cladophora, Lyngbya and certain other algae kill strains of bacteria (Pseudomonas and Mycobacterium) and also exhibit antiviral properties.

Certain cyanophycean members and those of Characeae possess larvicidal properties.

Laminaria saccharina (Phaeophyceae) has been reported to be used as cure for goitre, scrofula and syphilis, in addition to its being a source of iodine.

<u>Porphyra umbilicus</u> (Rhodophyceae) is also known to have some medicinal value.

The use of brown algae as a source of iodine and ultimately their use against goitre has long been known and discussed.

The following two quotations are often used by the men of ancient medicine in our area:

a) Saivalam sitalam snigdham santap vran-nasanam

-- Raj Nighantu

the algae are cold, contain fat, could be used against wounds and heat.

b) Saivalamtuvaram tiktam madhuram sitalam laghu. Snigdham dahtrisa pittaraktajvarharam param

-- Bhava Prakas

the algae are very efficient against thirst, body-temperature. They are soft and useful as sweet in taste.

In addition to the above points the later views about algae as drug plants is as follows:

Durvaaksatputikkumbhikplavsaivalam.
Jalen kvathitan pitam sukrameh haram param

-- Chakradatta

the algae and several other things may be used as decoction to help against excessive night-fall and semen-discharge.

This contains some indication on the way we could use the algae as drug plants. A water extract, which may later be warmed,is the right preparation.

Personal communication with quaks, vaidyas and people of Ayurveda has revealed that the algae are currently being used for the following ailments. It may be noted that these persons have no idea of the Latin binomial nomenclature, and correct identification of the plants becomes a big problem when dealing with them.

1. Unnatural semen discharge in youths at night is often a matter of concern. One part algae is boiled in four parts water till the volume has a shrink to one fourth. 250 ml of this decoction has to be taken by the patient twice a day for one week.
2. The above preparation has also been recommended in leucoria, and for enhancing clotting of blood.
3. Cold drink, or sherbat of the algae is recommended when a patient feels a burning sensation in the stomach. Unusual thirst may also be cured with this treatment.
4. In cases of high fever the algae may be used as paste over the forehead.
5. When urination is difficult, or has completely stopped, a drink made of algal extracts is very helpful.
6. Algae also possess some antiseptic properties, and may be used to enhance healing of wounds.
7. Piles may be treated or made less severe by the application of algal pastes over the wounds.

In our area the vaidyas often use Spriogyra, Ulva, Oedogonium and related forms for this purpose (personal communication from Pandit Professor Krishna Mohan Pathak, Ayurvedic College, Mohanpur House, Darbanga).

We have tried to verify some of the claims mentioned above on a modern scientific basis. Investigations were performed to assess the anti-viral properties of Spirogyra spp. and to isolate the proteinaceous inhibitors involved.

Dried powder of Spirogyra was mixed with a suspension of TMW (Tobacco mosaic virus) and used to inoculate the leaves of Synthi tobacco (Nicotiana tabacum). The controls were devised by mixing virus inoculum with distilled water. It was seen that the number of local lesions decreased by more than one half under the influence of Spirogyra as shown in the figure below. To save space we have not described the details of the experimental technique; they may be found in Misra (1977).

Further investigations revealed that the inhibiting principle was proteinaceous in nature. However, more details have to be worked out to understand the usefulness of algae as drug plants. But the large algal population in fresh and marine waters will be great boons if they could be effectively used for controlling diseases of man, animals and plants. Plant pathologists have isolated several naturally occurring substances that could be used in anti-microbial compounds, and the algae would add a further new class to them.

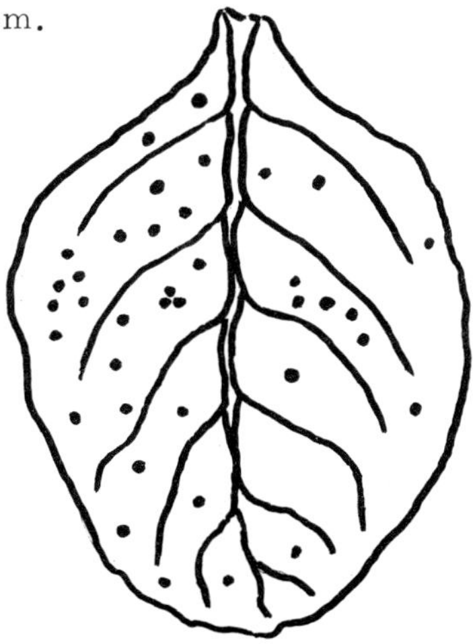

Xanthi tobacco leaf with local lesions produced by TMV.
Left half = TMV + buffer/water - right half = TMV + Spirogyra sap.
Addition of the algae reduces the number of local lesions, by partly inhibiting the virus, due to certain proteinaceous compounds.

A part of the work was completed in the laboratory of Prof. Dr. F. Nienhaus, Institute of Plant Diseases, University of Bonn, Federal Republic of Germany. Mrs. C. Mack was of great help in designing the experiments. Prof. Anantlal Thakur, Dr. Ramkaran Sarma, Prof. K. M. Pathak of Sanskrit University, Darbhanga were helpful in explaining the ancient books on Ayurveda.

Summary

The Indian ancient system of medicine, Ayurveda is highly developed and takes into account numerous forms of plant life. However, the use of algae for medicinal purposes is little known. There are records that *Laminaria sachharum* of Phaeophyceae and *Porphyra umbilicalis* and *P. vulgaris* of Rhodophyceae are used frequently. A systematic and biochemical account is not available for *Laminaria* while some work was done for *Porphyra* long ago by Chopra in 1933. *Porphyra* is rich in organic iodine. Vitamins B and C, folic acid and niacin are also found. *Laminaria*, being less worked out, is used more frequently, particularly against goitre, scrofula, and syphilis. Both the genera are well distributed along Indian seacoasts. Some fresh water green algae of different taxa are also in use. *Seratophyllum submersum* and *Vallisneria spiralis* (angiosperms) have been many times wrongly referred to as algae in the literature. Antiviral principles have been discovered in *Spirogyra*.

References

Böhtlingk, K., Rudolph, R.: Sanskrit Wörterbuch. Otto Zeller, Osnabrück, pp 306 (1966).

Chopra, R. N., Chopra, I. C., Varma, B. S.: Glossary of Indian Medicinal Plants. CSIR, New Delhi (1969).

Diksita, B.: Sidhanta Kaumudi. Vol. 2, pp 268, Banaras (1962).

Misra, A.: Virus inhibiting proteins in Mosses. Z. Pflanz.krankh. u. Pflanzenschutz, 84, 6 (1977).

Radhakantdeva, Raja: Shabdakalpadruma Vol. 5, pp 138 (1961).

Sharma, P. V.: Dravyaguna Vijnanam. Vol. 2, pp 677-678, Chaukhamba, Varanasi (1975).

Shrma, R. K., Dash, B.: Caraka Samhita. pp 80, Chaukhamba, Varanasi (1976).

Srinivasana, K. S.: Phycologia Indica (Icones of Indian Marine Algae), Botanical Survey of India, Calcutta (1973).

PHARMACEUTICAL STUDIES ON MARINE ALGAE IN JAPAN

Kazutosi Nisizawa

Department of Fisheries, College of Agriculture and Veterinary Medicine, Nihon University, Shimouma-3, Setagaya, 154 Japan

I. Introduction

It seems to be at least ten thousand years ago that a relationship between the ancient people living in Japan Archipelage and marine algae began to occur, because the remains of some marine algae such as Eisenia bicyclis, Sargassum fulvellum, Hizikia fusiformis and Undaria pinnatifida which are now generally a favorite food among Japanese have been found in the relics of these natives.[1] In Japanese math, it has been reported that marine algae have been offered to God on the festivals of Yamato Imperial Court which is said to have begun its reign between the third to fifth centuries A.D. In addition, a variety of marine algae have also been used as poetic subjects together with other plants and animals in "Manyoshu", the oldest collection of thirty-one syllable Japanese odes composed by the people of commonalty as well as aristocracy during about 400 years from 347 to 759 A.D.

Thus marine algae have become familiar material among the Japanese race in the course of Japanese history, and some of them have been fixed as usual food for so many centuries that the algae have even been believed useful or necessary for human longevity. It was, however, not until recently that this mysterious story began to acquire its theoretical basis, because most scientific research on marine algae from a pharmaceutical point of view has been carried out since around the middle of the 1950's. Of these investigations, the results obtained by some workers in Japan may be particularly noteworthy, and the main parts of them will be briefly introduced in this review.

II. Anthelmintics

Murakami et al. (1953)[2] isolated two substances of anthelmintic activity from a red alga, Digenea simplex, which were named α-L-kainic and α-L-allokainic acids. They were a stereoisomer of each other (1955).[3] The finding originated in a hint taken from the fact that people living on the western coast of Kyushu, Japan, have used this seaweed as a vermifuge. These authors determined the chemical structure of the acids as follows.

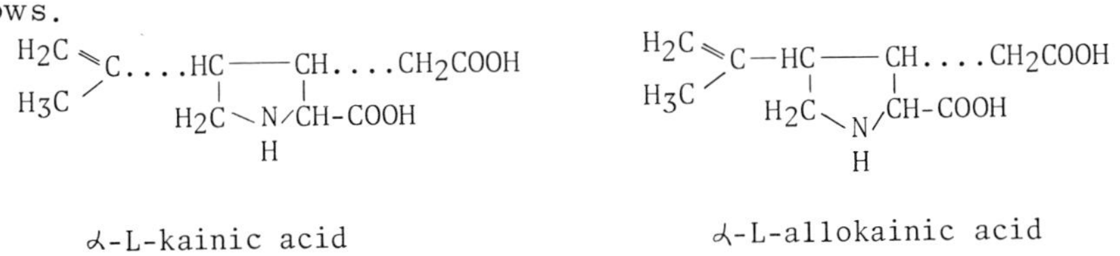

α-L-kainic acid α-L-allokainic acid

(3-Carboxymethyl-4-isopropenyl proline)

Marine Algae in Pharmaceutical Science
edited by H. A. Hoppe, T. Levring, Y. Tanaka

According to Sugimoto (1960),[4,5] doses of 20 mg of α-L-kainic acid to each of 43 first-year pupils of a primary school caused a decrease in the number of worm eggs excreted from 60.5% of the individuals tested and the rest was found to be free of the eggs. This efficacy of kainic acid was far larger than that of santonin used in the similar test on the fourth-year pupils.

On the other hand, a substance named domoic acid which showed a similar activity to kainic acid was isolated by Daigo (1959) from Chondria armata belonging to the same family as Digenea simplex.[6,7] Doses of 20 mg of domoic acid per day to each boy and girl between 10 and 15 years old, all of whom were excreting the worm eggs, brought about a negative result upon examination after 3 weeks. The name of this drug is connected with "domoi" which is the dialect in Yakushima, Kagoshima-ken, Japan, for Chondria armata.

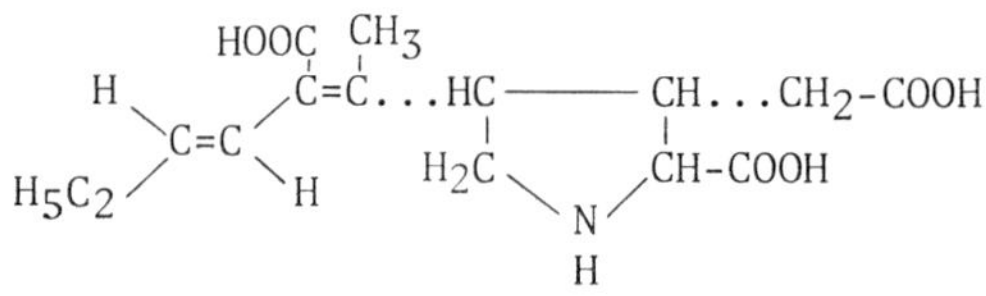

Domoic acid
(3-Carboxymethyl-4-(5-carboxymethylhexa-1, 3-dienyl) proline)

Part of the experimental results are shown in Table 1.

Table 1. Anthelmintic effect of domoic acid isolated from Chondria armata.

Name of subject	Sex	Age	Body weight (Kg)	Worm egg excretion before dose	Days after dose* and number of egg excreted						Worm egg excretion after dose
					1	2	3	4	5	Sum	
K.Y.	Male	10	25.3	+	1					1	-
M.I.	Female	10	31.1	+	1	**				1	-
T.S.	Female	15	49.3	+			1			1	-

* 20 mg of pure domoic acid were dosed per person. ** Thread worm

III. Antihypertensives

It was found by Takemoto et al. (1964)[8] that a basic amino acid effective in preventing or curing the hyperpiesia of mouse is present in Laminaria angustata. These authors succeeded in isolating it and determined its structure to be a methyl derivative of L-lysine.

The authors named this amino acid laminine in connection with the genus name, *Laminaria.* However, laminine was distributed besides *L. angustata* over the other various species of the Laminariaceae, although in small amounts, as shown in Table 2.[9] In addition, laminine was also isolated from *Heterochordaria abietina*, an edible small brown alga belonging to the Chordariaceae,[10] and now called *Analipus japonicus*.

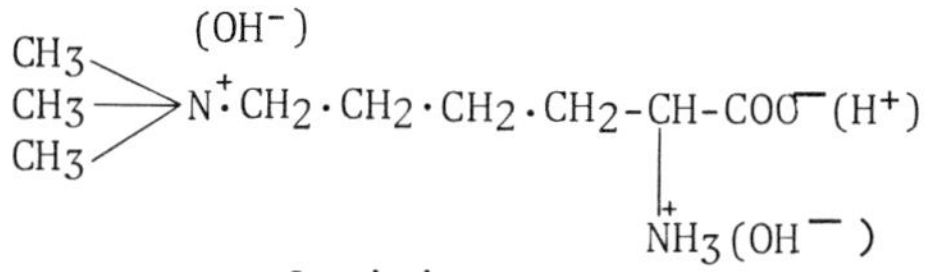

Laminine
((5-amino-5-carboxypentyl)-trimethyl ammonium hydroxide)

A precise pharmaceutical examination of laminine was later carried out by Ozawa et al. (1967).[11] The value of LD_{50} for mouse was estimated to be 394 mg/Kg body weight on intravenous injection and 2.98-3.57 g/Kg body weight on hypodermic injection Laminine suppressed heart beating *in vivo* as well as *in vitro*, and transitorily depressed blood pressure without change in pulse rate. Relaxation of smooth muscle of small intestine, blood vessels and trachea was also caused by this amino acid. It blocked the action of acetylcholine and histamine which are known to induce the contraction of smooth muscle.

Table 2. Distribution of laminine in various species of Laminariaceae.

Alga	Locality	Date of collection	Laminine μM per 1 g dry frond
Laminaria japonica	Muroran	Aug. 1963	0.07
L. japonica var. *ochotensis*	Rishiri	Ditto	0.10
L. fragilis	Muroran	Ditto	0.10
L. religiosa	Shyakotan	Ditto	0.04
L. angustata	Erimo	Ditto	0.34
L. angustata var. *longisima*	Kaigara	Ditto	0.11
L. cichoriodes	Nemuro	Ditto	0.13
L. yezoensis	Nosappu	Ditto	0.18
L. diabolica	Nemuro	Ditto	0.11
Kiellmaniella gyrata	Nemuro	Ditto	0.08
K. crassifolia	Muroran	Ditto	0.02
Costaria costata	Muroran	Ditto	0.04
Arthrothamnus bifidus	Nosappu	Ditto	0.06
Eisenia bicyclis	Shima	Apr. 1963	-
Eckloniopsis radicosa	Tanegashima	Aug. 1963	-
Ecklonia cava	Shimoda	Ditto	0.16
Undaria pinnatifida	Kesennuma	Feb. 1964	0.02
U. undarioides	Shima	Apr. 1962	0.01
Alaria crassifolia	Muroran	Aug. 1963	-

IV. Antibloodcoagulation and Antihyperlipemia Activities

In Japan, as early as 1941 Kimura investigated the fact that extracts from *Laminaria* showed an antibloodcoagulation activity[12] and also served as an activator of lypase.[13] In this connection, it is noteworthy that a patent for the preparation of fucoidan fractions which were most active in antibloodcoagulation and antilipemia was published in 1960 in Japan, Nakajima being the attorney.[14] The inventor was G.F. Springer in America and Ciba Ltd. was the applicant. The polysaccharides of average molecular weights, 5000-50000, fractionated with aqueous ethanol (45-67%) showed both of these activities.

These data are shown in Tables 3-5.

Table 3. Antibloodcoagulation tests for an extract from *Laminaria japonica**.

Grade of algal growth	Mature	1.2 cm of frond in width	Moderate growth	Mature
Coagulation time (min)				
Addition	780	1440	1440	1440
Without addition	18	12	13	13
Date of collection	August	June	July	July

* Algal powder was extracted with 40 times amount of physiological saline for 2 days. One milliliter of the extract was added to 1 ml of cardiac blood of a rabbit and the coagulation time measured.

Table 4. Anticoagulation activity of extracts from various species of Laminariaceae.

Species	*L. diabolica*	*L. longipedalis*	*L. religiosa*	*L. religiosa*	*Kiellmaniella gyrata*
Coagulation time (min)					
Addition	540	540	720	1440	450
Without addition	12	12	18	11	12
Date of collection	July	July	August	July	July

Table 5. Enhancement of pancreas lipase activity by various sterols. The activity of control assay is taken as unity.

	Amount of sterol added (mg)	3	4.8	6	18	24	30	60
	Unsaponifiable lipid fraction*	6.6	8.7	10.6	8.1	8.3	8.5	8.5
Sterol	Cholesterol	1.4	1.4	1.4	1.4	2.5	2.1	1.9
	Ergosterol	1.2	3.5	3.5	4.1	4.0	5.3	4.2
	Na-cholate	1.1	2.3	2.3	1.7	2.5	1.1	1.1

* From *L. japonica*

V. Depression of Cholesterol Level

When the rats of California University Strain were fed for 10 weeks with diets containing 5% of dried powder of green laver (*Enteromorpha compressa*), purple laver (*Porphyra tenera*) or sea tangle (*Laminaria japonica*) each and 1% of exogenous cholesterol, Kaneda et al.[15] found that the contents of plasma cholesterol of the sample rats were decreased significantly, particularly clearly with the diets containing the former two lavers. However, the levels of cholesterol in liver were not decreased by these diets. The results are shown in Table 6.

Table 6. Effect of dried powder of green and purple lavers and of a tangle on the cholesterol levels of plasma, liver and adrenal in rats.

Test group	Plasma cholesterol Total (mg%)	Plasma cholesterol Free (mg%)	Liver cholesterol Total (mg%)	Liver cholesterol Free (mg%)	Liver total lipids (mg/g)	Adrenal toal cholesterol (mg/g)
I: Without addition of cholesterol and bile salts	58.7±1.10*	19.4±0.91*	1.3±0.21*	1.2±0.19*	23.9± 0.78*	14.4
II: Addition of cholesterol (1%) and bile salts	123.6±1.36	27.7±0.96	48.4±0.90	3.6±0.31	148.0± 1.38	23.3
III: II+5% green laver	79.7±1.53	14.5±1.00	50.7±1.02	2.9±0.20	144.8± 1.78	25.3
IV: II+5% tangle	142.5±1.3	31.6±1.01	47.8±0.95	3.0±0.31	181.6± 1.41	20.1
V: II+5% purple laver	78.3±1.36	6.2±0.62	66.2±0.96	5.4±0.46	196.2± 1.61	31.7
VI: II+0.5% ergosterol	126.5±1.8	31.7±0.89	24.6±0.89	2.2±0.14	119.1± 1.21	19.5

* Standard deviation, $\sigma = \sqrt{\frac{\Sigma d^2}{N}}$

Basic diet consists of 63.26% of sucrose, 22% of casein, 5% of Celluflour, 4% of a salt mixture, 0.24% of choline-HCl, 0.5% of a vitamin mixture (casein, p-aminobezoic acid, inositol, tochopherol, ascorbic acid, B_1, Ca-pantothenate, niaci B_{12}, B_6, A, D, folic acid, menadion, and biotin), and 5% of cotton seed oil.

In contrast, diets containing 1% of either fucosterol or a certain unsaponifiable fraction extracted from *Sargassum ringoldianum* decreased both of the plasma and liver cholesterol levels of rats under similarly arising conditions. In particular, liver cholesterol decreased largely; roughly by half content of the control. These effects, however, were slightly lower than that by diets containing β-sitosterol in place of other additives mentioned above[16].

The hypocholesterolemic effect of algae-containing diets seemed to be due at least in part to excretion of a greater quantity of cholesterol than in rats fed without algal powder, although this fact is still unexplained. Possibly, the effective substances would combine with the cholesterol added to diet to lower the concentration of its absorbable portion.

Thereafter, Kaneda et al. succeeded in isolating the hypocholesterolemic substances from a green alga, *Monostroma nitedum* and *Porphyra tenera*.[17] They were found to be betaine group. The most effective one was identified as -homobetaine[18] isolated from the green laver.[19,20]

$(CH_3)_3\text{-}\overset{+}{N}\text{-}CH_2\text{-}COOH$

Glycine betaine

$(CH_3)_3\text{-}\overset{+}{N}\text{-}CH_2\text{-}CH_2\text{-}CH_2\text{-}COOH$

γ-Butylobetaine

$(CH_3)_3\text{-}\overset{+}{N}\text{-}CH_2\text{-}CH_2\text{-}COOH$

β-Homobetaine

$H_2CO\text{-}CH_2\text{-}CH_2\text{-}CH(N^+(CH_3)_3)\text{-}COOH$, with H_2C bonded to $HCOH$ bonded to H_2COH

0-1-Glyceryl homoserine betaine (Ulvaline)

When β-homobetaine was fed to rats, being mixed at 0.12% concentration with a given diet containing 1% cholesterol, the cholesterol level of rat plasma was lowered by nearly 70% of the control level, but all of the betaines tested were ineffective in the hypocholesterolemic activity toward liver.

On the other hand, it was observed by Tsuchiya and his colleagues[21] that a diet containing 5% of dried powder of *Heterochordaria abietina* lowered significantly (approximately 12%) blood cholesterol level of rats fed with a diet containing 1% cholesterol.

Table 7. Effect of Heterochordaria abietina (Analipus japonicus) on blood cholesterol level in rats.

	Feeding	Number in Experimental group	Weight gain (g/rat)	Whole blood cholesterol (mg/dl)
I:	Basic diet*	7	176.1	155.4+2.9
II:	I+cholesterol(1%)	7	170.7	173.0+3.5
III:	II-ether-soluble fraction	7	169.0	152.3+6.5
IV:	II+residual fraction	7	175.1	165.0+4.4

* Basic diet used here consists almost of the same composition as that by Kaneda et al.[15)]

The authors then extracted oil from this alga, and dosed 27 mg of them per animal per day. The similar effect to feeding with 5% agal powder was observed.

The result is shown in Table 7. Fucosterol showed also the same effect on plasma cholesterol level in chickens but not in rats. Carrageenan, agar and alginic acid were hypocholesterolemic agents. In particular, carrageenan was one of the excellent ones. Phaeophytin and phaeophoribide derived from chlorophyll of Analipus als showed a similar effect. The proteins bound with iodine compounds such as thyroxine, triiodo-, diiodo-, and monoiodotyrosines were better substances for preventing hypercholesterolemia in rats. In Fig. 1, the effect of some iodocompounds from brown algae on the blood cholesterol level of rat is shown.

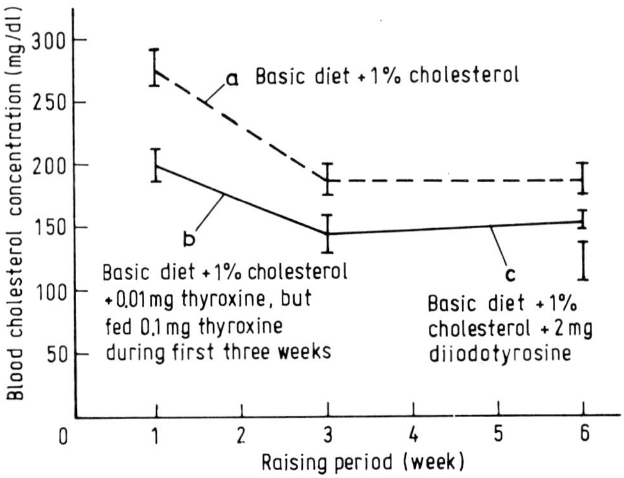

Fig. 1. Changes in blood cholesterol level of rats fed a diet containing iodoamino acid.

The iodine organic compounds have been found in various species of marine algae other than *Analipus japonica*, as shown in Table 8.[21]

Table 8. Iodoamino acid compounds in marine algae (mg% on a dry weight basis).

Iodoamino acid-I	*Heterochordaria abietina**	*Laminaria* sp.	*Undaria pinnatifida*	*Sargassum* sp.	*Sargassum thunbergii*	*Polyshiphonia urceolata*
Monoiodotyrosine-I	1.49	1.55	+	10.61	4.60	0.77
Diiodotyrosine-I	2.06	2.33	+	4.04	4.11	2.31
Triiodotyrosine-I	2.41	0.91	+	1.52	0.70	0.79
Thyroxine-I	2.84	1.30	+	1.90	0.73	0.52
Iodine-bound protein-I	2.14	0.67	+	5.56	1.86	0.79
Unknown 1-I	0.96	0.52	+	2.02	1.21	0.77
Unknown 2-I	2.93	1.38	+	6,70	9.44	0.77

Further studies on the hypocholesterolemic effect of alginate carried out by Tsuchiya et al.[22] showed that the sodium alginates of $\overline{DP}$'s 417 and 226 are most strong in this activity among the 4 kinds of alginate samples; commercial alginates and the sodium alginate of $\overline{DP}$ 13 and the above-mentioned alginate samples. The result is shown in Table 9.

Table 9. Effect of Na-alginates on plasma and fecal cholesterol levels and amounts of fecal alginates in rats.

Supplements to basal ration	Weight gain (g/rat)	Plasma cholesterol (mg/100 ml)	Fecal cholesterol (g/100 g dry matter)	Fecal alginate (g/100 g dry matter)
I: 1% of cholesterol	147	203.3±6.3*	7.6±0.35*	3.7±1.0*
II: I+3% of commercial Na-alginate	138	177.2±6.9	11.5±0.28	98.4±1.8
III: I+3% of purified commercial alginate, $\overline{DP}$ 417	137	158.1±15.2	12.1±0.28	110.3±5.8
IV: I+3% of purified commercial alginate, $\overline{DP}$ 226	148	154.7±13.3	12.0±0.46	107.3±5.5
V: I+3% Na-alginate obtained from commercial alginate, $\overline{DP}$ 13	152	210.1±14.0	10.3±0.54	67.2±6.1

* Standard deviation, $\sigma = \sqrt{\frac{\Sigma d^2}{N}}$.

VI. Antibiotics

In 1955, two fractions were prepared by Saito and Samejima[23] from various marine algae and tested for the antibiotic activity against *Bacillus subtilis*, *B. mesentericus*, *Micrococcus pyogenes* var. *aureus*, *Ptoteus vulgaris* and *Escherichia coli*. For example, the blades of a small marine red alga, *Symphyocladia gracilis* (or *S. latiuscula*) were extracted with 80% methanol. The extract was concentrated to a syrup, and divided into water soluble (Fraction I) and insoluble (Fraction II) portions. The two fractions were added to cultures of bacteria after appropriate dilution, and the growth patterns of bacteria were investigated, as shown in Table 10.

Table 10. Effect of Fractions I and II on the growth of various bacteria.

Extract from	Fraction I Dilution rate	B. subtilis 24 (hr)	B. subtilis 48 (hr)	E. coli 24 (hr)	E. coli 48 (hr)	Fraction II Dilution rate	B. subtilis 48 hr	B. mesentericus 48 hr	P. vulgalis 48 hr
Ulva pertusa	200	-	+	-	-	200	+	+	±
	400	+	+	+	+	2000	++	+	+
Monostroma sp.	200	+	+	+	+	200	+	+	±
	400	+	+	+	+	2000	++	++	+
Laminaria angustata	200	-	-	-	-	200	-	-	-
	400	-	+	-	±	2000	-	+	-
	1000	+	+	-	+				
Undaria pinnatifida	200	±	+	-	+	200	-	-	±
	400	+	+	+	+	2000	-	+	+
Rhodomela larix	200	-	-	-	-	200	-	-	-
	400	-	+	-	±	2000	-	+	±
	1000	+	+	-	+				
Digenea simplex	200	+	+	+	+	200	+	+	±
	400	+	+	+	+	2000	+	+	+
Symphyocladia gracilis	200	Not tested				200	-	-	-
	400	Not tested				2000	-	-	±

In contrast to Fraction I's from Monostroma sp. and Digenea simples, those from Laminaria angustata and Rhodomela larix showed some inhibitory activities against bacterial growth. Fraction II's from L. angustata, R. larix and S. gracilis were also more inhibitory than others. The water-soluble inhibitory substance migrated with a Rf value of 0.35 on paper chromatography (the solvent system, Butanol: acid: Water, 4:1:1 v/v), and the substances were different from inorganic bromine and iodine compounds. It was found by another experiment that a brominated phenolic compound extracted from Fraction II obtained from S. gracilis which was insoluble in petroleum ether but soluble in ether, showed a higher antibiotic activity against B. mesentericus than those against P. vulgaris and E. coli, although the compound seemed slightly different from a brominated phenolic compound isolated from R. larix by Mautner et al.[24]

Recently, Kurata investigated the antibiotic activity of a series of brominated phenolic compounds isolated from various species of red algae belonging to the Rhodomelaceae,[25] whose chemical structures are shown in Fig. 2.

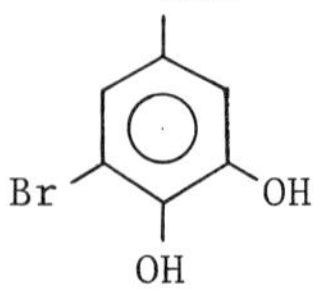

(I)

3-bromo-4,5-dihydroxybezaldehyde

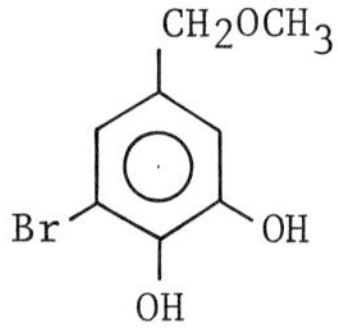

(II)

3-bromo-4,5-dihydroxybenzyl methyl ether

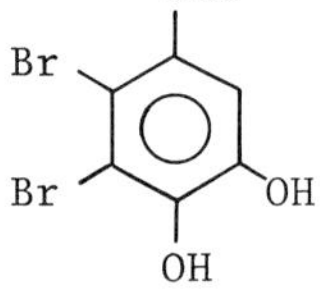

(III)

2,3-dibromo-4,5-dihydroxybezaldehyde

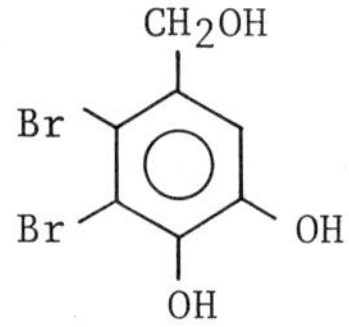

(IV)

2,3-dibromo-4,5-dihydroxybezyl alcohol (lanosol)

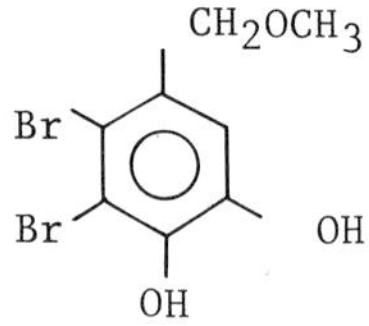

(V)

2,3-dibromo-4,5-dihydroxybenzyl methyl ether (lanosol methyl ether)

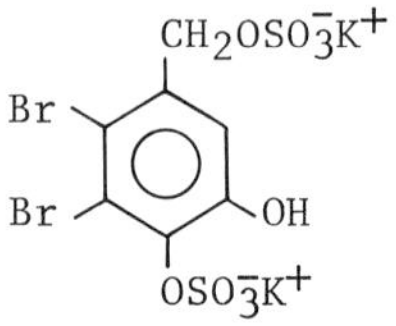

(VI)

4,5-disulfate dipotassium

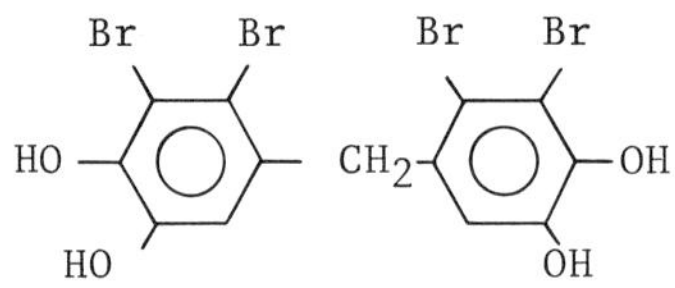

(VII)

2,2',3,3'-tetrabromo-4,4',5,5'-tetrahydroxydiphenyl methane

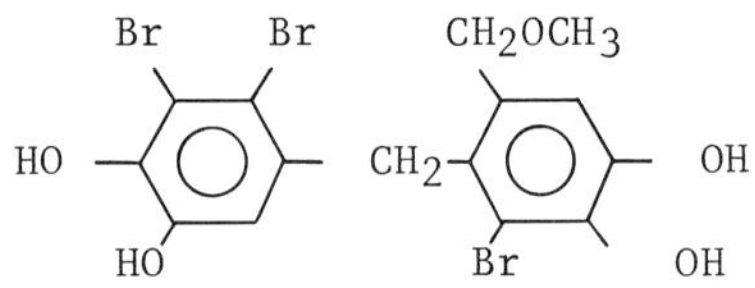

(VIII)

2,2',3-tribromo-3',4,4',5-tetrahydroxy-6-methoxymethyl

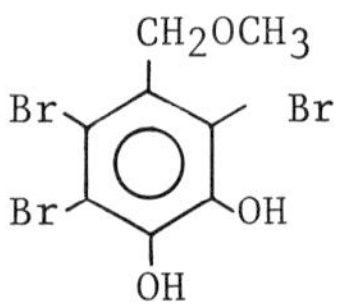

(IX)

2,3,6-tribromo-4,5-dihydroxybezylmethyl ether

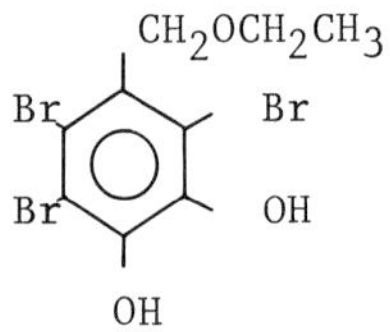

(X)

2,3,6-tribromo-4,5-dihydroxybezylethyl ether

From **Polysiphonia** urceolata: I and II, **P. morrowii: I,** **Rhodomela larix**: III, IV, V, VI, VII, and VIII, R. subfusa: III, IV, V, and VI, Odonthalia corymbifera: V, VI and VII, Symphyocladia latiuscula: IX, and X were isolated.

Fig. 2. Chemical structures of algal bromophenol compounds of antibiotic activity and their origin.

Bacteria and fungi were cultured on a commercial agar medium (manufactured by Nissei Co.) and on a potato-dextrose agar medium (manufactured by Nissei Co.), respectively. To 20 ml of these media, were added 10 ml of dimethylformamide dissolving the compounds from I to X individually, and the growth of microbes to be tested was compared with that of standard microbes obtained from the culture collections of the Central Research Institute of Ueno Pharmaceutical Company. Part of the results obtained by Kurata is summerized in Table 11.

Table 11. Antibiotic activities of various bromophenolic compounds against bacteria and fungi. The antibiotic activity is expressed by an approximate minimum amount (g/ml) to inhibit microbial growth.

Microorganism	Bromophenolic compounds shown in Fig. 2									
	I	II	III	IV	V	VI	VII	VIII	IX	X
Staphylococcus aureus, 20	20	29	41	42	79	106	32	22	66	15
Pseudomonas aeruginospae	20	29	41	42	79	106	32	22	66	15
Proteus vulgaris HX-19	20	29	41	42	79	106	32	22	66	15
Salmonella pullorum	20	29	41	42	79	106	32	22	66	15
Escherichia coli 8057	20	29	41	42	79	106	32	22	66	15
Klebsiella pneumoniae ST 101	20	29	41	42	79	106	32	22	66	15
Saccharomyces uvarum 7016	20	29	41	42	79	106	32	22	66	15
Pichia membranaefaciens K-17	20	29	41	42	79	106	32	22	66	15
Schizosaccharomyces pombe diner 0-77	20	29	41	42	79	106	32	22	66	15
Aspergillus niger	20	29	41	42	79	106	32	22	66	15
Fusarium moniriforme IAM 5062	20	29	41	42	79	106	32	22	66	15
Trichoderma T-1 ATCC 9645 IAM 5061	20	29	41	42	79	106	32	22	66	15
Pennicillium citrinum 2125	20	29	41	42	79	106	32	22	66	15

VII. Antitumor Activity

Some examinations for the possible antitumor activity of algal extracts have recently been attempted in Japan. Yamamoto and Nagumo reported that a fraction of funoran from Gloiopeltis furcata, a sulfated galactose of carrageenan type,[26] and a hot water-extract from Sargassum fulvellum are curative for Sarcoma-180 solid tumor.[27] Thereafter, Nakazawa et al.[28] examined the antitumor activity of aqueous extracts from a variety of marine algae including green, brown and red algae by the intraperitoneal injection of varying concentrations. For example, a 4.4 mg dose of the extracted material from Dictiopteris undulata per mouse caused mean survival days of 37.3 (96% increase in life span) while its LD_{50} value was more than 35.0 mg per mouse.

Then, the authors partially purified the crude extracts from Sargassum horneri, S. tortile and S. hemiphyllum first by dialysis against water and divided them into the inner (DI) and outer (DO) substances. The inner ones were then fractionated by salting out with ammonium sulfate into precipitate (DIP) and supernatant (DIS). The outer ones were extracted with methanol after lyophilization, and the residue was fractionated by gel filtration on Sephadex G-25(fine) into DO-I-VI fractions. Their antitumor activity was then tested, revealing that each fraction inhibited in different degrees the growth of Ehrlich ascites carcinoma cells and Sarcoma-180 cells which were trasplanted subcutaneously and intraperitoneally in mice. With Fraction DO-I, a dose of 75g/mouse/day x6 caused the prolongation of life span for 75% of the animals tested. Similar results were found with DIP and DIS whose carbohydrate contents were determined to be 50% and 70% glucose, respectively, by the phenol-sulfuric acid method.

Most of the fractions from S. horneri showing an antitumor activity were fucose-containing polysaccharides with 20 to 30% of ester sulfate.[29] In the studies using mice, a range of doses from 125 to 750 g of the polysaccharide fractions/mouse/day was found to be effective not only to the ascites tumor ($3x10^5$ cells/mouse) and solid tumor ($2x10^6$ cells/mouse) of Ehrlich ascites carcinoma, but also to the solid tumor ($2x10^6$ cells/mouse) of Sarcoma-180. As to the possible mechanism of this efficacy, the authors have assumed that the polysaccharides show host-mediated effect through the direct cytotoxity to activate the transplanted immunity, because of causing an increase in the number of white blood cells, a change in the serum electrophoretic pattern, and an inhibition of the tumor growth by administration of this serum in mice, to which solid tumor had been subcutaneously transplanted.

It was found also that a polysaccharide fraction precipitated with 4 times amounts of ethanol from a slightly yellowish extract of a green marine alga, Codium pugniformis is active against both the ascites tumor of Ehrlich carcinoma and the solid tumors produced by Ehrlich carcinoma and Sarcoma-180.[30] An acid hydrolysate of this polysaccharide contained galactose, arabinose and mannose but a small amount of amino acids. Possibly, the polysaccharide may be a sulfated galactan, judging from the result obtained by Love and Percival from Codium fragile polysaccharide.[31]

Another similar study was carried out by Ito and Sugimura on an active substance extracted from Sargassum thunbergii.[32] Starting from the routine extraction of the brown alga with hot water, the purification of the effective substance was carried out by repeating precipitation with ethanol. The active substance contained galactose, xylose, fucose, glucose, mannose and glucuronic acid in addition to a small amount of sulfate. Then, each group of 10 mice was inoculated with Ehrlich ascites of $2x10^5$ cells and raised with 20 or 10 mg of the active fraction (F-1)/Kg/dayx10 to measure the survival rate of mice. As compared with the control tests, in which all mice died at 14 days after the inoculation, 100 or 40% of mice survived, respectively, by treatment with these amounts of active fraction.

VIII Selective Removal of Poisonous Heavy Metals

There are few studies in Japan similar to those of Skoryna and his associates on the selective binding capacity of alginic acid and other acid polysaccharides of seaweeds.[33,34,35]

On an assumption that part of radioactive strontium once absorbed into the body would be excreted into the digestive tract through bile and digestive fluid and consequently they would be prevented from reabsorption by alginate fed to the experimental animals together with routine diet, Ichikawa[36] attempted experiments to demonstrate the assumption using cat and rat as the subject. These animals which had been injected with some amounts of radiostrontium (^{89}Sr) intravenously or intraperitoneally were fed a diet containing Na-alginate for a given period, and then a given amount of pilocarpine was injected to promote the secretion of bile and other digestive fluid. After being fed Na-alginate for a further given period, remaining radiostrontium in body was determined. However, no remarkable decrease in its content was observed. Possibly, as has been stated by this worker, too, this is due to the fact, that the amount of radiostrontium excreted into digestive tract is so small that the alginate fedding has resulted in inefficancy. In fact, the amount of radioactive strontium excreted into feces was found in his experiment to be far smaller than that into urine.

IX. Enzyme activator

It was found by Komura et al.[37] that phytol and a terpene isolated from Hizikia fusiformis stimulated the activity of pancreatic esterase when soluble substrates such as methyl butylate and monopropionin as well as triacetin were used as substrate. The esterase activity enhanced by phytol was inhibited by addition of bile salts, lecithin, Triton X-100 or urea.

X. Vitamins and Related Compounds

The content of vitamins and related substances in seaweeds were widely examined by Kanazawa and Kakimoto,[38] and by Kanazawa.[39] Among red algae, Porphyra tenera contains various vitamins relatively in abundance, and thus provitamin A, thiamine (B_1), riboflavin (B_2), pyridoxine (B_6), nicotinic acid, folic acid, biotin, lipoic acid, cobalamine (B_{12}), choline and ascorbic acid(C) have been found in this laver. Furthermore, it was found that the following vitamins and related substances are particularly rich in some red algae: B_2, folic acid and choline in Gelidium amansii; B_{12} in Chondrococcus nipponicus; folic acid in Grateloupia ramosissima; B_2 and folic acid in Gloiopeltis tenax; provitamin A and B_{12} in Gracilaria gigas; B_1, nicotinic acid, pantothenic acid, folic acid, biotin, lipoic acid and inositol in Gracilaria textorii; B_2, nicotinic acid, pantothenic acid and lipoic acid in Chondrus ocellatus; pantothenic acid in Lomentaria catenata; and nicotinic acid, pantothenic acid, folic acid and B_{12} in Laurencia okamurai.

Among green algae, only a few species were examined for vitamin content. The examples are biotin in Ulva pertusa, B_2, folic acid in Monostroma nitidum, and provitamin A, nicotinic acid, biotin B_{12} and choline in Enteromorpha linza.

In brown algae, the following algae were the examples: nicotinic acid, folic acid and lipoic acid in Spatoglossum pacificum; choline, B_{12}, lipoic acid in Colpomenia sinusa; choline and inositol in Myelophycus caespitosum; folic acid in Hydroclathrus clathratus; provitamin A and and nicotinic acid in Laminaria sp.; biotin in Ecklonia cava; provitamin A in Hizikia fusiformis; folic acid in Sargassum fulvellum; pantothenic acid, biotin and inositol in Sargassum thunbergii.

In addition, it was reported by Watanabe[40] that Undaria pinnatifida contains a substance which cures artificial nicotine poisoning. White rats previously poisoned with nicotine were fed powder of this brown alga and recovered in as healthy a condition as before poisoning. The substance was found in unsaponifiable fraction extracted with ether from the algal fronds, but no further structural study was carried out on it thereafter.

XI. Uses for Manure, Animal Feed and Food for Human Health

Recently in Japan, seaweed meal has begun to be used widely for manure in agriculture and a food good for human health in addition to feed for domestic animals. A commercial product, Algit, is one of the examples. The Algit and related products are being imported from Norway and they are the powdered algal bodies of Ascophyllum nodosum. It is particularly noteworthy that Algit has been reported to increase the proliferation of useful soil bacteria.[41]

On the other hand, powder of brown algae such as Hizikia fusiformis and Undaria pinnatifida grown in Japan has been manufactured in an industrial scale as a food in aiming at the possible promotion of human health, and named Alkalon[42] and Viva Natural,[43] respectively. Most recently, alginate has been used, being named Mosanon, as an agricultural medicine for protection of tobacco plant from the infection of tobacco mosaic virus. It has been reported that an appropriate concentration of Mosanon was greater in efficacy than any other sugar and polysaccharide tested.[44]

References

1) Miyashita, A.: The Marine Algae (Kaisoo), pp. 315, Hosei University Press, Tokyo, Japan (1974) (in Japanese).

2) Murakami, S., Takemoto, T., and Shimizu, Z.: Studies of the effective principles of Digenea simplex, I. Separation of the effective fraction by liquid chromatography, II. On the digeneaside. J. Pharm. Soc. Japan, 73: 1026-1029 (1953) (in Japanese with English summary).

3) Murakami, S., Takemoto, T., Zheng, R.-C., Daigo, K., and Takagi, S.: Studies of the effective principles of Digenea simplex, X. On α-allokainic acid. J. Pharm. Soc. Japan, 75: 1252-1255(1955) (in Japanese with English summary).

4) Sugimoto, Y.: Takeda Products Manual 1960, p.170-171, Takeda Industrial Parmaceutical Co., Osaka, Japan (1960) (in Japanese).

5) Takagi, M.: Advances in Pharmacology in Japan (ed. by Tokida, J. and Hirose, H.), p.321-325, VEB Gustav Fisher Verlag, Jena. (1975)

6) Daigo, K.: Studies on the constituents of Chondria armata, 1. Detection of the anthelmintic constituents, 2. Isolation of an anthelmintic constituent, 3. Constitution of domoic acid. J. Pharm. Soc. Japan, 79: 350-360 (1959) (in Japanese with English summary).

7) Takemoto, T.: On the domoic acid: A new anthelmintic principle. Kagaku (Chemistry), 14: 59-62 (1959) (in Japanese).

8) Takemoto, T., Daigo, K., and Takagi, N.: Studies on the hypotensive constituents of marine algae, I. A new basic amino acid "laminine" and the other basic constituents isolated form Laminaria angustata. J. Pharm. Soc. Japan, 84: 1176-1179 (1964) (in Japanese with English summary).

9) Takemoto, T., Daigo, K., and Takagi, N.: Studies on the hypotensive constituents of marine algae, III. Determination of laminine in Laminariaceae. J. Pharm. Soc. Japan, 85; 37-40 (1965) (in Japanese with English summary).

10) Takemoto, T., Takagi, N., and Daigo, K.: Studies of the hypotensive constituents of marine algae, IV. Amino acid constituents of Heterochordaria abietina. J. Pharm. Soc. Japan, 85: 843-845 (1965) (in Japanese with English summary).

11) Ozawa, H., Gomi, Y., and Otsuki, I.: Pharmacological studies on laminine monocitrate. J. Pharm. Soc. Japan, 87: 935-939 (1967) (in Japanese with English summary).

12) Kimura, J.: Studies on the substances of antibloodcoagulation activity in marine algae, IV. in Laminaria japonica. Hokkaido Igaku Zasshi (Acta Medica Hokkaidonensia), 19: 427-436 (1941) (in Japanese).

13) Inagaki, N.: Studies on the chemical utilization of marine algae. Shokuryo no Kagaku (Food Sciences), 1: 194-195 (1947) (in Japanese).

14) Nakajima, N.: Preparation of a new polysaccharide. Patent Gazette of Japanese Patent Agency, Patent Application Number, 3098:1-5 (1960) (in Japanese).

15) Kaneda, T., Tokuda, S., and Arai, K.: Studies of the effects of marine products on cholesterol metabolism, I. The effects of edible seaweeds. Bull. Jap. Soc. Sci. Fish., 29: 1020-1023 (1963)(in Japanese with English summary).

16) Kaneda, T., Arai, K.: Studies on the effects of marine products on cholesterol metabolism in rats, III. The effects of lipids isolated from Sargassum ringoldianum. Bull. Jap. Soc. Sci. Fish., 30: 589-593 (1964) (in Japanese with English summary).

17) Abe, S., Takeda, F., and Kaneda, T.: Studies of the effects of marine products on cholesterol metabolism in rats, VII. The influence of dried purple laver on cholesterol levels of organs and cholesterol excretion in rats. Bull. Jap. Soc. Sci. Fish., 33: 1050-1056 (1967) (in Japanese with English summary).

18) Abe, S., and Kaneda, T.: Studies of the effect of marine products on cholesterol metabolism in rats, IX. Effect of betaines on plasma and liver cholesterol levels. Bull. Jap. Soc. Sci. Fish., 39: 391-393 (1973) (in English).

19) Abe S., and Kaneda, T.: Studies on the effect of marine products on cholesterol metabolism in rats, VIII. The isolation of hypocholesterolemic substances from green laver. Bull. Jap. Soc. Sci. Fish., 39: 383-389 (1973) (in English).

20) Abe, S., and Kaneda, T.: Studies on the effect of marine products on cholesterol metabolism in rats, XI. Isolation of a new betaine, Ulvaline, from a green laver, *Monostroma nitidum* and its depressing effect on plasma cholesterol levels. Bull. Jap. Soc. Sci. Fish., 41: 567-571 (1975) (in English).

21) Tsuchiya, Y.: Comparative hypocholesterolemic activities of marine algae. Proceedings of the VIth International Seaweed Symposium (ed. by Margalef, R.), pp. 747-757, Subsecretaria de la Marina Mercante, Madrid, Spain (1969).

22) Ito, K., and Tsuchiya, Y.: The effect of algal polysaccharides on the depressing of plasma cholesterol levels in rats. Proceedings of VIIth International Seaweed Symposium (ed. by Nisizawa, K. et al.), pp. 558-561, University of Tokyo Press, Tokyo, Japan (1972).

23) Saito, K., and Sameshima M.: Studies on the antibiotic action of algal extracts. Part 3. J. Agr. Chem. Soc. Japan, 29: 427-429 (1955) (in Japanese with English summary).

24) Mautner, H.G., Gardner, G.M., and Pratt, R.: Antibiotic activity of seaweed extracts, II. *Rhodomela larix*. J. Am. Pharm. Assoc., 42: 294-296 (1953).

25) Kurata, K.: Antibiotic activity of bromophenolic compounds isolated from red algae. Annual Meeting of the Japanese Society of Scientific Fisheries, Shimonoseki, October. Abstracts, p. 106 (1976) (in Japanese).

26) Hirase, S., and Watanabe, K.: Fractionation and structural investigation of funoran. Proceedings of VIIth International Seaweed Symposium (ed. by Nisizawa, K. et al.), pp. 451-454, University of Tokyo Press, Tokyo, Japan (1971).

27) Yamamoto, I., and Nagumo, S.: Tumor-inhibiting effect of tangles and _Sargassum_ against Sarcoma-180 solid tumor. Annual Meeting of the Japanese Society of Cancer. Abstracts, p 1225 (1973) (in Japanese).

28) Nakazawa, S., Kuroda, H., Abe, F., Nishino, T., Otsuki, M., and Umezaki, I.: Antitumor effect of water-extracts from marine algae (I). Chemotherapy, _22_: 1435-1442 (1974) (in Japanese with English summary).

29) Nakazawa, S., Abe., F., Kuroda, H., Kohno, K., Higashi, T., and Umezaki, I.: Antitumor effect of water-extracts from marine algae (II). _Sargassum horneri_ (Turner) C. Agardh, Chemotherapy, _24_: 443-447 (1976) (in Japanese with English summary).

30) Nakazawa, S., Abe. F., Kuroda, H., Kohno, K., Higashi, T., and Umezaki, I.: Antitumor effect of water-extracts from marine algae (III). _Codum pugniformis_ Okamura. Chemotherapy, _24_: 448-450 (1976) (in Japanese with English summary).

31) Love, J., and Percival, E.: The polysaccharides of the green seaweed _Codum fragile_. Part II. The water-soluble sulfated polysaccharides. J. Chem. Soc. 3338-3345 (1964).

32) Ito, H., and Sugiura, M.: Antitumor polysaccharide fraction from _Sargassum thnbergii_. Chem. Pharm. Bull., _24_: 1114-1115 (1976).

33) Skoryna, S. C., Paul, T. M., and Edward, D.W.: Studies on inhibition of intestinal absorption of radioactive strontium, 1. Prevention of absorption from ligated intestinal segments. Canada. Med. Ass. J., _91_: 285-288 (1964).

34) Skoryna, S.C., Hong, K.C., and Tanaka, Y.: The effects of enzymic degradation products of alginates on intestinal absorption of radiostrontium. Proceedings of VIIth International Seaweed Symposium (ed. by Nisizawa, K. et al.), pp. 605-607, University of Tokyo Press, Tokyo, Japan (1971).

35) Tanaka, Y., Waldron-Edward, D., and Skoryna, S.C.: Studies on inhibition of intestinal absorption of radioactive strontium, VII. Relationship of biological acivity to chemical composition of alginates obtained from North American seaweeds. Canad. Med. Ass. J., 99: 169-175 (1968).

36) Ichikawa, R.: Behavior of alginic acid toward radioactive strontium. Kagaku to Seibutsu (Chemistry and Biology), 7: 208-211 (1969) (in Japanese).

37) Komura, T., Nagayama, H., and Wada, S.: Stimulation of pancreatic esterase activity by phytol isolated from Hizikia fusiformis. Agr. Chem. Soc. Japan, 49: 149-155; and 157-162(1975) (in Japanese with English summary).

38) Kanazawa, A., and Kakimoto, D.: Studies of the vitamins of seaweeds, I. Folic acid and folinic acid. Bull. Jap. Soc. Sci. Fish., 24: 573-577 (1958)(in Japanese with English summary).

39) Kanazawa, A.: Studies on the vitamin B-complex in marine algae, I. On vitamin contents. Memoirs Fac. Fish. Kagoshima Univ., 10: 38-69 (1961).

40) Watanabe, Y.: Harmful effects of nicotine can be prevented by intake of Undaria. Oral Presentation at the General Meeting of Nippon Dietetical and Foodtechnology Society (cited from the Reference 5) (1968).

41) Proceedings of the IIIrd Symposium on the effects of Algit on agriculture in Japan (ed. by Niikura, H.), pp. 1-44, All Japan Algit-Agriculture Promoting Conference, Tabuse, Yamaguchi-ken, Japan, (1976) (in Japanese).

42) Hara, T.: A guide to Alkalon. Sasalon and Alkalon, pp. 17-22, Taishū Igakusha (Popular Medicine Co.), 32-17, Oyamahigashi Itabashi-ku, Tokyo, Japan (1971) (in Japanese).

43) Tooyama, F.: A guide to the management of health. Kaiso Shigen Kaihatsu Ltd. (Marine Algal Resources Developing Co. 29-8, Shinjuku-1, Shinjuku-ku, Tokyo, Japan (1976) (in Japanese).

44) Yamashina, Y.: A review of Mosanon. Reports of the Agricultural Chemicals Division, Mitsubishi Kasei Industrial Co., 5-2, Marunouchi-2, Chiyoda-ku, Tokyo, Japan (1976) (in Japanese).

Part II. Articles on Special Constituents of Marine Algae

ANTIBIOTIC SUBSTANCES FROM MARINE FLORA

M. Aubert, J. Aubert, M. Gauthier
I.N.S.E.R.M.
1, avenue Jean-Lorrain, 06300 Nice (France)

For a long time, a great number of medicaments have been extracted from the sea and from sea plants and animals. Here we shall limit ourselves to a discussion of substances produced by marine flora and micro-organisms which act antibacterially against saprophyte or pathogenous sea-floor micro-organisms.

Such antibacterial activity has frequently been observed, but these substances have only rarely been used for pharmaceutical purposes, although they have considerable antibiotic properties; they have a wide range of activity and are liable to be effective against stocks of pathogenous micro-organisms that are resistant to classic antibiotics.

In the following we shall discuss possibilities offered in this field by seaweeds, planktonic microalgae and marine bacteria.

I - Substances produced by Seaweeds

The production of inhibitory substances by seaweeds was noted as early as 1917 by Harder (33). Since then, numerous studies have enabled us to find and extract a considerable number of antibacterial substances from marine algae of all three groups: Rhodophyceae, Phaeophyceae and Chlorophyceae. Tables I - VI give a bibliographic summary of studies on seaweeds that have been found to have antibacterial properties against germs in the soil.

Marine Algae in Pharmaceutical Science
edited by H. A. Hoppe, T. Levring, Y. Tanaka

Table I - Seaweeds with antibacterial action on soil bacteria

Algal species	Nature of isolated compounds	Bibliographical references
Rhodophyceae		
Porphyra tenera	fatty acids carbonyl terpenes	Katayama 1962 (41)
Grateloupia	not identified	Kamimoto 1955 (39)
Callophyllis	"	Pratt et al. 1951 (52)
Gloepeltis	"	Kamimoto 1955 (39)
Cystoclonium	acrylic acid	Roos 1957 (54)
Gracilariopsis	"	Allen and Dawson 1960 (2)
Iridophycus	"	Pratt et al. 1951 (52)
Plocamium	"	Kamimoto 1955 (39)
Chrysymenia	not identified	Kamimoto 1955 (39)
Halosaccion	"	Pratt et al. 1951 (52)
Gelidium	not identified	Kamimoto 1955 (39)
Ceramium rubrum	sulphur compounds	Ikawa 1973 (35)
Ceramium	not identified	Roos 1957 (54)
Delesseria	"	Roos 1957 (54), Hornsey and Hide 1974 (34)
Falkenbergia rufolanosa	bromic compounds	Codomier et al. 1977 (20), Burkholder et al. 1960 (15)
Phycodrys	not identified	Olesen et al. 1963 (50), Roos 1957(54)
Murrayella	"	Burkholder et al. 1960 (15)
Amansia	"	Burkholder et al. 1960 (15)
Chondria	"	Burkholder et al. 1960 (15)

Table II - Seaweeds with antibacterial action on soil bacteria (cont.)

Algal species	Nature of isolated compounds	Bibliographic references
Rhodophyceae (cont.)		
Asparagopsis armata	bromic compounds	Codomier et al. 1977 (20)
	brominated compounds	Hornsey and Hide 1974 (34)
Bonnemaisonia asparagoides	not identified	Hornsey and Hide 1974 (34)
Bonnemaisonia hamifera	"	"
Gracilaria verrucosa	"	"
Phyllophora crispa	"	"
Phyllophora membranifolia	"	"
Chondrus crispus	"	"
Dilsea carnosa	"	"
Gloiosiphonia capillaris	"	"
Callocolax neglectus	"	"
Antithamnion cruciatum	"	"
Antithamnion plumula	"	"
Callithamnion tetragonum	"	"
Callithamnion tetricum	"	"
Sphondylothamnion multifidum	"	"
Membranoptera alata	"	"
Bostrychia scorpiodes	"	"
Brongniartella byssoides	"	"
Chondria dasyphylla	"	"
Halopitys incurvus	"	"
Odonthalia dentata	"	"
Pterosiphonia thuyoides	"	"
Rhodomela confervoides	"	"
Chondria oppositiclada	terpene compounds	Fenical and Sims 1974

Table III - Seaweeds with antibacterial action on soil bacteria (cont.)

Algal species	Nature of isolated compounds	Bibliographical references
Rhodophyceae (cont.)		
Wrangelia	not identified	Burkholder et al. 1960 (15), Olesen et al. 1963 (50)
Digenea	fatty acids	Katayama 1962 (41)
	carbonyl	
	fractions of terpenes	
Polysiphonia	acrylic acid	Chesters and Stott 1956 (18), Roos 1957 (54) Sieburth and Conover (1964) (21) (57)
Polysiphonia nigrescens	not identified	Hornsey and Hide 1974 (34)
Polysiphonia nigra	"	"
Polysiphonia urceolata	"	"
Pterocladia capillacea	"	Khaléafa et al. 1975 (42)
Rhodomela	bromophenol	Mautner et al. 1953 (47), Sameshima and Saito 1955 (56), Roos 1957 (53)
Laurencia hybrida	not identified	Hornsey and Hide 1974 (34)
Laurencia obtusa	"	Burkholder et al. 1960 (15)
Laurencia pinnatifida	"	"
Laurencia glandulifera	bromic compounds	Irie et al. 1968 (36)
Laurencia nipponica	"	"
Laurencia intermedia	"	"
Spyridia	not identified	Olesen et al. 1960 - 1963 (50)
Rytiphlaea tinctoria	bromic compounds	Chevelot-Magueur et al. 1976 (19)

Table IV - Seaweeds with antibacterial action on soil bacteria (cont.)

Algal species	Nature of isolated compounds	Bibliographical references
Phaeophyceae		
Scytosiphon lomentaria +)	not identified	Khaléafa et al. 1975 (42)
Ectocarpus	"	Roos 1957 (54)
Dictyota	not identified	Allen, Dawson 1960(2) Starr et al. 1962(60) Olesen et al.1963(50) Hornsey and Hide 1974 (34)
Padina	not identified	Allen and Dawson 1960 (2) Burkholder et al. 1960 (15)
Dictyopteris	"	Olesen et al. 1963 (50) Hornsey and Hide 1974 (34)
Desmarestia	sulphuric acid	Roos 1957 (54)
Desmarestia aculeata	not identified	Hornsey and Hide 1974 (34)
Desmarestia ligulata	"	"
Chordaria flagelliformis	not identified	Hornsey and Hide 1974 (34)
Alaria esculenta	"	"
Cystoseira baccata	"	"
Cystoseira tamariscifolia	"	"
Cystoseira fimbriata +)	"	Khaléafa et al. 1975 (42)
Chorda	not identified	Roos 1957 (54)
Halopteris scoparia +)	"	Khaléafa et al. 1975 (42)

+) This species also has antifungal properties.

Table V - Seaweeds with antibacterial action on soil bacteria (cont.)

Algal species	Nature of isolated compounds	Bibliographic references
Phaeophyceae (cont.)		
Laminaria	acrylic acid	Saito and Sameshima 1955 (56)
	carbonyl, terpenes	Chesters and Stott 1956 (18)
		Roos 1957, Katayama 1962 (41)
Laminaria digitata	not identified	Hornsey and Hide 1974 (34)
Laminaria saccharina	"	"
Macrocystis	not identified	Pratt et al. 1951 (52)
Egregia	"	"
Postelsia	"	"
Undaria	"	Saito and Samashima 1955 (56)
		Kamimoto 1955 (39)
Ascophyllum nodosum	"	Vacca and Walsh 1954 (62)
Fucus	"	Roos 1957 (54)
Pelvetia	Acrylic acid	Chesters and Stott 1956 (18), Challenger et al. 1956 (17), Pratt et al. 1951 (52)
Halidrys	"	Chesters and Stott 1956 (18), Challenger et al. 1956 (17)
		Hornsey and Hide 1974 (34)
Sargassum	fatty acids, carbonyl terpenes	Katayama 1962 (41), Conover and Sieburth 1962 (21), Kamimoto 1955 (39)
Cryptophyceae		
Hemiselmis	not identified	Duff, Bruce and Antia 1966 (25)
Rhodomonas	"	"

Table VI - Seaweeds with antibacterial action on soil bacteria (cont.)

Algal species	Nature of isolated compounds	Bibliographic references
Chlorophyceae		
Codium	fatty acids, carbonyl and fractions terpenes	Katayama 1962 (40), Kamimoto 1955 (39)
Codium fragile	not identified	Hornsey and Hide 1974 (34)
Codium tomentosum	"	"
Derbesia	fatty acids, carbonyl and fractions of terpenes	Allen and Dawson 1960 (2)
Enteromorpha	"	Roos 1957 (54), Allen and Dawson 1960 (2), Starr et al. 1962 (60), Katayama 1962 (40), Challenger et al. 1963 (17)
Ulva	fatty acids, carbonyl and acrylic terpenes	Pratt et al. 1951 (52), Roos 1957 (54), Challenger et al. 1957 (17), Katayama 1962 (40), Kamimoto 1955 (39), Hornsey and Hide 1974 (34)
Cladophora +)	not identified	Khaléafa et al. 1975 (42), Starr et al. 1962 (60), Hornsey and Hide 1974 (34)
Cladophoropsis	"	Allen and Dawson 1960 (2)
Spongomorpha	acrylic acid	Roos 1957 (54), Challenger et al. 1957 (17)
Pithophora	not identified	Klein 1964 (43)
Bryopsis plumosa	"	Hornsey and Hide 1974 (34)

+) This species also has antifungal properties.

Without wishing to go into details of all the research carried out in this field, we shall restrict ourselves to considering those studies that have led to identification of chemical compounds responsible for antibiotic activity. These are primarily fatty acids, terpenes, carbonyls and bromophenol compounds.

1. Volatile compounds

In 1953 Katayama first isolated fatty acids, terpenes and carbonyls from Rhodophyceae (*Porphyra tenera*, *Digenia simplex*), Phaeophyceae (*Laminaria* and *Sargassum*) and Chlorophyceae (*Enteromorpha*, *Codium fragile*, *Ulva*). These substances are active at very low concentrations on *Escherichia coli* and *Staphylococcus aureus* 209 P (in dilutions of 2000 and 3000 they inhibit growth of *Escherichia coli* and *Staphylococcus aureus* 209 P, while just fractions of carbonyls and terpenes have a paralytic toxic effect on some annelids).

Among the fatty acids, acrylic acid seems to play an ecological role in some marine spheres. This has been attested in seaweeds by several authors since 1951, who have found it in Rhodophyceae (*Polysiphonia*, *Tridophycus*) (Pratt et al. 1951), in Phaeophyceae (*Laminaria*, *Pelvetia* and *Halidrys*) and in Chlorophyceae (*Ulva*, *Spongomorpha*, *Enteromorpha*).

It is interesting to note that according to Challenger et al. (17) acrylic acid results from hydrolysis of the dimethyl-B-propriothetin produced by these algae, which sets free dimethyl sulphur and nascent acrylic acid; the latter, according to Sieburth, possesses strong antibiotic properties with a wide antibacterial range, activity increasing in a more acid medium (reaching an activity maximum at pH 5-6). That is, acrylic acid can only have ecological effects close to the alga by which it was produced, as upon contact with sea water (alkaline pH) it is transformed into an antibacterially inactive compound. This property differentiates it from other fatty acids, terpenes and carbonyls which are widespread in the sea and whose influence may be felt on the scale of the ocean or may even have an effect on exchanges between water and the atmosphere, as has been shown by some studies on aerosols (M. Aubert et al. (6)). More recently Fénical & Sims (1974), working with Chondria oppositiclada, isolated an antibiotic substance which proved to be a cyclopropane containing a sesquiterpene.

2. Bromic compounds

These were found in Rhodophyceae by Mautner et al. (1953) (47), later by Saito and Shameshina (56) and by Roos (1957) (54).

These authors all carried out research on the thallus of Rhodomela larix. They found a wide range of antibacterial activity (towards Staphylococcus aureus, Bacillus megathorium, Escherichia coli, Pseudomonas aeruginosa, Mycobacterium tuberculosis, Mycobacterium avium, Mycobacterium phlei, Aerobacter aerogenes, Klebsella pneuminiae, Salmonella gallinarium, Saccharomyces pestorianus, Candida albicans and Trichochyton rubrum).

This chemically-oriented research showed the presence of bromic and phenolic compounds.

Later work by Teste et al. carried out at C.E.R.B.O.M. on the Rhodophycea Rytiphlaea tinctoria emphasized the antibacterial action of this alga on Staphylococcus aureus 209 P and was especially directed at establishing the chemical nature of the active substance. Isolation of the latter was achieved from the water extracted from the alga (19).

The data collected seem to indicate that the compound obtained from the pigment contains bromic compounds.

More recently, bromic and brominated compounds have been found by Teste et al., in Falkenbergia rufolanosa and Asparagopsis armata. Their antibacterial properties were demonstrated against Staphylococcus aureus and Escherichia coli (20).

3. Sulphur compounds

Working with Ceramium rubrum, Ikawa et al. (35) isolated a sulphuric compound with a high degree of antibiotic activity against Bacillus subtilis.

II - Substances produced by planktic algae

Generally speaking, antibacterial activity in connection with phytoplankton was first discovered "in situ" in comparative studies of sea-floor and planktic bacterial populations.

a) Sieburth's observations (1959):
- on the lack of various types of germs in the intestines of Antarctic penguins, which enabled him to show the antibiotic effect of a planktic microalga, Phaeocystis poucheti (58), found in the shrimps eaten by the penguins;
- on the correlation of seasonal variations in the anticoliform activity of the water of the Bay of Narangansett with the efflorescence of the Diatom Skeletonema costatum (Sieburth and Pratt 1962) (57).

b) Aubert's studies (1965) on the antibiotic properties of sea water and the presence of certain phytoplanktic forms; his studies have shown an important correlation between the presence of Rhizosolenia alata and the antibacterial properties of the sea water in which it is found (6).

Since then a number of studies have led to the discovery of antibacterial properties in water extracted from phytoplanktic organisms of several families (Diatoms, Dinoflagellata, Cyanophyta, Chrysophyta); some micro-organisms may contain two different substances, as in the case of Asterionella japonica or Chaetoceros lauderi (7).

Table VII - Planktic algae with antibacterial action on soil bacteria (bibliographical data)

Planktic species	Nature of isolated compounds	Bibliographical references
Diatoms:		
Fragillaria prinata	peptides	Berland et al. 1972 (10); Sieburth et al. 1964 (57);
Skeletonema costatum	fatty acids	Aubert et al. 1968 (5); Ackman 1968 (1)
Thalassiosira decipiens	fatty acids	Aubert et al. 1968 (5)
Thalassiosira nana	"	Steemann Nielsen 1955 (61)
Bacteriastrum elegans	"	Aubert et al. 1968 (5)
Chaetoceros socialis	"	"
Chaetoceros peruvianus	"	"
Chaetoceros lauderi+	polysaccharides	Aubert et al. 1968 (5)
	fatty acids	Pesando and Gheho 1977
Lithodesmium undulatum	not identified	Aubert et al. 1968 (5)
Licmophora abbreviata	not identified	Aubert et al. 1968 (5)
Gyrosigma spenceri	"	"
Thalassiothrix frauenfeldi	"	"
Asterionella japonica	nucleosides fatty acids	Aubert et al. 1977 (6)
Nitzschia seriata	not identified	Aubert et al. 1968 (5)
Nitzschia longissima	"	"
Nitzschia ascicularis	"	"
Rhizosolenia alata	"	"
Bacillaria paradoxa	"	"

+ This diatom also has antifungal properties.

Table VIII - Planktic algae with antibacterial action on soil bacteria (cont.)

Panktic Species	Nature of isolated compounds	Bibliographical references
Diatoms (cont.):		
Asterionella notata+	not identified	Gauthier 1969 (30)
Navicula incerta	"	Aubert et al. 1970 (unpubl.)
Cyclotella nana	"	Duff et al. 1966 (25)
Phaeodactylum	"	"
Dinoflagellata:		
Zooxanthellae	terpene and carbohydrates	Burkholder et al. 1958 (16)
Gonyaulax tamarensis	"	"
Prorocentrum micans	"	"
Cyanophyta:		
Hydrocoleus	terpene and	Burkholder et al. 1960 (15)
Lyngbia	carbohydrates	"
		Aubert et al. 1968 (5)
Trichodesmium erythraeum	terpene and carbohydrates	Ramamurphy 1970 (53)
Chrysophyta:		
Prymnesium parvum	terpene and carbohydrates	Aubert et al. 1968 (5)
Coccolithus	terpene and	Duff et al. 1966 (25)
Isochrysis	carbohydrates	"
Monochrysis	"	"
Phaeocystis	acrylic acid	Sieburth 1960 (57)
Dunaliella	not identified	Duff et al. 1966 (25)
Stichochrysis immobilis	"	Berland et al. 1974 (10)
Tetraselmis	not identified	Duff et al. 1966 (25)

+ This diatom also has antifungal properties.

Tables VII and VIII show all the studies carried out on the antibiotic activity of phytoplanktic organisms; it has only been possible to isolate a very small number of substances.

Before considering the chemical nature of the active substances, we would like to emphasize the pharmaceutical potential of some cellular extracts we have tested.

They often have a very wide antibacterial range, and are likely to act on germs that are resistant to classic antibiotics. We have observed:

- bactericidal effects of cellular extract of *Asterionella japonica* on 8 stocks of resistant pathogenic staphylococci from the Roger Bellon laboratory (stocks: CEB, GOO, BED, SIN, GIR, PES, DIG, HET);
- antibacterial activity of 11 cellular extracts of phytoplankton on 21 anaerobic germs from the Institut Pasteur in Paris that were resistant to usual antibiotics (4).

1. Carbohydrate compounds

The studies carried out at C.E.R.B.O.M. on antibiotic fractions of some cellular extracts have enabled some of them to be isolated and identified. One of the active fractions of *Chaetoceros lauderi* is an acid polysaccharide with a high molecular weight, which might include a uronic group (Pesando, personal communication).

2. Protein compounds

Berland et al. (10) have isolated an antibiotic substance which proved to be a peptide from a Diatom (*Fragillaria prinata*) and from a Chrysophycea (*Stichochysis immobilis*).

Elsewhere, working with a cellular extract from *Asterionella japonica*, Aubert et al. (7) have shown that one of the active fractions of the extract which has a low molecular weight (400-600) and contains ribose could be a nucleoside.

3. Fatty acids

a) Another active fraction has been isolated from the cellular extract of Asterionella japonica. This was found to have remarkable bactericidal properties after exposure to light, and was especially active on Sarcina lutea, also on Staphylococcus aureus 209 P, Clostridium tetami, Clostridium botulinum, Clostridium speticum, Clostridium histolyticum (7).

This fraction has been the object of more detailed studies at C.E.R.B.O.M. by D. Pesando on the isolation, purification and structure of the lipid substance. It could be identified as a fatty acid whose structure was studied by ultra-violet and infrared spectrophotometry, R.N.N. and mass spectrography. The results of these observations indicate that it would appear to be a C_{20} fatty acid with 5 malonic double bonds in positions 5, 8, 11, 14 and 17 (5) (51).

b) Similarly, another active fraction of Chaetoceros lauderi was found to be a fatty acid with the same bactericidal properties after exposure to light (Gauthier, Bernard, Aubert).

c) Ackman et al. also found the active fraction of Skeletonema costatum to be a fatty acid (1).

d) Among the other fatty acids that have been identified, acrylic acid was isolated by Sieburth from the Chrysophycea Phaeocystis poucheti (58). In the section dealing with substances produced by seaweeds we mentioned the conditions under which acrylic acid is formed and those under which it possesses antibiotic qualities (acid environment, formation by enzymatic process).

e) Finally, it should be observed that Burkholder et al. noted the presence of terpenes in Zooxanthellae (16).

III - Antibiotic substances produced by marine micro-organisms

Although they are not algae, these micro-organisms may conveniently be studied with them, as they are traditionally classed among the vegetal species and belong to the marine microflora.

As early as 1947 the description by Rosenfeld and Zobell (55) of stocks of marine bacteria capable of producing antibiotics showed the importance of the possibility of isolating from their marine habitat certain micro-organisms capable of synthesizing substances that were active against terrestrial microflora and against germs pathogenic to Man. This was primarily an ecological discovery, and the use of such bacteria was only considered within the frame of endogenic microbic antagonism in the marine world; so far, they have not been the object of any direct medical application.

However, a study of the work carried out in this area since then shows that research has followed two quite different directions:

- on the one hand, many authors have attempted a global evaluation of the proportion of these germs in natural marine populations. A few examples may be given here, which give a quantitative statement of the facts: 9 of the 56 stocks studied by Rosenfeld and Zobell were active against bacteria of the genera _Proteus_, _Salmonella_ and _Shigella_; more recently, Krassilnikova (1961) (44) described the antibacterial properties of 27 stocks on 326 stocks of bacteria isolated at different depths. Finally, Baam et al. have shown that of 60 stocks isolated from samples taken from different marine zones, 51 were antibiotically active on _Salmonella typhosa_ and _Staphylococcus aureus_;
- on the other hand, some research has been carried out in an attempt to define more clearly the antibacterial range of the stocks that have been isolated and the chemical nature of the inhibitory substances in operation. This work has been primarily concerned with marine bacteria and actinomycetes.

1. Antibiotic properties of marine bacteria

Marine bacteria produce various substances with a varying degree of inhibitory action on three types of terrestrial and amrine organisms: on other bacteria, on lower fungi (or yeasts) and on viruses.

a) Antibacterially active substances:
These may be globally divided into two groups according to their molecular weight (Table IX):
- substances with a low molecular weight, which are closer to antiseptics than to "sensu strictu" antibiotics. They are essentially bromopyrrole compounds which accumulate in the environment of the germs by which they are produced (Corpe 1962) (21) and are generally autotoxic. They have the following chemical structure:

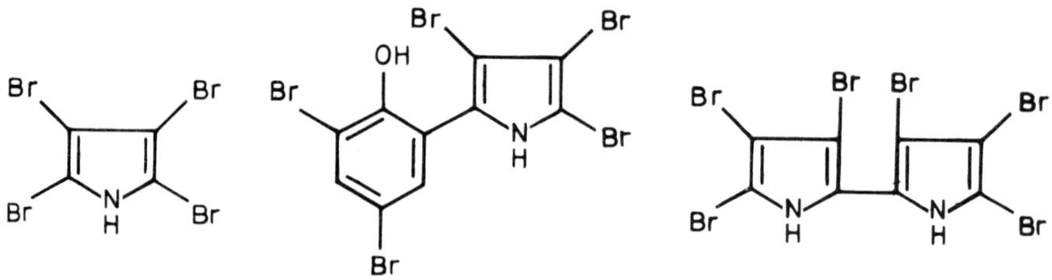

These products are not very specific and act in high doses on a large number of terrestrial and marine grampositive and gramnegative germs. They are, however, of considerable ecological importance and are the only known examples of bromic compounds synthesized by marine-organisms.
- macromolecular inhibitors, including certain polyanionic antibiotics and a large number of exocellulary lytic enzymes.

Polyanionic antibiotics, which could be regrouped under the special heading "alteromycines", are produced by a group of bacteria that are widely distributed in the marine environment and belong to the genus Alteromonas. They are negatively charged molecules with a fairly high molecular weight (5000 to more than 200000) containing a high proportion of acid polysaccharides as well as other compounds whose nature is not yet known. They essentially inhibit grampositive bacteria, but they are more active on germs with a low degree of catalytic or peroxide activity (Gauthier 1976) (28). On contact with them, affected bacteria release hydrogen peroxide (H_2O_2), whose lethal action may be controlled by respiratory enzymes.

The exocellulary lytic enzymes are produced by very numerous sea-water bacteria. The nature of their activity is not specified; they affect most terrestrial germs, which are more sensitive to them than marine bacteria (Jannasch 1968) (37). These enzymes are most often produced by bacterial symbionts on algae, especially by Myxobacteria (*Cytophaga*).

A final mention should be made of the inhibitory activity of various marine bacterial stocks which produce antibiotic substances whose chemical nature has not been ascertaines; these are sometimes highly specified in their action on other micro-organisms. For example, Sieburth (1967) (59) demonstrated the agglutination and inhibition of arthrobacters by certain substances produced by species of the genus *Pseudomonas*. This antagonism was observed in the natural habitat and confirmed "in vitro" using pure stocks.

b) Substances with antifungal properties:
The few studies available in this area were carried out by Buck et al. and Mitchell et al. (21, 22). They show that the effects of marine bacteria on yeasts and lower fungi (terrestrial and marine) are due to lytic enzymes which are relatively specific to parietal parts of these micro-organisms.

c) Substances with antiviral properties:
Magnusson, Gundersen et al. (1967) (45) demonstrated deactivation of the polio virus by the secretions of a germ similar to *Vibrio marinus*. The authors found that these bacteria restore the antiviral properties of sea

Table IX - Bibliographical data relating to the antibiotic properties of marine bacteria

Productive	Chemical nature of the antibiotic products	Inhibited species	Authors
Pseudomonas bromoutilis	2-(2' hydroxy-3', 5'-dibromophenol)-3, 4, 5-tribromopyrrole	gram +ve and-ve terrestrial bacteria	Burkholder et al. 1966 (14)
	tetrabromopyrrole hexabromo-2, 2'-bipyrrole	gram +ve and -ve terrestrial and marine bacteria	
	n-propyl 4-hydroxybenzoate	Photobacterium phosphoreum	
"Chromobacterium sp."	4-hydroxybenzaldehyde	gram +ve and -ve terrestrial and marine bacteria	Andersen et al. 1974 (3)
Alteromonas luteoviolaceus	bromopyrroles	gram +ve and -ve terrestrial and marine bacteria	Gauthier & Flatau 1976 (28)
Alteromonas rubra Alteromonas citrea Alteromonas luteoviolaceus Alteromonas sp.	polyanionic complexes (polysaccharides and	gram +ve and -ve terrestrial bacteria	Gauthier 1976 (27) Ballester 1976 (9)
Chaetomium globosum	n-acetyl hexosaminidase	gram +ve and -ve bacteria (Staphylococcus, Micrococcus, Bacillus)	Imada et al. 1973 (37)
epiphyte marine bacteria from algae (not specified)	proteases, polysaccharidases, esterases, etc.	various marine and terrestrial bacteria	Corpe & Winters 1972 (22)

Pseudomonas sp.	exocellulary lytic enzymes	*Escherichia coli*, Azetobacter	Mitchell & Nevo 1964 (48)
Vibrio sp.	not known	Pseudomonas aeruginosa	Doggett 1968 (24)
Pseudomonas	not known	*Arthrobacter*	Sieburth 1967 (59)
Pseudomonas piscidia	not known	*Rhodotorula minuta*	Buck & Meyers 1966 (13)
Pseudomonas sp.	not known	various marine and terrestrial yeasts	Buck et al. 1963 (11)
agarolytic marine bacteria	exocellulary lytic enzymes	various terrestrial Oomycetes *Pythium debaryanum*	Mitchell & Wirsen 1968 (49)

water that has previously been deactivated by heating, but that the bacteria only develop this property after incubation at a low temperature (4° - 12° C).

2. Antibiotic properties of marine actinomycetes

There are several studies concerning the synthesis of antibiotics by actinomycetes on land or in freshwater, but surprisingly little research has been acrried out on similar properties of these micro-organisms in the sea. In 1958 Grein and Meyers (30) did, however, indicate the antibacterial activity of 70 stocks of actinomycetes (Nocardia, Streptomyces, Micromonospora) from 200 stocks taken from marine sediments. These germs had inhibitory action on both grampositive and gramnegative bacteria.

These results, although condensed, indicate that it would be interesting to carry out a systematic study of such organisms in this field.

Conclusion

Marine flora and micro-organisms have been shown to be capable of synthesizing or giving off numerous biologically active substances.

Among these we have mentioned only those with bactericidal properties, especially those which are effective against saprophyte or pathogenic seafloor germs which are sometimes resistant to classic antibiotics used in therapy.

Here it is interesting to note that certain water cellular extracts from Diatoms (Asterionella notata, Asterionella japonica) have been used in therapeutic tests in dermatology for localised treatment of infectious skin diseases and of metabolic lesions. The first results seem promising (Aubert & Emery).

The substances isolated from Diatoms (A. japonica and Chaetoceros lauderi) would also appear to be entirely non-toxic (in the pharmaceutical sense), as was shown by a series of experiments carried out in our laboratory in collaboration with Professor Cier (Lyon, France).

In fact, research on the direct toxicity of the raw extract showed no mortality up to a dose close to 1 g/kg.

Research on sublethal toxicity of the total extract from Asterionella japonica demonstrated that daily doses of 100 mg/kg administrated intra-peritoneally to mice for three consecutive weeks produced no toxic effects. Furthermore, histological examination showed no effect on liver or kidneys.

However, it should be emphasized that although many studies have so far proved successful, they have to the best of our knowledge not been the object of chemical synthesis for a rational utilisation in human therapy.

References

1. Ackman, R. G., Cocher, C. S., McLachlan, J.: Marine Phytoplancton fatty acids. J. Fish. Biol. Canada, 25 (8) 1603-1620 (1968).
2. Allen, M. B., Dawson, E. Y.: Production of antibacterial substances by benthictropical algae. J. Bacteriol., 79, 459-460 (1960).
3. Andersen, R. J., Wolfe, M. S., Faulkner, D. J.: Autotoxic antibiotic-production by a marine Chromobacterium. Mar. Biol., 27, 281-285 (1974).
4. Aubert, J., Gambarotta, J. P.: Etude de l'Action antibactérienne d'espèces phytoplanctoniques marins vis-à-vis de germes anaérobies. Rev. Intern. Oceanogr. Méd., 25, 39-47 (1972).
5. Aubert, M., Aubert, J., Gauthier, M.: Pouvoir auto-épurateur de l'eau de mer et substances antibiotiques produites par les organismes marin. Rev. Intern. Océanogr. Méd., 10, 137-208 (1968).
6. Aubert, J., Bely, O., Breitmayer, J. P.: Transfert de micro-organismes et de substances d'origine biologique par les aérosols marins. Presse. therm et clim., 114e année (1) 51-53 (1977).

7. Aubert, M., Aubert, J., Gauthier, M., Pesando, D., Daniel, S.: Origine et Nature des substances antibiotiques dans le milieu marin. VIe Partie - Étude biochimique des substances antibactériennes extraites d'Asterionella japonica (Cleve). Rev. Intern. Océanogr.

8. Baam, B. R., Gandhi, N. M., Freitas, Y. M.: Antibiotic activity of marine-micro-organism. Helgol. wiss. Meeresunters., 13 (1-2), 181-185 (1966).

9. Ballester, M. Y.: Purification partielle d'un antibiotique produit par une bactérie marine (Alteromonas NCMB 1890), Thèse spécialité, Université Aix-Marseille, 94 p. (1976).

10. Berland, B., Bonin, D., Maestrini, S.: Etude expérimentale de l'influence de facteurs nutritionnels sur la production du phytoplancton de Méditerranée. Thèse Sc. Naturelles. Université Aix-Marseille, 239 p. (1974).

11. Buck, J. D., Ahearn, D. G., Roth, F. J., Meyers, S. P.: Inhibition of yeasts by a marine bacterium. J. of Bact., 85 (5), 1132-1135 (1963).

12. Buck, J. D., Meyers, S. P.: Antiyeast activity in marine environment. I. Ecological consideration. Limnol. and Oceanogr., 10 (3), 385-391 (1965).

13. Buck, J. D., Meyers, S. P.: In vitro inhibition of Rhodotorula minuta by a variant of the marine bacterium Pseudomonas piscida. Helgo. wiss. Meeresunters., 13 (1-2), 171-180 (1966).

14. Burkholder, P. R., Pfister, R. M., Leitz, F. H.: Production of a pyrrole antibiotic by a marine bacterium. Appl. Microb., 14 (4), 649-653 (1966).

15. Burkholder, P. R., Burkholder, L. M., Almodovar, L. R.: Antibiotic Activity of Some Marine Algae of Puerto-Rico. Botanica Marina, 2, 149-156 (1960).

16. Burkholder, P. R., Burkholder, L. M., Almodovar, L. R.: Antimicrobial Activity of Horny Corals. Science, 127, 1174-1175 (1958).

17. Challenger, F., Bywood, P. Thomas, P., Hayward, B.: Studies on biological methylation. Arc. Biochem. Biophy., 69, 514-523 (1957).

18. Chesters, G. C. C., Stott, J. A.: Production of antibiotic substances by seaweeds. Pro. 2nd Intern. Seaweed Symp. Trondhjem, 1955, 49-53 (1956).

19. Chevolot-Magueur, A. M., Cave, A., Potier, P., Teste, J., Chiaronia, Riche, Cl.,: Composés bromés de Rytiphlea tinctoria (Rhodophycea). Phytochemistry, Vol. 15, 767-771 (1976.

20. Codomier, L., Bruneau, Y., Combaut, G., Teste, J.: Etude biologique et chimique d'Asparagopsis armata et de Falkenbergia rufolanosa (Rhodophycées bonnemaisoniales). C. R. Acad. Sc. Paris, 284, Série D, 1163-1165 (1977).

21. Conover, J. T., Sieburth, J. Mc N.: Effect of Sargassum distribution on itdepibiota and antibacterial activity. Bot. Mar., 6, 147-157 (1964).

22. Corpe, W. A., Winters, H.: Hydrolytic enzyms of some periphytic marine bacteria. Can. J. Microbiol. 18, 1483-1490 (1972).

23. Corpe, W. A.: Extracellular accumulation of pyrroles in bacterial cultures. Appl. Microbiol., 11, (2) 145-150 (1963).

24. Doggett, R. G.: New anti Pseudomonas agent isolated from a marine Vibrio J. of Bact., 95, (5), 1972-1973 (1968).

25. Duff, D. C. B., Bruce, D. L., Antia, N. D.: The antibacterial activity of marine planktonic aglae. Canad. Journ. Bact., 12, 877-884 (1966).

26. Gauthier, M. J.: Antagonisme et synergie dans l'antibiose chez certaines bactéries marines. Rôle des enzymes respiratoires dans ces phénomènes. Rev. Intern. Océanogr. Méd., 26, 65-83 (1972).

27. Gauthier, M. J.: Modification of bacterial respiration by a macromolecular polyanionic antibiotic produced by amarine. Alteromonas. Antimicrob. Agents Chemother., 9, (3), 361-366 (1976).

28. Gauthier, M. J., Flatau, G.: Antibiotic activity of marine violet pigmented Alteromonas with special reference to the production of bromisated compounds. Canad. J. Microbiol., 22, (6), 1-9 (1976).

29. Gauthier, M. J.: Propriétés antibactériennes des micro-organismes marins. Thèse Sciences, (Nice, 116 p.) (1970).

30. Gauthier, M. J.: Activité antibactérienne d'une Diatomée marine: Asterionella notata (Grein). Rev. Intern. Océanogr. Méd., 15-16, 71-103 (1969).

31. Grein, A., Meyers, S. P.: Growth characteristic and antibiotic production of actinomycetes isolated from littoral sediments and materials suspended in seawater. J. Bacteriol., 76, (5), 457-463 (1958).

32. Guncersen, K., Brandberg, A., Magnusson, S., Lycke, E.: Characterisation of a marine bacterium associated with virus inactivus capacity. Acta path. microbiol. scandinav., 71, 281-286 (1968).

33. Harder, R., Oppermann, A.: Über antibiotische Stoffe bei den Grünalgen Stichoccus bacillaris und Protosiphon bomyoides. Arch. Microbiol., 19, 398-401 (1953).

34. Hornsey, I. S., Hide, D.: The production of antimicrobial compounds by British marine Algae. I. Antibiotic producing marine Algae. Br. phycol., 9, 353-361 (1974).

35. Ikawa, M., Thomas, I. M., Buckley, J., Uesbel, J.: Sulfur and the toxicity of the red algae Ceramium rubrum et Bacillus subtilis. J. of Phycology, 9, (3), 302-304 (1973).

36. Irie, T. M., Izawa, M., Kurosawa, E.: Tetrahedron. Lett. 2091 (1968).

37. Imada, A., Nakahama, K., Igarasi, S., Isono, M.: A bacteriolytic enzyme from Chaetonium globossum, a marine - isolate. Arch. Microbiol., 91, 41-54 (1973).

38. Jannasch, H. W.: Competetive elimination of Enterobacteriaceae from seawater. Appl. Microbiol., 16, (10), 1616-1618 (1968).

39. Kamimoto, K.: Studies on the antibacterial substances extracted from seaweeds on the growth of some pathogenic bacteria. Japon. J. Bacteriol., 10, 897-902 (1955).

40. Katayama, J.: Volatile constituents. In: Physiology and Biochemistry of Algae, pp. 467-473. Ed. by R. A. Lewin. Acad. Press Inc. New York (1962).

41. Katayama, J.: Chemical studies on volatile constituents of seaweeds. Antibacterial action of Enteromorpha sp. Bull. Japan Soc. Sci. Fisheries, 22, 248-250.

42. Khaleafa, A. F., Kharboushman, M. A. M., Metwalli, A., Moshen, A. F., Serwi, A.: Antibiotic (Fungicidal) Action from extracts of some seaweeds. Bot. Mar., Vol. XVIII, 163-165 (1975).

43. Klein, S.: New antituberculosis substance. Bot. Mar., Vol. VI, (314), 248-251 (1964).

44. Krassilnikova, E. N.: Propriétés antibiotiques de micro-organismes marins isolés à différentes profondeurs. Mikrobiologia, 30, 653-791 (1961).

45. Lewin, R. A.: Antibiotics from algae. In: Physiology and Biochemistry of Algae, 841-842 (1962).

46. Magnusson, S., Gundersen, K., Brandberg, A., Lycke, E.: Marine bacteria and their possible relation to the virus inactivation capacity of seawater.Acta path. microbiol. scnadinav., 71, 274-280 (1967).

47. Mautner, H. G., Gardner, G. M., Pratt, R.: Antibiotic Activity of Seaweeds Extracts. J. Am. Pharm. Assoc. Sci., Ed., 43, 294-296 (1953).

48. Mitchell, R., Nevo, Z.: Decomposition of structural polysaccharides of bacteria by marine micro-organisms. Nature 205, (4975) 1007-1008 (1964).

49. Mitchell, R., Wirsen, Z.: Lysis of non marine fungi by marine micro-organisms. J. of Gen. Microbiol., G.B, 52, (3), 335-351 (1968).

50. Olesen, P. E., Maretzki, A., Almodovar, A. L.: An investigation of antimicrobial substances from marine algae. Bot. Mar., Vol. VI F. 3/4, 224-231 (1964).

51. Pesando, D.: Etude chimique et structurale d'une substance lipidique antibiotique produite par une Diatomée marine, Asterionella japonica. Rev. Intern. Océanogr., Méd., 25, 49-69 (1972).

52. Pratt, R., Mautner, H., Gardner, G. M., Sha Y., Dufrenoy, J.: Report on antibiotic activity of seaweeds extracts. J. Amer. Pharm. Assoc., 40, 575-579 (1951).

53. Ramamurphy, V. D.: Antibacterial activity of the marine blue-green algae Trichodesmium erythream in the gastro-intestinal contents of the sea gull Laurus brumicephalus. Marine Biol., 6, 74-76 (1970).

54. Roos, H.: Untersuchungen über das Vorkommen antimikrobieller Substanzen in Meeresalgen. Kiel. Meeresforsch., 13, 41-58 (1957).

55. Rosenfeld, D. W., Zobell, C. E.: Antibiotic production by marine micro-organisms. J. of Bact., 54, 393-398 (1947).

56. Saito and Sameshina, J.: Studies on antibiotic action of algae extracts. J. Agr. Chem. Soc. Japan., 29, 427-430 (1955).

57. Sieburth, J. Mc. N.: Antibacterial substances produced by marine algae. Development in Industrial Microbiology, Vol. 5, 123-134 (1964).

58. Sieburth, J. Mc. N.: Acrylic Acid an "Antibiotic" principale in Phaeocystis Blooms in Antarctic waters. Sciences, 132, 676-677 (1960).

59. Sieburth, J. Mc. N.: Inhibition and agglutination of Arthrobacter by Pseudomonas. J. Bacteriol., 93, (6), 1911-1916 (1967).

60. Starr, I. J., Dieg, E. F., Church, K. K., Allen, M. B.: Antibacterial and antiviral activities of algal extracts studied by acridine orange staining. Texas Report Biol. Med., 20, 271-278 (1962).

61. Steeman Nielsen, E.: The production of antibiotics by plankton algae and its effect upon bacterial activities in the sea. Deep-sea research, 3 (Suppl.) 281-286 (1955).

62. Vacca, D. D., Walsh, R. A.: The antibacterial extract obtained from an extract of Ascophyllum nodosum. J. Am. Pharm. Assoc. (Sci. ev) 43, 24-26 (1954).

ANTI-VIRAL PROPERTIES OF ALGAL POLYSACCHARIDES AND RELATED COMPOUNDS

D. W. Ehresmann, E. F. Deig, and M. T. Hatch
Naval Biosciences Laboratory, School of Public Health,
University of California, Berkeley, California 94720 USA

INTRODUCTION

Research directed toward the discovery and development of drugs which inhibit the replication of disease-causing mammalian viruses has only recently yielded agents with clinical promise. The success of vaccines in controlling many important viral diseases (e.g., poliomyelitis and small pox), coupled with the toxicity of many anti-viral compounds, lead to the concept that development of effective anti-viral drugs was both fruitless and unnecessary. However, the more recent recognition of a cellular anti-viral system (i.e., interferon) and the discovery of differences between the molecular biology of viral replication and normal cellular metabolism suggested that viral pathogenesis could be inhibited without concomitant disruption of normal cell function. Additionally, the realization that some viral diseases may be uncontrollable by immunization (e.g., diseases caused by the herpesviruses and rhinoviruses) has renewed interest in anti-viral chemotherapy.

Most drugs currently showing therapeutic promise against viral diseases, with the exception of interferon, are synthetic compounds. Therefore, one approach to the discovery of new anti-viral agents is to screen nonbiological compounds and to chemically modify known synthetic inhibitors. However, the existence of biological substances exhibiting anti-viral properties seems probable considering the diversity of compounds possessed by living organisms. Additionally, since viruses similar to

Marine Algae in Pharmaceutical Science
edited by H. A. Hoppe, T. Levring, Y. Tanaka

mammalian disease-producing agents have been found to infect species of many eukaryotic phyla, it is possible that lower organisms contain compounds functioning in defense against viral infection.

This review will focus on the current state of knowledge of anti-viral agents from marine algae. Although literature in this area is not extensive, recent investigations have indicated that these agents may be of therapeutic value in treating some common mammalian viral diseases. Inasmuch as the active component(s) are generally polysaccharides possessing inhibitory action similar to anionic polyelectrolytes (or polyanions) from non-algal sources, some relevant findings of polyanions (i.e., polyglucose sulfate, heparin and dextran sulfate) are also included to broaden the discussion wherever appropriate. The reader is referred to other more extensive treatment of the anti-viral activity of polyionic substances, which serve as background of previous work in this area. (Vaheri, 1964).

ANTI-VIRAL SUBSTANCES OF ALGAL ORIGIN AND RELATED COMPOUNDS

That marine algae possess anti-viral substances was first suggested by Gerber, *et al*. (1958) who observed that polysaccharides extracted from *Gelidium cartilagenium* and carrageenan (from *Chondrus crispus*) afforded protection for embryonated eggs against influenza B and mumps virus. Since *G. cartilagenium* is a common source of commercial agar, Gerber, *et al*. (1958) suggested that the anti-viral polysaccharide was probably the cause of the inhibitory action of plaque-formation observed with a variety of viruses, when an agar overlay was used in bioassay.

In 1965, Kathan discovered an inhibitor of bacterial and influenza neuraminidase in kelp meal (algal source not identified).

This substance inhibited the replication of influenza A (Lee and PR-8 strains) in embryonated eggs. Chemical characterization of partially purified inhibitor revealed the presence of both polysaccharide and protein.

The chemical nature and spectrum of inhibition of the anti-viral polysaccharide in agar was explored by a series of investigations (Takemori and Nomura, 1960; Takemoto and Liebhaber, 1961; Takemoto and Fabisch, 1964). Takemori and Nomura (1960) reported the inhibition of poliovirus by an aqueous extract of agar and Takemoto and Liebhaber (1961) identified the agar inhibitor of encephalomyocarditis virus (EMC) as a sulfated polysaccharide. In the later study, two other sulfated polyanions, polyglucose-SO_4 and heparin, were reported to inhibit EMC plaque formation. The anti-viral spectrum of the agar polysaccharide was also extended to herpes simplex virus by Takemoto and Fabisch (1964). Commercial agar contains both sulfated and nonsulfated polysaccharides (Guiseley, 1968) and the nonsulfated fraction (agarose) lacks anti-viral properties (Bergold and Mazzoli, 1968). Since these studies on the anti-viral polysaccharide in agar, a wide variety of biological and synthetic sulfated polyanions have been shown to inhibit the replication of many mammalian viruses (Table 1.).

With this background, an investigation of California marine algae for anti-viral compounds was initiated at the Naval Biosciences Laboratory (University of California). In 1974, Deig, *et al*. (1974) reported the inhibition of herpes simplex viruses types 1 and 2 in cell culture by aqueous extracts of *Cryptosiphonia woodii* and *Farlowia mollis*, both members of family Dumontiaceae (Rhodophyta). The inhibitor was subsequently detected in six additional members of Dumontiaceae, namely *Farlowia crassa*, *F. compressa*, *Constantinea simplex*, *Pikea pinnata*, and *Leptocladia conferta* (Ehresmann, *et al*., 1977). Recent studies indicate the inhibitor is a polysaccharide (Ehresmann, *et al*., 1975; Ehresmann, *et al*., 1977)

Table 1. Sulfated Polyanionic Inhibitors of Viral Replication

Polyanion	Virus Inhibited	Reference
Agar Polysaccharide	Poliovirus	Takemori and Nomura, 1960
	EMC	Takemoto and Liebhaber, 1961
	Herpes simplex	Takemoto and Fabisch, 1964
	Dengue	Schulze, 1964
Sulfated Polysaccharide from species of Dumontiaceae (Rhodophyta)	Herpes simplex, types 1 and 2, Coxsackie B5	Deig *et al*.,1974 Ehresmann *et al*. 1977.
Dextran Sulfate	Herpes simplex	Nahmias *et al*., 1964
	EMC, ECHO, Coxsackie A9	Takemoto and Fabisch, 1964
Heparin	Herpes simplex	Nahmias and Kibrick, 1964; Vaheri and Cantell, 1963.
	Rous Sarcoma	Solomon *et al*. 1966.
Sulfated Amylopectin	Herpes simplex	Nahmias *et al*. 1964
Sulfated Amylose	Herpes simplex	Nahmias *et al*. 1964
Polyvinyl Sulfate	Herpes simplex	Nahmias *et al*. 1964
Polystyrene Sulfonate	Herpes simplex	Nahmias *et al*. 1964
Sulfated Glycosaminoglycan	Yellow Fever	Voss et al., 1974

with a high degree of sulfation (Hatch, et al., 1977). The degree of inhibition varied with algal source and was in the region of 2 to 4 Log_{10} reduction in virus titer.

MECHANISM OF ACTION

Nahmias et al. (1964) examined a variety of biological and nonbiological sulfated polyanions for inhibitory effect on herpes simplex virus. They found that the degree of inhibition varied with the degree of sulfation of the molecule with maximum effect obtained when one sulfate group per hexose unit is present in the polysaccharide. The inhibitory effect required a minimum molecular size since sulfopolyglucin, a highly sulfated molecule containing only 12 to 14 glucose residues per molecule, was a poor inhibitor. However, once a minimum size was reached, inhibition was not dependent upon molecular size. Additionally, the inhibitory effect was independent of molecular branching, type of glycosidic bond and type of sulfur bond (sulfonic or sulfate ester).

The finding that minute plaque-forming mutants of poliovirus (Takemori and Nomura, 1960) and EMC (Takemoto and Liebhaber, 1961) are inhibited by the sulfated agar polysaccharide whereas large plaque mutants are unaffected, suggests that a specific viral function is blocked by the inhibitor. This is supported by Germanov and Sokolov (1969) who were able to obtain twenty-eight spontaneous or hydroxylamine-induced mutants of herpes simplex virus sensitive to the agar polysaccharide.

Studies on the mode of action of inhibition have suggested that virus adsorption to susceptible cells is blocked by sulfated polyanions (Liebhaber and Takemoto, 1963; Schulze, 1964; Takemoto and Fabisch, 1964). The inhibition was prevented or reversed by the presence of the polycations DEAE-dextran and protamine which would bind to the polyanionic inhibitor. Similar characteristics have been established for the anti-herpes polysaccharide from species of Dumontiaceae (Hatch, et al., 1977). However, these investigations have not revealed if the inhibition is caused by the binding of the sulfated polyanion to the surface of the virion or cell. Virus adsorption to susceptible cell plasma membranes is an electrostatic phenomenon and requires a molecular affinity between the cell surface and virions (Fenner, et al., 1974). Areas of the cell surface showing this affinity are called viral receptors. It could be that the strongly negatively charged sulfated polyanions bind to positively charged molecules on the surface of the virion or cell and sterically inhibit adsorption. The positively charged molecule binding the polyanion might be near the cellular receptor or the receptor itself. Alternately the polyanion might bind near to or on the virion molecule which is complementary to the cell receptor. In this regard, it is of interest to note that Gibson and Roizman (1977) have found that the external envelope of herpes simplex virus contains the polyamine spermidine, a positively charged molecule which could bind sulfated polyanions.

ANIMAL AND HUMAN STUDIES

Gerber et al. (1958) reported that the intranasal administration of agar polysaccharides prevented death in 70% of mice receiving 100 LD_{50} units of murine pneumonia virus (PVM) provided the polysaccharide was given prior to virus inoculation. Lehel and Hadhazy (1966) found that heparin injected into rabbit skin prior to or at the same time as infection with a herpesvirus resulted in a reduction in the severity of virus-induced lesions. Museteanu et al. (1975) reported that sulfated glycosaminoglycans decreased the death rate in mice infected with yellow fever virus when the inhibitor was administered with virus. More recently, Deig, et al., (unpublished data) found that aqueous extracts of four members of family

Dumontiaceae protected infant mice against herpes simplex-induced encephalitis. The reduction in mortality afforded by the extracts was greater than 2 Log_{10}, when extracts were administered two hours prior to, or simultaneous with virus. Additional studies with the Dumontiaceae inhibitor in other herpes simplex disease models are currently being conducted in our laboratory and by investigators of the National Institute of Health's Anti-viral Substances Program.

The dependence of *in vivo* effectiveness on early administration of the inhibitor is consistent with the concept that sulfated polyanions block virus adsorption and suggests that the clinical value of these agents may be limited to a prophylactic regimen. However, with many disease-causing viruses, the spread of virions from infected cells to adjacent non-infected cells occurs primarily extracellularly. Even with the herpesviruses, a cell-associated group, significant extracellular spread occurs (Watson, 1973). The anti-viral polyanions therefore may have therapeutic potential in treating certain established viral disorders, particularly limited diseases such as herpes simplex dermatitis, by reducing the severity of infection. Alternatively, combination of relatively non-toxic sulfated polyanions with the toxic intracellular anti-viral drugs (e.g., 9-β-D-Arabinofuranosyl Adenine and 5-Iodo-2'-deoxyuridine) in a therapeutic preparation, may allow a reduction in the level of the toxic agent. These possibilities need exploration.

The only sulfated polyanion which has received clinical examination in humans is the sulfated glycosaminoglycan L5. Voos *et al*. (1977), reported that when L5 was evaluated against zoster infections (herpesvirus) by direct application to the lesions, healing of vesicles was facilitated and the course of the disease was shortened. This investigation encourages continued evaluation of sulfated glycosaminoglycan and other sulfated polyanions as potential human anti-viral therapeutics.

ACKNOWLEDGMENTS

The investigations reported from the author's laboratory were supported by the Office of Naval Research and by the Sea Grant College Program, Grant No. R/MP-12, University of California.

REFERENCES

1. Bergold, G.H., Mazzoli, R.: Plaque-formation by arboviruses. J. Gen. Virol. 2, 273-284 (1968).

2. Deig, E.F., Ehresmann, D.W., Hatch, M.T., Riedlinger, D.J.: Inhibition of herpesvirus replication by marine algae extracts. Antimicrob. Agents Chemother. 6, 524-525 (1974).

3. Ehresmann, D.W., Deig, E.F., Hatch, M.T., Riedlinger, D.L.: Characterization of the anti-herpesvirus activity in extracts of marine algae. Abstracts of the Annual Meeting of the American Society for Microbiology, A7 (1975).

4. Ehresmann, D.W., Deig, E.F., Hatch, M.T., DiSalvo, L.H., Vedros, N.A.: Antiviral substances from California marine algae. J. Phycol. 13, 37-40 (1977).

5. Fenner, F., McAuslan, B.R., Mims, C.A., Sambrook, J., White, D.O.: The Biology of Animal Viruses. Second edition. Academic Press, N.Y. p. 178 (1974).

6. Gerber, P., Dutcher, J.D., Adams, E.V., Sherman, J.H.: Protective effect of seaweed extracts for chicken embryos infected with influenza B or mumps virus. Proc. Soc. Exp. Biol. Med. 99, 590-593.(1958).

7. Germanov, A.B., Sokolov, M.I.: Selection and genetic studies on mutants of herpes simplex virus sensitive to sulfated agar polysaccharides. Arch. ges. Virus forsch. 28, 77-84 (1969).

8. Gibson, W., Roizman, B.: Compartmentalization of spermine and spermidine in the herpes simplex virion. Proc. Nat. Acad. Sci. USA, 68, 2818-2821 (1971).

9. Guiseley, K.B.: Seaweed colloids. In Encyclopedia of Chemical Technology. John Wiley and Sons, Inc. Vol. 17, pp. 763-784 (1968).

10. Hatch, M.T., Ehresmann, D.W., Deig, E.F., Vedros, N.A.: Further studies on the chemical composition and an initial in vivo evaluation of antiviral material in extracts of macroscopic algae. J. Phycol. Supplement to Vol. 13, 28 (1977).

11. Kathan, R.H.: Kelp extracts as antiviral substances. Ann. N.Y. Acad. Sci. 130, 390-397 (1965).

12. Lehel, F., Hadhazy, G.: Effect of heparin on herpes simplex virus infection in the rabbit. Acta microbiol. Hung. 13, 197-204 (1966).

13. Liebhaber, H., Takemoto, K.K.: The basis for the size differences in plaques produced by variants of encephalomyocarditis (EMC) virus. Virology 20, 559-566 (1963).

14. Museteanu, C., Voss, H., Rossner, R.: Sulfated glycosaminoglycans as virus inhibitors: Inhibitory effect of glycosaminoglycan polysulfates on yellow fever virus 17D in animal experiments. Zentralbl. Bacteriol. orig. A. 230, 298-305 (1975).

15. Nahmias, A.J., Kibrick, S.: Inhibitory effect of heparin on herpes simplex virus. J. Bacteriol. 87,1060-1066 (1964).

16. Nahmias, A.J., Kibrick, S., Bernfeld, P.: Effect of synthetic and biological polyanions on herpes simplex virus. Proc. Soc. Exp. Biol. Med. 115,993-996 (1964).

17. Schulze, I.T.: Reversible inhibition of type 2 dengue virus by agar polysaccharide. Virology 22, 79-90 (1964).

18. Solomon, J.J., Glatt, K.A., Okazaki, W.: Inhibitory effect of heparin on rous sarcoma virus. J. Bacteriol. 92, 1855-1856 (1966).

19. Takemori, N., Nomura, S.: Mutation of polioviruses with respect to size of plaque. II. Reverse mutation of minute plaque mutant. Virology 12, 171-184 (1960).

20. Takemoto, K.K., Liebhaber, H.: Virus-polysaccharide interactions. I. An agar polysaccharide determining plaque morphology of EMC virus. Virology 14, 456-462 (1961).

21. Takemoto, K.K., Fabisch, P.: Inhibition of herpes virus by natural and synthetic acid polysaccharides. Proc. Soc. Exp. Biol. Med. 116, 140-144 (1964).

22. Vaheri, A., Cantell, K.: The effect of heparin on herpes simplex virus. Virology 21, 661-662 (1963).

23. Vaheri, A.: Heparin and related polyionic substances as virus inhibitors. Acta Path. microbiol. scand. Suppl. 171, 1-97 (1964).

24. Voss, H., Sensch, K.H., Panse, P.: Sulfated glycosaminoglycans as virus inhibitors: The effect of sulfated glycosaminoglycans in different virus-host systems as demonstrated by the plaque inhibition test. Zentralbl. Bacteriol. orig. A. 229, 1-36 (1974).

25. Voss, H., Pohle, H.D., Museteanu, C.: Sulfated glycosaminoglycans as virus inhibitors: Clinical aspects of zoster with the proposal of a new treatment. Zentralbl. Bacteriol. orig. A 237, 1-34 (1977).

26. Watson, D.H.: Replication of the viruses - morphological aspects. In The Herpesviruses, A.S. Kaplan (ed.) Academic Press, N.Y. pp. 133-159 (1973).

ANTIBIOTICS FROM ALGAE

K.-W. Glombitza
Institut für Pharmazeutische Biologie, Universität,
D - 53oo Bonn, Germany

Since the discovery of autantagonistic light sensitive metabolites in the culture media of the Cyanophycea Nostoc punctiforme (9o) many species of algae have been investigated for antialgal, antibacterial, antifungal, and antiviral properties. It turned out that antimicrobial metabolites are present in all alga classes. Very large numbers of active species were found in the Ulvales (Chlorophyceae), in the Bonnemaisoniaceae and Rhodomelaceae (Rhodophyceae), and in the Fucales (Phaeophyceae). Investigations were carried out with culture media, homogenates of thallus, whole thallus pieces or extracts, which had been prepared with lipophilic or hydrophilic solvents or with the eluotropic series. Results were reviewed in detail by Wolters (239), Sieburth (214), Krzywicka (137) and Baslow (14). Recently a large variety of marine algae were screened again for antibiotic activities (7,62,63,68,95-97,133,134,189 and others). In the present review, however, the author will quote exclusively papers which contribute in a notable way to the structural elucidation of inhibiting compounds.

Until the mid sixties final purification and establishment of structures had been achieved in only a few cases. In the last five years an increasing number of reports on new substances has been published (47).

There were frequent discussions in the mid fifties as to whether or not chlorophylls, or poorly defined chlorophyllides were responsible for the inhibiting effects of culture liquids, extracts or sea water (6,16,17,21,33,106). In the meantime, it seems rather evident that this group of substances are antibiotically inactive. Upon illumination, however, they undergo rapid destruction, turning into inhibitors themselves, or catalyzing the formation of inhibiting compounds (162).

Marine Algae in Pharmaceutical Science
edited by H. A. Hoppe, T. Levring, Y. Tanaka

CHLORELLIN, UNSATURATED FATTY ACIDS AND THEIR DERIVATIVES

The first successful studies for identification of algal inhibitors were carried out with culture solutions of various unicellular Chlorophyceae, especially Chlorella. The antialgal and antibacterial activities increased with the age of the cultures, and they were greater in illuminated than in dark cultures (1o4,1o5,1o7,136,151,19o,192,195,196,198,23o,234,235). An active fraction isolated from such cultures (191) was called Chlorellin by Pratt et al. (194). It has an inhibiting effect on gram-positive and gram-negative bacteria. In spite of many efforts, up to now no definite structure proposal for one or more of the effective principles has been published. They are supposed to be a mixture of fatty acids, in which the more unsaturated compounds are converted by photooxidation into antibiotically active derivatives (222).

Similar observations were made with culture solutions and extracts of other algal species (Ochromonas danica (1), Nitzschia spec. (13), Stichococcus mirabilis, Protosiphon botryoides (91), Chlamydomonas reinhardii (197)). However, as there is no correlation between the surface activity of the culture solution and the antibiotic activity, Pratt (193) has rejected this theory of Spoehr et al. (222,223). According to several other reports (for example 2,3) and the exact investigations of Pesando (187) with extracts from the diatom Asterionella japonica, the photo-oxidation products of a eikosapentaenic acid (1) develop a strong antibiotic activity after short illumination. Therefore, the participation of derivatives of highly unsatu-

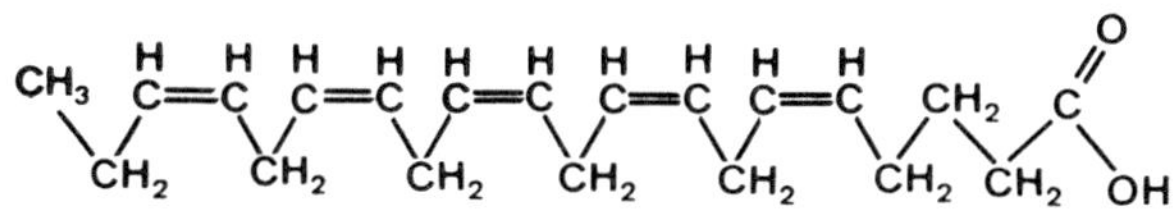

(1)

rated fatty acids in the antibiosis of algal cultures and extracts seems to be proved. More recently additional evidence was given by Olfers-Weber (181), that the inhibitors in question of the fresh water Chlorophycea Hydryodictyon reticulatum are also such derivatives of unsaturated fatty acids.

A derivative of such fatty acids formed in vivo might be present in Chondriol (2) from Chondria oppositiclada (53).

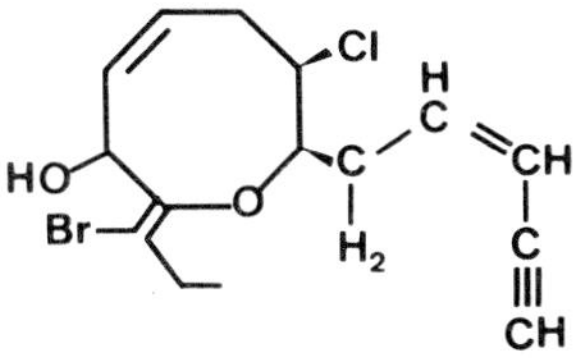

Chondriol (2)

It is a cyclic ether, whose antibiotic activity was found to be comparable with those of prepacifenol and debromlaurinterol (219).

ACRYLIC ACID

Acrylic acid is the first antibiotically active compound from algae which was unambiguously identified. It was detected by Sieburth (211,212) in Phaeocystis poucheti. It occurs partly free, or as salts, partly bound to dimethylsulfide as the antibiotically inactive thetine (4, dimethyl-ß-carboxyethyl sulfonium compound), which may be split enzymatically or, in vitro,

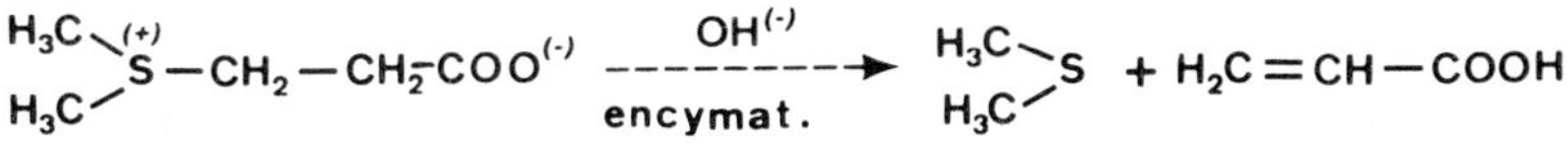

(4) (3)

with diluted alcaline reagents to dimethyl sulfide and 3 (4,5,25,27,3o,31, 1oo,1o1). Taylor et al. (233) detected acryloyl choline in Amphidinium carteri. Acrylic acid is excreted in low concentration into sea water by Phaeocystis (84). According to Obata et al. (175), the formation of dimethyl sulfide from 4 is perhaps partly due to microorganisms. 3 inhibits especially gram-positive, and to a lower degree also gram-negative organisms (212,213,224). 3, 4 or dimethyl sulfide where found in many alga species in the meantime (Tab. 1). Quite often, however, the active concentrations of free acrylic acid, or acrylic acid produced from 4 are too small to explain the antibiotic activities found with thalli or extracts.

Table 1, Occurence of acrylic acid in algae (citations of literature)

	Dimethylsulfide	Acrylic acid	Propiothetin
CYANOPHYCEAE			
Oscillatoria agardii	83		83
Phytoplankton	83		83
DINOPHYCEAE			
Amphidinium carteri [x]	lol/236/loo		lol/236/loo
Gyrodinium cohnii	lol/loo		lol/loo
Glenodinium spec.	loo		loo
HAPTOPHYCEAE			
Phaeocystis poucheti		211/212	
Syracophaera carterae [x]	236/5		5/236
EUGLENOPHYCEAE			
Aspasia longa		loo	
Euglena gracilis		loo	
BACILLARIOPHYCEAE			
Cyclotella cryptica	5		5
Cyclotella nana	loo		
Skeletonema costatum [x]	5		5
Phaeodactylum tricornutum	5		5
CHLOROPHYCEAE			
Haematococcus pluvialis	loo		loo
Nannochloris oculata	loo		
Chlamydomonas	loo		loo
Tetraselmis spec.	5		5
Ulothrix spec.	3o		
Enteromorpha intestinalis [x]	25/3o/64	64/68	25/3o/64
Enteromorpha prolifera	175		
Enteromorpha compressa	3o		
Enteromorpha spec.	125/127	125/126/127	
Ulva lactuca [x]	3o/64/236	64/68/lo9/236	64
Ulva linza [x]	64	64/68	64
Ulva pertusa	125/127/175	125/126/127	
Acrosiphonia centralis [x]	3o	68	64
Acrosiphonia arcta [x]	64	64	64
Acrosiphonia sonderi [x]	64	64	64
Oedogonium	3o		
Cladophora rupestris	3o/64	64/68	64
Codium difforme [x]		68	
Codium fragile	64/122/125	64/68/125/126/127	64
Codium tomentosum	64	64	64
Spongomorpha lanosa	64	64	64
Monostroma undulatum	64	64	64
Monostroma spec.	125		
RHODOPHYCEAE			
Nemalionales			
Asparagopsis armata		68	
Bangiales			
Porphyra tenera	174		174
Porphyra spec.	127	127	

Table 1 , continued

	Dimethylsulfide	Acrylic acid	Propiothetin
RHODOPHYCEAE			
Gigartinales			
Furcellaria fastigiata		68	
Calliblepharis jubata		68	
Sphaerococcus coronipofol		68	
Gracilaria foliifera		68	
Gracilaria verrucosa		68	
Chondrus crispus		68/225	
Gigartina acicularis		68	
Gigartina stellata	3o	68	
Cystoclonium purpureum		225	
Plocamium vulgare		225	
Cryptonemiales			
Dilsea carnosa		68	
Dumontia incrassata		68	
Rhodymeniales			
Lomentaria clavellosa		68	
Ceramiales			
Callithamnium tetricum		68	
Ceramium rubrum	3o	225	
Griffithsia flosculosa		68	
Digenea simplex	127	127	
Delesseria sanguinea		68/225	
Hypoglossum woodwardii		68	
Laurencia hybrida		68	
Laurencia obtusa		68	
Laurencia pinnatifida		68/225	
Polysiphonia brodiaei ˣ		68	
Polysiphonia elongata ˣ		68	
Polysiphonia nigra ˣ		68	
Polysiphonia fruticulosa		68	
Polysiphonia lanosa ˣ	25/27/86	25/27/68	25/27
Polysiphonia nigrescens ˣ	25	25/68	
Polysiphonia thuyoides ˣ		68	
Polysiphonia urceolata		225	
Rhodomela confervoides		68/225	
PHAEOPHYCEAE			
Chorda filum		(68)	
Saccorhiza polyschides		68	
Alaria esculenta		(68)	
Clodostephus verticillatus		(68)	
Dictyota dichotoma		(68)	
Fucus serratus		68	
Bifurcaria bifurcata		(68)	
Cystoseira tamariscifolia		(68)	
Halidrys siliquosa	3o	(68)	
Sargassum spec.	127	127	
Pelvetia canaliculata	3o		
Laminaria spec.	127	127	

ˣAlgae are rich in acrylic acid.

Numbers show the paper with positive results. Numbers in (..) mean, the occurence is not unambiguously proven.

HALOGENATED ALIPHATIC COMPOUNDS

In the highly volatile fractions of the genera *Asparagopsis*, *Bonnemaisonia* and *Ptilonia*, belonging to the Bonnemaisoniaceae, a great variety of halogenated alkanes (5-24), saturated (31-59) and unsaturated ketones (6o-78), aldehydes (29), alcohols (26-28), epoxides (3o) and halogenated derivatives of acetic and acrylic acids (82-87, isolated by their esters), have been detected recently. A notable antibiotic activity against Bacillus subtilis, Staphylococcus, Fusarium and Vibrio was shown for the halogenated heptanones (55-59; (221); MIC o.1 - 1 mg/ml), and some haloacetones and halobutanones (49). Extracts of *Asparagopsis* sp., *Bonnemaisonia hamifera* and its tetrasporophyte *Falkenbergia* turned out to be exceptionally active (22, 95,68,18o); in solutions of raw substance mixtures, free iodine is developed very quickly (153). Upon addition of sodium thiosulfate to the extracts no more inhibition is detected in the agar diffusion test (88).On the other hand, as the diffusion of soluted iodine (J_3^-) is very slow and causes strong discolorations, the large zones of inhibition could to be due to iodine, split off subsequently, or to organic iodine complexes (88).

TERPENES

Many substances discussed so far are steam-volatile. A part of them, therefore, was found in distillates of various algae, as reported by Katayama (111-13o and 161). In addition to 3 he found short chain saturated fatty acids,alkanes and terpenes. He determined an antibiotic activity not only with the low molecular acids, but also - although often rather weak - in the terpene containing fractions (127,129). Substances concerned are terpenes as cineol, linalool, geraniol, limonene, pinene, myrcene, ocimene and similar compounds, also widely distributed in higher plants. In the meantime, these compounds were detected in other algae also (6o,1o8,146,147,2o7,22o,231, 237,241,242). As they are usually present in extremely small quantities, and have low biological activities (129), they cannot be made responsible in general for the antibiotic effects of algae and alga preparations.

Especially in the class Rhodophyceae, a great variety of unusual, mostly halogenated, and sometimes aromatic terpenes were found in the meantime

Halogenated Aliphatic Compounds

Alkanes, Alkenes

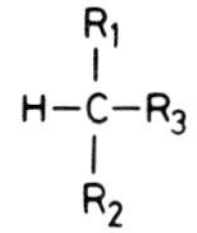

No	R_1	R_2	R_3	Lit.
5	J	H	H	5
6	Br	Br	H	3
7	Cl	J	H	5
8	Br	J	H	5
9	J	J	H	3
10	Cl	Cl	Cl	5
11	Cl	Cl	Br	5
12	Cl	Br	Br	3,2,5,4
13	Cl	Br	J	3,2
14	Br	Br	Br	2,3,4,5,9
15	Br	Br	J	3,2,5
16	Br	J	J	3,2
17	J	J	J	3,2

No		Lit.
18	CCl_4	5
19	CBr_4	3,2,5
20	$Br-CH_2-CH_2J$	3

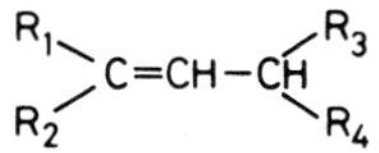

No	R_1	R_2	R_3	R_4	Lit.
21	Br	Br	Cl	Cl	3,2
22	Br	Br	Br	Cl	3,2
23	Br	Br	Br	Br	3,2
24	Br	J	Br	Br	3,2

Alcohols, Aldehydes, Epoxides

No	
25	JCH_2-CH_2OH

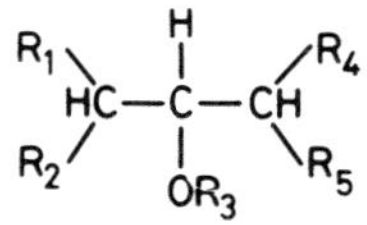

No	R_1	R_2	R_3	R_4	R_5	Lit.
26	Br	H	H	Br	Br	3
27	Br	H	Ac	Br	Br	3
28	Br	Br	Ac	Br	Br	3

No		Lit.
29	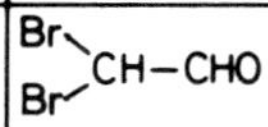	3
30	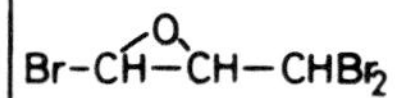	3

Carboxylic acid derivatives

No		Lit.
79	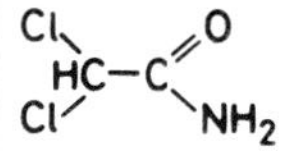	8
80	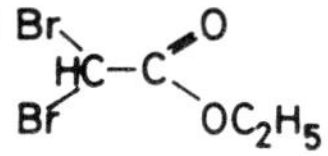	5,4
81	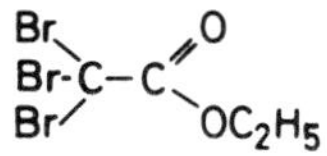	5,4
82	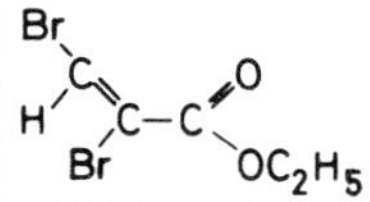	5,4

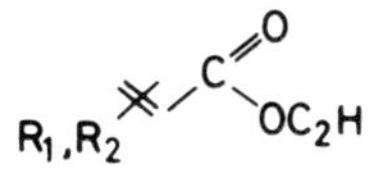

No	R_1	R_2	Lit.
83	J	H	4
84	Br	Cl	5
85	Br	Br	5,4
86	Br	J	4
87	J	J	4

Detected in:

Asparagopsis taxiformis

1.) Fenical 1974

2.) Burreson et al. 1975

3.) Burreson et al. 1976

4.) McConnell & Fenical 1977

Asparagopsis armata

5.) McConnell & Fenical 1977

9.) Codomier et al 1977

Bonnemaisonia hamifera

6.) Siuda et al. 1975

Ptilonia australasica

7.) Kazlauskas et al. 1977

Marginisporum aberrans

8.) Ohta & Takagi 1977

Falkenbergia rufolanosa

9.) Codomier et al 1977

Ketones

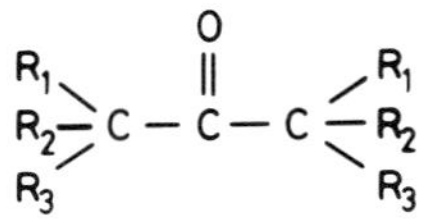

No	R_1	R_2	R_3	R_4	R_5	R_6	Lit.
31	Br	H	H	H	H	H	3
32	J	H	H	H	H	H	2
33	Cl	H	H	Cl	H	H	5
34	Cl	H	H	Br	H	H	1,5,9
35	Cl	H	H	J	H	H	5
36	Br	H	H	Br	H	H	3,1,9
37	Br	H	H	J	H	H	3
38	Br	Br	H	H	H	H	3
39	Cl	Cl	H	Cl	H	H	5
40	Cl	Cl	H	Br	H	H	5
41	Cl	Br	H	Cl	H	H	5
42	Cl	Br	H	Br	H	H	1,3,5,9
43	Br	Br	H	Cl	H	H	1,5,9
44	Br	Br	H	Br	H	H	1,5,9
45	Br	Br	H	J	H	H	5,3
46	Cl	Cl	H	Cl	Cl	H	5
47	Cl	Br	H	Cl	Cl	H	5
48	Cl	Br	H	Cl	Br	H	5
49	Br	Br	H	Cl	Cl	H	5
50	Br	Br	H	Cl	Br	H	1,5,9
51	Br	Br	H	Br	Br	H	1,3,5,9
52	Br	Br	Cl	H	H	H	3
53	Br	Br	Br	H	H	H	3
54	Cl	Cl	Cl	Cl	Cl	Cl	3

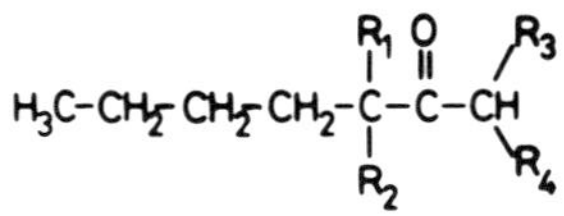

No	R_1	R_2	R_3	R_4	Lit.
55	Br	H	Br	H	6
56	Br	H	Br	Br	6
57	Br	Br	Br	H	6
58	Br	Br	J	H	6
59	Br	Br	Br	Br	6

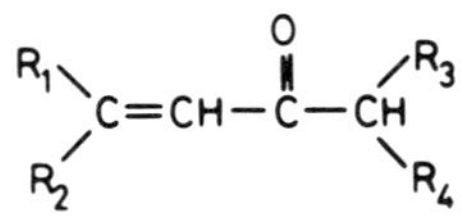

No	R_1	R_2	R_3	R_4	Lit.
60	Br	Br	H	H	3
61	Br	Br	Br	H	3
62	Br	Br	J	H	3
63	Br	Br	Br	Br	3
64	Br	Br	Br	Cl	3
65	Br	Cl	Br	Br	3
66	Cl	Cl	Br	Br	3
68	Br	Cl	Br	Cl	3

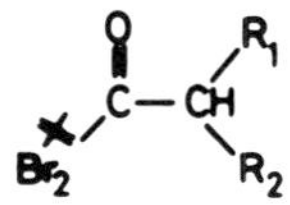

No	R_1	R_2	Lit.
69	H	H	2
70	Cl	H	2
71	Br	H	2
72	J	H	2
73	Cl	Br	2
74	Br	Br	2,5

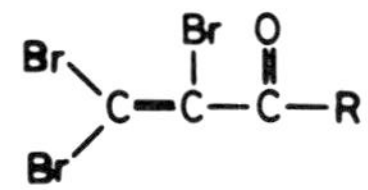

No	R	Lit.
75	$-CH_2Br$	7
76	$-CH=CH-CBr_2-CH_2-CH_3$	7
77	$-CHBr-CH_2-CH_2-CH_2-CH_3$	7
78	$-CBr_2-CH_2-CH_2-CH_2-CH_3$	7

(review : 5o). In very few cases only, substances were checked for their antibiotic activity. The results published up to now show that the phenolic halogenated terpenes (laurinterol, 88, (26,98) and isolaurinterol, 9o (99))are more active, especially against staphylococci and mycobacteria, than the unhalogenated (debromlaurinterol, 89), or the non-aromatics (prepacifenol, 91; cycloeudesmol, 92, (219) and 3-bromo-3-chloromethyl-4,6-dichloro-1-methyl cyclohexane-1-carboxaldehyde (166)).

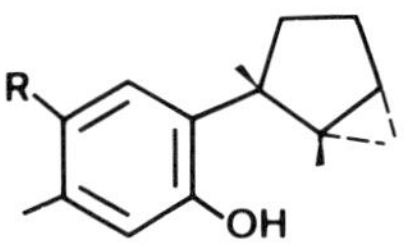

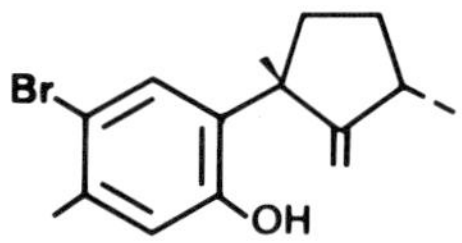

Laurinterol R = Br (88)

Debromlaurinterol R = H (89)

Isolaurinterol (90)

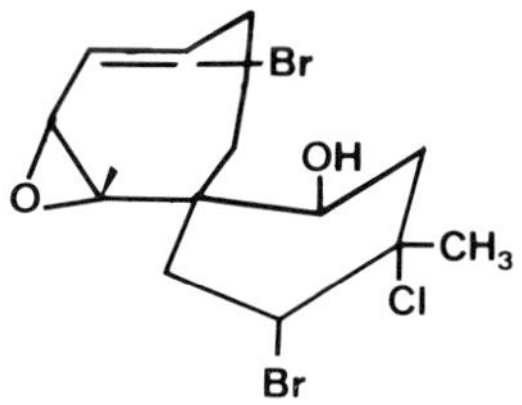

Prepacifenol (91)

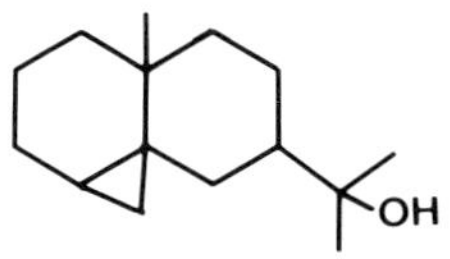

Cycloeudesmol (92)

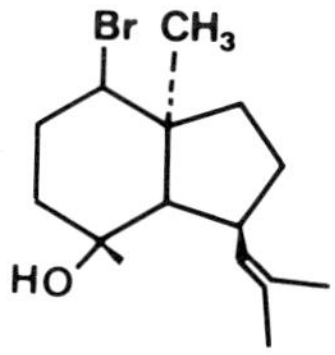

Oppositol (93)

A moderate antibiotic activity was also found with the halogenated sesquiterpene oppositol (93) from Laurencia subopposita (87).
It is open up to now, whether the antibiotic activity found in Dictyotales (22,68,95,96,18o) is due to the diterpenes like pachydictyol (94, Pachydictyon coriaceum (92); Dictyota hespera (155); D.dichotoma v. implexa (48))as was supposed by Robertson & Fenical (2o2) or by dictyol A (95) and B (96, D. dichotoma var. implexa (46,158)), or by acetoxycrenulatin (97, D. crenulata, P. coriaceum (155)).

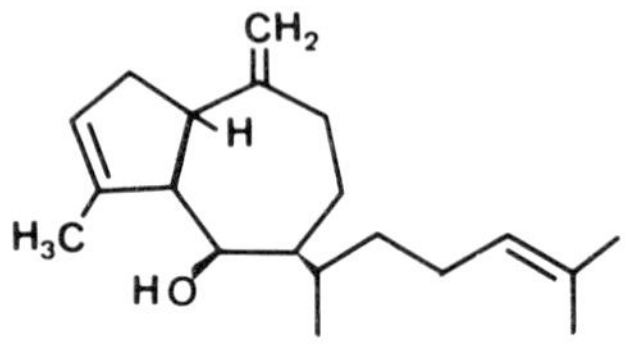

Pachydictyol A (94)

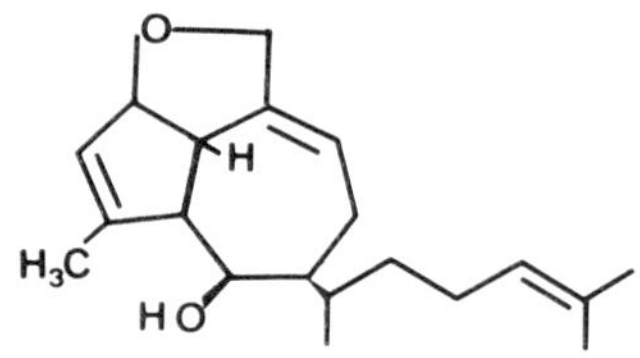

Dictyol A (95)

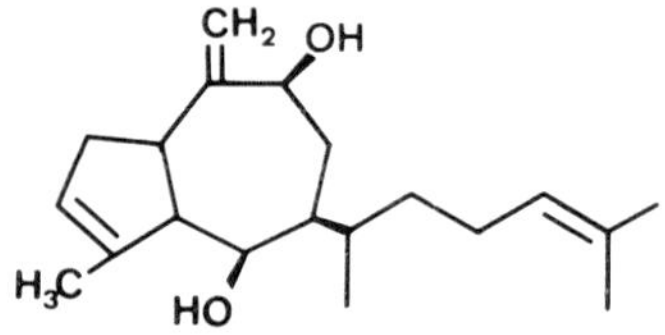

Dictyol B (96)

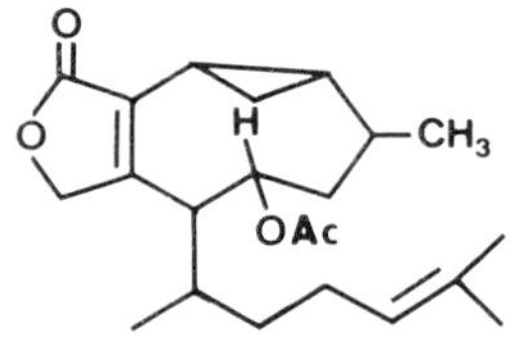

Acetoxycrenulatin (97)

The thallus surface of Delisia fimbriata, belonging to the Bonnemaisoniaceae, is nearly free of bacterial and other epiphytes (188).

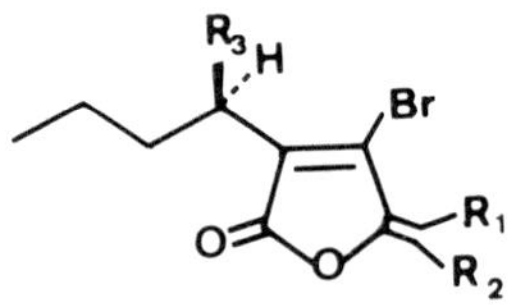

Fimbrolides

No	R_1	R_2	R_3
98	Br	H	H
99	H	Br	H
100	Br	Br	H

Hydroxyfimbrolides

No	R_1	R_2	R_3
101	Br	H	OH
102	H	Br	OH
103	J	H	OH
104	H	J	OH
105	Cl	H	OH
106	H	Cl	OH
107	Br	Br	OH

Acetoxyfimbrolides

No	R_1	R_2	R_3
108	Br	H	OAc
109	H	Br	OAc
110	J	H	OAc
111	H	J	OAc
112	H	Cl	OAc
113	Br	Br	OAc

The strong antibiotic activity of dichloromethane extracts of this alga (132) is due to the fimbrolides (88-1oo), hydroxyfimbrolides (1o1-1o7) and acetoxyfimbrolides (1o8-1o9), isolated by (132,188).

SULFUR CONTAINING HETEROCYCLIC COMPOUNDS

In Chondria californica, a species belonging to the Rhodomelaceae, various peculiar heterocyclic compounds were found : cyclic polysulfides and their oxidation products, several thiepanes (114-119) and thiolanes (12o-121), which are antibiotically active (24o).

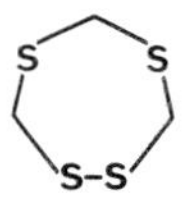
(114)

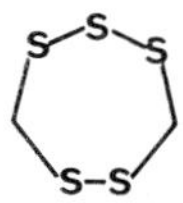
Lenthionin
(115)

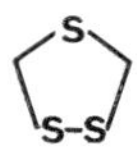
(116)

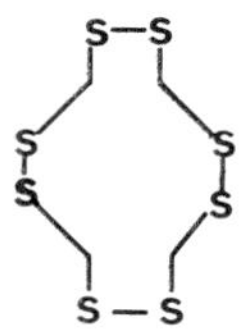
(117)

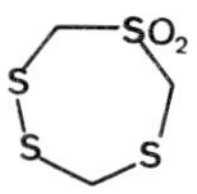
(118)

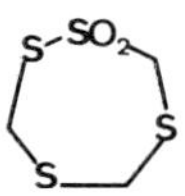
(119)

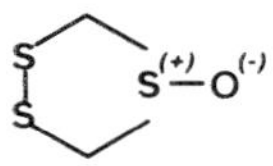
(120)

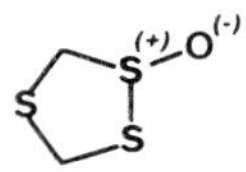
(121)

Lenthionin (115) and thiepane (114) had been found previously in the mushroom Lentinus edulis (163,164). Similar cyclic polysulfides are present also in Dictyopteris species (2o3).

PHENOLIC INHIBITORS

The antimicrobial properties of free phenols are well known. Therefore, phenols are frequently used as antiseptics. Among phenolic substances detected in algae are p-cresol (122), found by Katayama (125) in Chlorophyceae, Rhodophyceae and Phaeophyceae and 3,5-dinitroguaiacol (123), recently detected in the Corallinacea Marginisporum aberrans (179), and

(122)

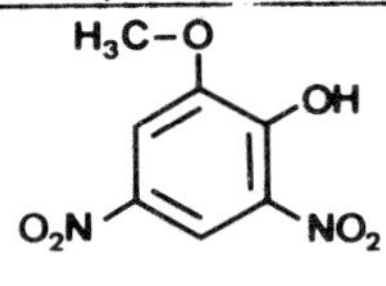
(123)

the "lignin"-like substances, which possibly protect the cell wall of a Staurastrum spec. against bacterial decay (85). A rather large group of red algae contain brominated and unbrominated derivatives of the p-hydroxybenzyl alcohols (124) and 3,4-dihydroxybenzylalcohols.

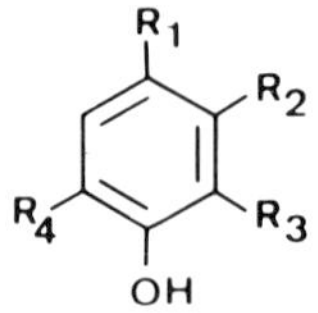

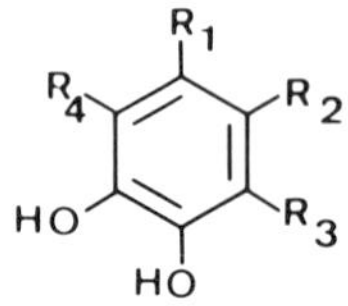

No	R_1	R_2	R_3	R_4
124	CH_2OH	H	H	H
125	CH_2OH	H	Br	H
126	CH_2OH	H	Br	Br
127	CH_2OH	Br	Br	Br
128	CH_2OCH_3	H	Br	H
129	CH_2OCH_3	H	Br	Br
130	$CH_2OC_2H_5$	H	Br	Br
131	CHO	H	H	H
132	CHO	Br	Br	Br
133	CH_2COOH	H	Br	Br
134	$CH_2C(=O)COOH$	H	Br	Br

No	R_1	R_2	R_3	R_4
135	CH_2OH	H	Br	H
136	CH_2OH	Br	Br	H
137	CH_2OH	Br	Br	Br
138	CH_2OCH_3	H	Br	H
139	CH_2OCH_3	Br	Br	H
140	$CH_2OC_2H_5$	H	Br	H
141	$CH_2OC_2H_5$	Br	Br	H
142	$CH_2OC_3H_7$	Br	Br	H
143	CHO	H	Br	H
144	CHO	Br	Br	H

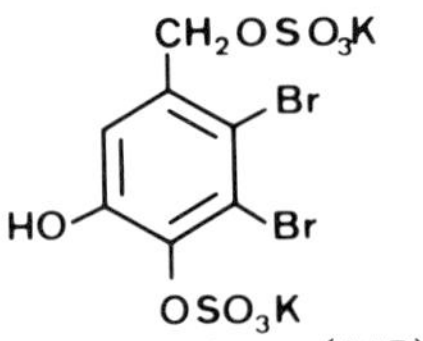

(145)

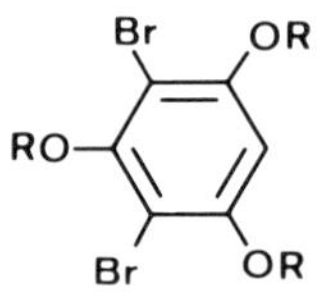

$R=CH_3$ (146)

(see(34))

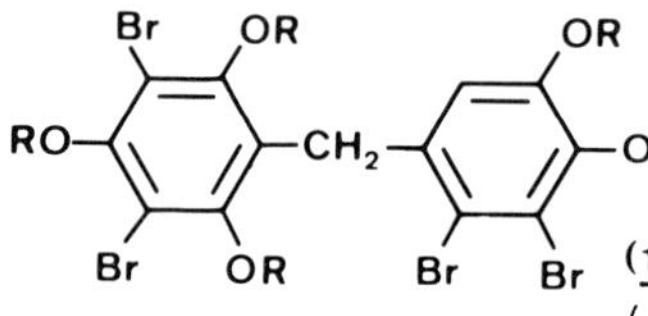

(147), $R=CH_3$

(see (34))

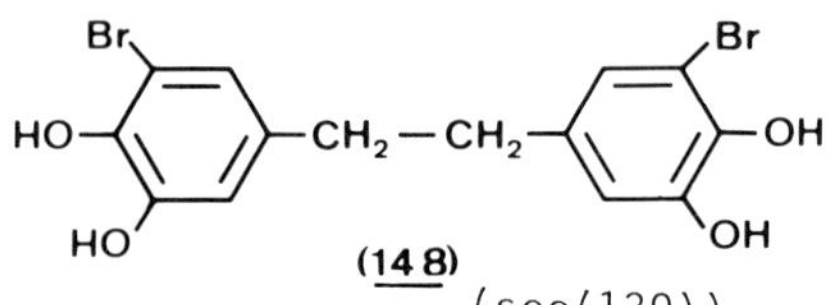

(148)

(see(139))

Table 2 , Occurence of bromophenols in algae (citations of literature)

No	Seaweed / Substance :	124	125	126	127	128	129	13o	131	132	133	134	135	136
	CYANOPHYCEAE													
1)	Calothrix brevissima			182	182					182				182
	PHAEOPHYCEAE													
2)	Fucus vesiculosus			183										183
	RHODOPHYCEAE													
	Gigartinales													
3)	Cystoclonium purpureum													
	Cryptonemiales													
4)	Corallina officinalis													184
	Ceramiales													
	R h o d o m e l a c.													
5)	Polysiphonia brodiaei		8o	184/8o									184/8o	184/8o
6)	P. urceolata		8o	8o			139						8o/184	184
7)	P. fastigiata (= P.lanosa)			8o									8o/226	8o
8)	P. nigrescens			8o/184									8o	8o/184
9)	P. morrowii													
1o)	P. fruticulosa		8o											8o
11)	P. elongata		8o											8o
12)	P. nigra		8o											8o
13)	P. violacea		8o											8o
14)	P. thuyoides		8o										8o	8o
15)	Pterosiphonia complanata													
16)	Rhodomela larix													238
17)	Rhodomela confervoides (= Rh.subfusca)			39/184										39/8o/184
18)	Symphocladia gracilis													
19)	Vidalia volubilis													
2o)	Lenormandia prolifera													2o4
21)	Odonthalia corymbifera													14o
22)	Odonthalia dentata			39/184		184	184	184					184	39/184
23)	Halophitys incurvus (=H.pinastroides)		8o								32[b]	32[b]	8o	
24)	Rhytiphlaea tinctoria													34[b]
	D a s y a c e a e													
25)	Dasya pedicellata var. stanfordiana	52							52					
	D e l e s s e r i a c.													
26)	Phycodrys rubens													
	C e r a m i a c e a e													
27)	Antithamnion plumula													184
28)	Ceramium rubrum													184

[a] Structure later corrected, [b] after methylation with diazomethane

Table 2, continued

No	137	138	139	14o	141	142	143	144	145	unidentified with $-SO_3^-$	unidentified without-SO_3^-	No
1)												1)
2)					183							2)
3)								184				3)
4)												4)
5)			184				184	184	8o			5)
6)		139	184				8o/139/184	184		186	186	6)
7)	8o	226	226	226	226	8o/226	8o/226	8o/226	79/8o/93[a]	9/1o/37/142/15o/185/186	11/186	7)
8)			184			8o	184	8o	8o			8)
9)							2o5					9)
1o)								8o	8o	186		1o)
11)							8o	8o	8o	9		11)
12)								8o	8o	9		12)
13)								8o	8o			13)
14)								8o	8o			14)
15)										9		15)
16)			131/238					131	238		152	16)
17)	8o		138/184					8o/138/184	8o/138			17)
18)											2o6	18)
19)										141		19)
2o)												2o)
21)			14o						14o		14o	21)
22)			184		184		184	184				22)
23)										12/141/186	186	23)
24)					34[b]					9/186	186	24)
25)												25)
26)			184									26)
27)			184									27)
28)			184									28)

and the dimerisation products (148 (139)), or condensation products with dibromophloroglucinol (147 (34)).

Owing to their tendency for crystallization and their easy detection, the presence of some of these compounds was shown rather early (9-11,37,141,142, 15o,2o6). But it took a long time until the first definite structure proposal was presented for 3-bromo-4,5-dihydroxybenzaldehyde (143) from Polysiphonia morrowii (2o5). In the meantime, 25 different substances were identified. Among these, however, the four benzylmethyl ethers (128,129,138,139), the three benzylethyl ethers (13o,14o,141), and possibly a benzylisopropyl ether (142) might be isolation artifacts, which may be formed very easily during extraction in the weakly acidic medium from the respective benzyl alcohols (125,126,135,136), and the short chain alcohol used as solvent. As pointed out by Weinstein et al. (238), the aldehydes (e.g.144), might be artifacts, but they might be also part of a biosynthetic pathway.

For the free phenols antibiotic activities with MIC down to 5o µg/ml against bacteria (Tab. 3 (8o)),and a LD_{50} of about 7 - 7o µg/ml against

Table 3, Minimum inhibition concentrations of some bromophenols (µg/ml)(8o)

Compound	Bacillus subtilis	Escherichia coli	Sarcina pelagia	Serratia marinorubra	Vibrio phytoplankton
135	50	>150	100	80	100
143	280	200	300	380	180
136	120	160	190	350	200
144	100	150	100	>530	150
126	100	420	400	>600	430

Seed: 10^4 cells/ml

unicellular algae (Tab. 4 (8o)), were determined. Up to 2o mg/ml the sulfuric acid ester (145) does not show any antibacterial activity against Bacillus subtilis and Escherichia coli (8o,185). Low concentrations of lanosol (136, $4 \times 1o^{-5}$ - $1o^{-6}$ M, 58), or other similar phenols (59) show even a growth stimulating effect on Polysiphonia urceolata and Goniotrichum alsidii. No conclusive comment on the physiological or ecological significances of the bromophenols is possible up to now. The amounts of these compounds in the genus Polysiphonia are in the range from o.o9 % (relative

Tab.4 LD_{50} of some phenols against microalgae (157)

	Phloroglucinol	1,4-Benzoquinone	5-Bromo-3,4-dihydroxy-benzaldehyde (143)	2,3 Dibromo-4,5-dihyroxy-benzylalkohol (136)
Skeletonema costatum	>0,1; <0,3	>0.005	<0.03	>0.03; <0.1
Cryptomonad 3-C	>0,3; <0,5	>0.005; <0.018	>0.03; <0.1	>0.03; <0.1
Monochrysis lutheri	>0,1; <0,3	>0.025; <0.050	>0.03; <0.1	>0.03; <0.1
Dunaliella tertiolecta	>1,0	>0.05; <0.1	>0.3	>0.3
Amphidinium carteri	>0,5; <1,0	>0.1; <0.15	>0.15; <0.3	>0.1; <0.15
Olisthodiscus sp.	>0.3; <0.5	>0.005; <0.018	<0.03	<0.03
Porphyridium sp	>0.5; <1.0	>0.2	>0.15; <0.3	0.15

to dry weight) in P. fruticulosa, up to 1.7 % in Polysiphonia nigrescens, and 1.9 % in P. lanosa (8o). It doesnot seem unlikely that in certain cases at the cell membrane or at the external surface of the cell walls local concentrations may be acheived, which may account for an inhibitory effect on potential bacterial or algal epiphytes.

ISOPRENYLATED AND BROMINATED HYDROQUINONES

Isoprenylated and brominated hydroquinones were isolated from the Chlorophycea Cymopolia barbata (94). It is possible but not evident, that these substances are responsible for the antibiotic activities of this seaweed,

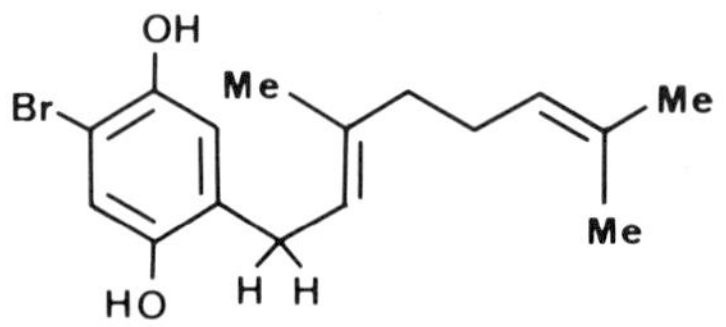

Cymopol (146)

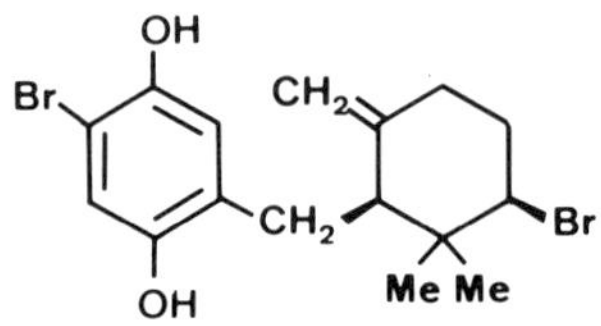

Cyclocymopol (147)

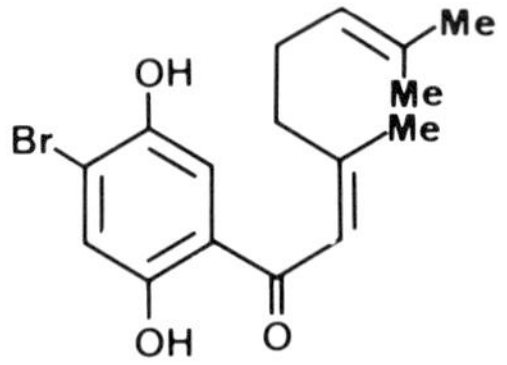

Isocymopolone (148)

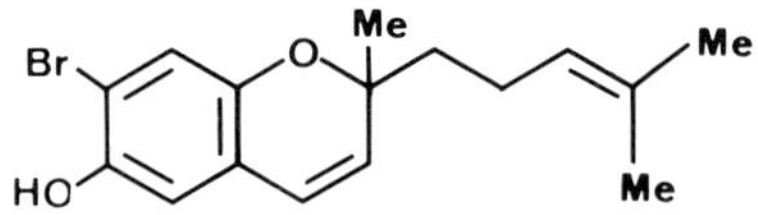

Cymopochromenol (149)

stated earlier (167,168). Whilst in cymopol (146), cyclocymopol (147), isocymopolone (148) and cymopochromenol (149) a monoterpenoid moiety forms the side chain, a diterpenoid side chain is present in non halogenated analogous substances from Dictyopteris undulata : zonarol (15o , (54)), chromazonarol (151), isochromazonarol (152, (51)) and the analogous zonaroic acid (153, (35)).

PHENOLS IN PHAEOPHYCEAE

When thallus fragments of numerous brown algae or their extracts were examined for antibiotic activities with the agar diffusion test, comparatively small inhibition zones were measured in most cases (62,88). Frequently these inhibition zones have a brownish colour, or upon spraying with solutions of diazonium salt, they show reactions typical of phenols.

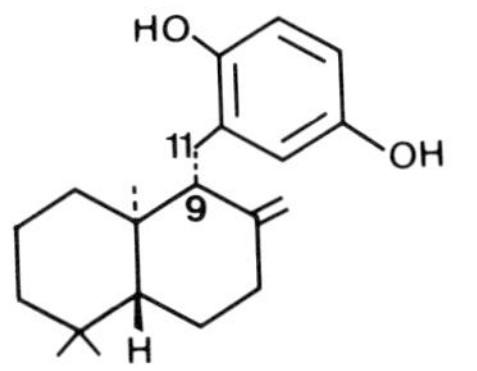

Zonarol (150)

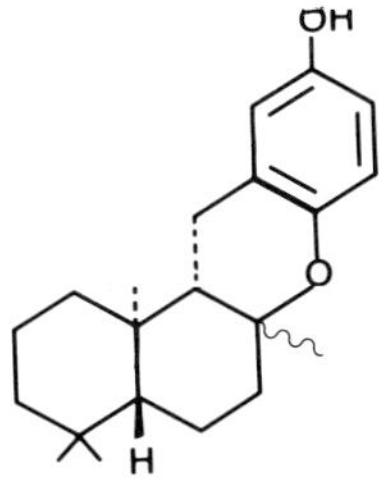

Chromazonarol (151)

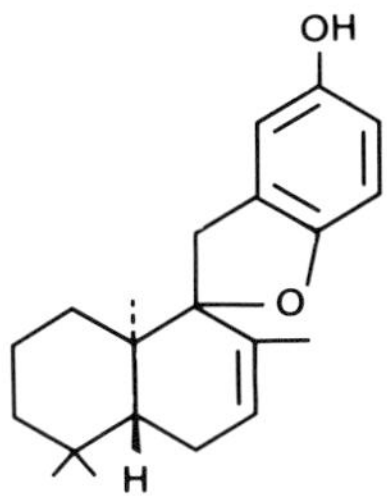

Isochromazonarol (152)

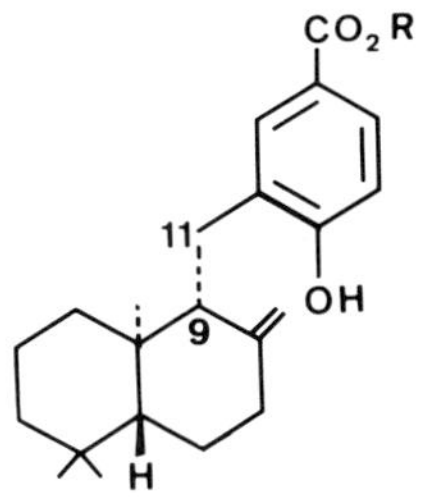

Zonaroic acid (153)

However, the occurence of phenolic tannin-like substances in brown algae has been known since the studies of Crato (42) on Chaetopteris plumosa, and thereafter has been confirmed frequently. Obviously the tannins are contained in vesicle-like organelles, which are called sometimes "physodes", sometimes "vesicles", or "Hansteensche Fucosankörner" (for review see :

199, 237a). They have been found not only in nearly all series of Phaeophyceae (listed by 237a), but also in various other algal classes, e.g. in the Heterochloridales (Xanthophyceae) (41). On the other hand, by no means all structues called "physodes" or "vesicles" contain tannins. In many cases, however, definitely positive proof resulted from specific colour reactions (43, and many others). Phenolic substances, sometimes in considerable amounts, are excreted by Phaeophyceae into the sea water (18,28,29,38,41, 43,55, 148,154,159,216-218). The origin of these substances and of the products arising from them ("Gelbstoff", (11o)) was not always realized at once.

In analogy to the physiological activities, well known from phenols and tannins of terrestrial plants (143), they are variously active in sea water and of outstanding ecological importance (156,38). First of all to be mentioned is their role as nonspecific enzyme inhibitors (44,45,8), which, although found several times (e.g. 177,1o2), was not always realized clearly. Antibacterial activities were noticed not only in the author's group, but also by Conover & Sieburth (38), with the tannin like polyphenols from Sargassum against Vibrio and Pseudomonas. Seshadri & Sieburth (2o9) stated that the Phaeophyceae tannins inhibit the development of yeast (Rhodotorula & Candida). Density variations of the yeast cover on the thallus surface of brown algae might be due to seasonal variations of phenol production and excretion. According to Jensen (1o3) the composition of bacterial populations on the surfaces of algae is also controlled by seasonal variation. Not all species of bacteria are inhibited, but some of them are even able to destroy the phenols (1o3). The growth rate of unicellular algae such as Olisthodiscus, Dunaliella and Monochrysis also is decreased by phenols (5o % decrease of growth, with about 15-1oo µg/ml, 156,145). Multicellular algae such as Ulva, Enteromorpha and Fucus react to phenols with inhibited or abnormal growth (165,227,228). Hydroids are at least as sensitive as bacteria to phenols of Phaeophyceae, whereas worms (215,229), copepods and pygnogonids tolerate higher concentrations (38,188a). Furthermore, fertilization and development of Echinodermata eggs and the growth of Balanus and Mytilus are disturbed by Fucus extracts containing phenols (19,2o,38,44). For fish larvae of Pleuronectes platessa, the LD 5o is about o,32 mg/l (217).

Although Crato established early (42,43) by means of more or less specific colour reactions that the content of the physodes "consists of rather complicated phenol like compounds, and that a constant constituent of them is phloroglucinol", although this was confirmed by various investigators (see 199,237a), and although phloroglucinol was identified in alkaline degradation products of extracts from Fucus vesiculosus (156), Sargassum ringgoldianum (176), and Ecklonia cava (232), it was not until 1973, that Glombitza et al. (77) reported on the unambiguous detection of free phloroglucinol in 17 algal species (Tab. 5).

Table 5 , Occurence of free phloroglucinol in Phaeophyceae (77)

Species		Species	
Sphacelariales		Cystoseira tamariscifolia (Hudson) Papenfuss	+ + +
Cladostephus spongiosus (Hudson) C. Agardh	+	Cystoseira baccata (Gmelin) Silva	+
Cladostephus verticillatus (Lightfoot) Lyngbye	+	Cystoseira granulata C. Agardh	+
Dictyotales		Cystoseira discors C. Agardh	+ +
Dictyota dichotoma (Hudson) Lamouroux	+	Cystoseira myriophylloides Sauvageau	(+)
Dictyopteris membranacea (Stackhouse) Batterson	?	Bifurcaria bifurcata Ross	(+)
Fucales		Halidrys siliquosa (L.) Lyngbye	(+)
Fucus vesiculosus L.	+	Laminariales	
Fucus serratus L.	(+)	Chorda filum (L.) Stackhouse	+
Fucus spiralis L.	+	Laminaria hyperborea (Gunner) Foslie	?
Himanthalia elongata (L.) S.F. Gray	+ +	Laminaria ochroleuca La Pylaie	+
		Saccorhiza polyschides (Lightfoot) Batterson	+

+ + +	$\geqq$ 0,5% rel. to dry weight
+ +	nearly 0,1%
+	$\leqq$ 0,1% detectable without enrichment
(+)	only detectable after enrichment
?	detection dubious even after enrichment

In most cases it accounts for only a small percentage of the total phenol fraction, the bulk of which consists of high molecular dehydropolymerisates of phloroglucinol. The higher molecular fraction has the properties of tannins (phlorotannin, (2o8)). Probably the authors describing a lignine from phaeophyceae have isolated phlorotannins, which often give the same reactions like lignine (in Durvillea (2o1); in Fucus vesiculosus (82) and (149)). The phloroglucinol rings are bound in the type of oligoaryls (fucols, (4o,65,72,2oo)), or oligoaryl ethers (phlorethols (76,69, e.a. see table 6).

Fucols

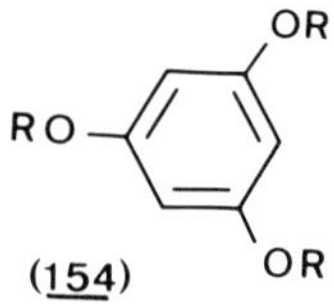

(154)

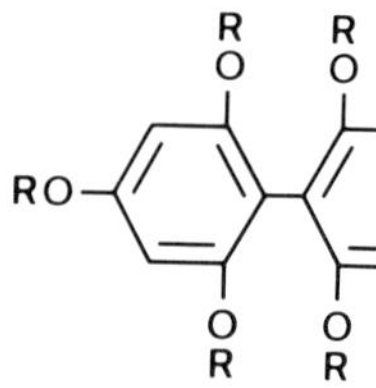

(155)

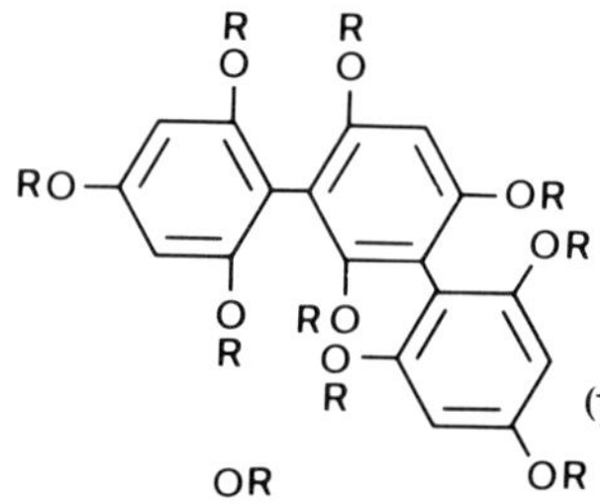

(156)

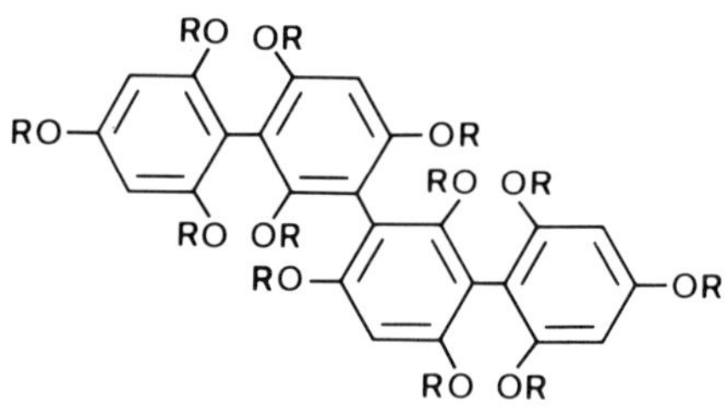

R= H_3CCO

(157)

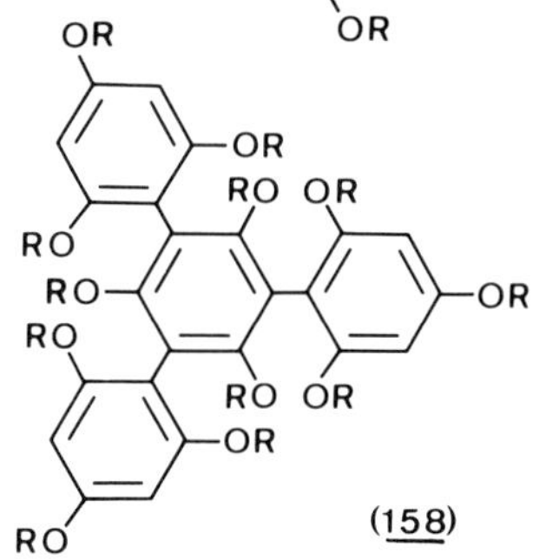

(158)

In some molecules diaryl as well as diaryl ether linkages might also be present at the same time (fucophlorethols (4o,65,66,71 e.a. see table 6)).

Fucophlorethols

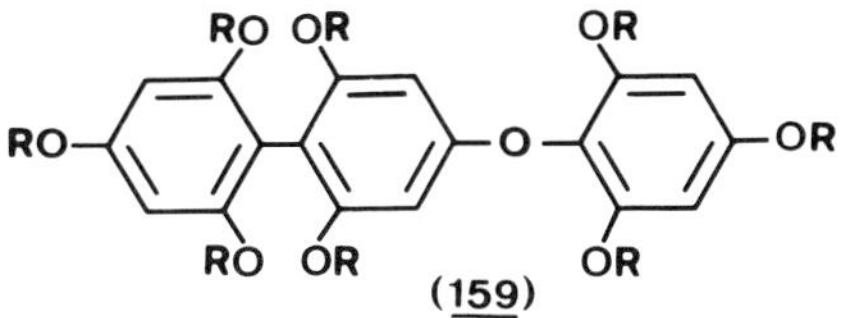

(159)

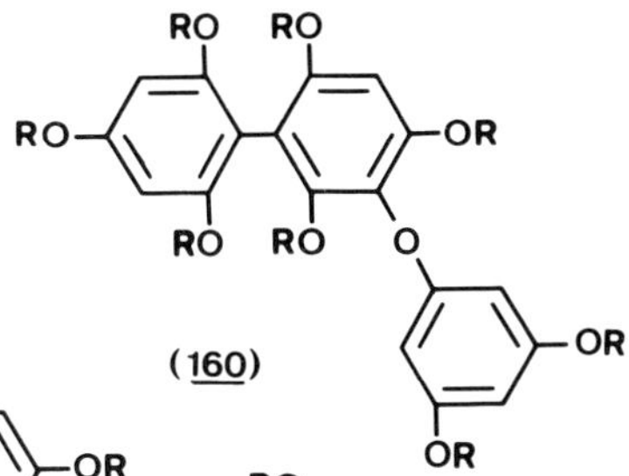

(160)

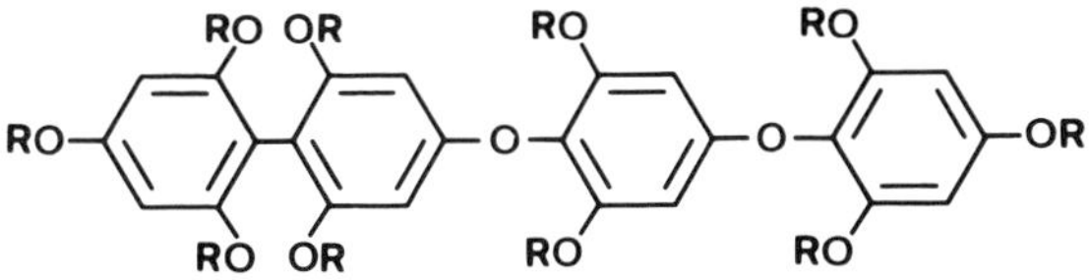

(161)

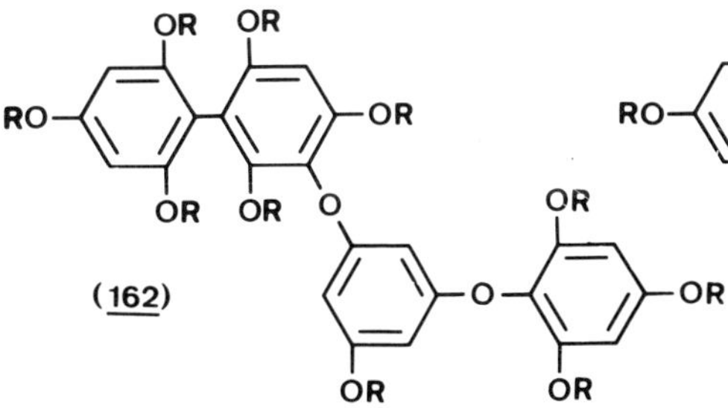

(162)

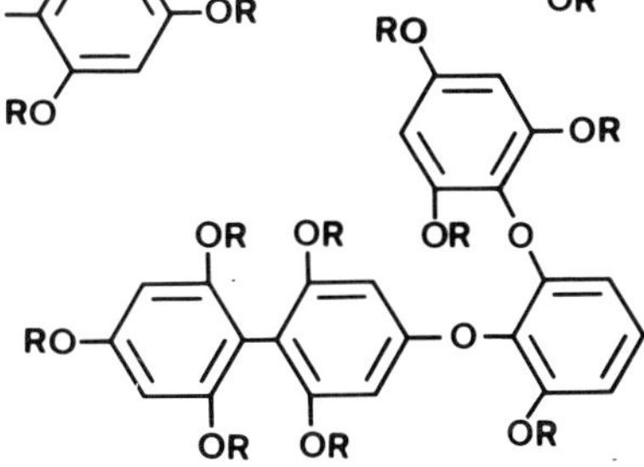

(163)

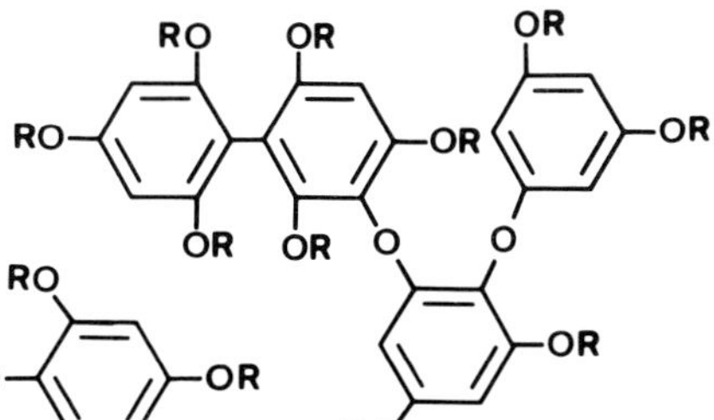

(164)

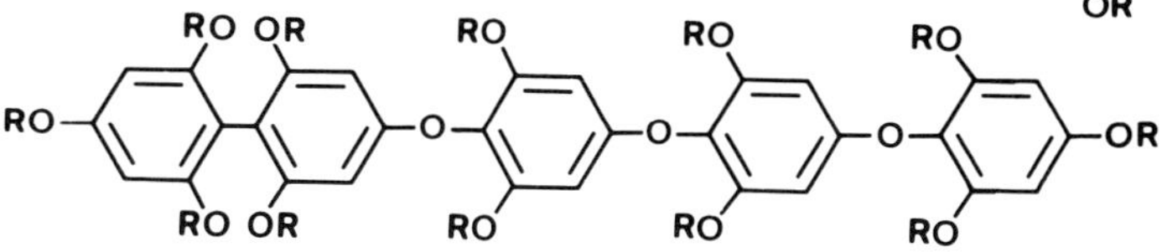

R = H_3CCO

(165)

The substances may be chlorinated (chlorinated fucophlorethols (7o)). In the molecules of fuhalols additional HO-groups are attached (78,2o8, e.a. see table 6). Most of the compounds, especially the fuhalols, are sensitive to oxidation. In many cases, therefore, their isolation was possible only after acetylation or methylation. In addition to the substances whose structures were definitely proven, a number of further homologues were isolated, in which only some structur elements, but not the precise

Fuhalols

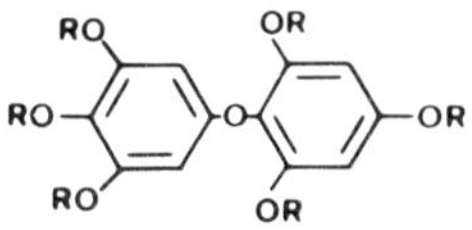

(166)

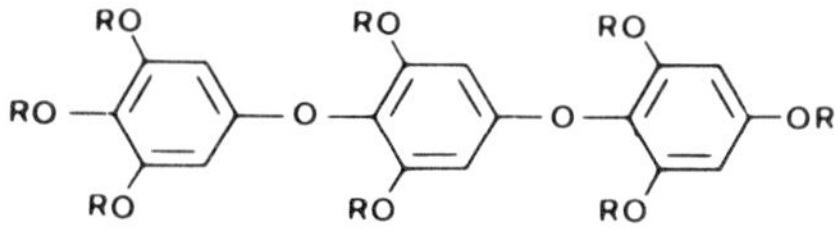

(167)

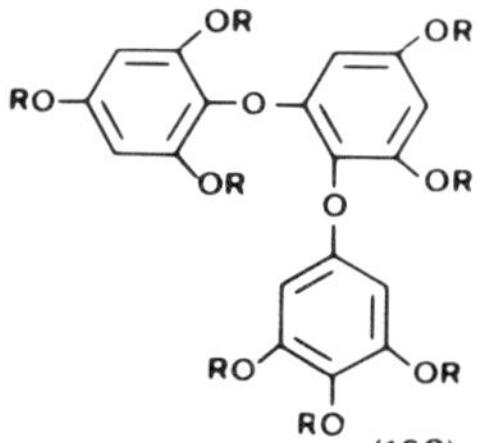

(168)

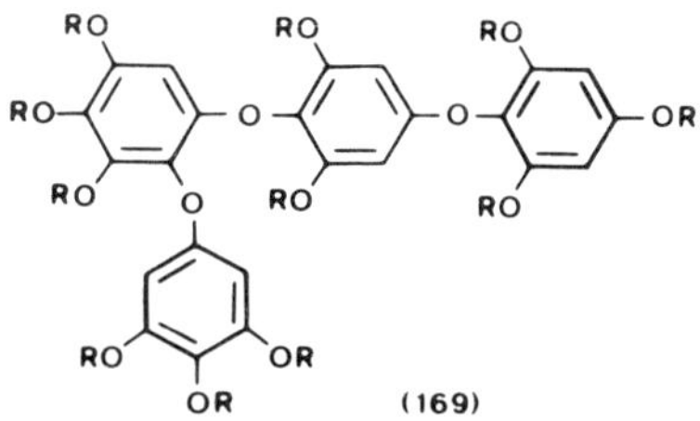

(169)

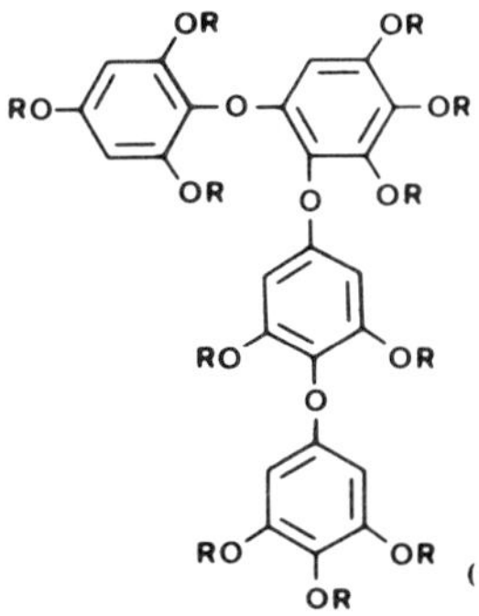

(170)

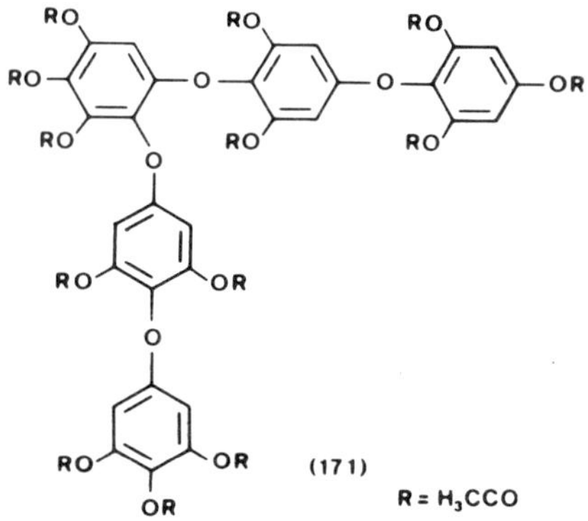

(171)

R = H_3CCO

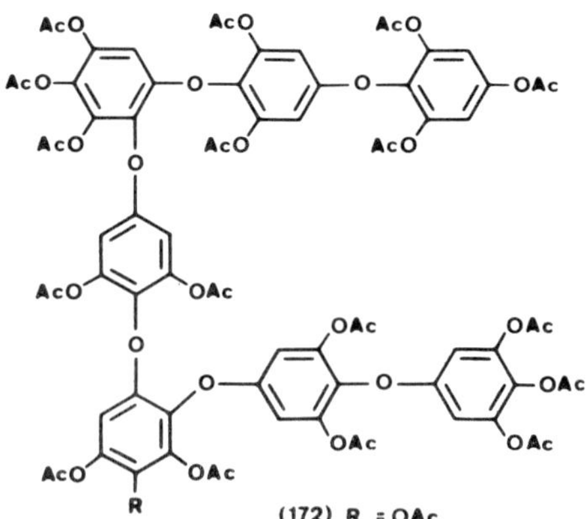

(172) R = OAc

(173) R = H

Table 6 , Occurence of phlorotannins in Phaeophyceae (citations of lit.)

	Substance × \ Seaweed	No	Dictyota dichotoma	Laminaria ochroleuca	Fucus vesiculosus	Himanthalia elongata	Cystoseira tamariscifolia	Cystoseira baccata	Bifurcaria bifurcata	Halidrys siliquosa	Sargassum muticum	No
Fucols	Difucol	155	67		65/72/2oo	73		81	75	61		155
	Trifucol	156			65/72	(73)			75			156
	Tetrafucol A	157			65/72	(73)						157
	Tetrafucol B	158			65/72							158
Fucophlorethols	Fucophlorethol A	159			4o/65/2oo							159
	Fucophlorethol B	16o		135				81				16o
	Fucodiphlorethol A	161			65/71							161
	Fucodiphlorethol C	162		135								162
	unnamed	163			4o							163
	Fucodiphlorethol B	164						81				164
	Fucotriphlorethol	165			65/71							165
Fuhalols	Bifuhalol	166					76	81	74	61	56	166
	Trifuhalol A	167						(81)	75	78	56	167
	Trifuhalol B	168									56	168
	Tetrafuhalol A	169							75	61	56	169
	Tetrafuhalol B	17o								61		17o
	Pentafuhalol	171							75	61	56	171
	Heptafuhalol	172							75	61/2o8	56	172
	Desoxyheptafuhalol	173								61		173
Phlorethols	Diphlorethol	174	67	69		73	76		75	61		174
	Triphlorethol C	175		69								175
	Monochlordiphlorethol	176		7o								176
	Monochlortriphlorethol C	177		7o								177

In (...) : Structure proposal not yet definitely proven.
× Usually isolated by the acetates.

Phlorethols

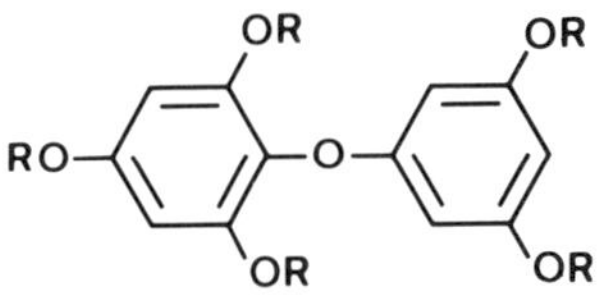

(174)

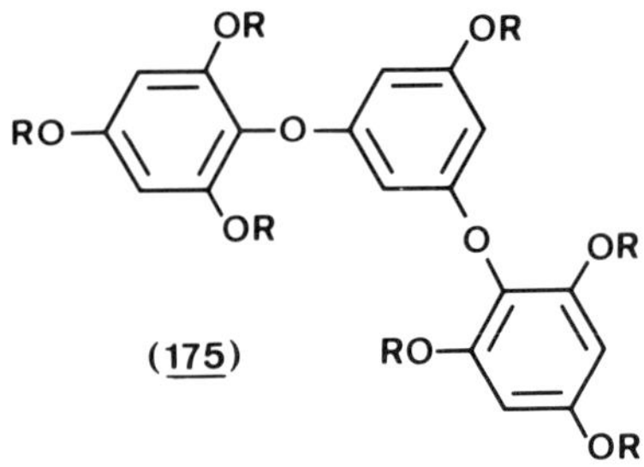

(175)

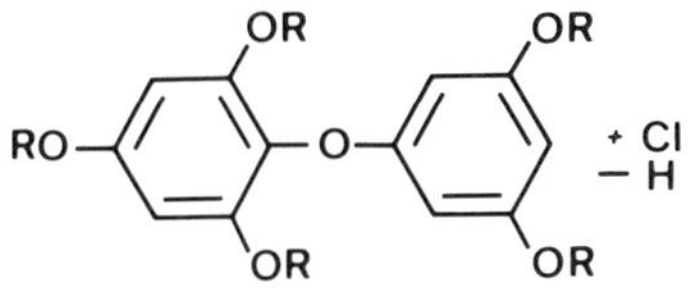

(176)

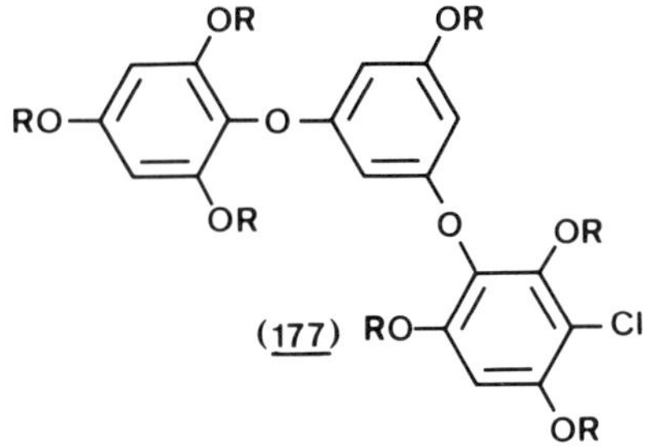

(177)

R = H_3CCO

arrangement of the rings, could be established. With several Phaeophyceae, e.g. Ascophyllum nodosum, it was not yet possible to enrich low molecular constituents. Nevertheless, it was shown that in this alga, phenol ethers with two free OH-groups in each benzene ring are present (57).
In addition to phlorotannins, gallotannins were also found in algae, in the Conjugata Spirogyra arcta (169-173). Their antimicrobial properties are virtually identical with those of gallotannins from higher plants.

Since the detection of acrylic acid from Phaeocystis poucheti by Sieburth (211,212), many antimicrobially active compounds from algae were identified. For a great number of active principles as Scenedesmin, Pandorin, Chonalgin, Sarganin, etc. (listed : 14,137,214,239), no exact data are available so far. This also holds true for additional inhibitors detected in recent years, e.g. :

The antifungally active goniodomin ($C_{43}H_{58}O_{11}$), which was isolated from the Dinoflaggelata Goniodoma (21o);

or

a toxic, dermatitis producing and antibacterially active principle of Lyngbya majuscula (16o);

or

two inhibitors of bacteria from the Chrysophyceae Stichochrysis immobilis, a peptide with MW 34oo and a component possibly belonging to the "Gelbstoffe" (15);

or

three inhibitors from the Chrysophycea Ochromonas malhamensis (89).

Antibiotic activity is not definitely proven for all substances mentioned in this paper. In some cases it seems likely by analogy reasoning. Up to now, no substance has been described in the literature, which seems to be worthy of clinical testing, or to justify hopes of therapeutic use in human medicine. Finally no definite statements on the ecological significance of the various substances are possible at this time.

References

1) Aaronson, S., Bensky, B. : Effect of aging of a cell population on lipides and during resistence in Ochroleuca danica. J.Protozool.14, 76-78 (1967).

2) Accorinti, J. : Competencia interespecifica e inhibidores antibacterianos de micro-algas marinas. Phyton (Buenos Aires) 21, 95-1o1 (1964).

3) Accorinti, J. : Versuche Hemmstoffe aus Scenedesmus obliquus zu isolieren. Phyton (Buenos Aires) 21, 1o3-o7 (1964).

4) Ackman, R.G., Tocher, C.S., McLachlan, J. : Dimethyl-ß-propiothetin determination by reactor gas-liquid chromatography, occurrence in algae and implications in fisheries. Proc.V[th].Intern.Seaweed Symp. Halifax, 235-42 (1965).

5) Ackman, R.G., Tocher, C.S. McLachlan J. : Occurrence of dimethyl-ß-propiothetin in marine phytoplankton. J.Fish.Res.Bd.Can. 23, 357-64 (1966).

6) Aubert, M. : Le comportement des bactéries terrigéenes en mer, relations avec le phytoplancton. Les Cahiers du CERBOM, Bd. 19 u. 2o, 1-278 (1965).

7) Aubert, J., Pesando D. : Action antibiotique d'extraits planctoniques vis-à-vis de germes anaérobies. Rev.Intern.Oceanogr.Med. 1o, 259-65 (1968).

8) Aufermann, B. : Vergleichende Untersuchungen über Verbreitung und Substratspektren einer unspezifischen Leucin-Decarboxylase bei marinen Rotalgen. Diss. Bonn (1976).

9) Augier, J. : La constitution chimique de quelques Floridées Rhodomélacées. Rev.Gen.Bot. 6o, 257-83 (1953).

1o) Augier, J., Henry, M.H. : Au sujet du brome dans les Rhodophycées. Bull.Soc.Bot.Fr. 97, 29-3o (195o).

11) Augier, J., Hoffmann, G. : Sucre et brome dans le Polysiphonia fastigiata (Roth.) Grev.Bull.Soc.Bot.Fr. 99, 8o-82 (1952).

12) Augier, J., Mastagli, P. : Sur un composé phénolique bromé extrait de l'algue rouge Halopitys incurvus. C.R.Acad.Sci. 242, 19o-91 (1956).

13) Badour, S.S., Gergis, M.S. : Cell division and fat accumulation in Nitzschia sp. grown in continuously illuminated mass culture. Arch. Mikrobiol. 51, 94-1o2 (1965).

14) Baslow, M.H. : Marine Pharmacology, Williams and Wilkinson & Co., Baltimore 8-13, 29-35, 43-44, 5o-65 (1969).

15) Berland, B.R., Bonin, D.J., Cornu, A.L., Maestrini, S.Y., Marino, J.P.: The antibacterial substances of the marine alga Stichochrysis immobilis (Chrysophyta). J.Phycol. 8, 383-92 (1972).

16) Blaauw-Jansen, G.: Chlorophyllide the probable precursor of a growth inhibitor. Nature 174, 312-13 (1954).

17) Blaauw-Jansen, G.: On the light induced transformation of chlorophyllide into a growth inhibiting substance. Proc.Ned.Acad.Wetenschap.Ser.C, 57, 498-5o6 (1954).

18) Bouck, G.B.: Fine structure and organelle association in brown algae. J.Cell.Biol. 26, 523-37 (1965)

19) Braham, J.M.: Inhibition of fertilization by Fucus extracts. Diss. Florida State Univ. (1963), Univers.Microfilms, Ann Arbor 63-6347.

2o) Braham, J.M., Metz, C.B.: Nature and action of the fertilization inhibitor from Fucus. Biol.Bull. 116, 3o5 (196o).

21) Bruce, D.L., Duff, D.C.B., Antia, N.J.: The identification of two antibacterial products of the marine planktonic alga Isochrysis galbana. J.Gen.Microbiol. 48, 293-98 (1967).

22) Burkholder, P.R., Burkholder, L.M., Almodõvar, L.R.: Antibiotic activity of some marine algae of Puerto Rico. Bot.Mar. 1/2, 149-156 (1959/61).

23) Burreson, B.J., Moore, R.E., Roller, P.: Haloforms in the essential oil of the alga Asparagopsis taxiformis. Tetrahedron Lett. 473-76 (1975).

24) Burreson, B.J., Moore, R.F., Roller, P.: Volatile halogen compounds in the alga Asparagopsis taxiformis (Rhodophyta). J.Agric.Food Chem. 24, 856-61 (1976).

25) Bywood, R., Challenger, F.: The evolution of dimethyl sulphide by Enteromorpha intestinalis. Isolation of dimethyl-ß-carboxyethyl sulphonium chloride from the alga. Biochem.J. 53, XXVI (1953).

26) Cameron, A.F., Ferguson, G., Robertson, J.M.: Laurencia natural products. II. Crystal structure and absolute stereochemistry of laurinterol acetate, a bicyclo-(3,1,o) hexane derivative. J.Chem.Soc.B.Chem.Commun. 692-93 (1969).

27) Cantoni, G.L., Anderson, D.G.: Enzymatic cleavage of dimethylpropiothetin by Polysiphonia lanosa. J.Biol.Chem. 222, 171-77 (1956)

28) Chadefaud, M.: Sur les physodes des Phéophycées. R.C.Acad.Sci. 194, 1675-77 (1932)

29) Chadefaud, M.: Le Cytoplasme des algues vertes et des algues brunes, ses éléments figurés et ses inclusions. 1er Thèse, Paris (1935).

3o) Challenger, F., Bywood, R., Thomas, P., Hayward, B.J.: Studies on biological methylation. XVII. The natural occurrence and chemical reactions of some thetins. Arch.Biochem.Biophys. 69, 514-23 (1957).

31) Challenger, F., Simpson, M.I.: Studies on biological methylation. XII. A precursor of the dimethyl sulphide evolved by Polysiphonia fastigiata. Dimethyl-2-carboxy-ethylsulphonium hydroxides and its salts. J.Chem.Soc. 1591-97 (1948),

32) Chantraine, J.M., Combaut, G., Teste, J.: Phenols bromés d'un algue rouge, Halopytis incurvus : Acides carboxyliques. Phytochemistry 12, 1793-96 (1973).

33) Chesters, C.G.C., Stott, J.A.: The production of antibiotic substances by seaweeds. 2nd Intern.Seaweed Symp. 49-54 (1955).

34) Chevolot-Magueur, A.-M., Cave, A., Potier, P., Teste, J., Chiaroni, A., Riche, C.: Composés bromés de Rytiphlea tinctoria (Rhodophyceae). Phytochemistry 15, 767-71 (1976).

35) Cimino, G., De Stefano, S., Fenical, W., Minale, L., Sims, J.: Zonaroic acid from the brown seaweed Dictyopteris undulata (which is the same as zonarioides). Experientia 31, 125o-51 (1975).

36) Codomier, L., Bruneau, Y., Combaut, G., Teste, J.: 'Etude biologique et chimique d'Asparagopsis armata et de Falkenbergia rufolanosa (Rhodophycées, Bonnemaisoniales). C.R.Acad.Sci.Ser.D 1163-65 (1977).

37) Colin, H., Augier, J.: Un glucide original chez les Floridées du genre Polysiphonia : le d-mannoside du glycérate de sodium. C.R.Acad. Sci. 2o9, 145o (1939).

38) Conover, J.T., Sieburth, J. McN. . : Effect of tannins excreted from Phaeophyta of planktonic survival in tide pools. Proc.Intern. Seaweed Symp. Vth., 99-1oo (1965).

39) Craigie, J.S., Gruenig, D.E.: Bromophenols from red algae. Science 157, 1o58-59 (1967).

4o) Craigie, J.S., McInnes, A.G., Ragan, M.A., Walter, J.A.: Chemical constituents of the physodes of brown algae. Characterization by ^{1}H and ^{13}C nuclear magnetic resonance spectroscopy of oligomers of phloroglucinol from Fucus vesiculosus (L.). Can.J.Chem. 55, 1575-82 (1977).

41) Craigie, J.S., McLachlan, J.: Excretion of colored ultraviolet-absorbing substances by marine algae. Can.J.Bot. 42, 23-33 (1964).

42) Crato, E.: Die Physode, ein Organ des Zellenleibes. Ber.Dtsch.Bot.Ges. 1o, 295-3o2 (1892).

43) Crato, E.: Morphologische und mikrochemische Untersuchungen über die Physoden. Bot.Z. 51, 157-95 (1893).

44) Esping, U.: A factor inhibiting fertilization of sea urchin eggs from extracts of the alga Fucus vesiculosus. I. The preparation of the factor inhibiting fertilization. Ark.Kemi 11, 1o7-15 (1957).

45) Esping, U.: II. The effect of the factor inhibiting fertilization on some enzymes. Ark.Kemi 11, 117-27 (1957).

46) Fattorusso, E., Magno, S., Mayol, L., Santacroce, C., Sica, D., Amico, V., Oriente, G., Piattelli, M., Tringali, C.: Dictyol A and B, two novel diterpene alcohols from the brown alga Dictyota dichotoma. J.Chem.Soc.Chem.Commun., 575-76 (1976).

47) Faulkner, D.J.: Interesting aspects of marine natural products chemistry. Tetrahedron 33, 142-51 (1977).

48) Faulkner, D.J., Ravi, B.N., Finer, J., Clardy, J. : Diterpenes from Dictyota dichotoma. Phytochemistry 16, 991-93 (1977).

49) Fenical, W. : Polyhaloketones from the red seaweed Asparagopsis taxiformis. Tetrahedron Lett., 4463-66 (1974).

5o) Fenical, W. : Halogenation in the Rhodophyta. A review. J.Phycol. 11, 245-59 (1975).

51) Fenical, W., McConnell, O. : Chromazonarol and isochromazonarol, new chromanols from the brown seaweed Dictyopteris undulata (zonarioides). Experientia 31, 1oo4-oo5 (1975).

52) Fenical, W., McConnell, O. : Simple antibiotics from the red seaweed Dasya pedicellata var. stanfordiana. Phytochemistry 15, 435 (1976).

53) Fenical, W., Sims, W., Radlick, P. : Chondriol, a halogenated acetylene from the marine alga Chondria oppositiclada. Tetrahedron Lett., 313-16 (1973).

54) Fenical, W., Sims, J.J., Squatrito, D., Wing, R.M., Radlick, P. : Zonarol and isozonarol fungitoxic hydroquinones from the brown seaweed Dictyopteris zonarioides. J.Org.Chem. 38, 2383-35 (1973).

55) Fogg, G.E., Boalch, G.T. : Extracellular products in pure cultures of a brown alga. Nature 181, 189-9o (1958).

56) Forster, M. : Diss. Bonn 1978 (in Vorbereitung).

57) Fretheim, K. : Konstitutsjonsstudia av phenoliske forbindelsar fra Ascophyllum nodosum. Diss. Trondheim Norway (1974).

58) Fries, L. : Growth stimulating effects of the bromophenol, Lanosol, on red algae in axenic culture. Experientia 29, 1436-37 (1973).

59) Fries, L. : Growth stimulation of axenic red algae by simple phenolic compounds. J.Exp.Mar.Biol.Ecol. 15, 1-9 (1974).

6o) Fujiwara, T. : Flavor of seaweeds. Koryo 92, 35-44 (1969).

61) Geisler, C. : Diss. Bonn (1977).

62) Glombitza, K.-W. : Antimikrobielle Substanzen in Algen. Habilitationsschrift Bonn D 5 (1968).

63) Glombitza, K.-W. : Antibakterielle Inhaltsstoffe in Algen. 1. Helgoländer wiss. Meeresunters. 19, 367-84 (1969).

64) Glombitza, K.-W. : Antimikrobielle Inhaltsstoffe in Algen. 2. Das Vorkommen von Acrylsäure in verschiedenen Meeresalgen. Planta Med. 18, 21o-221 (197o).

65) Glombitza, K.-W. : Highly hydroxylated phenols from Phaeophyceae. In VIIIth Intern.Seaweed Symp. Bangor, Wales (1974) in press.

66) Glombitza, K.-W. : Highly hydroxylated phenols of the Phaeophyceae. In Faulkner, D.J. and Fenical, W.H., Marine natural products chemiestry. Nato Conf.Ser.IV : Marine Sci., Plenum Press New York/ London (1977).

67) Glombitza, K.-W., Geisler, C., Eckhardt, G. : Antibiotica aus Algen. 21. Phlorotanninvorstufen aus Dictyota dichotoma. Phytochemistry 16, 2o35-6 (1977).

68) Glombitza, K.-W., Hansen,C. : in Vorbereitung (1977).

69) Glombitza, K.-W., Koch, M., Eckhardt, G. : Antibiotica aus Algen. 14. Polyhydroxyphenyläther aus Laminaria ochroleuca. Phytochemistry 15, 1o82-83 (1976).

7o) Glombitza, K.-W., Koch, M., Eckhardt, G. : Antibiotica aus Algen. 17. Chlorierte Phlorethole aus Laminaria ochroleuca. Phytochemistry 16, 796-97 (1977).

71) Glombitza, K.-W., Rauwald, H.-W., Eckhardt, G. : Antibiotica aus Algen. 18. Fucophlorethole, Polyhydroxyoligophenyläther aus Fucus vesiculosus. Planta Med. 32, 33-45 (1977).

72) Glombitza, K.-W., Rauwald, H.-W., Eckhardt, G. : Antibiotica aus Algen. 12. Fucole, Polyhydroxyoligophenyle aus Fucus vesiculosus. Phytochemistry 14, 14o3-o5(1975).

73) Glombitza, K.-W., Rauwald, H.-W., Eckhardt, G. : Antibiotica aus Algen. 19. Phlorotannine aus Himanthalia elongata. Phytochemistry 16, 1614 (1977).

74) Glombitza, K.-W., Rösener, H.U. : Antibiotica aus Algen. 11. Bifuhalol, ein Diphenyläther aus Bifurcaria bifurcata. Phytochemistry 13, 1245-47 (1974).

75) Glombitza, K.-W., Rösener, H.U., Koch, M. : Antibiotica aus Algen. 15. Polyhydroxyoligophenyle und Phenyläther aus Bifurcaria bifurcata. Phytochemistry 15, 1279-81 (1976).

76) Glombitza, K.-W., Rösener, H.U., Müller, D. : Antibiotica aus Algen. 13. Bifuhalol und Diphlorethol aus Cystoseira tamariscifolia. Phytochemistry 14, 1115-16 (1975).

77) Glombitza, K.-W., Rösener, H.U., Vilter, H., Rauwald, W. : Antibiotica aus Algen. 8. Phloroglucin aus Braunalgen. Planta Med. 24, 3o1-o3 (1973).

78) Glombitza, K.-W., Sattler, E. : Antibiotica aus Algen. 1o. Trifuhalol, ein neuer Triphenyläther aus Halidrys siliquosa. Tetrahedron Lett. 43, 4277-8o (1973).

79) Glombitza, K.-W., Stoffelen, H. : 2,3-Dibrom-5-hydroxybenzyl-1',4-disulfat (Dikaliumsalz) aus Rhodomelaceen. Planta Med. 22, 391-95 (1972).

8o) Glombitza, K.-W., Stoffelen, H., Murawski, U., Bielaczek, J., Egge, H.: Antibiotica aus Algen. 9. Bromphenole aus Rhodomelaceen. Planta Med. 25, 1o5-14 (1974).

81) Glombitza, K.-W., Wiedenfeld, G., Eckhardt.G. : Antibiotica aus Algen. 2o. Niedermolekulare Phlorotannine aus Cystoseira baccata. Arch. Pharm., in press (1978).

82) Gol'man, L.P., Mikhaseva, M.F., Reznikov, V.M. : Infrared spectra of lignin preparations of pteridophytes and algae. Dokl.Akad.Nauk B. SSR 17, 1o31-33 (1973).

83) Granroth, B., Hattula, T. : Formation of dimethyl sulfide by brackish water algae and its possible implication for the flavor of Baltic herring. Finn.Chem.Lett., 148-5o (1976).

84) Guillard, R.R.L., Hellebust, J.A. : Growth and the production of extracellular substances by two strains of Phaeocystis poucheti. J.Phycol. 7, 33o-38 (1971).

85) Gunnison, D., Martin, A. : Basis for the resistance of several algae to microbial decomposition. Appl.Microbiol. 29, 729-38 (1975).

86) Haas, P. : The liberation of methyl sulphide by seaweed. Biochem.J. 29, 1297-99 (1935).

87) Hall, S.S., Faulkner, D.J., Fayos, J., Clardy, J. : Oppositol, a brominated sesquiterpene alcohol of a new skeletal class from red alga, Laurencia subopposita. J.Am.Chem.Soc. 95, 7187-89 (1973).

88) Hansen, C. : Untersuchungen über die Verbreitung antibiotischer Eigenschaften und quantitative Bestimmung der antibiotisch wirksamen Acrylsäure bei marinen Algen. Diss. Bonn (1976).

89) Hansen, J.A. : Antibiotic activity of the chrysophyte Ochromonas malhamensis. Physiol. Plant. 29, 234-38 (1973).

9o) Harder, R. : Ernährungsphysiologische Untersuchungen an Cyanophyceen, hauptsächlich dem endophytischen Nostoc punctiforme. Z.Bot. 9, 145-242 (1917).

91) Harder, R., Oppermann, A. : Über antibiotische Stoffe bei den Grünalgen Stichococcus und Protosiphon botryoides. Arch.Mikrobiol. 19, 398-4o1 (1953).

92) Hirschfeld, D.R., Fenical, W., Lin, G.H.Y., Wing, R.M., Radlick, P., Sims, J.J. : Marine natural products. VIII. Pachydictyol A, an exceptional diterpene alcohol from the brown alga Pachydictyon coriaceum. J.Am. Chem.Soc. 95, 4o49-5o (1973).

93) Hodgkin, J.H., Craigie, J.S., McInnes, A.G. : The occurrence of 2,3-dibrombenzylalcohol-4,5-disulfate dipotassium salt in Polysiphonia lanosa. Can.J.Chem. 44, 74-78 (1966).

94) Hoegberg, H.E., Thomson, R.H., King, T.J. : The cymopols, a group of prenylated bromohydroquinones from the green calcareous alga Cymopolia barbata. J.Chem.Soc. Perkin Trans. 1, 1696-7o1 (1976).

95) Hornsey, I.S., Hide, D. : The production of antimicrobial compounds by British marine algae. 1. Antibiotic-producing marine algae. Br.phycol.J. 9, 353-61 (1974).

96) Hornsey, I.S., Hide, D. : The production of antimicrobial compounds by British marine algae. II. Seasonal variation in production of antibiotics. Br.phycol.J. 11, 63-67 (1976).

97) Hornsey, I.S., Hide, D. : The production of antimicrobial compounds by British marine algae. III. Distribution of antimicrobial activity within the algal thallus. Br.phycol.J. 11, 175-81 (1976).

98) Irie, T., Suzuki, M., Kurosawa, E., Masamune, T. : Laurinterol and debromlaurinterol, constituents from Laurencia intermedia. Tetrahedron Lett., 1837-4o (1966).

99) Irie, T., Izawa, M., Kurosawa, E., Masamune, T. : Laurinterol, debromlaurinterol and isolaurinterol constituents of Laurencia intermedia Yamada. Tetrahedron 26, 3271-77 (197o).

1oo) Ishida, Y. : Physiological studies on evolution of dimethyl sulfide from unicellular marine algae. Mem.Coll.Agr., Kyoto Univ. 94, 47-82 (1968).

1o1) Ishida, Y., Kadota, H. : Participation of dimethyl-ß-propiothetin in transmethylation reaction in Gyrodinium cohnii. Nippon Suisan Gakkaishi 34, 699-7o5 (1968).

1o2) Jacobi, G. : Enzyme systems. In : Physiology and biochemistry of algae. Lewin, R.A. Academic Press, New York (1962).

1o3) Jensen, A. : personal communication (1976).

1o4) Jørgensen, E.G.: Growth inhibiting substances formed by algae. Physiol.Plant. 9, 712-26 (1956).

1o5) Jørgensen, E.G. : Effect of cell extracts of algae on the growth of Bacillus subtilis. Carnegie Inst.Washington, Yearb. 348-52 (196o).

1o6) Jørgensen, E.G. : Antibiotic substances from cells and culture solutions of unicellular algae with special reference to some chlorophyll derivatives. Physiol.Plant. 15, 53o-45 (1962).

1o7) Jørgensen, E.G., Steemann-Nielsen, E. : Effect of filtrates from cultures of unicellular algae on the growth of Staphylococcus aureus. Physiol.Plant. 14, 896-9o8 (1961).

1o8) Juettner, F. : ß-Cyclocitral and alkanes in Microcystis (Cyanophyceae). Z. Naturforsch., C : Biosci. 31 C, 491-95 (1976).

1o9) Kahn, V. : Glycine as methyl donor in dimethyl-ß-propiothetin synthesis. J.Exp.Bot. 15, 225-31 (1964).

11o) Kalle, K. : Fluorescenz und Gelbstoff im Bottnischen und Finnischen Meerbassin. Dtsch. Hydrogr.Z. 2, 117-24 (1949).

111) Katayama, T. : Chemical studies on volatile constituents of seaweed. 2. On volatile acids of Ulva pertusa K. . Bull.Jpn.Soc.Sci.Fish. 19, 793-97 (1953).

112) Katayama, T. : Chemical studies on volatile constituents of seaweed. 3. On the terpenes of volatile constituents of Ulva pertusa K. . Bull.Jpn.Soc.Sci.Fish. 21, 412-15 (1955).

113) Katayama, T. : Chemical studies on volatile constituents of seaweed. 4. On the carbonyl compounds of volatile constituents of Ulva pertusa. Bull.Jpn.Soc.Sci.Fish.21, 416-19 (1955).

114) Katayama, T. : Chemical studies on volatile constituents of seaweed. 5. On volatile constituents of Enteromorpha sp.. Bull.Jpn.Soc. Sci.Fish. 21, 42o-24 (1955).

115) Katayama, T. : Chemical constituents of seaweed. 6. On volatile constituents of Sargassum spec.,Bull.Jpn.Soc.Sci.Fish. 21, 425-28 (1955).

116) Katayama, T. : Volatile constituents of seaweeds. 7. Flavor components and other characteristics of Porphyra tenera. Nippon Suisangaku Kaishi 22, 244-47 (1956).

117) Katayama, T. : Chemical studies on volatile constituents of seaweeds. 1o. Antibacterial action of a seaweed (Enteromorpha). Bull.Jpn.Soc. Sci.Fish. 22, 248-5o (1956).

118) Katayama, T. : Chemical studies on volatile constituents of seaweeds. 11. Anthelmintic action of the volatile compounds of seaweed. Bull.Jpn.Soc.Sci.Fish.22, 251-52 (1956).

119) Katayama, T. : Chemical studies on volatile constituents of seaweeds. 12. Effect of the volatile components of seaweeds on the muscle contraction of earthworms. Bull.Jpn.Soc.Sci.Fish. 22, 253-56 (1956).

12o) Katayama, T. : Chemical studies on volatile constituents of seaweeds. 9. On the volatile constituents of Codium fragile (Suringar) Hariot. J.Fac.Fisheries Anim. Husbandry, Hiroshima Univ. 2, 67-76 (1958).

121) Katayama, T. : Chemical studies on volatile constituents of seaweeds. On volatile constituents of Digenia simplex. Bull.Jpn.Soc.Sci. Fish. 24, 2o5-o8 (1958).

122) Katayama, T. : Chemical studies on volatile constituents of seaweed. 13. On volatile constituents of Laminaria spec.. Bull.Jpn.Soc. Sci.Fish. 24, 346-54 (1958).

123) Katayama, T. : The volatile constituents of seaweed. 14. On the volatile constituents of Laminaria. Nippon Suisan Gakkaishi 24, 925-32 (1959).

124) Katayama, T. : Chemical studies on volatile constituents of seaweed. 15. On volatile constituents of Laminaria spec.. Bull.Jpn.Soc. Sci.Fish. 26, 818-22 (196o).

125) Katayama, T. : Chemical studies on volatile constituents of seaweed. 16. Their phylogenetic and biochemical significance. Bull.Jpn.Soc. Sci.Fish. 27, 75-84 (1961).

126) Katayama, T. : Biochemical significance of the existence of acrylic acid in seaweeds. Nippon Sorui Gakkai 12, 14-19 (1964).

127) Katayama, T. : Volatile constituents of algae. 2o. Pharmacological action of volatile constituents and biochemical significance of the existence of acrylic acid. Kagoshima Daigaku Suisan Gakubu Kiyo 13, 58-72 (1964).

128) Katayama, T. : Volatile constituents of algae. 21. On volatile constituents of Chlorella pyrenoidosa. Kagoshima Daigaku Suisan Gakubu Kiyo 15, 13-18 (1966).

129) Katayama, T., Nagai, I. : Chemical significance of the volatile components of spices in the food preservative viewpoint. 4. Structure and antibacterial activity of terpenes. Bull.Jpn.Soc.Sci.Fish. 26, 29-32 (196o).

13o) Katayama, T., Tomiyama, T. : Chemical studies on volatile constituents of seaweed. 1. The fractionation of the volatile constituents of Ulva pertusa. Bull.Jpn.Soc.Sci.Fish. 17, 122-27 (1951).

131) Katsui, N., Suzuki, Y., Kitamura, S., Irie, T. : 5,6-Dibromoprotocatechualdehyde and 2,3-dibromo-4,5-dihydroxybenzyl methyl ether. Tetrahedron 23, 1185-88 (1967).

132) Kazlauskas, R., Murphy, P.T., Quinn, R.J., Wells, R.J. : A new class of halogenated lactones from the red alga Delisea fimbriata (Bonnemaisoniaceae). Tetrahedron Lett., 37-4o (1977).

133) Khaleafa, A.F., Kharboush, M.A.M., Metwalli, A., Mohsen, A.F., Serwi, A. : Antibiotic (fungicidal) action from extracts of some seaweeds. Bot.Mar. 18, 163-65 (1975).

134) Kobbia, I.A., Zaki, D. : Biological evaluation of algal filtrates. Planta Med. 3o, 9o-92 (1976).

135) Koch, M. : Diss. in Vorbereitung, Bonn (1978).

136) Krzywicka, A. : Influence of time exposure to light on the activity of the antibacterial substance in Chlorella vulgaris. Bull.Acad.Pol. Sci. Cl II, 14, 5o9-12 (1966).

137) Krzywicka, A. : Growth inhibitors and antibacterial substances in algae. Postepy Mikrobiol. 8, 41-59 (1969).

138) Kurata, K., Amiya, T. : Chemical studies on constituents of marine algae - 2. Constituents of the red marine alga Rhodomela subfusa. Bull.Jpn.Soc.Sci.Fish. 41, 657-59 (1975).

139) Kurata, K., Amiya, T., Nakano, N. : 3,3'-Dibromo-4,4',5,5"-tetrahydroxybibenzyl, a new bromophenol from the red alga Polysiphonia urceolata. Chem.Lett., 821-22 (1976).

14o) Kurata, K., Amiya, T., Yabe, K. : Constituents of a red marine alga, Odonthalia corymbifera. Nippon Suisan Gakkaishi 39, 973-78 (1973).

141) Lamure, J. : Les constituents glucidiques de Vidalia volubilis et de Halopithys pinastroides. C.R.Acad.Sci. 218, 246-47 (1944).

142) Leman, A. : Sur un produit phénolique extrait des algues. Bull.Chem. Soc.France 11, 564-67 (1944).

143) Leifertová, I., Hejtmánková, N., Hlava, H., Kudrnáčová, J., Šantavý, F. : Antifungal and antibacterial effects of phenolic substances. A. study of the relationship between the biological activity and the constitution of the investigated compounds. Acta Univ. Palack.Olomucensis, 74, 83-1o1 (1975).

144) Le Touzé, M.H. : Contribution a l'étude histologiques des Fucacées. Rev.Gén.Bot. 24, 33-47 (1912).

145) Levring, T. : Culture experiments. The influence of algal and plankton extracts. K.Vet.O.Vitterh.Samh.Handl., F 6 Ser. B, 3, 11-18 (1945).

146) Liersch, R. : Über das ätherische Öl von Grünalgen. I. Die Öle der Gattung Chlorella. Arch.Microbiol. 1o7, 353-56 (1976).

147) Liersch, R. : Essential oil of green algae. III. The oils of some Chlorella mutants. Arch.Microbiol. 1o9, 321-22 (1976).

148) Loiseaux, S. : Ultrastructure of zoidogenesis in unilocular zoidocysts of several brown algae. J.Phycol. 9, 277-89 (1973).

149) Manskaya, S.M. : Lignin of the cell membran in the evolutionary order of plants. Khim.Ispol z.Lignina 3-11, Sergeev, V.N."Zinatne" : Riga, USSR (1974).

15o) Mastagli, P., Augier, J. : Étude de la structure de composé phénolique contenu dans Polysiphonia fastigiata. C.R.Acad.Sci. 229, 775-76 (1949).

151) Matusiak, K., Jaroszýńska, T., Kryzwicka, A. : Activity of antibacterial substance in Chlorella vulgaris and Chlorella pyrenoidosa at various stages of their development cycle and the influence of light on the process. Bull.Acad.Pol. Sci. Cl. II Ser.Sci.Biol. 13, 667-71 (1965).

152) Mautner, H.G., Gardner, G.M., Pratt, R. : Antibiotic activity of seaweed extracts. II. Rhodomela larix. J.Am.Pharm.Assoc. 42, 294-96 (1953).

153) McConnell, O., Fenical, W. : Halogen chemistry of the red alga Asparagopsis. Phytochemistry 16, 367-74 (1977).

154) McCully, M. : Histological studies on the genus Fucus. III. Fine structure and possible functions of the epidermal cells of the vegetative thallus. J.Cell.Sci. 3, 1-16 (1968).

155) McEnroe, F.J., Robertson, K.J., Fenical, W. : Diterpenoid synthesis in brown seaweeds of the family Dictyotaceae. In Faulkner, D.J., Fenical, W.H. : Marine natural products chemistry. Nato Conf. Ser.IV : Marine Science, Plenum Press New York/London (1977).

156) McLachlan, J., Craigie, J.S. : Algal inhibition by yellow ultraviolet-absorbing substances from Fucus vesiculosus. Can.J.Bot. 42, 287-92 (1964).

157) McLachlan, J., Craigie, J.S. : Antialgal activity of some simple phenols. J.Phycol. 2, 133-35 (1966).

158) Minale, L., Riccio, R. : Constituents of the digestive gland of the molluscs of the genus Aplysia.I. Novel diterpenes from Aplysia defilans. Tetrahedron Lett. 2711 (1976).

159) Moebus, K. : Jahresbericht der Biologischen Anstalt Helgoland (1972).

16o) Moikeha, S.N., Chu, G.W. : Dermatitis-producing alga Lyngbya majuscula in Hawaii. II. Biological properties of the toxic factor. J.Phycol. 7, 8-13 (1971).

161) Moore, R.E. : Volatile compounds from marine algae. Acc.Chem.Res. 1o, 4o (1977).

162) Moreth, C.M., Yentsch, C.S. : Role of chlorophyllase and light in the decomposition of chlorophyll from marine phytoplankton. J.Exp. Mar.Biol.Ecol. 4, 238-49 (197o).

163) Morita, K., Kobayashi, S. : Isolation and synthesis of lenthionine, an odorous substance of Shiitake, an edible mushroom. Tetrahedron Lett. 573 (1966).

164) Morita, K., Kobayashi, S. : Isolation, structure and synthesis of lenthionine and its analogs. Chem.Pharm.Bull. 15, 988 (1967).

165) Moss, B. : The transition from vegetative to fertile tissue in Fucus vesiculosus. Brit.Phycol.Bull. 3, 567-73 (1968).

166) Mynderse, J. : Halogenated monoterpenes from Plocamium cartilagineum. Diss. UC, San Diego (1975).

167) Nadal, N.G.M., Chapel, C.M.C., Rodrîguez, L.V., Perazza, J.R.R., Vera, L.T. : Antibiotic properties of marine algae. III Cymopolia barbata. Bot.Mar. 9, 21-25 (1966).

168) Nadal, Martînez, N.G., Rodrîguez, L.V., Casillas, C. : Isolation and characterization of sarganin complex, a new broad spectrum antibiotic isolated from marine algae. Antimicrob.Agents Chemother., 131-34 (1964).

169) Nakabayashi, T.,Hada, N. : Studies on tannins of Spirogyra arcta. 1.Isolation of tannins. J.Agric.Chem.Soc.Jpn. 28, 387-91 (1954).

17o) Nakabayashi, T., Hada, N. : Studies on tannins of Spirogyra arcta. 2. Investigation of the becoming of methyl gallate and the sugar in Spirogyra tannin. J.Agric.Chem.Soc. Jpn. 28, 788-91 (1954).

171) Nakabayashi, T. : Studies on tannins of Spirogyra arcta. 3. Estimation of tannin in Spirogyra arcta. J.Agric.Chem.Soc.Jpn. 28, 958-61 (1954).

172) Nakabayashi, T. : Studies on tannins of Spriogyra arcta. 4. Formation of alkyl gallate from tannic acid by fungous tannase. J.Agric.Chem.Soc. Jpn. 29, 161-65 (1955).

173) Nakabayashi, T. : Studies on tannins of Spyrogyra arcta. 5. On the structure of Spyrogyra tannin. J.Agric.Chem.Soc.Jpn. 29, 897-99 (1955).

174) Noda, H., Horiguchi, Y. : Studies on the flavor substances of "Nori", the dried laver Porphyra tenera. I. Dimethyl sulfide and dimethyl-ß-propiothetin. Bull.Jpn.Soc.Sci.Fish. 41, 481-86 (1975).

175) Obata, Y., Igarashi, H., Matano, K. : Studies on the flavor of marine algae. I. On the aromatic component of green algae. Bull.Jpn.Soc. Sci.Fish. 17, 6o-62 (1951).

176) Ogino, C., Taki, Y. : Studies on the tannins of brown alga. Sargassum ringgoldianum Harv. J.Tokyo Univ.Fish. 43, 1-5 (1957).

177) Ohmann, E. : Der Nachweis einiger Enzyme des Intermediärstoffwechsels in Algen des Golfes von Neapel. Pubbl.Staz.Zool.Napoli 33, 331-4o (1963).

178) Ohta, K., Takagi, M. : Halogenated sesquiterpenes from the marine red alga Marginisporum aberrans. Phytochemistry 16, 1o62-63 (1977).

179) Ohta, K., Takagi, M. : Antimicrobial compounds of the marine red alga Marginisporum aberrans. Phytochemistry 16, 1o85-86 (1977).

18o) Olesen, P.E., Maretzki, A., Almodovar, L.A. : An investigation of antimicrobial substances from marine algae. Bot.Mar. 6, 224-32 (1964).

181) Olfers-Weber, R. : Antibakterielle Substanzen aus der Süßwasseralge Hydrodictyon reticulatum. Diss. Bonn (1977).

182) Pedersén, M., DaSilva, E.J. : Simple brominated phenols in the blue-green alga Calothrix brevissima West. Planta 115, 83-86 (1973).

183) Pedersén, M., Fries, L. : Bromophenols in Fucus vesiculosus. Z.Pflanzenphysiol. 74, 272-74 (1975).

184) Pedersén, M., Saenger, P., Fries, L. : Simple brominated phenols in red algae. Phytochemistry 13, 2273-79 (1974).

185) Peguy, M. : A propos de l' activité présumé d' une solution aqueuse du corps phénolique bromé et sulfoné extrait du Polysiphonia lanosa (Linné) Tandy. C.R.Acad.Sci. 252, 2134-15 (1961)

186) Peguy, M. : Quelques essais de séparation par chromatographie sur papier des corps phénoliques bromés extraits de certaines Rhodomelacées. Proc.Inter.Seaweed Symp. IVth. Biarritz 1963, 366-69 (1964).

187) Pesando, D. : Etude chimique et structurale d' une substance lipidique antibiotique produite par une diatomée marine : Asterionella japonica. Rev.Intern.Océanogr.Med. 25, 49-69 (1972).

188) Pettus, J.A., Wing, R.M., Sims, J.J. : Marine natural products XII. Isolation of a family of multihalogenated gamma-methylene lactones from the red seaweed Delisea fimbriata. Tetrahedron Lett. 41-44, (1977).

188a) Pinter, P. : Bryozoa-algal associations in Southern California waters. Bull.S.Calif.Acad.Sci. 68, 199-218 (1969).

189) Porzi, G., Bertazzoni, E.M. : Attivitá antibatterica e presenza de mercurio in dodici specie di alghe marine. Boll.Soc.Ital.Biol.Sper. 51, 1591-95 (1975)

19o) Pratt, R. : Influence of the size of the inocolum on the growth of Chlorella vulgaris in freshly prepared culture medium. Am.J.Bot. 27, 52-56 (194o).

191) Pratt, R. : Studies on Chlorella vulgaris. V. Some properties of the growth inhibitor formed by Chlorella vulgaris. Am.J.Bot. 29, 142-48 (1942).

192) Pratt, R. : Studies on Chlorella vulgaris. IX. Influence on growth of Chlorella of continuous removal of chlorellin from culture solution. Am.J.Bot. 31, 418-421 (1944).

193) Pratt, R. : Studies on Chlorella vulgaris. XI. Relation between surface tension and accumulation of Chlorella. Am.J.Bot. 35, 634-37 (1948).

194) Pratt, R., Daniels, T.C., Eiler, J.B., Gunnison, J.B., Kumler, W.D., Oneto, J.F., Strait, L.A., Spoehr, H.A., Hardin, G.J., Milner, H.W., Smith, J.H.C., Strain, H.H. : Chlorellin, an antibacterial substance from Chlorella. Science 99, 351-52 (1944).

195) Pratt, R., Fong, J. : Studies on Chlorella vulgaris. II. Further evidence that Chlorella cells form a growth-inhibiting substance. Am.J.Bot. 27, 431-36 (194o).

196) Pratt, R., Oneto, J.F., Pratt, J. : Studies on Chlorella vulgaris. X. Influence of the age of the culture on the accumulation of Chlorellin. Am.J.Bot. 32, 4o5-o8 (1945).

197) Proctor, V.W. : Studies of algal antibiosis using Haematococcus and Chlamydomonas. Limnol.Oceanogr. 1, 125-39 (1956).

198) Provasoli, L. : Nutritional aspects of algal growth. Proc. Can. Soc. Plant Physiol. 6, 126-27 (1965).

199) Ragan, M.A. : Physodes and the phenolic compounds of brown algae. Composition and significance of physodes in vivo. Bot.Mar. 19, 145-54 (1976).

2oo) Ragan, M.A., Craigie, J.S. : Physodes and the phenolic compounds of brown algae. Isolation and characterization of phloroglucinol polymers from Fucus vesiculosus (L.). Can.J.Biochem. 54, 66-73 (1976).

2o1) Ripa, R., Borquez, L.M. : Lignin in the phaeophycean alga, Durvillea antarctica. Sciencia (Valparaiso) 34, 36-38 (1967).

2o2) Robertson, K.J., Fenical, W. : Pachydictyol-A epoxide, a new diterpene from the brown seaweed Dictyota flabellata. Phytochemistry 16, 1o71-73 (1977).

2o3) Roller, P., Au, K., Moore, R.E. : Isolation of S-(3-Oxoundecyl)-thioacetate, Bis-(3-oxoundecyl) disulphide, (-)-3-Hexyl-4,5-dithiacycloheptanone, and S-(trans-3-oxoundec-4-enyl)thioacetate from Dictyopteris. Chem.Commun. 5o3 (1971).

2o4) Saenger, P., Pedersén, M., Rowan, K.S. : Bromo-compounds of the red alga Lenormandia prolifera. Phytochemistry 15, 1957-58 (1976).

2o5) Saito, K., Ando, Y. : A bromphenolic compound from the red alga Polysiphonia morrowii. Harv.. Nippon Kagaku Zasshi 76, 478-79 (1955).

2o6) Saito, K., Sameshima, M. : Studies on the antibiotic action of algae extracts. J.Agric.Chem.Soc.Jpn. 29, 427-3o (1955).

2o7) Sakevich, A.I., Atanasova, R.S., Mashanova, N.S. : Component composition of essential oil in bluegreen algae Microcystis aeruginosa. Dopov.Akad.Nauk.Ukr.RSR Ser.B 34, 164-66 (1972).

2o8) Sattler, E., Glombitza, K.-W., Wehrli, F.W., Eckhardt, G. : Antibiotica aus Algen. 16. Mitt. Polyhydroxyphenyläther aus der Phaeophycee Halidrys siliquosa. Tetrahedron 33, 1239-44 (1977).

2o9) Seshadri, R., Sieburth, J.M. : Cultural estimation of yeasts on seaweeds. Appl. Microbiol. 22, 5o7-12 (1971).

21o) Sharma, G.M., Michaels, L., Burkholder, P.R. : Goniodomin, a new antibiotic from a Dinoflagellate. J.Antibiot. 21, 659-64 (1968).

211) Sieburth, J.M. : Acrylic acid, an "antibiotic" principle in Phaeocystis blooms in antarctic waters. Science 132, 676-77 (196o).

212) Sieburth, J.M. : Antibiotic properties of acrylic acid, a factor in the gastrointestinal antibiosis of polar marine animals. J.Bacteriol. 82, 72-79 (1961).

213) Sieburth, J.M. : Antibacterial compositions and methods of using same. US.P. 3 o82 15o dated 19.3.1963 (16.8.6o).

214) Sieburth, J.M. : Antibacterial substances produced by marine algae. Dev.Ind.Microbiol. 5, 124-34 (1964).

215) Sieburth, J.M., Conover, J. : Sargassum tannin, an antibiotic which retards fouling. Nature 2o8, 52-53 (1965).

216) Sieburth, J.M., Jensen, A. : Studies on algal substances in the sea. I. Gelbstoff (humic material) in terrestrial and marine waters. J.Exp.Mar.Biol.Ecol. 2, 174-89 (1968).

217) Sieburth, J.M., Jensen, A. : Studies on algal substances in the sea. II. The formation of Gelbstoff (humic material) by exudates of Phaeophyta. J.Exp.Mar.Biol.Ecol. 3, 275-89 (1969).

218) Sieburth, J.M., Jensen, A. : Studies on algal substances in the sea. III. The production of extracellular organic matter by littoral marine algae. J.Exp.Mar.Biol.Ecol. 3, 29o-3o9 (1969).

219) Sims, J.J., Donnell, M.S., Leary, J.V., Lacy, G.H. : Antimicrobial agents from marine algae. Antimicrob.Agents Chemother. 7, 32o-21, (1975).

22o) Sirenko, L.A., Sakevich, A.J. : Die Bildung von Terpenen bei Blaualgen. Dokl.Akad.Nauk SSSR 177, 959-6o (1967).

221) Siuda, J.F., VanBlaricom, G.R., Shaw, P.D., Johnson, R.D., White, R.H., Hager, L.P., Rinehart, K.L. : 1-Iodo-3,3-dibromo-2-heptanone, 1,1,3,3-tetrabromo-2-heptanone and related compounds from the red alga Bonnemaisonia hamifera. J.Am.Chem.Soc. 97, 937-38 (1975).

222) Spoehr, H.A., Smith, J.H.C., Strain, H.H., Milner, H.W., Hardin, G.J. : Fatty acid antibacterials from plants. Carnegie Inst.Washington Pub. 586, 1-67 (1949).

223) Spoehr, H.A., Smith, J.H.C., Strain, H.H., Milner, H.W., Hardin, G.J. : Chlorellin and similar antibiotics substances. Carnegie Inst. Washington Pub. 44, 66-71 (1945).

224) Stack, V.T. : Toxicity of alpha, beta-unsaturated carbonyl compounds to microorganisms. Ind.Eng.Chem. 49, 913-17 (1957).

225) Stoffelen, H. : Antibakterielle Eigenschaften in der Rotalgenfamilie der Rhodomelaceen, unveröffentlichte Details. Diss. Bonn D 5 (1973).

226) Stoffelen, H., Glombitza, K.-W., Murawski, U., Bielaczek, J., Egge, H.: Bromphenole aus Polysiphonia lanosa (L.) Tandy. Planta Med. 22, 396-4o1 (1972).

227) Suneson, S. : Über wachstumsfördernde Wirkung von Algenextrakten auf Ulva und Enteromorpha. Fysiogr.Sällsk.Förhandl. 12, 183-2o2 (1942).

228) Suneson, S. : Weitere Untersuchungen über wachstumsfördernde Wirkung von Algenextrakten auf Ulva latuca. Fysiogr.Sällsk.Förhandl. 13, 193-2o1 (1943).

229) Suyehiro, Y., Searashi, T. : The utilization of the inedable seaweeds. Bull. Tokyo Univ. Physiographical Sci.Res.Inst. I, 44-5o (1948).

23o) Swanson, C.A. : The effect of culture filtrates on respiration on Chlorella vulgaris. Am.J.Bot. 3o, 8-11 (1943).

231) Tabachek, J.L., Yurkowski, M. : Isolation and identification of blue-green algae producing muddy odor metabolites, geosmin, and 2-methyl-isoborneol, in saline lakes in Manitoba. J.Fish.Res.Board Can. 33, 25-35 (1976).

232) Takahashi, T. : Studies on Ecklonia cava. III. On black pigment. Tokyo Kogyo Shikensho Hokoku 26, 1-9 (1931).

233) Taylor, R.F., Ikawa, M., Sasner, J.J., Thurberg, F.P., Andersen, K.K. : Occurence of choline esters in the marine dinoflagellate Amphidinium carteri. J.Phycol. 1o, 279-83 (1974).

234) Telutchenko, M.M., Davydova, N.V., Fedorov, V.D. : Wechselbeziehungen zwischen Wasserpflanzen und Mikroorganismen. 2. Einfluß des Entwicklungszustandes der Protokokkenkulturen Chlorella vulg. und Scenedesmus obliquus auf die Hemmung des Mäusetyphuserregers Salmonella typhimurium. Naucnye dokl. vyssey skoly, Biol. Nauki 4, 157-63 (1962).

235) Telutchenko, M.M., Fedorov, V.D. : The interrelation between algae and bacteria in reservoirs. Byull.Moskov.Obshch.Ispytat.Prirody Otdel Biol. 67, 148-49 (1962).

236) Tocher, C.S., Ackman, R.G. : The identification of dimethyl propiothetin in the algae Syracosphaera carterae and Ulva lactuca. Can.J.Biochem. 44, 519-22 (1966).

237) Velev, T.K.: Isolierung, Trennung und Identifizierung der aus Grünalgen gewonnenen Konkrete. Diss. Univ. Sofia (1973).

237a)Vilter, H. : Gerbstoffhaltige Physoden der Braunalgen. Staatsarb., Bonn (1973).

238) Weinstein, B., Rold, T.L., Harrell, C.E., Burns III, M.W., Waaland, J.R. Reexamination of the bromophenols in the red alga Rhodomela larix. Phytochemistry 14, 2667-7o (1975).

239) Wolters, B. : Antibiotisch und toxisch wirksame Substanzen aus Algen und Moosen. Planta Med. 12, 85-99 (1964).

24o) Wratten, S.J., Faulkner, D.J. : Cyclic polysulfides from the red alga Chondria californica. J.Org.Chem. 41, 2465-67 (1976).

241) Zolotovich, G., Velev, T., Mikhailova, S., Dilov, Kh. : Odorous substances in cultivated microalgae. Dokl.Acad.Sel'skokhoz.Nauk.Bolg. 4, 455-55 (1971).

242) Zolotovich, G., Velev, C., Mikhailova, S. : Study of the composition of concretions from fresh water cultivated microalgae. Prilozhna Mikrobiol. 4, 69-86 (1974).

CHEMICAL CHARACTERIZATION AND THERAPEUTIC EVALUATION OF ANTI-HERPESVIRUS POLYSACCHARIDES FROM SPECIES OF DUMONTIACEAE.

M.T. Hatch, D.W. Ehresmann, and E.F. Deig

Naval Biosciences Laboratory, School of Public Health, University of California, Berkeley, California 94720 U.S.A.

INTRODUCTION

Recently we found an antiviral compound in marine red algae that may have some usefulness in the control of herpes simplex virus (HSV) infections (6). The HSV problem is of considerable public health importance, particularly in view of the fact that type 1 viruses cause acute respiratory illness, recurrent oral lesions, herpes keratitis (an eye infection responsible for up to 18,000 cases of blindness in the USA every year) and meningitis; whereas viruses comprising type 2 initiate venereal disease and are under suspicion as one cause of cervical cancer (3, 5, 17). Although several antiviral agents appear to have a potential application in the treatment of herpesvirus infections of man (24), prevention of recurrences of herpetic keratitis, oral lesions and venereal infections attributable to HSV is still a critical unsolved problem (1, 16, 20). Moreover, increasing numbers of individuals are currently suffering from exaggerated herpetic diseases that can be life threatening or can maim severely (2). Continued exploration is therefore necessary to discover new and clinically useful drugs active against HSV. However, it is noteworthy that the selection of a candidate antiviral for clinical testing in man is made only after numerous preclinical evaluations including: (a) identification in terms of accepted chemical nomenclature of active chemical moieties, (b) examination for possible unique physicochemical characteristics of the viral inhibitor, (c) demonstration of selective antiviral activity in at least one *in vivo* and one *in vitro* model system and (d) determination of

Marine Algae in Pharmaceutical Science
edited by H. A. Hoppe, T. Levring, Y. Tanaka

drug-host interactions, particularly with respect to viral replication, cell metabolism and toxicity of the active component or its degradation products at potentially effective dosage levels.

In our earlier studies (6, 10), we presented evidence that the selective anti-herpesvirus material extracted from red algae consisted of large molecules which were excluded from Sephadex G-50 and G-200 columns. The active substance was stable to heat and a wide range of pH. It was not partitioned into common lipid solvents but could be precipitated in the presence of methanol, ethanol or acetone. Spectroscopic observations of active crude extracts from one source, *Constantinea simplex* Setchell, produced a broad peak at wavelengths from 260 to 280 nm. The cumulative evidence suggested that the active agent consisted of a major polysaccharide fraction of unknown homogeneity bound in some way to a minor protein or nucleic acid fraction. This paper is concerned with further chemical characterization of the antiviral substance as well as results of our initial experiments to determine the *in vivo* efficacy of crude and partially purified algal extracts in a herpesvirus-induced mouse encephalitis model.

MATERIALS AND METHODS

Source of Seaweed and Preparation of Extracts

Specimens of *Farlowia mollis* (Harv. & Bail.) Farlow & Setchell, *Cryptosiphonia woodii* Ag. and *C. simplex* were harvested during low tides from the same coastal areas of northern California previously reported (6, 10). The seaweeds were packed in ice and transported back to the laboratory, where they were thoroughly washed with sea water and mechanically homogenized (either before or after storage at -60 C) in citrate-phosphate buffer (pH 7.0) as described (6), unless noted otherwise. Polysaccharides were extracted, as reported previously (10) using the method outlined by Su and Hassid (29). Finally, the residue was recovered by filtration and suspended in water. The antiviral fraction was then removed from aqueous solution by ethanol precipitation after which the precipitate was usually washed and dried by lyophilization. In some instances, the viral inhibitor was partitioned using water-saturated phenol, as outlined previously (10), or was selectively precipitated with concanavalin A (Con A), a lectin from the jack bean (*Canavalia ensiformis*).

Chemicals and Enzymes

All chemicals were reagent grade unless specified otherwise. Con A (3X crystallized) was purchased from Miles Laboratories, Inc., Elkhart, Ind. Ribonuclease A (RNase) and pronase (*Streptomyces griseus* protease) were from Calbiochem, La Jolla, Ca. Heparin (sodium salt) and heparin (ammonium salt) were from Abbott Laboratory, North Chicago, Ill., and Scientific Products Div., American Hospital Supply Corp., McGaw Park, Ill., respectively. Sodium dextran sulfate and diethylaminoethyl-dextran (DEAE-dextran) were from Pharmacia Fine Chemicals Inc., Piscataway, N.J. Bio-Gel A-0.5 was obtained from Bio-Rad Laboratories, Richmond, Ca. Kappa, lambda and iota carrageenans, as well as PA and PC unialgans, were supplied by Donald W. Renn (Marine Colloids, Inc., Rockland, Me.).

In Vitro Antiviral Assay

The presence of anti-herpesvirus substances in algal extracts or fractions eluted from columns was usually detected and quantitated as described previously (6, 10), using cell cultures infected with serially diluted HSV-1 and HSV-2 in the presence or absence of algal material. The protective effect of a given extract or fraction was termed the index of protection (IOP), defined as the $\log_{10}$ difference between $TCID_{50}$ titers of the virus in treated and untreated control cells.

In order to rapidly detect antiviral substances in primary

extracts of algae or subsequent fractions eluted from Bio-Gel columns, we mixed a constant amount of extract (usually reconstituted from the lyophilized state in 0.2 M phosphate buffer pH 5.8, 6.5 or 7.6), or a given eluate, with an equal volume of Con A suspended in phosphate buffer at a specified pH to a final concentration of 140-180 µg/ml. The mixture was left undisturbed at room temperature for 2 h, after which Con A-inhibitor aggregates were removed by low speed centrifugation. The residual amount of unreacted Con A in solution was then detected quantitatively by a Fluorolog spectrofluorometer (SPEX Industries, Inc., Metuchen, N.J.).

In Vivo Test System

NAMRU strain mice (11) were used throughout this study. The animals were usually housed as litters of 8 animals with their mothers in Isocap cages until they were two to four days old at which time they were inoculated by the intraperitoneal (ip) route. Each mouse in a given group received 0.03 ml of a suspension containing either algal extract, HSV or both virus and extract. The treatment schedules utilized in these studies consisted of: (a) administering a given extract and subsequently (usually 2 or 24 h) injecting serially diluted virus by the ip route, (b) challenging groups of mice with given virus dilutions and 2 or 24 h later, administering a single dose per animal of algal extract or (c) administering both extract and serially diluted virus at the same time. Groups of 8 animals were routinely held as placebo-treated, virus-infected controls. Citrate-phosphate buffer served as the placebo solution. The antiviral effect of given extracts was determined by comparing the number of survivors, after 3 weeks of infection of extract-treated and placebo-treated control animals. The LD_{50} titer of treated and control groups of mice was determined (26) and antiviral action was expressed as IOP_m values, where the subscript m differentiates between animal mortality and *in vitro* IOP values.

RESULTS

Chemical Composition of Antiviral Material

In a preliminary chemical analysis of active extracts derived from *F. mollis* (10), we showed that RNase and DNase, when added to extracts at a final concentration of 10 µg/ml, failed to significantly affect antiviral activity. However, trace amounts of nucleic acids were found in the extracts.

Table 1. Stability to selective chemical treatment of anti-herpesvirus polysaccharides derived from *Constantinea simplex*.

Extract Treatment	Virus Titer[1] ($TCID_{50}$/0.2 ml) HSV-1	IOP[2]
RNase, DNase	$10^{4.0}$	2.5
Phenol	$10^{4.3}$	2.2
Pronase	$10^{5.0}$	1.5
None	$10^{4.3}$	2.2
Untreated cells (Control)	$10^{6.5}$	
Con A purified extract	$10^{4.0}$	2.5

1. In cell cultures exposed to treated extracts and challenged with HSV-1, as described in the text.
2. Index of Protection = $\log_{10}$ difference between $TCID_{50}$ values of HSV-1 in treated and control cells.

Results summarized in Table 1 provide strong evidence that active site(s) on antiviral molecules derived from *C. simplex* were also not composed of nucleic acids, although small amounts of nucleic acids may have been present at non-active sites or as contaminants. In this study, the concentration of RNase and DNase used was increased to 20 μg/ml. Even so, nuclease-treated extracts elicited a strongly protective response against HSV-1, as 2.5 IOP values were routinely observed. Similar results were obtained when cells treated with these extracts were challenged with HSV-2.

A second phase of this study was to ascertain if proteins were prominent in the structure of the viral inhibitor. We extracted C. simplex with acetate buffer (pH 6.0) and then added 20 μg/ml pronase (final concentration). After a reaction time of 16 h, excess enzyme was removed by dilution and antiviral material was partitioned using the liquified phenol method described previously (10). Residual phenol was removed with bentonite and diethyl ether. As a control, we partitioned antiviral material by the phenol method in the absence of pronase. As shown in Table 1, extracts without pronase produced an IOP of 2.2 whereas pronase treatment reduced the IOP to 1.5. This observation suggests that the active antiviral substances may indeed contain some protein molecules. However, additional studies are needed to substantiate this possibility.

With the knowledge that Con A reacts with both HSV-1, HSV-2, and the HSV-induced antigen (7, 15), an attempt was made to selectively agglutinate the antiviral material in extracts from C. simplex as Con A-inhibitor complexes. When 180 mg Con A suspended in phosphate buffer was reacted with 40 ml of purified material (reconstituted from the lyophilized state at a concentration of 5 mg/ml), a precipitate was formed. The precipitated complex was recovered and shown to dissociate after 24 h in solution containing α-methyl-D-glucoside, after which antiviral material was recovered from solution by ethanol precipitation. The IOP of the lectin-purified material was 2.5 compared to an IOP of 2.2 with the untreated extract (Table 1). The finding that the purified product was only slightly more protective than the crude extract suggests that chemical degradation or dissociation and possibly loss of specific reactive sites on the antiviral molecule occurred as a consequence of the purification procedure; or that activity of the antiviral agent was associated with a limited number of reactive sites on inhibitor molecules per se or on the viral genome. The finding that Con A selectively removed the viral inhibitor from solution indicated that the antiviral molecule contained branch chain polysaccharides with terminal non-reducing α-D-glucopyranosyl, α-D-mannopyranosyl or β-D-fructofuranosyl residues (28).

Table 2. Analyses for the constituent anions and cations in polysaccharides from *Constantinea simplex*.

	Percent	Milliequivalents/100 g	
H_2O	9.95		
ester-SO_4^-	20.08	209.0	
free $SO_4^=$	0.37	7.7	227.6 mEq. anions
pyruvic acid	0.96	10.9	
Cl^-	0.0		
$PO_4^{\equiv}$	0.0		
Na^+	1.90	82.6	
K^+	3.02	77.2	244.5 mEq. cations
Ca^{++}	0.36	18.0	
Mg^{++}	0.81	66.6	
Ash	17.54		

Further analysis of the antiviral polysaccharide polymer from *C. simplex* resulted in the data of Table 2. As can be seen, a number of anions and cations were identified. Of particular interest was the finding of large amounts of sulfate anions (20%) in the ester form. However, there was a noticeable excess of 16.9 mEq. of cations, which is a rather large discrepancy (6.9% based on cations). This discrepancy could be accounted for by the presence of another anionic moiety, such

Table 3. Antiviral activity of carrageenans, unialgans, and Constantinea simplex extract in cell culture.

Challenge virus	Index of Protection (IOP)*					
	Carrageenans			Unialgans		
	Kappa	Lambda	Iota	PA	PC	C. simplex
Herpes simplex Type 1	2.5	2.0	2.0	0	2.3	3.5
Herpes simplex Type 2	2.2	2.0	2.0	0.5	2.5	ND
Adeno - 7	2.7	2.0	3.2	0.2	3.0	0.7
Coxsackie B_5	2.2	2.2	2.5	0	0	3.0
Echo - 6	3.0	3.5	3.3	0	0.5	0.3
Polio - 2	0.3	0.3	0	0	0	0
Rhino - 2	0	0	1.2	0	0.9	0.9
Vaccinia	0.7	0.5	1.0	0.7	0.5	0
Vesicular stomatitis	2.0	0.9	2.0	0	2.4	0.2
Western equine encephalitis	3.5	3.7	3.5	0	3.7	0

*IOP = $\log_{10}$ difference between $TCID_{50}$ values of virus in treated and control cells.

as carboxylic acid residues. Nonetheless, the data indicated that the polysaccharide contained approximately 1 ester sulfate for each 2 hexose moieties and a mere scattering of pyruvate (about 1 for every 35-40 hexoses).

Antiviral Properties of Carrageenans and Unialgans

Data presented in Table 3 compare the antiviral activity of extracts from C. simplex with other sulfated polysaccharides from marine sources including kappa, lambda and iota carrageenans and unialgans. The carrageenans all were active against HSV-1 and HSV-2, with IOP values varying from 2.0 to 2.5. These values were lower than those found for cells treated with the extract from C. simplex (i.e., an IOP of 3.5). In addition, Table 3 shows that the 3 carrageenans were also active against the adeno, Coxsackie, echo, and western equine encephalitis viruses, and comparatively inactive against polio, rhino and vaccinia viruses. Kappa and iota carrageenans were also active against vesicular stomatitis virus whereas lambda carrageenan was inactive. The PC unialgan showed significant activity against HSV (both types), adeno, vesicular stomatitis and western equine encephalitis viruses whereas PA unialgan was inactive against all of the viruses tested. In contrast, the specificity of the antiviral substances from C. simplex was more limited (Table 3) in that this material was active only against HSV and Coxsackie viruses. It is important to mention that thus far only extracts from C. simplex have demonstrated inhibitory activity against Coxsackie B_5 viruses. Extracts from other species of the family Dumontiaceae have been significantly active only against HSV.

Antiviral Action of Polyanionic Substances

The suspected polyanionic nature of algal sulfated polysaccharides prompted a comparative study with sulfated polyanions from other sources known to be active against HSV. We found that heparin (sodium salt) exerted significant antiviral action giving an IOP of 2.8 toward HSV (Table 4). When heparin was tested as an ammonium salt, the IOP toward this virus was reduced to a barely significant level of 2.0. However, not all polyanions tested were active. Dextran sulfate failed to show significant antiviral action (e.g., the IOP was < 2.0). On the other hand, kappa, lambda and iota carrageenans, unialgan PC and sulfated polysaccharides derived from C. simplex and F. mollis all demonstrated a level of anti-herpesvirus activity similar to the polyanion heparin under the test conditions used. Results summarized in Table 4 also show that the antiviral fraction in these materials was lost when the polycation DEAE-dextran was added. In some instances, the inhibitory action was completely negated whereas in others it was present, but reduced to insignificant levels. Certain of these experiments were carried out with both HSV-1 and HSV-2 and no significant difference in results were noted.

Table 4. Comparison of anti-herpes activity in sulfated polyanions with material from marine sources. Effect of DEAE-dextran interaction.

Test material treating cells	Index of Protection (IOP)* for HSV-1	
	Before DEAE-dextran	After DEAE-dextran
Kappa carrageenan	2.5	1.6
Lambda carrageenan	2.0	0.8
Iota carrageenan	2.0	1.6
Heparin/Na salt	2.8	1.0
Heparin/NH_4 salt	2.0	0
Dextran sulfate	0.5	0.5
C. simplex extract	3.5	1.3
F. mollis extract	3.0	1.0
PC Unialgan	2.3	0.0

*IOP = $\log_{10}$ difference between $TCID_{50}$ values of HSV-1 in treated and control cell cultures.

Con A as a Tool for Detecting Anti-herpesvirus Substances in Test Solution

The practicality of using the selective saccharide-binding property of Con A as a tool for rapid detection, and possible quantitative estimation of algal polysaccharides in extracts was examined. First, an analytical procedure for detection and quantitation of Con A molecules in solution was sought. We found that the lectin suspended in phosphate buffer (pH 6.5) displayed fluorescent properties with an excitation wavelength of 296 nm and emission wavelength of 340 nm. The analytical curve (Fig. 1) was linear between about 20 µg/ml and 180 µg/ml. Because Con A in solution below pH 6.0 is known to form a molecular species made up of two protomers whereas at pH 6.7 and above it may consist of 4 protomers (9, 27), analytical

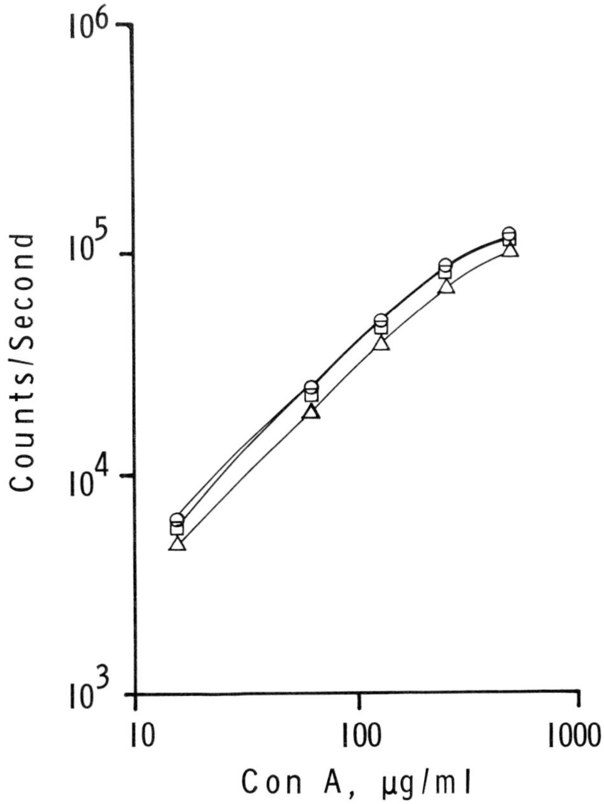

Fig. 1. Comparison of fluorescent analytical curves for concanavalin A in phosphate buffer: O, pH 7.6; □, pH 5.8, △, pH 6.5 (excitation at 296 nm and emission measured at 340 nm).

curves for Con A in buffer above and below pH 6.5 were determined. Since data for Con A dimers and tetramers were reasonably similar (Fig. 1), in all subsequent analytical studies we used buffered solutions of Con A at pH 7.6.

Next, the presence and relative quantities of Con A binding material in algal extracts was assessed by measuring the Con A left in the supernatant after allowing a specified amount of lectin to react with material in algal extracts as described

Table 5. Anti-herpesvirus activity of Cryptosiphonia woodii extracts in infant mice.

Time (hr) between infection and treatment (1)	Mortality(2) from HSV-1		
	Treated group	Untreated (control)	IOPm(3)
(—)2	$10^{3.0}$	$10^{5.2}$	2.2
0	$10^{2.7}$	$10^{5.0}$	2.3
2	$10^{5.3}$	$10^{5.6}$	0.3
24	$10^{5.5}$	$10^{6.0}$	0.5

(1) A negative number indicates treatment preceded virus.

(2) LD_{50} values.

(3) Index of Protection = $\log_{10}$ difference between LD_{50} values for treated and control groups. Significant antiviral activity = IOPm $\geq$ 2.0.

in the Materials and Methods section. Results of preliminary studies showed that extracts from C. woodii and C. simplex reduced the level of Con A in solution from 225 µg/ml to 110 µg/ml and 145 µg/ml, respectively. When these same extracts were evaluated in cell culture the IOP values, respectively, were 3.3 and 2.4. Numerous eluates from Bio-Gel columns were also tested for presence of antiviral material in this manner. Those found to contain active substances by the cell culture method also effectively reduced the level of Con A in the test solution. Additional details of this work will be reported elsewhere.

Therapeutic Evaluation of Anti-herpesvirus Material in Extracts from C. woodii.

Table 5 presents a summary of data from the assay of soluble primary extract from *C. woodii* for antiviral activity toward type 1 HSV in infant mice. Before use in this work, the extract was shown to interfere with the replication of the virus in cell culture. It can be seen that antiviral activity *in vivo* occurred if the extract was given before or at the same time as the mice were infected. Under these conditions, the algal material significantly protected infant mice from fatal encephalitis initiated by HSV. Mice infected for as short a period as 2 hours before treatment were not protected. Mice injected with extract only showed no mortality or other symptoms of a toxic response.

DISCUSSION

Results of the present investigation support our earlier claim (10) and provide strong evidence suggesting that anti-herpesvirus substances found in extracts from marine red algae (family Dumontiaceae) consist of carbohydrate moieties of unknown homogeneity possibly associated with nitrogen-containing molecules (i.e., nucleic acids or proteins) at specific sites on the macromolecule. This hypothesis is supported by results showing: (i) active components could be partitioned in aqueous solution containing phenol and the addition of pronase to purified extracts diminished antiviral activity; (ii) some physical and chemical properties, including % N_2, stability of the inhibitor over a wide pH range, exclusion from Sephadex and Bio-Gel columns and precipitation by boiling or cold ethanol, methanol and acetone, and finally (iii), selective interaction of Con-A with antiviral molecules. The capability of the lectin to form precipitable aggregates with specific inhibitor moieties suggests that at least a portion of the active molecules consist of branched, rather than linear polysaccharides, because Con-A precipitates disaccharide moieties with multireaction sites (13). Whether the lectin is bound preferentially to α-linked sugar residues having the D-arabino configuration (23) or polyelectrolyte structures (8) remains to be determined. Such interactions could involve sulfate groups of algal polysaccharides, or more likely, if present, nonreducing terminal α-linked N-acetyl-D-glucosamine residues.

Although evidence presented in Table 2 shows sulfate reactive groups ($-SO_3^-$) were linked with active polysaccharides, the uniqueness of the compound was shown by comparative studies with sulfated polysaccharides from other sources (i.e., kappa, lambda and iota carrageenans, unialgans, agar and heparin). The algal polysaccharides were specifically active against HSV, except in one case where extracts from C. simplex also interfered with the replication of Coxsackie B_5 viruses; whereas other sulfated polysaccharides tested produced an inhibitory effect against a wide spectrum of viruses. This latter finding is in agreement with those of other investigators who found that many RNA- and DNA-containing viruses are sensitive to sulfated polysaccharides from synthetic or natural sources (12, 31). Thus, our knowledge regarding the diversity of antiviral activity of sulfated polysaccharides in general and the seemingly novel selective interaction between algal antivirals, and herpesvirus in particular, is still incomplete. This raises a number of questions regarding the influence of those distinct structural features, constituent monosaccharides and amino acids which make up active macromolecules of the algal sulfated polysaccharide, and, in addition, the sequence of events leading to specific anti-herpesvirus effects which remain to be explored. It is also relevant to determine whether antiviral substances extracted from all 8 species of Dumontiaceae have identical structural configurations and chemical groups at active sites or whether we are dealing with a number of different but closely related compounds which perchance have specific sites sufficiently similar so they can function in the same manner as a class of unique anti-herpesvirus inhibitors. Finally, the important question as to whether some common mechanism of action exists among sulfated polysaccharides from all sources in that they either interfere with virus adsorption and penetration processes without being consumed, interact chemically or physically directly with virus particles to form virus-inhibitor complexes, induce protective cellular responses, or a combination of all of these, remains to be established.

Evidence obtained to date suggests that those sulfated polysaccharides extracted from Gelidium cartilaginium (the common source of domestic agar) inhibit the development of a number of viruses or virus variants including encephalomyocarditis (EMC) viruses (19, 30). The sulfated agar polysaccharides interact directly with the virus particle and cause a decrease in their rate of attachment to cultured cells. It has also been shown that the degree of virus inhibition is dependent on the extent of sulfation of the sulfated agar polysaccharide. In addition, it is known that sulfated polyanions (i.e., sodium dextran sulfate) can gain entrance into L cells (19). So an intracellular effect on viral multiplication attributable to sulfated polysaccharides cannot be entirely ruled out.

Comparison of the anti-herpesvirus effect of algal sulfated polysaccharide with some sulfated polyanions (see Table 4) also suggests that algal and other sulfated polysaccharides examined are ionic and heparinoid in nature, as the cationic polymer DEAE-dextran effectively removed the antiviral effect in all instances. Heparinoids are compounds having properties similar to heparin (25), a naturally-occurring sulfated polysaccharide. Thus, it is tempting to speculate that mechanisms thought to account for the inhibitory effect of this polyelectrolyte or polyanion on HSV, and the antiviral action of algal sulfated polysaccharides are reasonably similar. Heparin is characterized intrinsically as having strong negative charge molecules (31). It is generally supposed that the fixed distance between charges permit the formation of electrostatic bonds between the negatively charged polyanion and oppositely charged sites on the virion surface (14, 22, 31). The formed virion-heparin complex is incapable of penetrating into the susceptible cell and is noninfectious but, upon dissociation, the virus regains its infectivity (21, 31). It is worthy of mention that algal polysaccharide-HSV complexes have also been observed in our laboratory and found to be non-infectious. Further details of this work are to be published elsewhere. The above simplified explanation, however, provides little insight into the unique specificity of our algal polysaccharide towards HSV. We consider it quite likely that factors in addition to electrostatic attachment of the virus to inhibitor molecules, resulting in prevention of adsorption and penetration of HSV particles to host cells, are involved in conferring the observed antiviral property of algal polysaccharide-protein compounds. Our results with the nucleases suggest, but are not necessarily conclusive, that components of nucleic acids do not play an important role in conferring specificity to algal antiviral molecules. Albeit single-stranded polynucleotides are sensitive to RNase, double-stranded polynucleotides are comparatively resistant (18). If double-stranded polynucleotides, which are known inducers of interferon (18) are indeed present at specific sites on the macromolecule, the possibility that antiviral activity of algal substances is associated with a highly specific HSV-induced interferon (32) must be considered. Experiments to explore this possibility are in progress.

The remarkable similarities between heparin and algal antiviral compounds in inhibition of the infectious process is also noteworthy. Data presented in Table 5 show that the administration of algal antiviral material to infant mice before or at the same time HSV was administered afforded a significant protective response, whereas no such protection was evident when mice were given a single inoculation of algal material 2 h after the infectious process was underway. Similarly, it has been previously reported (22) that the addition of heparin and HSV to cultured cells at the same time, prevented most of the virus from replicating. However, addition of heparin to cultured cells 3 h after infection with the virus also had no protective effect.

Results of a preliminary study evaluating the ability of the antiviral material to prevent or cure herpes-induced skin lesions in adult mice have indicated that this material could also prevent but not cure lesions. The chief problems encountered with this animal model related to the nature of the infection and the manner of treatment. The virus in skin lesions was able to spread to uninfected cells without being exposed to the extracellular environment. Therefore, high molecular weight antiviral material, unable to penetrate cell membranes, may have been unable to block the migration of virus to uninfected cells. In addition, treatment with unpurified material meant that insufficient amounts of active substance may have reached susceptible cells in the zone of infection. Further studies to define mechanisms of action of algal antivirals and to improve their efficacy *in vivo* should clarify this point.

ACKNOWLEDGMENTS

This investigation was supported by the Office of Naval Research and by the University of California, Sea Grant College Program, Grant No. R/MP-12. The authors wish to thank their collaborators Drs. Kenneth B. Guiseley and Donald W. Renn (Marine Colloids, Inc.) for providing the research data presented in Table 2, and Kirk Wilhelmsen for excellent technical assistance.

REFERENCES

1. Adams, H.G., Benson, E.A., Alexander, E.R., Vontver, L.A., Remington, M.A., Holmes, K.K.: Genital herpetic infection in men and women: Clinical course and effect of topical application of adenine arabinoside. J. Infect. Dis. Suppl. 133, A151-A159 (1976).

2. Alford, C.A., Jr., Whitley, R.J.: Treatment of infections due to herpesvirus in humans: A critical review of the

state of the art. J. Infect. Dis. Suppl. 133, A101-A108 (1976).

3. Aurelian, L.: Possible role of herpesvirus hominis, type 2, in human cervical cancer. Federation Proc. 31, 1651-1659 (1972).

4. Becker, Y.: Antiviral drugs. Mode of action and chemotherapy of viral infections of man. In: Monographs in Virology, Melnick, J.L. (ed.). pp. 13-23. Basel - Munchen-Paris-London-New York-Sydney: S. Kargev, 1976.

5. Davis, D.J.: Measurements of the prevalence of viral infections. J. Infect. Dis. Suppl. 133, A3-A5 (1976).

6. Deig, E.F., Ehresmann, D.W., Hatch, M.T., Riedlinger, D.J.: Inhibition of herpesvirus replication by marine algae extracts. Antimicrob. Agents Chemother. 6, 524-525 (1974).

7. De Leon, M.P., Hessle, H., Cohen, G.H.: Separation of herpes simplex virus-induced antigens by concanavalin A affinity chromatography. J. Virol. 12, 766-774 (1973).

8. Doyle, R.J., Woodside, E.E., Fishel, C.W.: Protein-polyelectrolyte interactions. The concanavalin A precipitin reaction with polyelectrolytes and polysaccharide derivatives. Biochem. J. 106, 35-40 (1968).

9. Edelman, G.M., Cunningham, B.A., Reeke, G.N., Jr., Becker, J.W., Waxdal, M.J., Wang, J.L.: The covalent and three-dimensional structure of concanavalin A. Proc. Natl. Acad. Sci. USA 69, 2580-2584 (1972).

10. Ehresmann, D.W., Deig, E.F., Hatch, M.T., DiSalvo, L.H., Vedros, N.A.: Antiviral substances from California marine algae. J. Phycol. 13, 37-40 (1977).

11. Garber, E.D., Hauth, F.C.: A new mutation with asymmetrical expression in the mouse. J. Heredity 41, 122-124 (1950).

12. Germanov, H.B., Sokolov, M.I.: Selection and genetic studies on mutants of herpes simplex virus sensitive to sulfated agar polysaccharides. Archiv für die gesamte Virusforschung 28, 77-84 (1969).

13. Goldstein, I.J., Hollerman, C.E., Smith, E.E.: Protein-carbohydrate interaction: II. Inhibition studies on the interaction of concanavalin A with polysaccharides. Biochemistry 4, 876-883 (1965).

14. Hochberg, E., Becker, Y. : Adsorption, penetration and uncoating of herpes simplex virus. J. Gen. Virol. 2, 231-241 (1968).

15. Ito, M., Barron, A.L. : Inactivation of herpes simplex virus by concanavalin A. J. Virol. 13, 1312-1318 (1974).

16. Kaufman, H.E., Meyer, R.F., Laibson, P.R., Waltman, S.R., Nesburn, A.B., Shuster, J.J.: Human leukocyte interferon for the prevention of recurrences of herpetic keratitis. J. Infect. Dis. Suppl. 133, A165-A168 (1976).

17. Klein, R.J.: Pathogenetic mechanisms of recurrent herpes simplex virus infections. Arch. Virol. 51, 1-13 (1976).

18. Kleinschmidt, W.J.: Biochemistry of interferon and its inducers, Ann. Rev. Biochem. 41, 517-542 (1972).

19. Liebhaber, H., Takemoto, K.K.: The basis for morphologic differences in plaques produced by variants of encephalomyocarditis (EMC) virus. Bacteriol. Proc. 62, 139 (1962).

20. Myers, M.G., Oxman, M.N., Clark, J.E., Arndt, K.A.: Therapy of local herpesviral infections. Photodynamic inactivation in recurrent infections with herpes simplex virus. J. Infect. Dis. Suppl. 133, A145-A150 (1976).

21. Nahmias, A.J., Kibrick, S.: Inhibitory effect of heparin on herpes simplex virus. J. Bacteriol. 87, 1060-1066 (1964).

22. Nahmias, A.J., Kibrick, S., Bernfeld, P.: Effect of synthetic and biological polyanions on herpes simplex virus. Proc. Soc. Exp. Biol. Med. 115, 993-996 (1964).

23. Poretz, R.D., Goldstein, I.J.: An examination of the topography of the saccharide binding sites of concanavalin A and of the forces involved in complexation. Biochemistry 9, 2890-2896 (1970).

24. Prusoff, W.H., Goz, B.: Chemotherapy - Molecular aspects. In: The Herpesviruses, Kaplan, A.S. (ed.). pp. 641-663. New York and London: Academic Press, 1973.

25. Pulver, R.: Heparinoide. Chemie und tier-experimentelle Prufung. Chemotherapia (Basel) 3: 388-414 (1961).

26. Reed, L.J., Muench, H.: A simple method of estimating fifty per cent endpoints. Amer. J. Hyg. 27, 493-497 (1938).

27. Sharon, N., Lis, H.: Lectins: Cell-agglutinating and sugar-specific proteins. Science 177, 949-959 (1972).

28. So, L.L., Goldstein, I.J.: Protein-carbohydrate interaction XVII. The effect of polysaccharide molecular weight on the concanavalin A-polysaccharide precipitation reaction. J. Immunol. 102, 53-57 (1969).

29. Su, J.C., Hassid, W.Z.: Carbohydrates and nucleotides in the red algae Porphyra perforata. I. Isolation and identification of carbohydrates. Biochemistry 1, 468-474 (1962).

30. Takemoto, K.K., Liebhaber, H.: Virus-polysaccharide interactions. 1. An agar polysaccharide determining plaque morphology of EMC virus. Virology 14, 456-462 (1961).

31. Vaheri, A.: Heparin and related polyionic substances as virus inhibitors. Acta Path. Microbiol. Scand. 62, Suppl. 171, 1-98. (1964).

32. Valle, M.J., Jordan, G.W., Haahr, S., Merigan, T.C.: Characteristics of immune interferon produced by human lymphocyte cultures compared to other human interferons. J. Immunol. 115, 230-233 (1975).

AMINES IN ALGAE

H. Kneifel
GSF, Abteilung Algenforschung und Algentechnologie,
Bunsen-Kirchhoff-Str. 13, D-4600 Dortmund 1, W.Germany

1. INTRODUCTION

Amines occur in all living organisms. Reviews have been published which deal with different aspects of amines occurring in plants (67, 68, 69, 70). In comparison with other plants, however, relatively little is known about amines in algae. Publications in this field contain mostly analytical data, with the functions of amines in algae being rarely understood. Amines not only play an important role in cell metabolism; they frequently also exhibit interesting pharmaceutical activities. Therefore, analysis of amines in living organisms can also offer the chance of finding interesting new natural products of commercial value.

Amino acids are the most investigated and best known of the naturally occurring amino compounds and are usually treated separately. In this review the term "amines" refers to those amino compounds which do not contain carboxylic groups or groups with similar or stronger acidity (e.g. sulfo or phosphono groups).

2. METHODS OF AMINE ANALYSIS

The lack of the acidic group makes amines less polar than corresponding amino acids and favors the application of different analytical methods. While amino acids are almost always determined quantitatively by ion-exchange chromatography ("amino acid analyzer"), separation after derivatization and methods offered by gas chromatography can be advantageous with amines. In order to interpret published analytical data, it is important to consider the methods of amine analysis applied.

Marine Algae in Pharmaceutical Science
edited by H. A. Hoppe, T. Levring, Y. Tanaka

Some methods are only able to detect relatively volatile amines, other methods cannot distinguish between closely related amines. The following pages contain a short summary of methods of amine extraction and separation used in the analysis of algae.

2.1. Methods of Extraction

Amines have to be extracted from the algal sample before they can be analyzed. Aliphatic amines have pK-values of 9 - 12. They are, therefore, soluble in aqueous solvents at pH <8 and can be extracted with organic solvents at pH >12. (Aromatic amines are weaker bases and have lower pK-values). Polyamines require relatively polar solvents for their extraction from aqueous solutions, but even with n-butanol extraction of spermine is difficult. Quarternary amines are even more difficult to transfer into organic solvents unless they contain hydrophobic residues. Extraction of amines from algal samples is, therefore, usually performed in acidic aqueous solutions. Diluted (10%) acetic acid has the advantage that the solvent can easily be removed from the extract by evaporation, diluted trichloroacetic acid prevents contamination of the extract with proteins. Ten per cent hydrochloric acid hydrolyzes ester bonds and is used to analyze total (bound + esterified) choline and ethanolamine.

An additional extraction step is sometimes required in order to concentrate the extracted amines selectively. It is usually performed by extraction with organic solvents (e.g. n-butanol) at high pH. (Alkaline solutions of amines should be kept under nitrogen to avoid oxidation). When analyzing amines as derivatives this step can often be omitted.

A different approach is the extraction of amines by steam distillation. This methods yields relatively pure fractions but its application is limited to volatile amines.

2.2 Methods of Separation

Gas chromatography (GC). Analysis of underivatized and derivatized amines can be performed by GC. Due to their interaction with the support material, chromatographic peaks of free amines often show tailing. This tailing can be reduced by alkalization of the stationary phase (12, 15).

Many of the published GC determinations of amines from biological materials are based on methods in which the amines were derivatized. Conversion of the amines into acyl, trimethylsilyl, dinitrophenyl derivatives and isothiocyanates for GC separations was reviewed by Drozd (14). Giumanini et al. converted polyamines into permethyl derivatives (19). Trifluoroacetylation of amines seems to be the method of choice (27, 52, 54). Formation of these relatively volatile derivatives is achieved by reaction of trifluoroacetanhydride with the dried extract. Trifluoroacetamides are better resolved by GC than free amines and also show more characteristic patterns in mass spectrometry (56, 79). However, partial loss of the most volatile trifluoroacetamides during the preparation of the sample and, especially in trace analysis, high degree of contamination with non-amino compounds are disadvantages of this method. It should be mentioned that tertiary amino groups generally cannot be derivatized for analysis and that quarternary amino compounds are not volatile enough to be determined by GC. Analysis of amines by GC can be made more selective by using a nitrogen-specific detector (N-detector) and more sensitive by using the electron capture detector in the separation of derivatives of amines containing halogens.

Liquid chromatography (LC). LC methods for the separation and analysis of free and derivatized amines were reviewed by Deyl (11) and by Seiler (64). Paper (PC) and thin-layer chromatography (TLC) were the first methods to be used in amine analysis of algae (72). Amines can be detected on the chromatograms by different spray reagents. Ninhydrin, the most frequently used reagent, reacts with primary and secondary amines.

With columns, ion-exchange chromatography (IEC) is mostly used (18, 46). Amines can be separated by the amino acid analyzer if stronger buffers and/or higher pH values are used.

Derivatization of amines with the fluorogenic dansyl group has been studied mostly by Seiler and Wiechmann. The dansylation reaction is performed by reacting an aqueous solution of the amines with an excess of dansyl chloride (5-dimethylamino-1-naphthalenesulfonyl chloride) at pH 8 - 10 in a mixture of water and acetone. An excess of dansyl chloride is destroyed after the reaction by addition of proline. Dansyl chloride reacts readily with primary and secondary amines, at higher pH also with phenols, imidazoles, and even with some alcohols. After the reaction the dansylamides are extracted with benzene or ethyl acetate and separated by TLC (silica gel or polyamide sheets) (62) or by high-performance liquid chromatography (HPLC; silica gel (1) or reverse phase (37, 48, 65)). Similar results can be obtained by using bansyl chloride (5-dibutyl-amino-1-naphthalenesulfonyl chloride) (63).

The use of the dansyl reagent in amine analysis has several advantages: a) high sensitivity with fluorescence detection; b) relatively high selectivity for primary and secondary amino groups; c) dansyl amides are less polar than free amines and are better separated by LC; d) the derivatization can be performed in aqueous solution, and by subsequent extraction of the dansyl derivative into organic solvents selective concentration can be achieved. Disadvantages are: a) dansyl derivatives of ammonia and dimethylamine are always observed as by-products of the derivatization procedure and these compounds, therefore, cannot be determined by this method; b) other side reactions are: demethylation of tertiary methyl-amino compounds (77, 78) and oxidation of certain compounds by dansyl chloride (61).

At present, methods with pre-column derivatization and post-column reaction seem to give comparable results in LC. If amines are present in sufficient concentrations the amino acid analyzer technique is advantageous, in trace analysis the dansylation technique appears to be preferable. The resolution of the chosen method should be good enough to separate the amines well. Dansyl derivatives of the homologues of spermidine are not sufficiently separated by TLC and this has led to misinterpretation of analytical results in the literature.

Other methods. High voltage electrophoresis has been used to separate polyamines (32). Separation of the triamines is pH-dependent with spermidine and norspermidine not being separated at pH 4.5 and 2.5. The optimum for the separation of these two amines is at pH 3.2.

Analysis of choline is difficult using LC methods. It can be analyzed by precipitation with Reinecke's salt or by biological methods (8, 30). Trimethylamine can be analyzed by GC or by a colorimetric method after oxidation to trimethylamine oxide (17).

3. PRIMARY AND SECONDARY AMINES

3.1. Monoamines

A detailed report on monoamines in seaweeds was published by Steiner and Hartmann (72). They analyzed volatile amines of 28 different seaweeds by PC after extraction of the amines by alkaline steam distillation. The authors detected the following monoamines by reaction with ninhydrin and co-chromatography with standards: methylamine (MEA), dimethylamine (DMA), ethylamine (ETA), propylamine (PRA), isobutylamine (iBA), isoamylamine (iAA) 2-phenylethylamine (PEA), and 3-mercaptomethyl propylamine (MMP). Of these amines the last one (fig.1) was

$$H_3C-S-CH_2-CH_2-CH_2-NH_2$$

Fig.1 3-methylmercapto propylamine (MMP)

detected for the first time in nature. Its exact structure caused some confusion in the literature. In their first publication the authors described it as 2-MMP and it was as such quoted in some review articles (4, 59). The correct structure was published later (22). Of the two structures the occurrence of the 3-isomer can be expected as it can be derived from methionine by decarboxylation. No data were given by the authors on the concentrations of these amines in algae. The structure of MMP was determined from 4mg of the compound isolated from 2kg of fresh Desmarestia aculeata.

The formation of the amines (probably with the exception of MEA) can be attributed to a highly active leucine carboxy-lyase. This enzyme, which was purified and characterized by Hartmann, shows highest activity with L-leucine, but is also active with other amino acids (23, 24, 25). It could be isolated from about 50% of all Rhodophytae tested, but was not found in Phaeophytae and Chlorophytae. The enzyme requires pyridoxal-5-phosphate as coenzyme. The function of the leucine carboxy-lyase could perhaps be related to the fact that only extracellular leucine is converted into isoamylamine. Bound to acidic groups on the outside of the algal cells, the amine might provide some protective functions. This hypothesis is supported by the observation that plants in the highest stadium of development have highest leucine carboxy-lyase activity, and plants with reduced activity are usually those with epiphytic diatoms (26).

MEA and DMA were also found in algae by other authors (37, 38, 53, 54). These amines, together with trimethylamine and other compounds are responsible for the odor of decaying algae. They are also produced following attack by bacteria. ETA was found in Scenedesmus acutus (54). Usually, concentrations of volatile monoamines in freshwater algae are low.

In addition to PEA the aromatic amines tyramine (TYA; e.g. in Chondrus crispus, Polysiphonia urceolata and Laminaria saccharina) and dopamine (DOA; in Monostroma fuscum) were detected in seaweeds (fig.2) (38, 72; 70, 76).

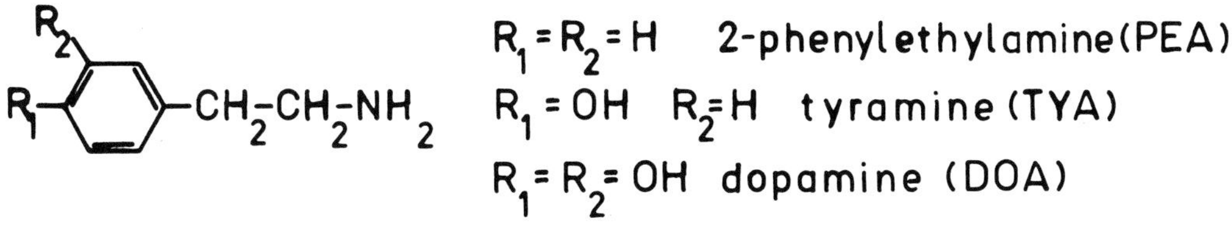

Fig.2 Phenylethylamine and its derivatives in algae

Fresh samples of Monostroma fuscum, which is a popular sea vegetable in East Asian countries, contained up to 3% DOA. It is probably formed by an enzyme which decarboxylates both D- and L-dihydroxyphenylalanine (dopa) but neither tyrosine nor phenylalanine. DOA is one of the so-called catecholamines which show physiological activity in higher animals. In plants catecholamines are precursors in the biosynthesis of some alkaloids and indole derivatives. PEA and TYA have also been found in Scenedesmus acutus (54).

Histamine (HIA; fig.3) is the decarboxylation product of histidine. It has strong pharmacological activity. HIA was detected in drum-dried Euglena from a sewage treatment pond in Israel (40).

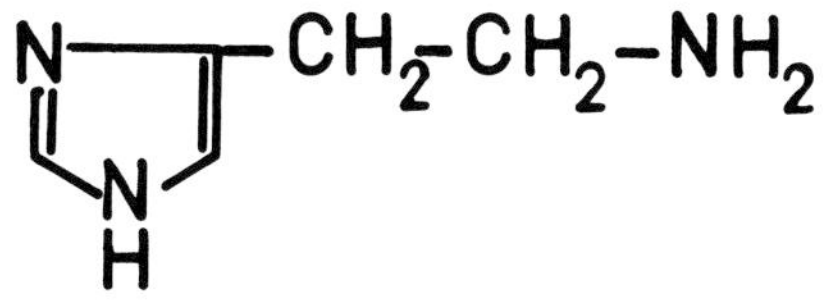

Fig.3 Histamine

Ethanolamine (EOH) is known to occur in plants as phosphatide, bound to phosphoric acid (fig.4). Presence of free EOH in algae, therefore, may be due to hydrolysis of phosphatides during the isolation procedure. To obtain reliable data on the concentrations of free EOH in algae the pH value of the extractant is crucial. At too low or too high pH values phosphatides would be hydrolyzed, at neutral pH extraction of EOH and other amines would be incomplete due to the ionic interaction of amines with acidic components of the algae. Usually, the total EOH content is determined by analysis of hydrolyzed samples.

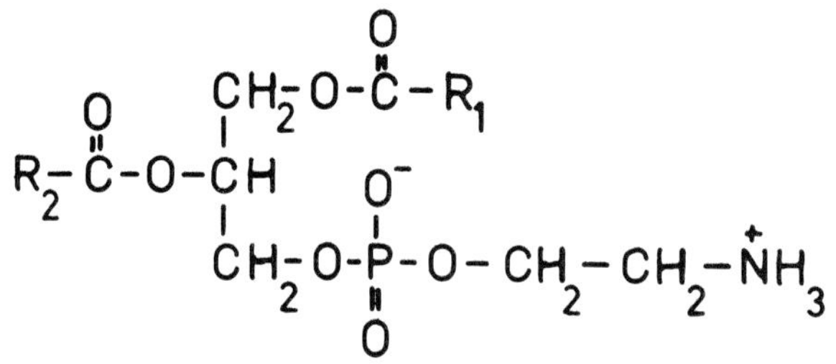

Fig.4 Phosphatidyl ethanolamine

Kanazawa et al. studied the content of EOH in Chlorella ellipsoidea *) (34). (Extraction: 1N $HClO_4$, then n-butanol from 0.5N NaOH; analysis: IEC. Under these conditions partial hydrolysis of phosphatides is possible). In synchronous cultures the concentrations of free EOH changed during the cell cycle (fig.5). D_n, D-L, L_1, L_2, L_3 and L_4 denote phases in the cell cycle with D_n to L_3: light phase (light is turned off at L_3) and L_4 to D_n maturing phase (dark phase). The concentration of EOH more than doubles in the dark phase.

Free EOH was also detected in algae by other authors using different methods (37, 53, 54). (Extraction: 0.8 - 1N $HClO_4$ or 10% acetic acid; analysis: dansyl - TLC (53), dansyl - HPLC - rev. phase (37), TFA - GC (54). Results v. appendix). In most freshwater algae EOH occurred in the highest concentrations relative to other monoamines. Spirulina, a blue-green alga,

*) Names of algae were taken from the publications reviewed. See also appendix.

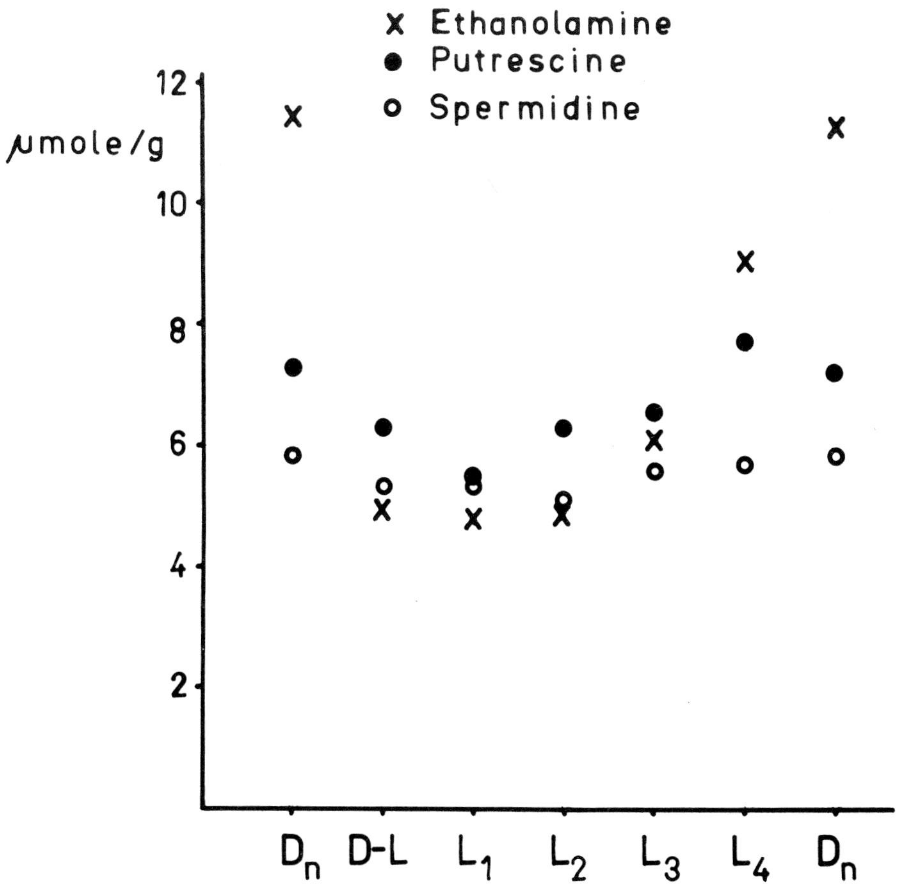

Fig.5 Concentrations of amines in Chlorella during the cell cycle (34)

had only trace amounts of EOH. This fact is in good agreement with the reported low values for choline in other blue-green algae. Choline and EOH have in one respect similar roles in biochemistry as they are both components of phosphatides.

3.2. Diamines and Polyamines

The diamine putrescine (PUT; 1,4-diaminobutane) and the polyamines spermidine (SPD) and spermine (SPM) have been detected in practically all organisms investigated. They are associated with nucleic acids, stabilizing their secondary structure and having stimulatory effects on their biosynthesis; they stimulate protein biosynthesis in vivo and in vitro, are involved in the biochemistry of cell division (44), and they are active in various other physiological systems. Their enhanced concentrations in urine of patients suffering from some kinds of cancer has attracted attention in clinical chemistry (16, 55). The physiological role and the occurrence of diamines and polyamines has been reviewed (2, 7, 74).

In comparison to these three amines other di- and polyamines have received only minor attention. Cadaverine (CAD), the decarboxylation product of lysine, has been detected in several plants (69). Much less is known about the occurrence and the biochemical role of the homologues of SPD and SPM (v. fig.6). Norspermidine (NSD) *) has been found in some plant viruses**) (32), in the thermophilic bacterium Caldariella acidophila (9) and in the white shrimp Penaeus setiferus (73); norspermine (NSM) in thermophilic bacteria (50) and in the above-mentioned white shrimp (73). Homospermidine was found to occur in

---

*) The prefix nor- indicates a homologue with one lacking $-CH_2-$ group, homo- a homologue with an additional $-CH_2-$group. The prefix sym- (sym-homospermidine) can be added to distinguish from the asymmetrical compound (asym-homospermidine). If not otherwise indicated the symmetrical homologues are discussed. Other names for norspermidine are: N-(3-aminopropyl)-1,3-diaminopropane, bis(3-aminopropyl)-amine, 3,3'-diaminodipropylamine, dipropylene triamine, 1,7-diamino-4-azaheptane; names for the other polyamines can be formed correspondingly. The trivial names caldine were given to norspermidine by De Rosa (9) and thermine to norspermine by Oshima (50). The names norspermidine, norspermine and homospermidine are preferred by the author as they show the structural relationship with spermidine and spermine.

**) These results could not be confirmed by other authors (5).

$H_2N-(CH_2)_4-NH_2$	$H_2N-(CH_2)_5-NH_2$
Putrescine (PUT)	Cadaverine (CAD)
$H_2N-(CH_2)_3-NH-(CH_2)_4-NH_2$	$H_2N-(CH_2)_3-NH-(CH_2)_4-NH-(CH_2)_3-NH_2$
Spermidine (SPD)	Spermine (SPM)
$H_2N-(CH_2)_3-NH-(CH_2)_3-NH_2$	$H_2N-(CH_2)_3-NH-(CH_2)_3-NH-(CH_2)_3-NH_2$
sym-Norspermidine (NSD)	sym-Norspermine (NSM)
$H_2N-(CH_2)_4-NH-(CH_2)_4-NH_2$	$H_2N-(CH_2)_3-NH-(CH_2)_5-NH_2$
sym-Homospermidine (HSD)	asym-Homospermidine

Fig.6 Diamines and polyamines of biochemical importance

relatively high concentrations in the leaves of the sandalwood tree, Santalum album (42), and in legume root nodules (71). Derivatives of HSD, sym-N,N,N',N'-tetramethyl-N-palmitoyl HSD and the corresponding N-palmitenoyl compound, which have been isolated from the plant Solanum tripartitum, had significant tumor-inhibiting activity (41). Asymmetrical HSD has been found in a mutant of the bacterium Escherichia coli (13).

Determinations of PUT and SPD in Chlorella were performed by Kanazawa et al. (34). (Extraction: 1N $HClO_4$; hydrolysis:

6N HCl; analysis: IEC, ninhydrin). In addition to EOH, PUT and SPD were the only major amines present. The change in PUT concentrations during the cell cycle was analogous with EOH but less obvious. SPD concentrations remained almost constant (v. fig.5).

Similar investigations by Rolle et al. with Chlorella and Scenedesmus from mass cultures were initiated by the fact that in Japan Chlorella extracts were used to stimulate the growth of lactic acid bacteria and that polyamines were believed to be the active factors (53). (Extraction: 0.8N $HClO_4$; analysis: dansyl - TLC). The average SPD concentration in Scenedesmus acutus was found to be 0.5 - 1.1mg SPD/g dry algae when grown in mass culture and about 2mg/g when grown in culture tubes at 30°C. Concentrations in Chlorella were somewhat higher (2 - 2.5mg/g). SPD concentrations decreased with age of the cultures and with decreasing temperatures. Urea as nitrogen source induced less SPD production than ammonium sulfate or potassium nitrate. Differences were in the order of 10 - 20%. (Later investigations showed that the determined "SPD" was actually the mixture of SPD homologues).

PUT, SPD and SPM in Euglena were determined by Bagni and Calzoni (3). The authors found about 37% of the total SPM content of the cells in the chloroplasts, 9% of the SPD and only 5% of the PUT. (Again, "SPD" was possibly a mixture of the homologues and "SPM" was probably NSM - see below).

Investigating extracts of Scenedesmus acutus by combined GC/MS Rolle et al. found the diamines CAD and piperazine in addition to the already established PUT and other monoamines (54). (Extraction: 10% acetic acid; evaporation; TFA - GC/MS). A group of peaks in the gas chromatogram (fig.7) had similar mass spectra. The compounds were identified as NSD, SPD and HSD by Kneifel et al. (36). It was the first time that homologues of SPD had been found in algae.

Screening algae for primary and secondary amines, best results were obtained with HPLC of dansylated extracts on reverse-phase

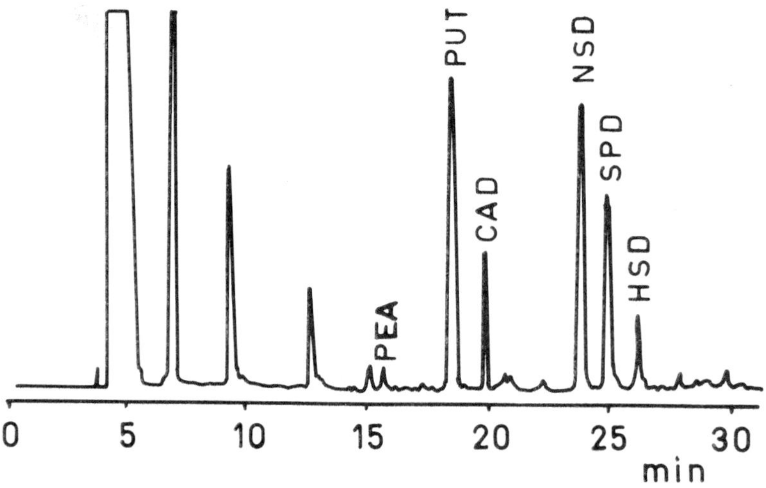

Fig.7 Gas chromatogram of a trifluoroacetylated extract of Scenedesmus acutus

columns (37, 38). (Extraction: 0.8N $HClO_4$). A typical chromatogram of a dansylated extract of Scenedesmus acutus can be seen on fig.8. Similar to the gas chromatogram, SPD is accompanied by its two homologues. The relatively smaller peak of NSD in GC compared to HPLC can perhaps be attributed to the poor extraction of this polar amine during the GC procedure.

HPLC investigations of other freshwater algae showed that di- and polyamines were always present. Of the diamines, PUT usually has high concentrations. CAD can sometimes be found in minor amounts, its concentration rises enormously with decay of algal cultures, especially after bacterial attack. 1,3-diaminopropane (DAP) is rare but it was found in the green microalga Neospongiococcum saccatum in very high concentrations (38). Of the polyamines, SPD (resp. NSD) occurs in relatively high concentrations. SPM is less prevalent and sometimes not detectable. Highest concentrations of SPM were found in Spirulina. This occurrence of SPM in Spirulina is interesting as SPM is usually not found in procaryotic organisms (7).

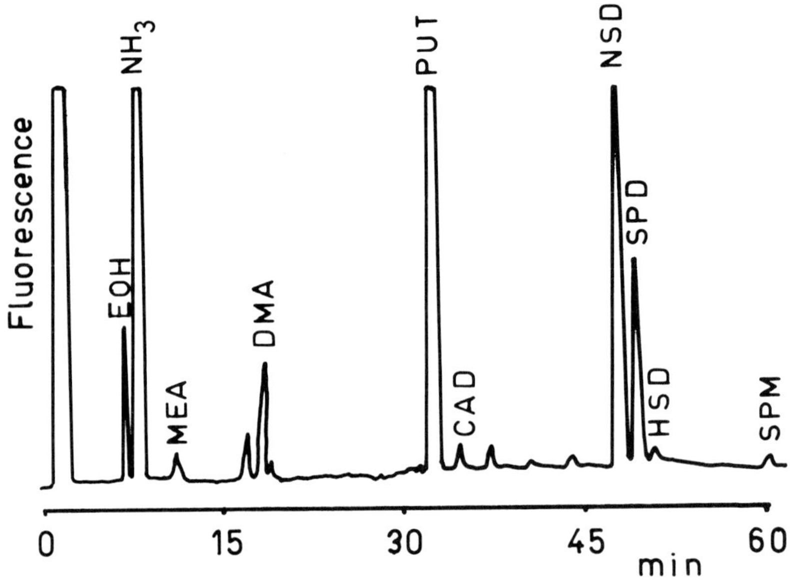

Fig.8 HPLC chromatogram of a dansylated extract of *Scenedesmus acutus*

The tetraamine of *Euglena gracilis* was first reported to be SPM (3, 37). Later investigations by Kneifel and Schuber showed that it was actually NSM and that SPM occurred only in traces (39). This was the first demonstration of the occurrence of a homologue of SPM in algae and it showed again that algae are unique in having such a variety of triamines and tetraamines. In several seaweeds (*Blidingia sp.*, *Enteromorpha*, *Codium fragile* and *Rhodomela confervoides*) HPLC chromatograms of dansylated extracts contained peaks which can be attributed to NSM.

While NSM is relatively rare in algae, homologues of SPD have been detected in various algae (fig.9). NSD occurs in high concentrations in green microalgae, especially in *Scenedesmus*, and in *Euglena*. All strains of *Scenedesmus* investigated so far had higher concentrations of NSD than of SPD. *Spirulina platensis* (Cyanophyta), on the other hand, did not contain more

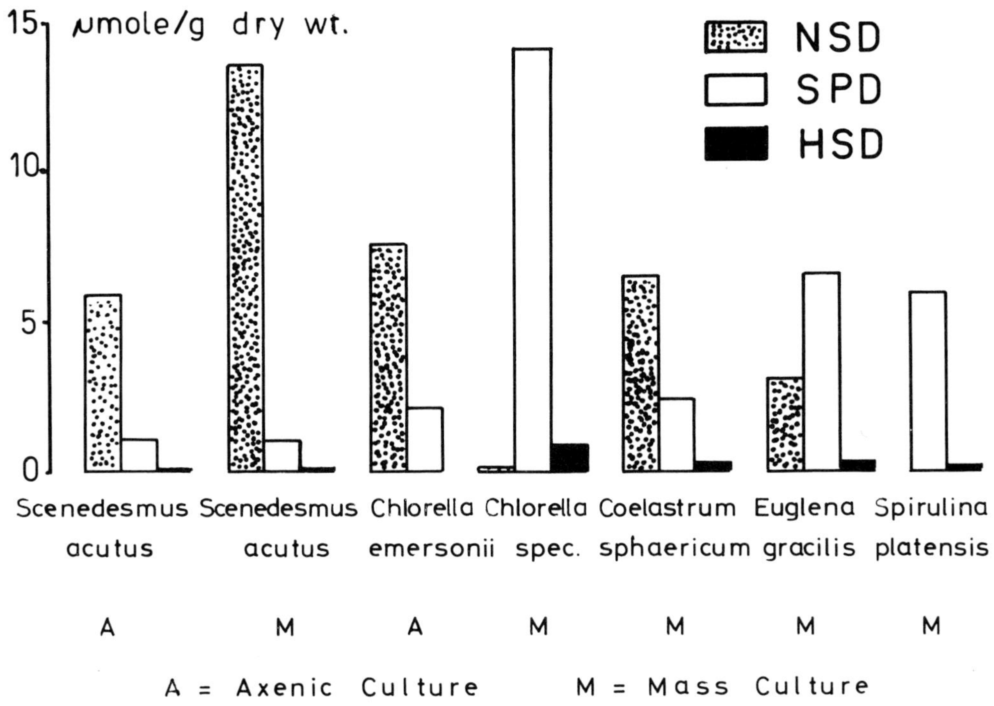

Fig.9 Homologues of SPD in freshwater algae

than traces of NSD. HSD concentrations were low in all algae investigated.

From these results it can be seen that homologues of SPD are much more widely distributed in plants than previously known. Some methods (e.g. TLC) can hardly distinguish between these homologues. T. Smith in his review on plant amines commented on HSD as follows (68): "It may hitherto have been confused with spermidine in chromatography". This could also be the case with NSD, and NSM may hitherto have been confused with SPM.

The occurrence of homologues of SPD and SPM in organisms raises two questions: a) function in relation to SPD and SPM and b) biosynthesis. So far, it can only be speculated whether SPD and SPM have in biochemistry the same functions as their

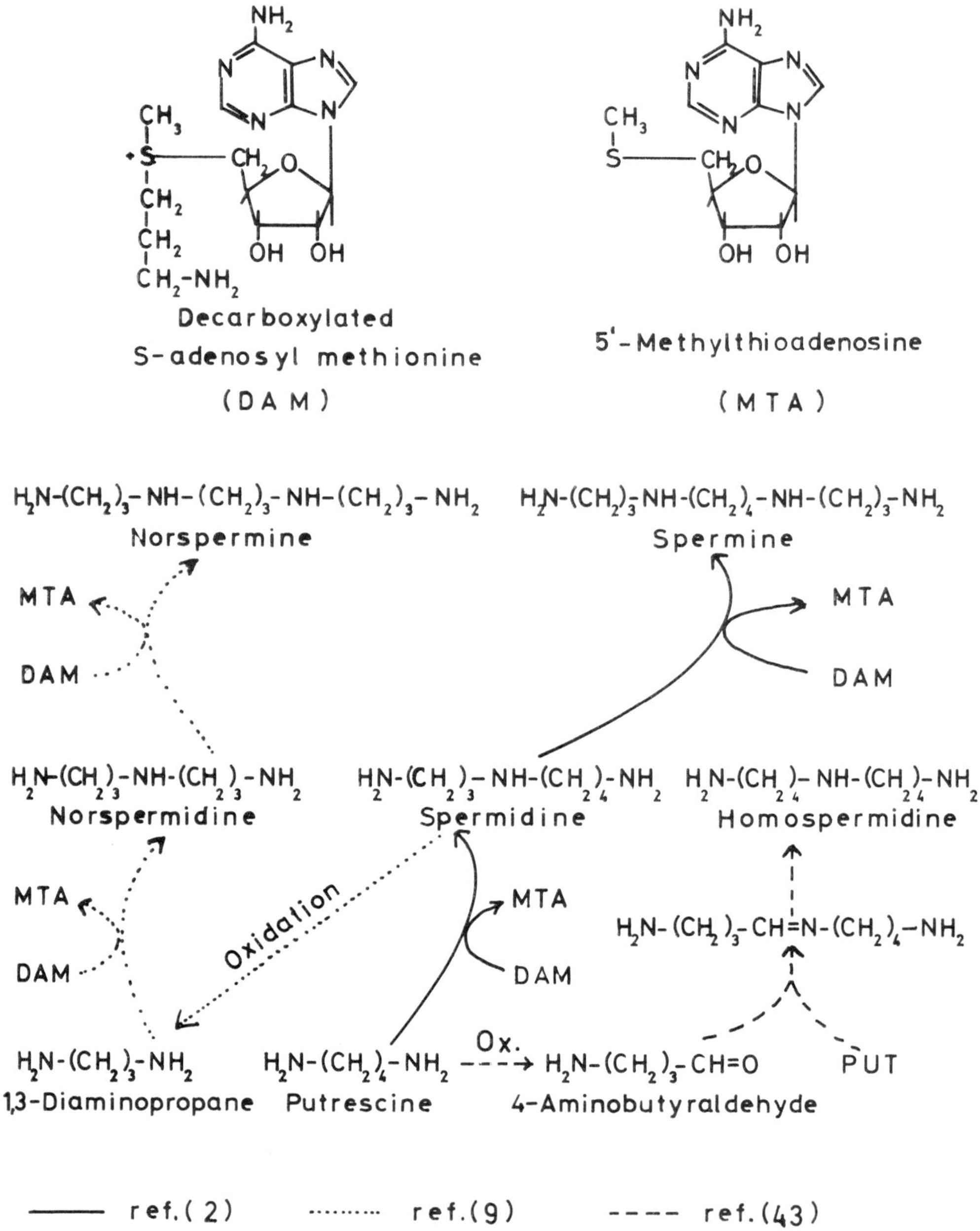

Fig.10 Biosynthesis of polyamines

homologues. There are no experimental data to answer this question.

Biosynthesis of these compounds seems to be related to that of SPD and SPM (fig.10). SPD is synthesized from PUT (C_4-unit), decarboxylated S-adenosylmethionine (DAM) provides the C_3-unit. Adding another aminopropyl residue to SPD yields SPM. Biosynthesis of NSD requires two C_3-units and can, therefore, not start directly from PUT. De Rosa postulated a mechanism for the biosynthesis of NSD (9). In his scheme DAP (C_3-unit) is supposed to react with DAM under formation of NSD which in turn is presumed to be the biological precursor of NSM. The low concentration of DAP in most algae and other organisms could be the result of a high turnover by the NSD-synthesizing enzyme system. While the diamines PUT and CAD are synthesized in plants from the corresponding amino acids ornithine (resp. from arginine via agmatine) and from lysine, DAP is produced by oxidation of SPD and SPM. This reaction, which is catalyzed by a diamine oxidase, is known to occur in several microorganisms and plants (9, 68).

Kuttan studied the biosynthesis of HSD in the leaves of the sandalwood tree (43). In his scheme PUT is the precursor for both sides of the symmetrical molecule. Deamination of PUT yields 4-aminobutyraldehyde which reacts with another molecule of PUT with formation of a Schiff's base. Reduction gives HSD.

4. TERTIARY AND QUARTERNARY AMINES

4.1. Trimethylamine

Trimethylamine (TMA) is the only tertiary amine occurring in higher concentrations in algae. Its occurrence in marine algae has been reported by several authors (17, 72). Freshwater algae, however, seem to contain only traces of it.

In marine algae and animals, TMA is found together with its oxide, trimethylamine oxide (TMAO). The formation of TMAO from TMA is thought to be a way of detoxification. The reversed reaction is brought about by reducing enzyme systems present in putrefying bacteria. The characteristic odor of decaying marine organisms can partly be attributed to this formation of TMA.

From the published data (v. appendix) it can be seen that marine green and red algae have somewhat higher TMA concentrations than brown algae (17).

4.2. Choline and Choline Derivatives

Choline (CHO) is widely distributed in all plant and animal cells, but is less common in bacteria. It is mostly found as part of phospholipids and sphingolipids which are phosphoesters of CHO with substituted glycerol attached to the phosphoric acid, and which are important parts of cell membranes. In plants, the most important of these compounds is lecithine (fig.11).

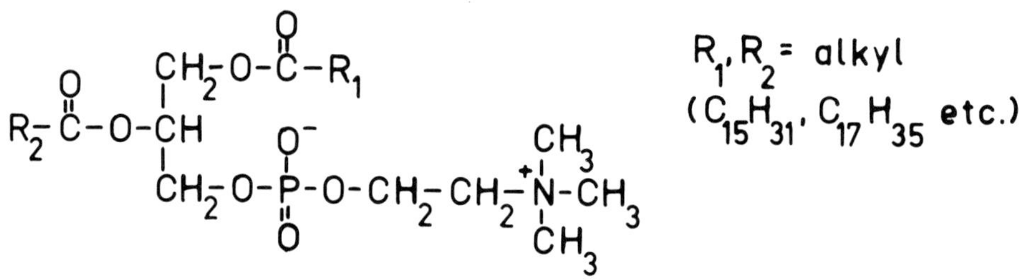

Fig.11 Phosphatidyl choline (lecithine)

Esters of CHO with carboxylic acids show neurophysiological activity and are toxic. Most important is acetylcholine which shows cholinergic activity.

CHO plays a role in metabolism as a methylating agent and is essential as part of the diet. A diet low in CHO causes fat accumulation in livers and hemorrhagic kidneys. Analysis of CHO and its derivatives in biological materials, therefore, is of interest a) in determining the nutritive value and b) in screening for toxic CHO esters.

Concentrations of CHO in algae were determined by Ikawa and Taylor (29, 30) and by Da Silva and Jensen (8). Of all algae investigated, blue-green algae had the lowest CHO content. It is interesting to note that bacteria, which are closely related to blue-green algae, also contain little CHO. Part of the CHO detected in other algae (v. appendix) occurs certainly in the form of phosphatides. The difficulties in determining free CHO in algal samples are the same as those encountered with EOH. Therefore, the total content of CHO is usually determined after hydrolysis. Da Silva and Jensen found a total of 2.6mg CHO per g dry *Ascophyllum nodosum* after hydrolysis and only 0.02mg/g after extraction with methanol. It is likely that the content of free CHO is higher, as methanol is probably not able to remove the strong base CHO from its ion pairs in the cell.

Highest CHO concentrations were found in the toxic dinoflagellate *Amphidinium carteri* (75). Of the 0.4% CHO (as chloride) about one quarter is present as phospholipids with the remaining 0.3% as water soluble compounds.Three derivatives were isolated and two of them identified: a)choline-O-sulfate and b)acrylylcholine. Acrylylcholine (fig.12) was active in reducing the heart beat of isolated *Mercenaria mercenaria* hearts. Acetylcholine shows the same type of activity but is 4000 times more active.

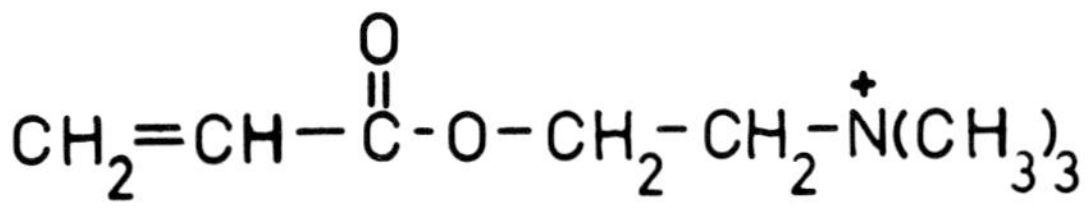

Fig.12 Acrylylcholine

4.3. Hordenine (N,N-dimethyltyramine)

Hordenine (fig.13) was detected by Güven et al. in the marine red alga *Phyllophora nervosa* (20, 21). It is similar to epinephrine in that it raises the blood pressure. Hordenine was the first alkaloid detected in an alga. It is, however, not a typical alkaloid as more recent definitions of alkaloids require the presence of a N-heterocyclic ring in the molecule.

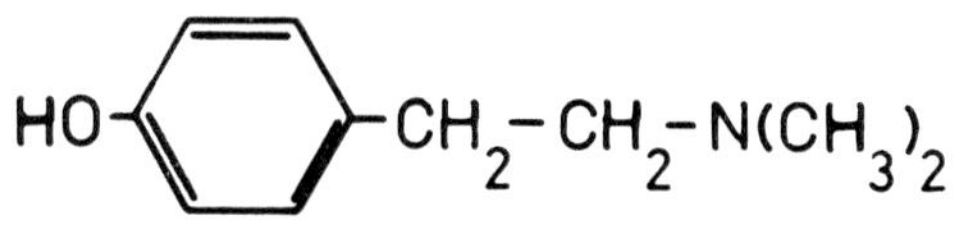

Fig.13 Hordenine

4. N-HETEROCYCLES

Of all amines from algae the most toxic ones are N-heterocycles: anatoxin-a, saxitoxin and gonyautoxin. Piperazine, found in trace amounts in Scenedesmus, has only little physiological activity.

4.1. Anatoxin-a

The blue-green alga Anabaena flos-aquae sometimes occurs in blooms in lakes, and ingestion of this alga can cause death to livestock and other animals. The poisonous compound, anatoxin-a, kills mice in 2 - 5min and was, therefore, called "Very Fast Death Factor". The death of the animals is preceded by gasps and tremors (10).

To isolate the toxin, A. flos-aquae was grown in cultures. The highest amount of the toxin was produced as the cultures matured (14 days) and part of it was excreted into the medium. The total amount of toxin could be obtained in the medium following lysis of the algal cells with Triton-X. The toxin was isolated best as follows: lyophilization of the concentrated culture liquid, extraction with 0.75% ethanol in chloroform, distribution of the evaporated extract between chloroform and water and final extraction of the toxin from the evapor. aqueous extract with 0.75% ethanol in chloroform.

The structure of anatoxin-a (fig.14) was determined by X-ray

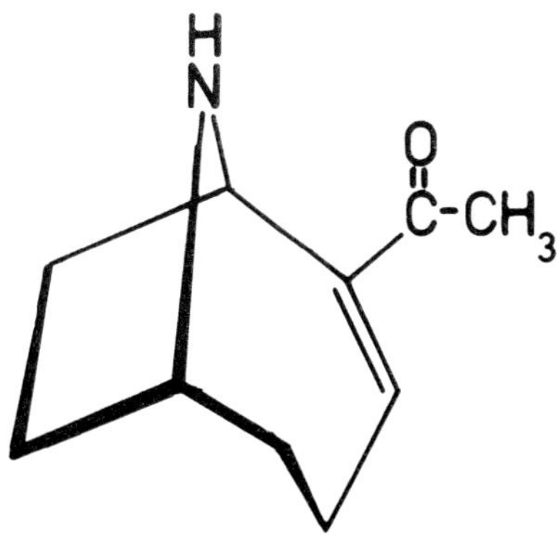

Fig.14 Anatoxin-a

crystallography of the N-acetyl derivative (28). The systematic name of anatoxin-a is 2-acetyl-9-azabicyclo[4.2.1]non-2-ene. The carbon-nitrogen skeleton of anatoxin-a is new for a natural product.

4.2. Saxitoxin and Gonyautoxins

For centuries it has been known that the bloom of certain red-colored dinoflagellates ("Red Tide") can cause mass mortality of fish (45, 59, 60). Around 1937 it became obvious that outbreaks of mussel poisoning are connected with these red tides. Clams and mussels can ingest the algae in the food chain and store the toxin for longer periods of time. This way otherwise edible clams and mussels become poisonous themselves and it is estimated that humans were poisoned in the USA in 1600 cases, causing 300 fatalities. Only in 1966, when saxitoxin was isolated from an axenic culture of *Gonyaulax catenella*, could it be shown that it was identical with the toxin from the Alaskan butter clam *Saxidomus giganteus* and from the mussel *Mytilus californianus*.

In order to isolate the toxin, it was extracted from poisonous clams with acidified ethanol, adsorbed on a weakly acidic ion-exchange resin, eluted with diluted acetic acid and finally purified on a column of acid-washed alumina with water/ethanol mixtures (57).

Elucidation of the structure was very tedious. In the course of this work several structures were proposed which were later abandoned. For the determination of the structure by X-ray crystallography crystals of the p-bromobenzenesulfonate salt of saxitoxin (58) and of the ethylketal (6) were used.

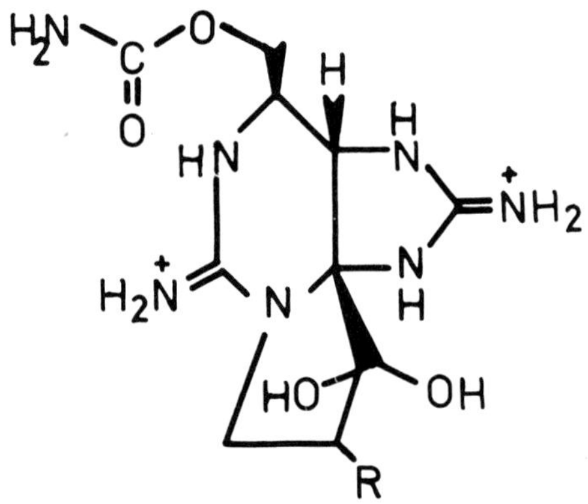

R = H Saxitoxin

R = α-OH Gonyautoxin-II

R = β-OH Gonyautoxin-III

Fig.15 Structures of saxitoxin and the gonyautoxins

Saxitoxin (fig.15) is the major poison of the toxic dinoflagellate Gonyaulax catenella which is found along the American west coast. The toxic dinoflagellate of the North Atlantic coasts of the United States and of Canada, Gonyaulax tamarensis, contains in addition to saxitoxin three other toxins. Two of the toxins, gonyautoxin-II and -III (fig.15), are epimeric hydroxysaxitoxins, the structure of the third toxin is still unknown (66). Gonyautoxin-II and -III were also found in a Japanese Gonyaulax species (51). The toxin of Aphanizomenon flos-aquae was reported to be identical with saxitoxin by

Jackim and Gentile (31), studies by Jüttner (33) gave different results.

Saxitoxin and the gonyautoxins are among the strongest poisons known in nature. 1mg can kill about 5500 mice, the lethal dose to man is estimated at about 0.3mg of toxin. The active part of the molecule is probably the guanidine moiety. The positively charged molecule interfers with the movement of the sodium ions through the membrane channels. Saxitoxin and gonyautoxin show similarity in structure (guanidine moiety) and physiological activity with tetrodoxin, the toxin of the Japanese pufferfish.

5. AMINES EXCRETED BY ALGAE

Algae are known to excrete soluble carbon compounds both in laboratory batch cultures and under more natural conditions. Excretion of nitrogenous compounds is mostly known from blue-green algae, less from other algae. Some of the excreted compounds seem to be peptides (49). Excretion of amines into algal cultures was studied by Herrmann and Jüttner (27). (Extr.: lyophilization of culture medium at low pH, sublimation after add. of 2N NaOH; analysis: TFA - GC). In addition to 2-aminopropane, 1-aminopropane was found in concentrations of 66μg/l (*Nitzschia*), 240μg/l (*Mycrocystis*) and 300μg/l (*Ankistrodesmus*). These concentrations are high enough to produce physiological activity and the excreted amines may, therefore, be of ecological importance. As the experiments were performed in axenic cultures it can be concluded that algae produce and excrete these amines. Furthermore, 1-aminobutane was detected during an algal bloom of *Chlamydomonas* in the water of the river Ruhr.

Large amounts of amines are produced from algae under anaerobic conditions by bacterial attack. Klein and Kneifel found mostly monoamines, high levels of PUT and CAD, but also homologues of SPD in culture media of *Scenedesmus* decomposed by bacteria (35). (Dansyl - HPLC). Compared to the amine concentrations within the algal cells, especially CAD and TYA showed high values in the culture medium.

Melnichenko et al. found TMA, DMA, ETA, MEA, and PRA in the culture medium of *Microcystis* incubated with bacteria (47). The excreted amines certainly contribute to the odor of water containing decaying algae, and may be of ecological importance.

6. SUMMARY

Amines, found in algae, can be classified in three groups:

a) Aliphatic monoamines are mostly derived from amino acids by decarboxylation. Concentrations of these amines are highest in marine red algae. Trimethylamine, which is mostly found in seaweeds, contributes to the odor of algae. Choline and ethanolamine, frequently found in other algae, are practically absent in blue-green algae.

b) Di- and polyamines are ubiquitous and are to be found at highest concentrations in green microalgae. Compared to other organisms, algae contain more complex patterns of di- and polyamines including homologues of spermidine and spermine.

c) Alkaloids are limited to very few examples, but saxitoxin and the gonyautoxins from dinoflagellates are among the strongest poisons known.

Acknowledgement

The author thanks the Bundesministerium für Forschung and Technologie for financial support of research on natural products in algae.

References

1. Abdel-Monem, M.M., Ohno, K.: Separation of the Dns derivatives of polyamines and related compounds by thin-layer and high-pressure liquid chromatography. J. Chromat. 107, 416-419 (1975).

2. Bachrach, U.: Function of naturally occurring polyamines. Academic Press, New York and London, 211 pages (1973).

3. Bagni, N., Calzoni, G.L.: Presence of polyamines in chloroplasts of Euglena and spinach leaves. IRCS Medical Science: Biochemistry 3, 263 (1975).

4. Baker, J.T., Murphy, V.(authors): Compounds from marine organisms. Handbook of marine science, CRC press, 226 pages (1976).

5. Beer, S.V., Kosuge, T.: Spermidine and spermine - polyamine components of Turnip Yellow Mosaic Virus. Virology 40, 930-938 (1970).

6. Bordner, J., Thiessen, W.E., Bates, H.A., Rapoport, H.: The structure of a crystalline derivative of saxitoxin. The structure of saxitoxin. J. Am. Chem. Soc. 97, 6008-6012 (1975).

7. Cohen, S.S.: Introduction to the polyamines. New Jersey, Prentice-Hall, Inc, (1971).

8. Da Silva, E., Jensen, A.: Benthic marine and blue-green algal species as a source of choline. J. Sci. Fd. Agric. 24, 855-861 (1973).

9. De Rosa, M., De Rosa, S., Gambacorta, A., Carteni-Farina, M., Zappia, V.: Occurrence and characterization of new polyamines in the extreme thermophile Caldariella acidophila. Biochem. Biophys. Res. Comm. 69, 253-261 (1976).

10. Devlin, J.P., Edwards, O.E., Gorham, P.R., Hunter, N.R., Pike, R.K., Stavric, B.: Anatoxin-a, a toxic alkaloid from Anabaena flos-aquae NRC-44h. Can. J. Chem. 55, 1367-1371 (1977).

11. Deyl, Z., Macek, K., Janak, J.(ed.): Liquid column chromatography. J. Chromat. Library, Vol.3, Elsevier Scient. Publ. Co., Amsterdam, 1175 pages (1975). Amines: 637-655.

12. Di Corcia, A., Samperi, R.: Gas chromatographic determination at the parts-per-million level of aliphatic amines in aqueous solution. Anal. Chem. 46, 977-981 (1974).

13. Dion, A.S., Cohen, S.S.: Polyamine stimulation of nucleic acid synthesis in an uninfected and phage-infected poly-

amine auxotroph of Escherichia coli K12. Proc. Natl. Acad. Sci. USA 69, 213-217 (1972).

14. Drozd, J.: Chemical derivatization in gas chromatography. J. Chromat. 113, 303-356 (1975).

15. Dunn, S.R., Simenhoff, M.L., Wesson, jr., L.G.: Gas chromatographic determination of free mono-, di-, and trimethylamines in biological fluids. Anal. Chem. 48, 41-44 (1976).

16. Durie, B.G.M., Salmon, S.E., Russell, D.H.: Polyamines as markers of response and disease activity in cancer chemotherapy. Cancer Res. 37, 214-221 (1977).

17. Fujiwara-Arasaki, T., Mino, N.: The distribution of trimethylamine and trimethylamine oxide in marine algae. Proc. VIIth Internat. Seaweed Symp. den Haag, 506-510 (1972).

18. Gehrke, C.W., Kuo, K.C., Ellis, R.L., Waalkes, T.P.: Polyamines - an improved automated ion-exchange method. J. Chromat. 143, 345-361 (1977).

19. Giumanini, A.G., Chiavari, G., Scarponi, F.L.: N-Permethylation of polyamines for gas chromatographic and mass spectrometric analyses. Anal. Chem. 48, 484-489 (1976).

20. Güven, K.C., Bora, A., Sunam, G.: About alkaloid content of marine algae. I. Hordenine from Phyllophora nervosa (D.C.) Grev. Eczacilik Bülteni (Turk. Bull. Pharmacy) 11, 177-184 (1969).

21. Güven, K.C., Bora, A., Sunam, G.: Hordenine from the alga Phyllophora nervosa. Phytochem. 9, 1893 (1970).

22. Hartmann, T.: 3-Methyl-mercaptopropylamine from the brown alga Desmarestia aculeata. Naturwiss. 55, 391 (1968).

23. Hartmann, T.: Leucine carboxy-lyase of marine Rhodophyceae: occurrence, distribution and some properties. Phytochem. 11, 1327-1336 (1972).

24. Hartmann, T.: Leucine carboxy-lyase of marine Rhodophyceae: 2. Properties of the enzyme from Polysiphonia urceolata. Biochem. Physiol. Pflanzen 163, 1-13 (1972).

25. Hartmann, T.: Leucine carboxy-lyase of marine Rhodophyceae. 3. Coenzyme specifity and activation by carbonylic

compounds. Biochem. Physiol. Pflanzen 163, 14-29 (1972).
26. Hartmann, T., Aufermann, B.: On the physiology of amine formation in the marine red alga Polysiphonia urceolata. Marine Biol. 21, 70-74 (1973).
27. Herrmann, V., Jüttner, F.: Excretion products of algae. Identification of biogenic amines by gas-liquid chromatography and mass spectrometry of their trifluoroacetamides. Anal. Biochem. 78, 365-373 (1977).
28. Huber, C.S.: The crystal structure and absolute configuration of 2,9-diacetyl-9-azabicyclo[4.2.1.]non-2,3-ene. Acta Cryst. B28, 2577-2582 (1972).
29. Ikawa, M., Borowski, P.T., Chakravarti, A.: Choline and inositol distribution in algae and fungi. Appl. Microbiol. 16, 620-623 (1968).
30. Ikawa, M., Taylor, R.F.: Choline and related substances in algae. Marine Pharmacognosy. Action of marine biotoxins at the cellular level. Acad. Press, New York and London, 203-240 (1973).
31. Jackim, E., Gentile, J.: Toxins of a blue-green alga: similarity to saxitoxin. Science 162, 915-916 (1968).
32. Johnson, M.W., Markham, R.: Nature of the polyamine in plant viruses. Virology 17, 276-281 (1962).
33. Jüttner, F.: Biochemistry of toxic "water blooms" of Aphanizomenon flos-aquae (L.) RALFS. Dr. thesis. Tübingen (1971).
34. Kanazawa, T., Yanagisawa, T., Tamiya, H.: Aliphatic amines occurring in Chlorella cells and changes of their content during the life cycle of the alga. Z. Pflanzenphysiol. 54, 57-62 (1966).
35. Klein, G., Kneifel, H.: Liberation of amines during the anaerobic decomposition of algae. Verhandl. Internat. Vereinig. theoret. u. angew. Limnol. 20, in press (1978).
36. Kneifel, H., Rolle, I., Paschold, B.: Amines of unicellular green algae, III. Identification of homologues of spermidine in the green alga Scenedesmus acutus 276-3a. Z. Naturforsch. 32c, 190-192 (1977).

37. Kneifel, H.: Investigations of natural products from algae. Chemiker-Z. 101, 165-168 (1977).
38. Kneifel, H., Meinicke, M., Soeder, C.J.: Analysis of amines in algae by high performance liquid chromatography. J. Phycol. 13, 36 (1977). (abstract).
39. Kneifel, H., Schuber, F.: Occurrence of norspermine in Euglena gracilis. Biochem. Biophys. Res. Comm., in prepar.
40. Kneifel, H.: unpublished.
41. Kupchan, S.M., Davies, A.P., Barboutis, S.J., Schnoes, H.K., Burlingame, A.L.: Tumor inhibitors, XLIII. Solapalmitine and solapalmitenine, two novel alkaloid tumor inhibitors from Solanum tripartitum. J. Org. Chem. 34, 3888-3893 (1969).
42. Kuttan, R., Radhakrishnan, A.N., Spande, T., Witkop, B.: sym-homospermidine, a naturally occurring polyamine. Biochem. 10, 361-365 (1971).
43. Kuttan, R., Radhakrishnan, A.N.: Studies on the metabolism of sym-homospermidine in sandal (Santalum album L.). Biochem. J. 128, 22P-23P (1972).
44. Mamont, P.S., Böhlen, P., McCann, P.P., Bey, P., Schuber, F., Tardif, C.: α-Methyl ornithine, a potent competitive inhibitor of ornithine decarboxylase, blocks proliferation of rat hepatoma cells in culture. Proc. Natl. Acad. Sci. USA 73, 1626-1630 (1976).
45. Martin, D.F., Martin, B.B.: Red tide, red terror. Effects of red tide and related toxins. J. Chem. Ed. 53, 614-617 (1976).
46. Marton, L.J., Heby, O., Wilson, C.B., Lee, P.L.Y.: An automated micromethod for the quantitative analysis of di- and polyamines utilizing a sensitive high pressure liquid chromatographic procedure. FEBS Lett. 41, 99-103 (1974).
47. Melnichenko, L.A., Goronowskij, I.T., Kulisch, A.F.: Determination of amines in the decomposition products of blue-green algae. Ukr. Chim. J. 40, 883-884 (1974).
48. Mietz, J.L., Karmas, E.: Chemical quality index of canned tuna as determined by high-pressure liquid chromatography.

J. Food Sci. 42, 155-158 (1977).

49. Newell, B.S., Dalpont, G., Grant, B.R.: The excretion of organic nitrogen by marine algae in batch and continuous culture. Can. J. Bot. 50, 2605-2611 (1972).

50. Oshima, T.: Thermine; a new polyamine from an extreme thermophile. Biochem. Biophys. Res. Comm. 63, 1093-1098 (1975).

51. Oshima, Y., Fallon, W.E., Shimizu, Y., Noguchi, T., Hashimoto, Y.: in Suisan Gakkaishi, in press. Ref.5 in Shimizu et al., (66).

52. Pailer, M., Hübsch, W.J.: Determination of primary and secondary amines as amides by gas chromatography on packed and capillary columns. Mh. Chemie 97, 1541-1553 (1966).

53. Rolle, I., Payer, R., Soeder, C.J.: Amines in unicellular green algae, 1. Spermidine content of Scenedesmus acutus (276-3a) and Chlorella fusca (211-8b). Arch. Mikrobiol. 77, 185-195 (1971).

54. Rolle, I., Hobucher, H.E., Kneifel, H., Paschold, B., Riepe, W., Soeder, C.J.: Amines in unicellular green algae, 2. Amines in Scenedesmus acutus. Analyt. Biochem. 77, 103-109 (1977).

55. Russell, D.H.: Increased polyamine concentrations in the urine of human cancer patients. Nature 233, 144-145 (1971).

56. Saxby, M.J.: The mass spectrometry of volatile derivatives II. N-alkyl trifluoracetamides. Org. Mass Spectrom. 2, 33-36 (1969).

57. Schantz, E.J., Mold, J.D., Stanger, D.W., Shavel, J., Riel, F.J., Bowden, J.P., Lynch, J.M., Wyler, R.S., Riegel, B., Sommer, H.: Paralytic shellfish poison. VI. A procedure for the isolation and purification of the poison from toxic clam and mussel tissues. J. Amer. Chem. Soc. 79, 5230-5235 (1957).

58. Schantz, E.J., Ghazarossian, V.E., Schnoes, H.K., Strong, F.M., Springer, J.P., Pezzanite, J.O., Clardy, J.: The structure of saxitoxin. J. Amer. Chem. Soc. 97, 1238-1239 (1975).

59. Scheuer, P.J.: Chemistry of marine natural products. Academic Press, 201 pages (1973).

60. Scheuer, P.J.: Marine toxins. Acc. Chem. Res. 10, 33-39 (1977).

61. Schulze, E., Neuhoff, V.: Oxidative side reactions during dansylation of SH-compounds. Hoppe-Seyler's Z. Physiol. Chem. 357, 225-231 (1976).

62. Seiler, N., Wiechmann, M.: TLC analysis of amines as their DANS-derivatives. Progress in thin-layer chromatography and related methods. Vol.I. Ann Arbor-Humphrey Sci. Publ., Ann Arbor, London, 94-144 (1970).

63. Seiler, N., Schmidt-Glenewinkel, T., Schneider, H.H.: 5-Di-n-butylaminonaphthalene-1-sulphonyl chloride - a new reagent for fluorescence labelling of amines, amino acids and peptides. J. Chromat. 84, 95-107 (1973).

64. Seiler, N.: Review. Chromatography of biogenic amines. I. Generally applicable separation and detection methods. J. Chromat. 143, 221-246 (1977).

65. Seiler, N., Knödgen, B., Eisenbeiss, F.: Determination of di- and polyamines by high-performance liquid chromatographic separation of their 5-dimethylaminonaphthalene-1-sulfonyl derivatives. J. Chromat. 145, 29-39 (1978).

66. Shimizu, Y., Buckley, L.J., Alam, M., Oshima, Y., Fallon, W.E., Kasai, H., Miura, I., Gullo, V.P., Nakanishi, K.: Structures of gonyautoxins II and III from the east coast toxic dinoflagellate Gonyaulax tamarensis. J. Amer. Chem. Soc. 98, 5414-5416 (1976).

67. Smith, T.A.: The occurrence, metabolism and functions of amines in plants. Biol. Rev. 46, 201-241 (1971).

68. Smith, T.A.: Review. Recent advances in the biochemistry of plant amines. Phytochem. 14, 865-890 (1975).

69. Smith, T.A., Wilshire, G.: Distribution of cadaverine and other amines in higher plants. Phytochem. 14, 2341-2346 (1975).

70. Smith, T.A.: Review. Phenethylamine and related compounds in plants. Phytochem. 16, 9-18 (1977).

71. Smith, T.A.: Homospermidine in Rhizobium and legume root nodules. Phytochem. 16, 278-279 (1977).
72. Steiner, M., Hartmann, T.: The occurrence and distribution of volatile amines in marine algae. Planta 79, 113-121 (1968).
73. Stillway, L.W., Walle, T.: Identification of the unusual polyamines 3,3'-diaminodipropylamine and N,N'-bis(3-aminopropyl)-1,3-propanediamine in the white shrimp Penaeus setiferus. Biochem. Biophys. Res. Comm. 77, 1103-1107 (1977).
74. Tabor, C.W., Tabor, H.: 1,4-diaminobutane (putrescine), spermidine, and spermine. Annual Rev. Biochem. 45, 285-306 (1976).
75. Taylor, R.F., Ikawa, M., Sasner, jr., J.J., Thurberg, F.P., Andersen, K.K.: Occurrence of choline esters in the marine dinoflagellate Amphidinium carteri. J. Phycol. 10, 279-283 (1974).
76. Tocher, R.D., Tocher, C.: Abstr. XI. Internat. Bot. Congr. p. 219 (1969). Ref. 19 in Smith (70).
77. Wiechmann, M.: Scope and limitations of the analytical use of dansyl chloride, I: The reaction of aromatic sulfonyl chlorides with aliphatic tertiary amines: the microanalytical aspects of the Hinsberg test. Hoppe-Seyler's Z. Physiol. Chem. 358, 967-980 (1977).
78. Wiechmann, M.: Scope and limitations of the analytical use of dansyl chloride, II: Formation of secondary dansyl amines by the reaction of dansyl chloride with N-oxides of aliphatic tertiary amino derivatives according to the Polonovski reaction. Hoppe-Seyler's Z. Physiol. Chem. 358, 981-984 (1977).
79. Zeman, A., Wirotama, I.P.G.: Identification of amines, II. Capillary gas-chromatography mass spectrometry (GC-MS) of trifluoroacetyl derivatives (TFA-amines). Z. Anal. Chem. 247, 158-163 (1969).

A P P E N D I X

Summary of published data on amines in algae

		MEA	DMA	TMA	ETA	PRA	iBA	iAA	MMP	PEA
Cyanophyta										
1. Anabaena flos-aquae		.	.	.	.	.	.	.	.	.
2. Anabaena variabilis		.	.	.	.	.	.	.	.	.
3. Aphanizomenon flos-aquae		.	.	.	.	.	.	.	.	.
4. Chroococcus cohaerens		.	.	.	.	.	.	.	.	.
5. Lyngbia sp.		.	.	.	.	.	.	.	.	.
6. Microcystis aeruginosa		–	.	–	.	+	.	.	.	.
7. Phormidium foveolarum		.	.	.	.	.	.	.	.	.
8. Phormidium luridum		.	.	.	.	.	.	.	.	.
9. Plectonema boryanum		.	.	.	.	.	.	.	.	.
		.	.	.	.	.	.	.	.	.
10. Spirulina platensis		1.05	.	.	–	.	–	–	.	–
Bacillariophyta										
1. Nitzschia actinastroides		+	.	–	.	+	.	.	.	.
Chlorophyta										
1. Acrosiphonia centralis	S	+	–	+	–	–	–	tr	–	–
2. Ankistrodesmus braunii*)		–	–	–	–	+	.	.	.	.
3. Chaetomorpha crassa	S	.	.	24.25	.	.	.	.	.	.
4. Chamaedoris orientalis	S	.	.	14.20	.	.	.	.	.	.
5. Chlorella ellipsoidea*)		.	.	.	.	.	.	.	.	.
6. Chlorella emersonii (fusca)*)		.	.	.	.	.	.	.	.	.
		+	+	.	.	.	.	.	.	.
7. Chlorella pyrenoidosa*)		.	.	.	.	.	.	.	.	.
8. Chlorella sp.		0.95	.	.	.	.	.	.	.	.
9. Cladophora rupestris	S	.	.	.	.	.	.	.	.	.
	S	+	–	+	–	–	–	tr	–	–
10. Codium divaricatum	S	.	.	18.28	.	.	.	.	.	.
11. Codium fragile	S	.	.	.	.	.	.	.	.	.
	S	.	.	19.28	.	.	.	.	.	.
	S	+	–	+	tr	–	+	+	–	–
12. Coelastrum sphaericum		0.79	.	.	.	.	.	.	.	.
13. Enteromorpha compressa	S	+	–	+	(+)	–	–	+	–	–
14. Enteromorpha linza	S	.	.	73.40	.	.	.	.	.	.
15. Enteromorpha rupestris	S	.	.	.	.	.	.	.	.	.
16. Monostroma fuscum	S	.	.	.	.	.	.	.	.	.
17. Neospongiococcum saccatum		.	.	.	.	.	.	.	.	.
18. Scenedesmus acutus		0.75	.	.	.	.	.	.	.	.
		+	+	.	.	.	.	.	.	.
		.	.	.	+	.	.	.	.	+
19. Ulva lactuca	S	.	.	.	.	.	.	.	.	.
	S	.	.	.	.	.	.	.	.	.
	S	+	–	+	–	–	–	–	–	–
20. Ulva pertusa	S	.	.	74.40	.	.	.	.	.	.
21. Ulva reticulata	S	.	.	35.30	.	.	.	.	.	.

*) Names of algae were taken from the literature reviewed. Some new names are generally accepted, e.g. Chlorella emersonii for Chl. fusca etc.

	EOH	PUT	CAD	NSD	SPD	HSD	SPM	CHO	other	ref.
1.	.	.	.	.	.	.	.	.	anatoxin-a	10
2.	.	.	.	.	.	.	.	<0.06		8
3.	.	.	.	.	.	.	.	0.96		30
4.	.	.	.	.	.	.	.	<0.06		8
5.	.	.	.	.	.	.	.	<0.06		30
6.	–	.	.	.	.	.	.	.		27
7.	.	.	.	.	.	.	.	<0.06		8
8.	.	.	.	.	.	.	.	<0.06		30
9.	.	+	.	.	+	.	–	.		6
	.	.	.	.	.	.	.	<0.06		30
10.	<0.05	0.3	–	<0.01	5.7	0.12	0.55	.		38

	EOH	PUT	CAD	NSD	SPD	HSD	SPM	CHO	other	ref.
1.	–	.	.	.	.	.	.	.		27

	EOH	PUT	CAD	NSD	SPD	HSD	SPM	CHO	other	ref.
1.	.	.	.	.	.	.	.	.		72
2.	–	.	.	.	.	.	.	.		27
3.	.	.	.	.	.	.	.	.		17
4.	.	.	.	.	.	.	.	.		17
5.	5-12	5-7	.		6		–	.		34
6.	4.3	10.7	0.12	7.6	2.1	<0.02	<0.02	.		38
	+	+	–	├──	10-15	──┤	–	.		53
7.	.	.	.	.	.	.	.	11.5		30
8.	4.0	7.5	–	0.15	14.1	0.83	–	.		37
9.	.	.	.	.	.	.	.	48.5		8
	.	.	.	.	.	.	.	.		72
10.	.	.	.	.	.	.	.	.		17
11.	.	.	.	.	.	.	.	56.2		8
	.	.	.	.	.	.	.	.		17
	.	.	.	.	.	.	.	.		72
12.	3.2	2.96	0.17	6.25	2.35	0.18	0.007	.		37
13.	.	.	.	.	.	.	.	.		72
14.	.	.	.	.	.	.	.	.		17
15.	.	.	.	.	.	.	.	21.7		8
16.	.	.	.	.	.	.	.	.	dopamine	76
17.	6.2	2.6	1.2	2.6	1.4	0.87	<0.05	.	DAP 23.4	38
18.	14.9	10.5	0.48	13.6	1.0	0.17	0.006	.		37
	+	+	–	├──	9-12	──┤	–	.		53
	+	+	+	+	+	+	.	.	TYA, piperazine	54
19.	.	.	.	.	.	.	.	0.13		30
	.	0.63	–	0.012	1.45	tr	0.04	.		38
	.	.	.	.	.	.	.	.		72
20.	.	.	.	.	.	.	.	.		17
21.	.	.	.	.	.	.	.	.		17

For explanation of abbreviations see last page.

		MEA	DMA	TMA	ETA	PRA	iBA	iAA	MMP	PEA
Chrysophyta										
1. Ochromonas danica		–	.	+	.	+	.	.	.	.
2. Prymnesium parvum		.	.	.	.	.	.	.	.	.
Dinophyta										
1. Amphidinium carteri		.	.	.	.	.	.	.	.	.
		.	.	.	.	.	.	.	.	.
2. Gonyaulax catenella		.	.	.	.	.	.	.	.	.
3. Gonyaulax tamarensis		.	.	.	.	.	.	.	.	.
Euglenophyta										
1. Euglena gracilis		.	.	.	.	.	.	.	.	.
		.	.	.	.	.	.	.	.	.
		4.9	.	.	.	.	.	.	.	.
		.	.	.	.	.	.	.	.	.
Phaeophyta										
1. Ascophyllum nodosum	S	.	.	.	.	.	.	.	.	.
	S	.	.	.	.	.	.	.	.	.
2. Chorda tomentosa	S	+	–	–	–	–	–	–	–	–
3. Chordaria flagelliformis	S	.	.	.	.	.	.	.	.	.
4. Chordata (?) flagelliformis	S	.	.	.	.	.	.	.	.	.
5. Desmarestia aculeata	S	.	.	.	.	.	.	.	.	.
	S	+	+	+	+	tr	+	++	++	+
6. Desmarestia viridis	S	.	.	14.62	.	.	.	.	.	.
	S	+	+	+	–	–	+	++	+	tr
7. Dictyopteris divaricata	S	.	.	9.76	.	.	.	.	.	.
8. Dictyosiphon foeniculaceus	S	.	.	.	.	.	.	.	.	.
9. Ectocarpus sp.	S	.	.	.	.	.	.	.	.	.
10. Eisenia bicyclis	S	.	.	25.71	.	.	.	.	.	.
11. Fucus edentatus	S	.	.	.	.	.	.	.	.	.
12. Fucus evanescens	S	.	.	.	.	.	.	.	.	.
13. Fucus serratus	S	.	.	.	.	.	.	.	.	.
	S	+	–	++	–	–	–	–	–	–
14. Fucus spiralis	S	.	.	.	.	.	.	.	.	.
	S	.	.	.	.	.	.	.	.	.
15. Fucus vesiculosus	S	.	.	.	.	.	.	.	.	.
	S	+	–	++	–	–	–	–	–	–
16. Halidrys siliquosa	S	+	–	+	–	–	–	–	–	–
17. Hizikia fusiformis	S	.	.	14.75	.	.	.	.	.	.
18. Laminaria agardhii	S	.	.	.	.	.	.	.	.	.
19. Laminaria digitata	S	.	.	.	.	.	.	.	.	.
	S	.	.	.	.	.	.	.	.	.
	S	+	(+)	–	–	–	(+)	(+)	–	–
20. Laminaria hyperborea	S	.	.	.	.	.	.	.	.	.
	S	+	(+)	–	–	–	(+)	(+)	–	–
21. Laminaria saccharina	S	.	.	.	.	.	.	.	.	.
	S	.	.	.	.	.	.	.	.	.
	S	+	(+)	–	–	–	(+)	(+)	–	–
22. Pelvetia canaliculata	S	.	.	.	.	.	.	.	.	.
23. Sargassum thunbergii	S	.	.	26.10	.	.	.	.	.	.

	EOH	PUT	CAD	NSD	SPD	HSD	SPM	CHO	other	ref.
1.	+	.	.	.	.	.	.	.	2-aminopropane	27
2.	.	+	.	.	+	.	+	.		6
1.	.	.	.	.	.	.	.	24.9		30
	.	.	.	.	.	.	.	.	acrylylcholine	75
2.	.	.	.	.	.	.	.	.	saxitoxin	58
3.	.	.	.	.	.	.	.	.	gonyautoxin II,III	66
1.	.	7.5	.	⊢—	0.82	—⊣	0.31	.		3
	.	.	.	.	.	.	.	5.1		30
	7.7	2.5	0.01	3.1	6.56	0.37	(4.18) →	.		37
	.	.	.	.	.	.	tr	.	NSM	39
1.	.	.	.	.	.	.	.	6.4-20.4		8
	.	.	.	.	.	.	.	<0.06		30
2.	.	.	.	.	.	.	.	.		72
3.	.	.	.	.	.	.	.	2.6		30
4.	.	.	.	.	.	.	.	36.4		8
5.	.	.	.	.	.	.	.	2.6		30
	.	.	.	.	.	.	.	.		72
6.	.	.	.	.	.	.	.	.		17
	.	.	.	.	.	.	.	.		72
7.	.	.	.	.	.	.	.	.		17
8.	.	.	.	.	.	.	.	8.3		8
9.	.	.	.	.	.	.	.	10.2		8
10.	.	.	.	.	.	.	.	.		17
11.	.	.	.	.	.	.	.	0.19		30
12.	.	.	.	.	.	.	.	0.32		30
13.	.	.	.	.	.	.	.	<0.6		8
	.	.	.	.	.	.	.	.		72
14.	.	.	.	.	.	.	.	<0.12		8
	.	.	.	.	.	.	.	0.26		30
15.	.	.	.	.	.	.	.	<0.06		30
	.	.	.	.	.	.	.	.		72
16.	.	.	.	.	.	.	.	.		72
17.	.	.	.	.	.	.	.	.		17
18.	.	.	.	.	.	.	.	2.6		30
19.	.	.	.	.	.	.	.	3-13		8
	.	.	.	.	.	.	.	4.5		30
	.	.	.	.	.	.	.	.		72
20.	.	.	.	.	.	.	.	2-5		8
	.	.	.	.	.	.	.	.		72
21.	.	.	.	.	.	.	.	0.1-2.6		8
	.	0.02	-	<0.002	0.01	-	0.03	.	TYA 0.16	38
	.	.	.	.	.	.	.	.		72
22.	.	.	.	.	.	.	.	0.6-5.7		8
23.	.	.	.	.	.	.	.	.		17

Phaeophyta (continued)		MEA	DMA	TMA	ETA	PRA	iBA	iAA	MMP	PEA
24. Scytosiphon lomentaria	S	.	.	22.89	.	.	.	.	.	.
	S	+	-	+	-	-	-	+	-	-
25. Undaria pinnatifida	S	.	.	14.51	.	.	.	.	.	.
Rhodophyta										
1. Ahnfeltia plicata	S	+	-	+	-	-	-	-	-	-
2. Ceramium rubrum	S	.	.	.	.	.	.	.	.	.
	S	.	.	.	.	.	.	.	.	.
	S	+	-	++	+	+	+	++	tr	+
3. Chondria crassicaulis	S	.	.	15.00	.	.	.	.	.	.
4. Chondrus crispus	S	.	.	.	.	.	.	.	.	.
	S	.	.	.	.	.	.	.	.	.
	S	+	-	+	-	-	-	(+)	-	-
5. Corallina officinalis	S	.	.	.	.	.	.	.	.	.
	S	+	-	+	-	-	-	tr	-	-
6. Cystoclonium purpureum	S	.	.	.	.	.	.	.	.	.
	S	+	-	+	+	+	+	++	+	+
7. Delesseria sanguinea	S	+	-	+	tr	-	+	+	+	tr
8. Dumontia incrassata	S	+	-	+	+	+	+	++	+	+
9. Euthora cristata	S	.	.	.	.	.	.	.	.	.
10. Furcellaria fastigiata	S	.	.	.	.	.	.	.	.	.
11. Gigartina stellata	S	.	.	.	.	.	.	.	.	.
12. Grateloupia elliptica	S	.	.	23.40	.	.	.	.	.	.
13. Grateloupia filicina	S	.	.	32.45	.	.	.	.	.	.
14. Grateloupia turuturu	S	.	.	22.40	.	.	.	.	.	.
15. Halosaccion ramentaceum	S	.	.	.	.	.	.	.	.	.
16. Odonthalia dentata	S	.	.	.	.	.	.	.	.	.
17. Phycodrys rubens	S	.	.	.	.	.	.	.	.	.
18. Phyllophora brodiaei	S	.	.	.	.	.	.	.	.	.
19. Phyllophora nervosa	S	.	.	.	.	.	.	.	.	.
20. Plocamium vulgare	S	+	-	+	-	-	-	(+)	-	-
21. Polyides rotundus	S	+	-	+	+	tr	+	++	+	+
22. Polysiphonia elongata	S	.	.	.	.	.	.	.	.	.
23. Polysiphonia fastigiata	S	.	.	.	.	.	.	.	.	.
24. Polysiphonia urceolata	S	.	.	.	.	.	12.5	11.4	.	.
	S	+	-	+	+	+	+	++	tr	+
25. Porphyra laciniata	S	.	.	.	.	.	.	.	.	.
26. Porphyra suborbiculata	S	.	.	42.85	.	.	.	.	.	.
27. Porphyra tenera	S	.	.	31.60	.	.	.	.	.	.
28. Porphyra umbilicalis	S	.	.	.	.	.	.	.	.	.
	S	+	-	-	-	-	-	-	-	-
29. Rhodomela confervoides	S	+	-	++	+	-	+	+	-	-
30. Rhodymenia palmata	S	.	.	.	.	.	.	.	.	.

Concentrations are in μmole/g dry wt.

Abbrev.: S seaweed; . not determined; - absent; + detected; ++ higher conc.; (+) doubtful; tr traces

MEA methylamine; DMA dimethylamine; TMA trimethylamine; ETA ethylamine; PRA propylamine; iBA isobutylamine; iAA isoamylamine

MMP 3-methylmercapto-propylamine; PEA phenylethylamine; EOH ethanolamine; PUT putrescine; CAD cadaverine; NSD norspermidine

	EOH	PUT	CAD	NSD	SPD	HSD	SPM	CHO	other	ref.
24.	.	.	.	.	.	.	.	.		17
	.	.	.	.	.	.	.	.		72
25.	.	.	.	.	.	.	.	.		17
1.	.	.	.	.	.	.	.	.		72
2.	.	.	.	.	.	.	.	6.4-14.7		8
	.	.	.	.	.	.	.	0.64		30
	.	.	.	.	.	.	.	.		72
3.	.	.	.	.	.	.	.	.		17
4.	.	.	.	.	.	.	.	7.7		30
	.	0.17	-	<0.002	0.009	-	0.023	.	TYA 0.16	38
	.	.	.	.	.	.	.	.		72
5.	.	.	.	.	.	.	.	0.64		8
	.	.	.	.	.	.	.	.		72
6.	.	.	.	.	.	.	.	1.9		30
	.	.	.	.	.	.	.	.		72
7.	.	.	.	.	.	.	.	.		72
8.	.	.	.	.	.	.	.	.		72
9.	.	.	.	.	.	.	.	1.3		30
10.	.	.	.	.	.	.	.	20.4		8
11.	.	.	.	.	.	.	.	14.0		8
12.	.	.	.	.	.	.	.	.		17
13.	.	.	.	.	.	.	.	.		17
14.	.	.	.	.	.	.	.	.		17
15.	.	.	.	.	.	.	.	5.1		30
16.	.	.	.	.	.	.	.	6.4		8
17.	.	.	.	.	.	.	.	6.4		30
18.	.	.	.	.	.	.	.	29.4		8
19.	.	.	.	.	.	.	.	.	hordenine	21
20.	.	.	.	.	.	.	.	.		72
21.	.	.	.	.	.	.	.	.		72
22.	.	.	.	.	.	.	.	1.3		30
23.	.	.	.	.	.	.	.	33.8		8
24.	.	<0.002	-	<0.002	0.017	-	0.038	.	TYA 1.22	38
	.	.	.	.	.	.	.	.		72
25.	.	.	.	.	.	.	.	7.7		8
26.	.	.	.	.	.	.	.	.		17
27.	.	.	.	.	.	.	.	.		17
28.	.	.	.	.	.	.	.	3.2		30
	.	.	.	.	.	.	.	.		72
29.	.	.	.	.	.	.	.	.		72
30	.	.	.	.	.	.	.	12.8		8

SPD spermidine
HSD homospermidine
SPM spermine
NSM norspermine
CHO choline
TYA tyramine
DAP 1,3-diaminopropane

It must be emphasized that culture conditions, age of cultures etc. have strong influence on content of amines in algae.

ANTIMICROBIAL AGENTS FROM MARINE RED ALGAE OF THE FAMILY BONNEMAISONIACEAE

Oliver J. McConnell and William Fenical
Institute of Marine Resources
Scripps Institution of Oceanography
La Jolla, California 92093

Abstract

Red marine algae of the family Bonnemaisoniaceae (Class-Rhodophyceae, Order-Nemaliales) are cosmopolitan in distribution. Representatives grow quite luxuriantly with few, if any, predators, both intertidally and sublittorally in polar, temperate, and tropical waters adjacent to all of the major continents of the world[1]. Early European investigators, who studied the distribution of this algae, also discovered that this family is characterized by specialized vesicular cells that contain unusually high amounts of halogens. Subsequently, these cells were shown to contain predominantly bromine, but in an unknown chemical form. Recent chemical investigations by us and others have revealed that the chemical form of the halogen in these specialized cells is mainly covalently bound bromine in secondary metabolic products. From antibiotic testing of these isolated and characterized compounds, we have been able to associate the reported antibiotic[2,3] and antimitotic[4] activity of crude extracts of these algae with these natural products and to evaluate their possible pharmaceutical applications.

Background

The presence of halogen in members of the Bonnemaisoniaceae has been recognized since the 1890's[5,6] when pressings of *B. asparagoides* on paper containing starch resulted in a blue stain characteristic of the presence of iodine. Kylin,[7-12] working with species of *Bonnemaisonia, Asparagopsis, Trailliella,* and *Falkenbergia,* associated the presence of iodine with ruptured vesicular cells, which are the specialized vegetative cells -- characteristic of the Bonnemaisoniaceae. Early workers could not completely explain this curious feature[13-15]. More recently, Wolk[16], using electron microprobe X-ray fluorescence spectroscopy, has shown that

Marine Algae in Pharmaceutical Science
edited by H. A. Hoppe, T. Levring, Y. Tanaka

vesicular cells of cultured *Trailliella* preferentially concentrate bromine and iodine to 30x and 3x their concentrations in adjacent vegetative cells. Wolk further showed that bromide was necessary in the culture medium for formation of the characteristic refractile vacuolar inclusions of these cells and for iodide oxidation. Thus the presence and association of iodine with ruptured vesicular cells was thought to be due to iodide oxidation. Von Stosch[17,18] showed with *Asparagopsis* that iodide was required for normal growth. Thus, the halogen content of members of the family Bonnemaisoniaceae plays an important role in the metabolic processes of these algae.

We feel that the halogenated components from algae of the family Bonnemaisoniaceae are, in great part, responsible for their survival against herbivores. From numerous field observations of *Asparagopsis* and *Bonnemaisonia* species, we have not found evidence of predation. These algae are very fine and delicate in form, hence they offer only slight physical defense against predators. From our experience in culturing these algae, we feel their growth rates are not sufficient to enable the algae to compete against rapacious herbivores -- especially in subtropical and tropical environments. Thus, the only conceivable protection for the algae is a chemical defense.

Since the halogenated compounds we will describe in this chapter contain the large amounts of bromine characteristic of the specialized vesicular cells, and because many of these compounds are capable of denaturing enzymes due to their abilities to alkylate numberous amino acid residues[19], these defensive substances are probably biosynthesized and stored in the vesicular cells. And, although the algae may halogenate the methyl ketone precursors in energy-providing biochemical transformations, the result is an array of compounds which renders the plant unpalatable and noxious to potential predators.

Since 1974[20], a great wealth of chemical information has been obtained regarding the structures of the halogenated metabolites in members of the family Bonnemaisoniaceae. There seems to be no question that these compounds are ecologically significant. In keeping with our interests in assessing both the *in vitro* and *in vivo* biological activities of algal metabolites, we have bioassayed many of the compounds from *Bonnemaisonia* and *Asparagopsis*. Reported herein are the results of those tests along with the results of similar testing, done in other laboratories, of metabolites from the related genera *Delisea* and *Ptilonia*.

Chemistry

Our chemical investigations of representatives of the family Bonnemaisoniaceae began with *A. taxiformis* (Delile) Trev. from the Gulf of California[20,21], and *A. armata* Harvey from La Escala, Spain[21]. The organic extract of lightly air-dried samples yielded quite a variety of halomethanes (1-4), haloacetaldehydes (5,6), haloacetones (7-24), and halobutenones (25-28, Fig. 1), while the pentane extract of algae stored in alcohol yielded additional halomethanes (29-33), and esters of haloacetic (34-36) and haloacrylic (37-42) acids (Fig. 2). Haloacetones and halobutenones were not detected. This strongly infers that the haloacetic and haloacrylic acids are in much greater abundance in the algae than the haloacetones and halobutenones. Chemical variation was noted in the relative proportions of 34-42 throughout the range of the Gulf of California (Fig. 3). Bromoform was the predominant constituent in all cases. The presence of chloroform (31) and carbon tetrachloride (32) in *A. armata* is quite exciting since these halocarbons, which are now recognized as atmospheric pollutants, are considered to be of a totally anthropogenic origin[22].

The structures of the *Asparagopsis* metabolites were, in general, initially assigned from combined gas chromatography-mass spectrometry (GC-MS) data. IR and NMR data were also quite informative, but final confirmations were made by comparison with commercially available standards or synthetic materials.

There are many possible origins for the compounds which have been observed in *Asparagopsis*. Acetone, which is also present in *Asparagopsis*[21], or its precursor, acetoacetate, appears to undergo halogenation reactions analogous to those observed for the fungal enzyme chloroperoxidase[23]. 1,1,1-Trihaloacetone derivatives react either chemically or enzymatically with base to yield haloforms and the corresponding acetic acids. Base-induced Favorsky rearrangements would yield the observed acrylic acids and would also explain the prevalence of the terminal olefin in these three-carbon acids (Fig. 4).

A. armata

cmpd		
1	-	$CHBr_2Cl$
2	-	$CHBr_3$
3	-	CBr_4
4	-	$CHBr_2I$
5	-	CH_2BrCHO
6	-	$CHBr_2CHO$

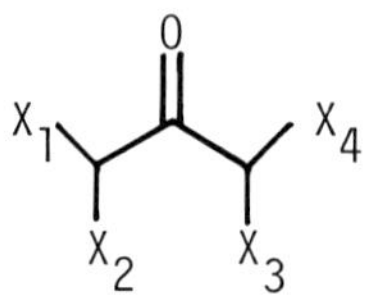

cmpd		X_1	X_2	X_3	X_4
15	-	H	Cl	H	Cl
16	-	H	Cl	Cl	Cl
17	-	Cl	Cl	Cl	Cl
18	-	Cl	Cl	H	Br
19	-	Cl	Br	H	Cl
20	-	Cl	Cl	Cl	Br
21	-	Cl	Cl	Br	Br
22	-	Cl	Br	Cl	Br
23	-	H	Cl	H	I
24	-	Br	Br	H	I

A. taxiformis

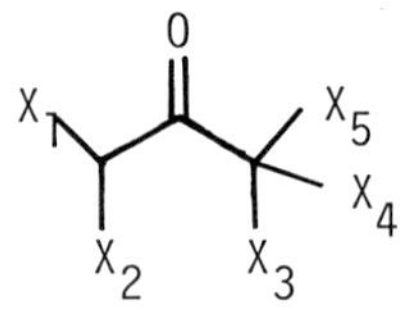

cmpd		X_1	X_2	X_3	X_4	X_5
7	-	Br	H	Br	H	H
8	-	Br	Br	Br	H	H
9	-	Br	Br	Br	Br	H
10	-	Br	H	Cl	H	H
11	-	Cl	H	Br	Br	H
12	-	Cl	Br	Br	H	H
13	-	Cl	Br	Br	Br	H
14	-	Br	Br	Br	Br	Br

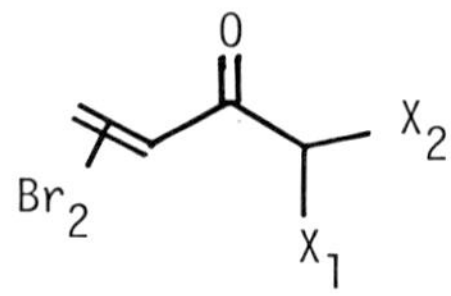

cmpd		X_1	X_2
25	-	Br	H
26	-	Cl	H
27	-	Cl	Br
28	-	Br	Br

Figure 1. Halogenated metabolites of *Asparagopsis* species. (air-dried plants)

Moore and his research group in Hawaii have reported a remarkably complete study on their local *A. taxiformis*[24-27]. In addition to the compounds we have found, the Hawaiian *Asparagopsis* contains the halomethanes (44-48), haloacetones (49-55), and halobutenones (56-63), as well as carbonyl diiodide (64), 2-iodoethanol (65), 2-iodo-1-bromoethane (66), halogenated isopropanols (67-85), halogenated 2-acetoxy-propanes (86-87), 1,3,3-tribromo-1-propene oxide (88), 1,1,3,3-tetrahalopropenes (89-92), 3,3-dibromoacrolein (93), halobutenols (94-101), haloacetamides (102-106), and haloacetic (107-113) and haloacrylic acids (114-118) (Fig. 5). These acids were obtained from aqueous extracts of the algae.

Wells and the research group at the Roche Institute have detected bromoform and a number of haloacetones (8-9, 11-13, 24,49,50) from the Australian *A. armata*[28].

Halomethanes:	CH_3I	CH_2ClI	$CHCl_3$	CCl_4	$CHBrCl_2$
	29	30	31	32	33

Haloacetic Acids:

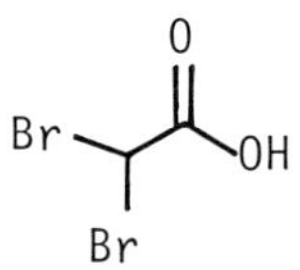
34

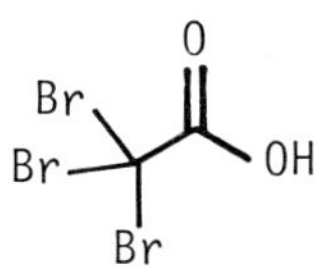
35

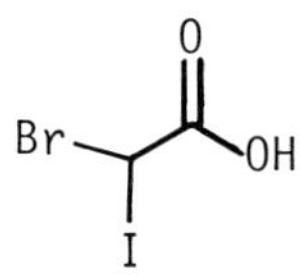
36

Haloacrylic Acids:

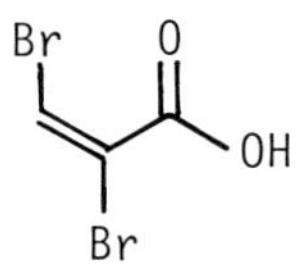
37

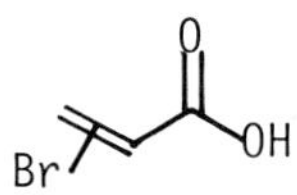
38

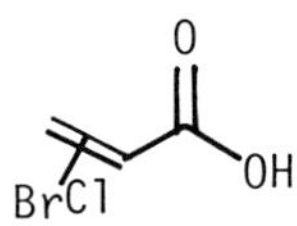
39

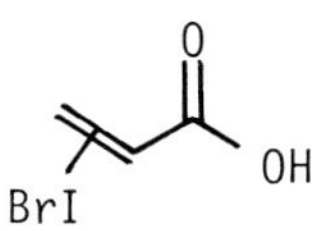
40

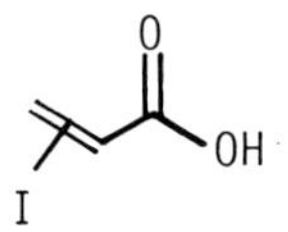
41

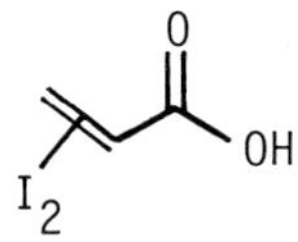
42

Figure 2. Halogenated methanes, acetic and acrylic acids from *A. taxiformis* and *A. armata*. The acids were isolated as ethyl esters following storage of the fresh algae in ethanol.

Falkenbergia rufolanosa, the alternate heteromorphic tetrasporophyte of *A. taxiformis*, which was collected along the Norwegian coast, contains large amounts of bromoform[29]. *F. rufolanosa* from Hawaii, on the other hand, has been reported to be devoid of volatile halogenated components[25].

We have chemically examined three species of *Bonnemaisonia*. *B. asparagoides* from La Escala, Spain, contains five chlorinated and brominated 1-octen-3-ones[30] (121-125) (Fig. 6).

Due to the limited availability of *B. asparagoides*, only 10 mg of total extract could be obtained. Recognition of compounds 121-125 in this extract was facilitated only by GC-MS. Subsequently, the total structures and stereochemistry of these compounds were confirmed by synthesis.

In contrast to the Mediterranean-derived species, *B. nootkana* from Monterey, California, provided exclusively seven and nine carbon metabolites[31], the major constituent being *trans*-1,3,3-tribromo-1-heptene oxide (126). Also present, but in minor amounts, are the brominated acids (127-130), brominated alcohols (131-134) and 1,1,3,3-tetrabromo-2-nonanone (135, Fig. 7). Similar to our work with *Asparagopsis*, the acids were detected as their esters following storage in alcohol, and their structures were determined by biomimetic Favorsky rearrangement of the appropriately brominated methyl ketone precursors. Base treatment of 135 and the seven-carbon homologue 1,1,3,3-tetrabromo-2-heptanone (136), yielded the branched acids, 130 and 129, respectively. Base treatment of 1,1,3-tribromo-2-heptanone (137) and -2-nonanone yielded the linear monobromo acids 127

Locale	Major Acids
Isla Angel de la Guarda 29° 35' NL	40
Isla Carmen 25° 59' NL	37,40
Cabo San Lucas 22° 80' NL	34-36

Figure 3. Chemical variation of *A. taxiformis* metabolites throughout the latitudinal range of the Gulf of California, Mexico.

and 128. We have recently confirmed the fatty acid origin of the major metabolite 126, by incorporation of $^{14}CH_3COO^-Na^+$ and U-[^{14}C]-$CH_3(CH_2)_{14}$-COO^-Na^+ using freshly collected algae from Monterey[32].

B. hamifera, collected in the Gulf of California, has previously been shown to contain the ketones 136 and 137 as well as 1,3-dibromo-2-heptanone (138), 1,3,3-tribromo-2-heptanone (139) and 3,3-dibromo-1-iodo-2-heptanone (140)[33] (Fig. 8). Re-examination of *B. hamifera* in our lab, after storage of the fresh plants in alcohol, revealed that 136 and 137 are less abundant than the acid 127 which was elucidated by GC-MS as its ethyl ester.

While *Asparagopsis* and *Bonnemaisonia* species are more cosmopolitan in their distribution, several other genera are known in this family and most have been examined chemically. Collections of *Delisea fimbriata* Lemour from both Australian[34] and Antarctic[35] waters have yielded a series of halogenated nine-carbon lactones (Fig. 9). Compounds 141-146 were found in both collections, while 147-156 were found only in the Australian collection. The structures of 141-146 were nicely elucidated by the Sims group using spectral data and X-ray analysis of the methanol addition product of 141. A complementary study, also well done, was completed by Wells using chemical degradation and spectroscopic data. It led

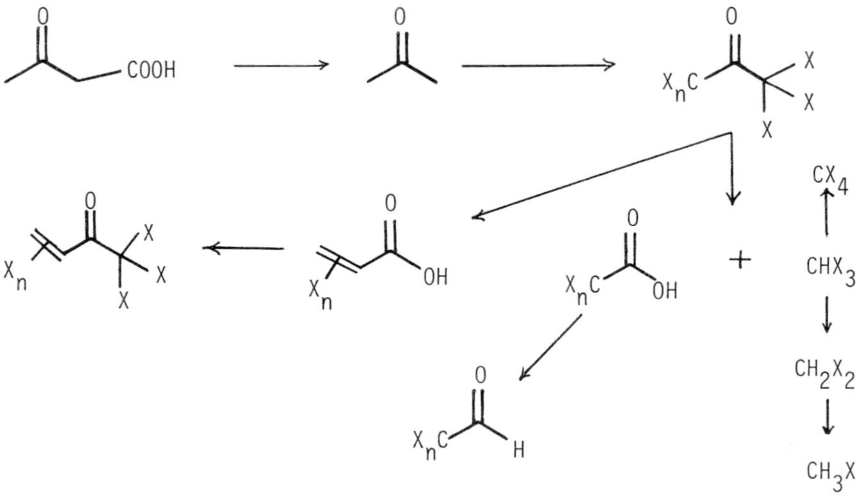

Figure 4. Possible biogenesis of the halogenated metabolites in *Asparagopsis*.

cmpd	
43 -	$CHBrI_2$
44 -	CHI_3
45 -	CHBrClI

cmpd	
46 -	CH_2Br_2
47 -	CH_2BrI
48 -	CH_2I_2

cmpd	
64 -	COI_2
65 -	ICH_2CH_2OH
66 -	$BrCH_2CH_2I$

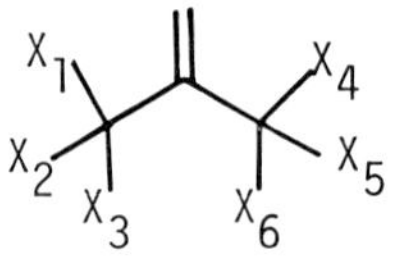

cmpd	X_1	X_2	X_3	X_4	X_5	X_6
49 -	H	H	H	H	H	Br
50 -	H	H	H	H	H	I
51 -	H	H	H	H	Br	Br
52 -	Br	H	H	H	H	I
53 -	H	H	H	Br	Br	Br
54 -	H	H	H	Br	Br	Cl
55 -	Cl	Cl	Cl	Cl	Cl	Cl

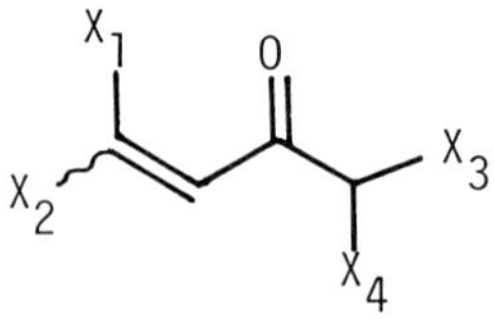

cmpd	X_1	X_2	X_3	X_4
56 -	Br	Br	H	H
57 -	Br	Br	H	Br
58 -	Br	Br	H	I
59 -	Br	Br	Br	Br
60 -	Br	Br	Br	Cl
61 -	Br	Cl	Br	Br
62 -	Cl	Cl	Br	Br
63 -	Br	Cl	Br	Cl

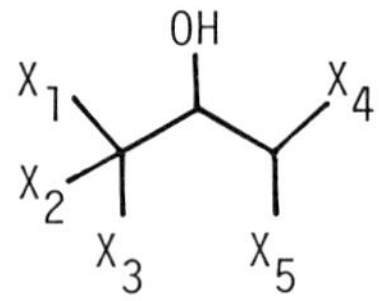

cmpd	X_1	X_2	X_3	X_4	X_5
67 -	H	Br	Br	H	H
68 -	H	Br	H	H	Br
69 -	H	Cl	H	H	I
70 -	H	Br	H	H	I
71 -	H	Br	Cl	H	Cl
72 -	H	Cl	Cl	H	Br
73 -	H	Br	Cl	H	Br
74 -	H	Br	Br	H	Cl
75 -	H	Br	Br	H	Br
76 -	H	Br	Cl	H	I
77 -	H	Br	Br	H	I
78 -	H	Cl	Cl	Cl	Cl
79 -	H	Br	Cl	Cl	Cl
80 -	H	Br	Br	Cl	Cl
81 -	H	Br	Br	Br	Cl
82 -	Br	Br	Br	H	Cl
83 -	Br	Br	H	Br	Br
84 -	Br	Br	H	Br	I
85 -	Br	Br	H	I	I

Figure 5. Additional *Asparagopsis* metabolites isolated from Hawaiian *A. taxiformis*.

cmpd

86 - $BrCH_2CH(OH_2)CHBr_2$

87 - $Br_2CHCH(OH_2)CHBr_2$

88 - $CHBrOCHCHBr_2$

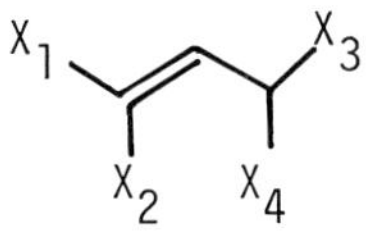

cmpd	X_1	X_2	X_3	X_4
89 -	Br	Br	Br	Br
90 -	Br	Br	Br	Cl
91 -	Br	Br	Cl	Cl
92 -	Br	I	Br	Br

93 -

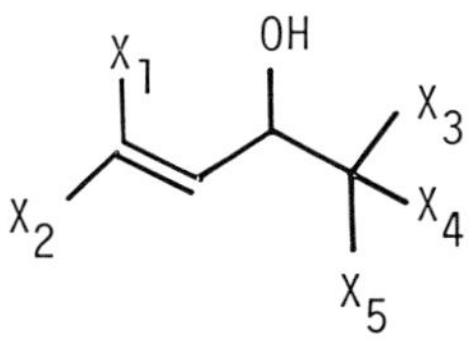

cmpd	X_1	X_2	X_3	X_4	X_5
94 -	Br	Br	H	H	Cl
95 -	Br	Br	H	H	Br
96 -	Br	Br	H	Cl	Cl
97 -	Cl	Cl	H	Br	Br
98 -	Br	Br	H	Br	Cl
99 -	Br	Cl	H	Br	Br
100 -	Br	Br	H	Br	Br
101 -	Br	Br	Br	Br	Br

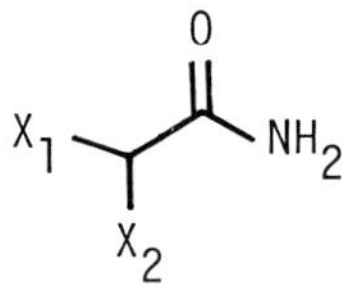

cmpd	X_1	X_2
102 -	Br	Cl
103 -	Br	Br
104 -	Cl	I
105 -	Br	I
106 -	I	I

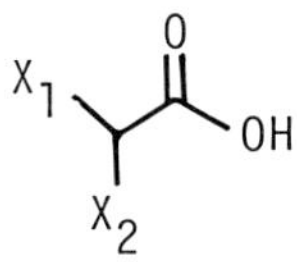

cmpd	X_1	X_2
107 -	Cl	H
108 -	Br	H
109 -	I	H
110 -	Cl	Cl
111 -	Br	Cl
112 -	Cl	I
113 -	I	I

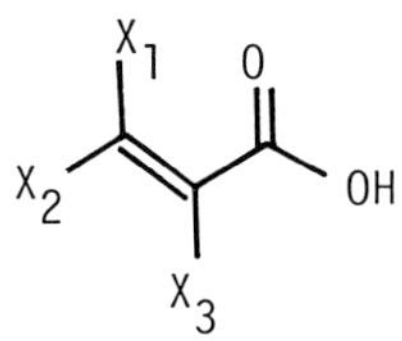

cmpd	X_1	X_2	X_3
114 -	Cl	H	H
115 -	Cl	Cl	H
116 -	Br	Br	H
117 -	Br	Br	Br
118 -	Br	Br	I

Figure 5. Continued.

to the structure elucidation of lactones 141-146 and 147-156. Also found by the Sims group in *D. fimbriata*[36] from Antarctica were five polyhalogenated 1-octen-3-ones (157-161), which are very similar to compounds 158-162 from *B. asparagoides*, and to those isolated by Wells from *Ptilonia australasica*[28] (159, 163-165, Fig. 10).

We believe that the polyhalo-1-octen-3-ones (121-125, 158-162, 164-165) from *B. asparagoides*, *D. fimbriata* and *P. australasica*, and the lactones (141-157) from *D. fimbriata*, have a common origin in acetate-derived precursors. Specifically, we suggest that polyhalogenated 2,4-nonadiones (with hydroxylation at C-6 for some of the lactones) undergo Favorsky reactions (Fig. 11). Base reaction, involving the C-2 carbonyl, could ultimately yield the halo-1-octen-3-ones. Reactions involving the C-4 carbonyl could yield a C-4 carboxylate-substituted n-octane skeleton which, through enolization of the remaining carbonyl toward C-1, could cyclize to give the *Delisea* lactones.

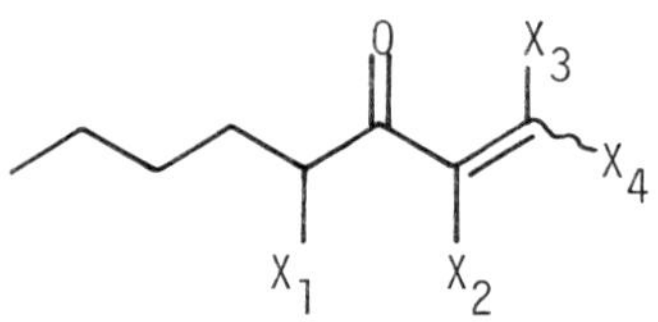

Naturally-Occurring Compounds

cmpd	X_1	X_2	X_3	X_4	E/Z
121 -	H	Cl	Br	Cl	1
122 -	Cl	Cl	Br	Cl	20
123 -	Br	Br	Cl	Cl	-
124 -	Cl	Br	Br	Cl	E
125 -	Br	Br	Br	Cl	E

Antibiotic Biotesting

compound(s)	mode	Pathogen A	B	C	D
121,122 (1:3 mix)	1	12(.25)	0(1.0)	10(.25)	0(1.0)
	2			2.2	
125 (25%X_1=H imp)	1	13(.25)	2(1.0)	15(.25)	0(1.0)
	2	1.5		3.0	
Synthetic, X_1=H,X_2=Cl/Br(3:2),X_3-Br,X_4=Cl	1	2(.25)	0(1.0)	7(.25)	0.1(1.0)
Synthetic, X_1=H,Br(2:1), X_2=X_3=X_4=Br	1	12(.25)	0(1.0)	10(.25)	0(1.0)
	2	12		2.2	

Figure 6. Halogenated metabolites isolated from *Bonnemaisonia asparagoides* and biotesting results. See Antimicrobial Testing section for explanation of bioassay results.

Compound		Mode	Pathogen A	Pathogen B	Pathogen C	Pathogen D
126	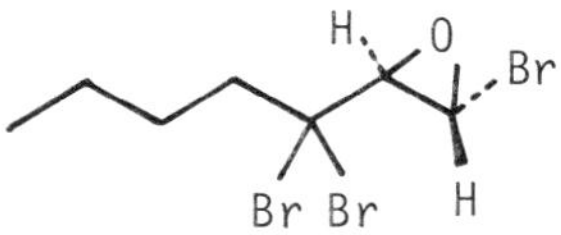	1	18(.25)	0(2.0)	37(.25)	1(1.0)
		2	3		500	
127	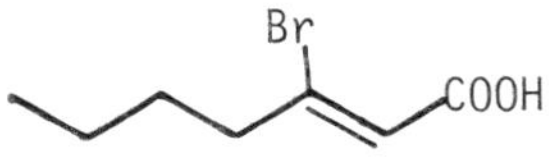	1	0(.25)	6(.25)	4(.25)	2(1.0)
128	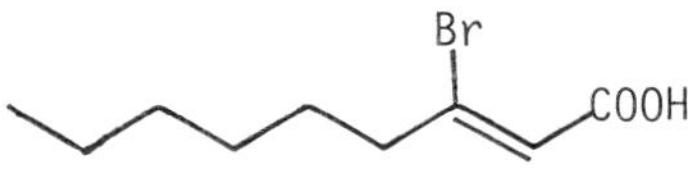	1	19(.25)	4(1.0)	5(.25)	3(1.0)
		2	15			
129	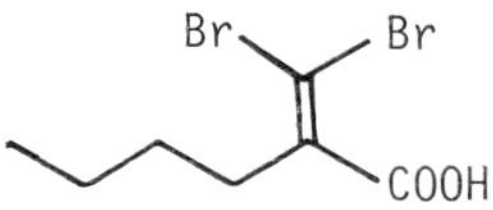	1	4(.25)	2(2.0)	0(.25)	1(1.0)
130	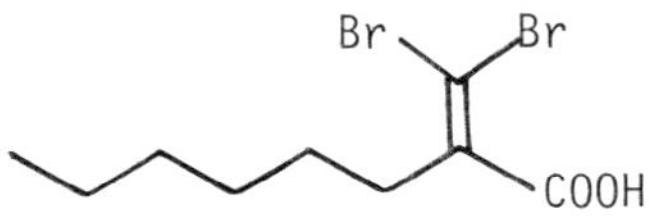	1	9(.25)	0(2.0)	0(.25)	2(1.0)
131	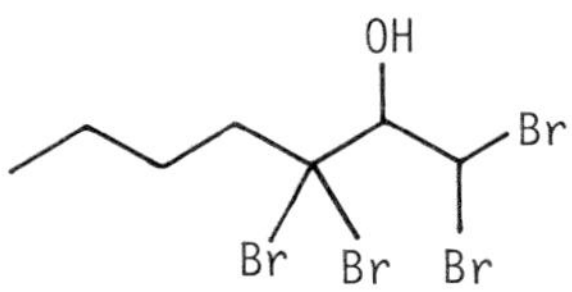	1	3(.25)	0(2.0)	0(.25)	2(1.0)
132	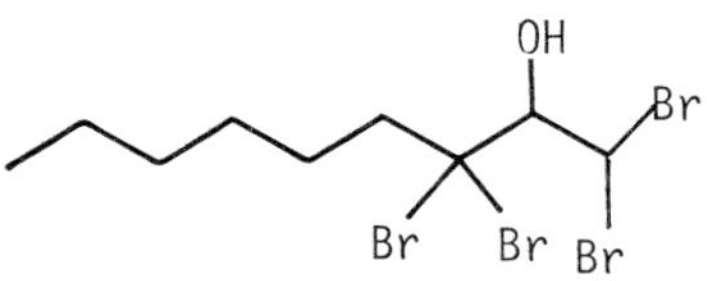	1	3(.25)	0(2.0)	0(2.0)	1(1.0)
133	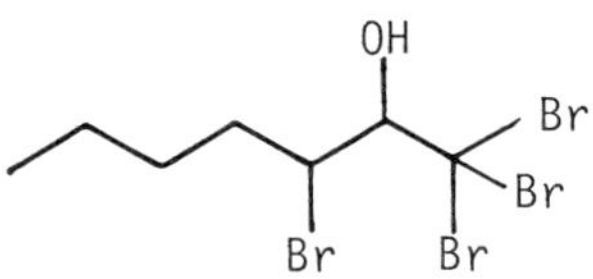					

Figure 7. Chemistry and biotesting results of *Bonnemaisonia nootkana* metabolites.

Compound		Mode	Pathogen A	B	C	D
134	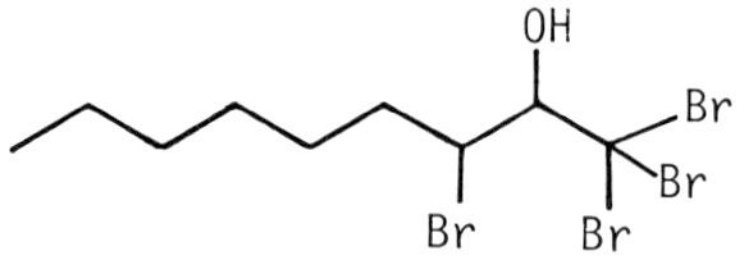					
135	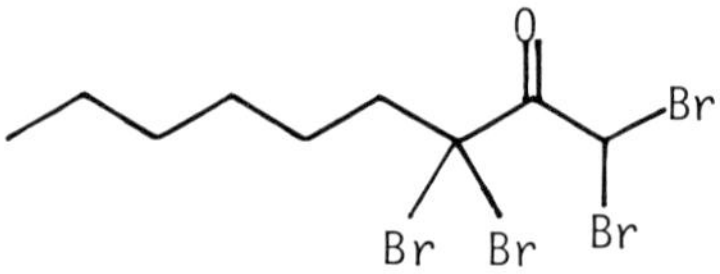	1	26(.1)	0(2.0)	29(.1)	1(1.0)
		2	.05		.05	

Figure 7. Continued.

Compound		Mode	Pathogen A	B	C	D
136	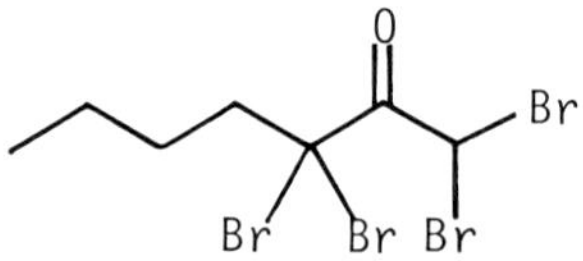	1	>36(.1)	3(.25)	34(.1)	1(1.0)
		2	37		25	
137	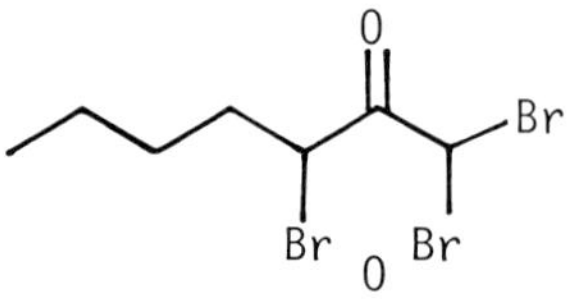	1	40(.25)	12(.25)	31(.25)	2(.1)
		2	5	300	50	
138	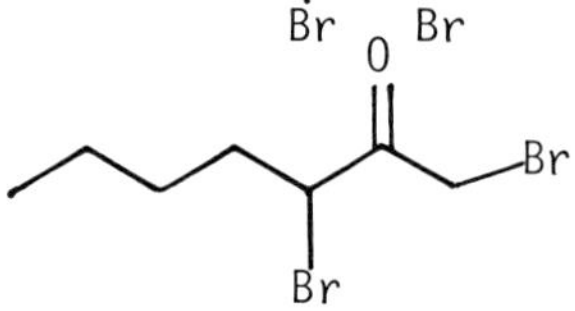					
139	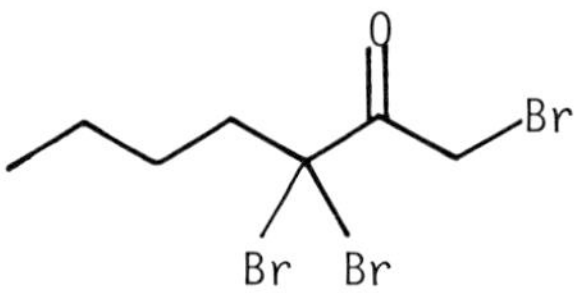					
140	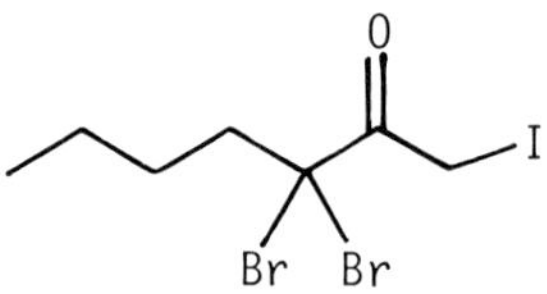					

Additional testing data on compound 136[33]

Pathogen	MIC (μg/ml)
I	100
J	100
K	500
L	500
A	1000
M	1000
N	1000

Figure 8. Chemistry and biotesting results with *Bonnemaisonia hamifera* metabolites.

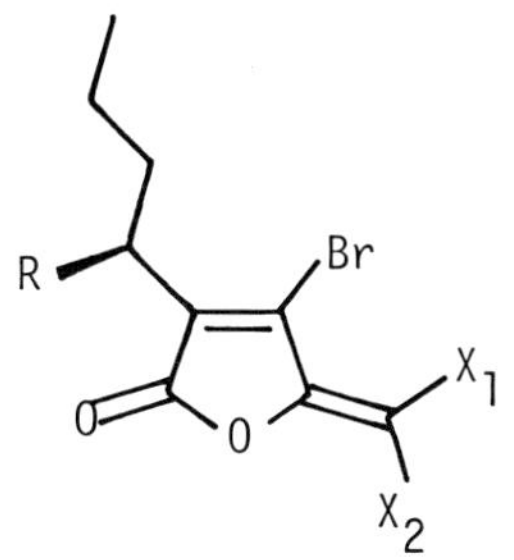

Compound	R	X_1	X_2	A	B	D	E	F	G	H	I
141	OAc	Br	H	1	10	80	80	5	10	1	1
142	OAc	H	Br								
143	OAc	I	H								
144	OAc	H	I								
145	OAc	H	Cl								
146	OAc	Br	Br								
147	OAc	Cl	H								
148	OH	Br	H	5	5	80	160	1	2	5	1
149	OH	H	Br								
150	OH	I	H								
151	OH	H	I								
152	OH	Cl	H								
153	OH	H	Cl								
154	OH	Br	Br								
155	H	Br	H								
156	H	H	Br	1	10			10	20	5	0.5
157	H	Br	Br								

(Pathogen - MIC (μg/ml); values are bracketed for compounds 141–142, 148–149 and 155–157.)

Figure 9. Chemistry and biotesting results for *Delisea fimbriata* metabolites.

Delisea fimbriata:

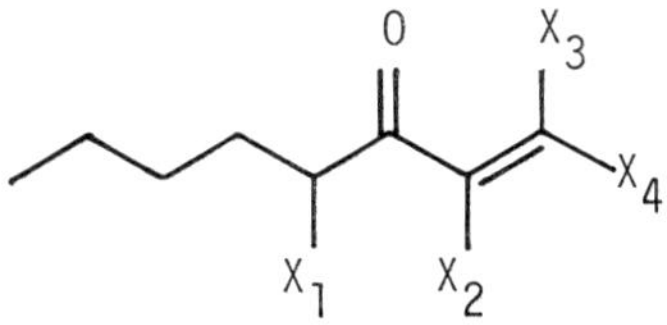

cmpd		X_1	X_2	X_3	X_4
158	-	H	Br	Br	Br
159	-	Br	Br	Br	Br
160	-	Cl	Br	Br	Br
161	-	H	I	Br	Br
162	-	Cl	I	Br	Br

Ptilonia australasica:

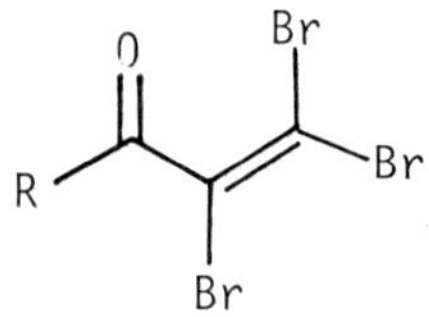

cmpd		R
163	-	$-CH_2Br$
164	-	$CH_3CH_2CBr_2CH\overset{t}{=}CH-$
165	-	$CH_3(CH_2)_3CBr_2-$
159	-	$CH_3(CH_2)_3CHBr-$

Figure 10. Halogenated metabolites from *Delisea fimbriata* (Antarctica) and *Ptilonia australasica*.

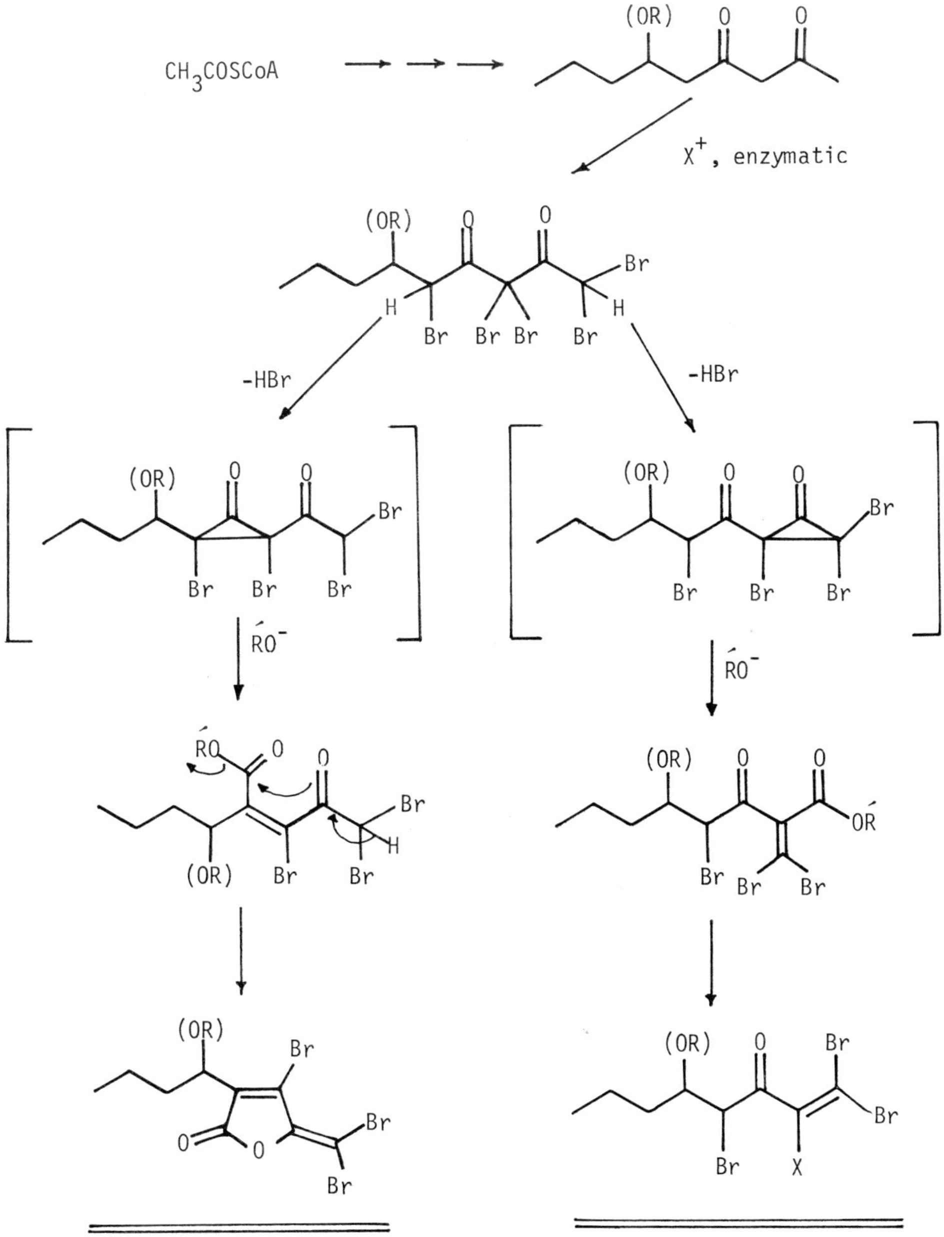

Figure 11. Proposed biogenesis for *Delisea*, *Ptilonia* and *Bonnemaisonia* compounds involving Favorsky rearrangements.

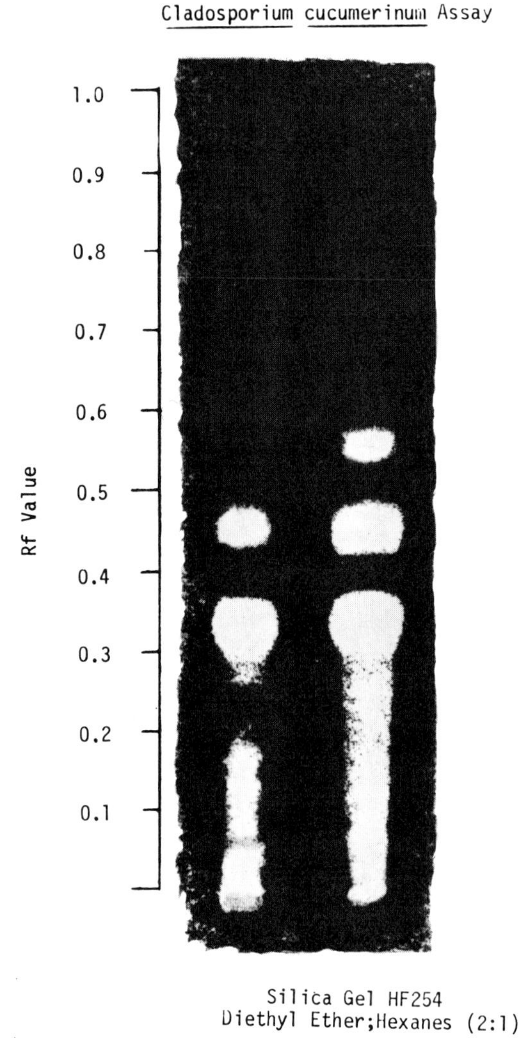

Figure 12. Thin-layer chromatographic antifungal bioassay of *Delisea fimbriata* crude extracts; left, CH_2Cl_2 extract; right, hexane extract.

Compound	% mix	Mode	A	B	C	D	E	F	G	H
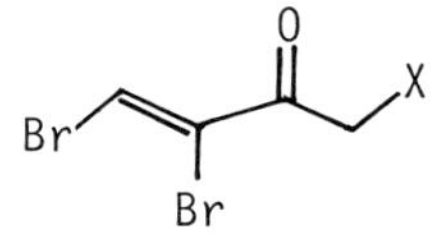										
166 - X = H (5:1 mix)		1	4(.25)	3(.25)	15(.25)	3(.25)				
25 - X = Br		2			30					
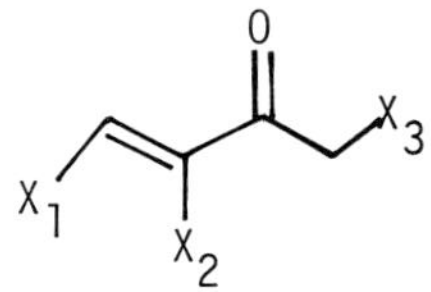										
167 - $X_1=X_2=Br, X_3=H$	5									
168 - $X_1=X_2=Cl, X_3=Br$	35	1	2(.25)	4(.25)	16(.25)	3(.25)				
169 - $X_1=X_3=Br, X_2=Cl$	20	2			30					
170 - $X_1=X_2=X_3=Br$	40									
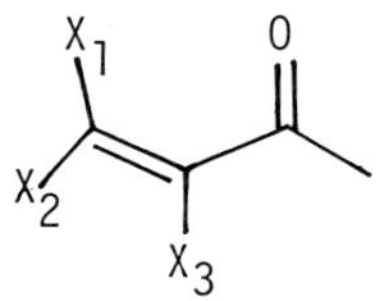										
171 - $X_1=H, X_2=X_3=Br$	28	1	1(.25)	3(.25	20(.25)	6(.25)				
172 - $X_1=X_2=Br, X_3=Cl$	14	2			2					
173 - $X_1=X_2=X_3=Br$	58									

Figure 13. Antimicrobial testing for the *Asparagopsis* compound 25 and for the synthetic analogs of *Asparagopsis* and *Delisea* compounds.

	Compound	Mode	Pathogen A	B	C	D	E	F	G	H
174	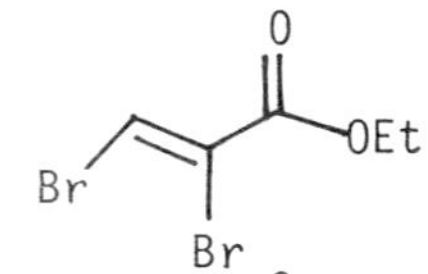	1	1(.25)		13(.25)	3(.25)				
		2			3					
175	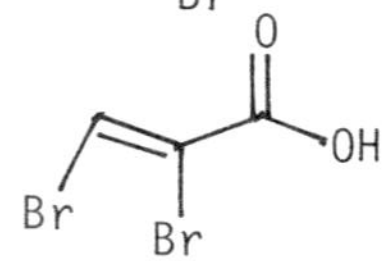	1	9(1.0)	1(1.0)	7(.5)	4(1.0)				
176	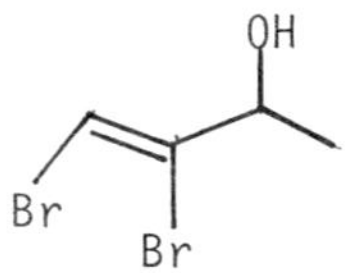	1			2(2.0)	1(2.0)				
177	- CBr_3CHO	1	6(.25)	4(.25)	5(.25)	4(1.0)				
	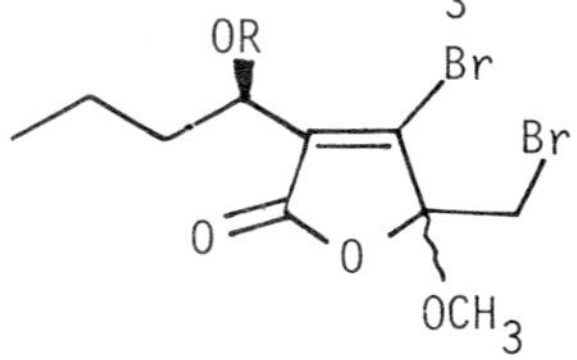									
178	- R = H	2	20	20	20		40		80	2
179	- R = Ac	2	80	160	80		40			80

Figure 13. Continued.

Compound		Mode	A	B	C	D	E	F	G	H
180	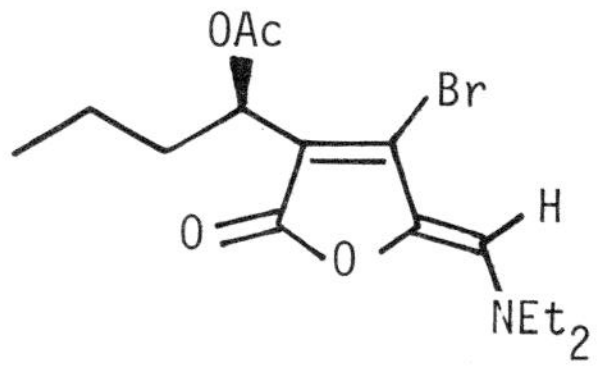	2	40							2
181	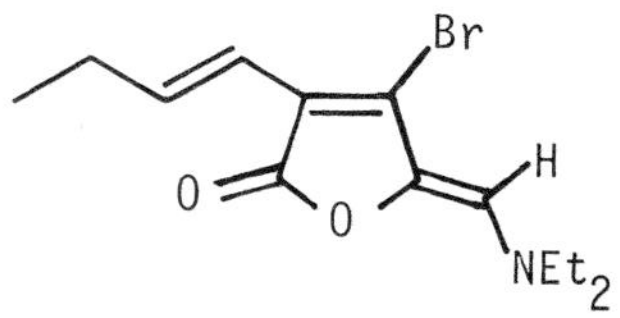	2	10							1
182	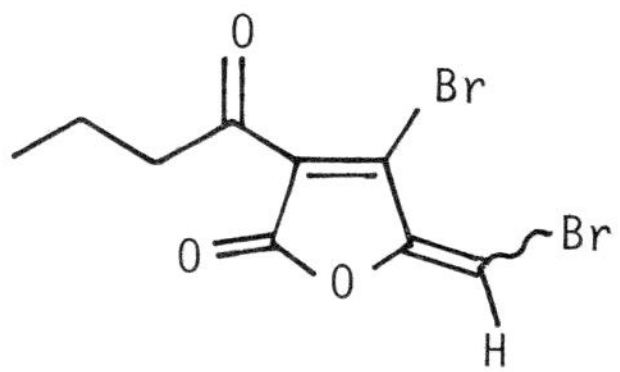	2	20	5	10		160		2	10

Figure 13. Continued.

Antimicrobial Testing

Of the one hundred and sixty-five compounds currently known from *Asparagopsis*, *Bonnemaisonia*, *Delisea* and *Ptilonia* species, less than half have been biotested and evaluated as antimicrobial agents. We have obtained antifungal and antibacterial testing results for nearly thirty of these compounds including several synthetic analogs. Testing results from the Roche group in Australia and from Professor Sims at the University of California, Riverside, with *Delisea* compounds, have also been obtained.

Antimicrobial testing results are tabulated along with the structures of the metabolites in figures 6-9, 12 and 13. In these figures the number refers to the mode of antibiotic testing employed. "1" is the agar plate method and positive activities are presented as zones of growth inhibition in millimeters. The value in parentheses is the disc loading concentration in milligrams. "2" refers to quantitative antimicrobial measurements using turbidity as a function of absorbance (550 nm), and data are given for minimum inhibitory concentrations (MIC) in micrograms per milliliter.

The qualitative antifungal test employed by the Sims group, using the plant pathogen *Cladosporium cucumerinum*[37], involves dispersing a spore suspension over a pre-developed thin layer chromatographic sheet and incubating (Fig. 12). The naturally dark color of *C. cucumerinum* and zones of white where antibiosis has occurred allow for a rapid analysis of the components of crude mixtures and extracts.

The methods employed in our testing have been described elsewhere[38]. The Roche group has employed methods similar to ours using slightly modified culture media[39]. Microorganisms selected by us were the human pathogens *Staph. aureus*, *E. coli*, *C. albicans* and the marine pathogen *Vibrio anguillarum*. The Roche group employed the same human pathogens, but also screened against *P. mirabilis*, *P. aeruginosa*, *C. neoformans*, *A. fumigatus* and *T. mentagrophytes*. Siuda *et al.*[33] have reported MIC values for the ketone 136 using the pathogens we selected as well as *M. apiospermum*, *G. sp.*, *S. pyogenes*, *D. pneumoniae*, *S. facecalis*, and *K. pneumoniae*. In the figures 6-9, 12 and 13, the bacterial and fungal pathogens have been abbreviated as A-N and these abbreviations correspond to the following complete list:

A. *Staphylococcus aureus* (Gram pos. bacteria)
B. *Escherichia coli* (Gram neg. bacteria)
C. *Candida albicans* (fungi)
D. *Vibrio anguillarum* (marine bacteria)
E. *Proteus mirabilis* (Gram neg. bacteria)
F. *Pseudomonas aeruginosa* (Gram neg. bacteria)
G. *Cryptococcus neoformans* (fungi)
H. *Trichophyton mentagrophytes* (fungi)
I. *Monosporium apiospermum* (fungi)
J. *Geotrichum* sp. (fungi)
K. *Streptococcus pyogenes* (Gram pos. bacteria)
L. *Diplococcus pneumoniae* (Gram pos. bacteria)
M. *Streptococcus facecalis* (Gram pos. bacteria)
N. *Klebsiella pneumoniae* (Gram neg. bacteria)

Generally, many of the metabolites tested here show formidable *in vitro* activities which favorably compare with existing antibiotics. In principal these natural products could find applications as new antimicrobial agents. It is important, however, to point out that *in vitro* tests frequently yield encouraging results which reflect general toxicity rather than selective antimicrobial activity. Additional and crucial evaluation of a new potential antibiotic must then be *in vivo* biotesting of animals such as mice to assess real pharmaceutical applications. The Roche group has evaluated several of the *Delisea*-derived compounds for both toxicity and *in vivo* activity. The lactones 155-157 and 141-142 show moderate toxicity with subcutaneous LD_{50} values of 12 mg/kg for the mixture 155-157 and 24 mg/kg for the mixture 141-142. Furthermore, subcutaneous injection of sublethal doses of these mixtures failed to counteract induced infections of *Staphylococcus aureus*. In the same bioassay the recognized antibiotic ampicillin gave 50% effective dose cures (ED_{50}) at levels of 0.5 mg/kg It must be concluded, therefore, that while effective in *in vitro* bioassays, the *Delisea* lactones have little or no *in vivo* activity and are useless as systemic antimicrobial agents. It would appear that the *Delisea* lactones are not absorbed and circulated to the site of infection or that they are rapidly excreted or metabolized to inactive compounds. Our chemical experience with these compounds suggests that they are effective alkylating agents.

It would seem that these metabolites would be rapidly deactivated by reaction with protein material.

A second application of an antimicrobial compound is as a topical antiseptic for abrasions and superficial wounds to prevent infection. Scientists at the Roche institute have assayed the *Delisea* lactone mixtures 155-157 and 141-142 in an assay with *S. aureus*, *E. coli* and *T. mentagrophytes* wound-infected mice. At the maximum tolerated levels of the lactones, cures of the topical infections could not be obtained. The most probable cause for deactivation of an antibiotic in an open infection is binding or deactivation by reaction with serum. Indeed, the Roche scientists showed a 500-1000 fold decrease in *S. aureus* activity by the lactones in *in vitro* assays in the presence of 50% human serum.

Antimicrobial agents can also be utilized in soaps, lotions, shampoos and creams as a general cleansing method for unbroken skin. However, since topically applied compounds will be absorbed to some extent through the skin, the nature and degree of toxicity of the compounds can severely limit their utility. The antimicrobials mentioned in this review would appear to be too toxic for this application. However, none have been tested and without specific evidence this potential application appears promising.

In summary, it is clear that red seaweeds of the family Bonnemaisoniaceae produce halogen-containing metabolites, most likely as a defensive strategy to gain a selective advantage against predation. Most of these metabolites appear to be general toxins (strong alkylating agents) which show impressive *in vitro* antimicrobial activities. Their utilization as pharmaceuticals, however, particularly as systemic antibiotics, would appear to be limited. One application of these compounds, which has yet to be explored, lies in their use as industrial disinfectants. Antibiotics are used in industry to cleanse surfaces and to prevent deterioration of organic substances. The major characteristic of an industrial antibiotic is that the compound exhibits broad spectrum antibiotic activity and be inexpensive to produce. The somewhat simple compounds produced by seaweeds of the family Bonnemaisoniaceae satisfy these criteria and this may well lead to their ultimate utilization in this application.

Acknowledgments

We wish to thank Professor Jim Sims, University of California, Riverside, for providing details of his bioassay techniques with certain *Delisea* compounds, and for providing Figure 12. We are especially grateful to Drs. J. L. Reichelt and R. J. Wells, Roche Research Institute of Marine Pharmacology, Sydney, Australia, for providing, in detail, their testing results with various *Delisea* metabolites and for their valuable suggestions.

References and Notes

1. Levring, T., Hoppe, H. A. and Schmidt, O. J.: Marine Algae; A Survey of Research and Utilization. Walter de Gruyter and Co., Berlin (1969).

2. Hornsey, I. S. and Hide, D.: The production of antimicrobial compounds by British marine algae I. antibiotic-producing marine algae. Br. Phycol. J., 9, 353-361 (1974).

3. Burkholder, P. R., Burkholder, L. M. and Almodóvar, L. R.: Antibiotic activity of some marine algae of Puerto Rico. Bot. Marina II, 149-156 (1960).

4. Starr, T. J., Kajima, M. and Piferrer, M.: Mitotic anomalies in tissue-cultured cells treated with extracts derived from marine algae. Texas Rep. Biol. Med., 24, 208-221 (1966).

5. Robertson, D.: *Bonnemaisonia asparagoides* C. Ag. that gave a blue stain to paper. Trans. Nat. Hist. Soc. Glasgow, 4, 172 (1896).

6. Golenkin, M.: Algologische notizen; 1. das vorkommen von freiem jod bei *Bonnemaisonia asparagoides*. Bull. Soc. Impér. Natur. Moscou, 8, 257-270 (1894).

7. Kylin, H.: Über die Blasenzellen einiger florideen und ihre Beziehung zur Abspaltung von Jod. Arch. för Bot., 14, 1-13 (1915).

8. Kylin, H.: Über *Spermothamnion roseolum* (Ag.) Pringsh und *Trailliella intricata* Batters. Bot. Not., 83-92 (1916).

9. Kylin, H.: Über die Blasenzellen der florideen. Bot. Not., 275-285 (1927).

10. Kylin, H.: Über *Falkenbergia hilldebrandii* und ihre Beziehung zur Abspaltung von Jod. Bot. Not., 233-254 (1928).

11. Kylin, H.: Über das vorkommen von jodiden, bromiden und jodidoxydasen bei dem meer-salgen. Hoppe-Seyler's Zeitschrift für Physiologische Chemie, 50-84 (1929).

12. Kylin, H.: Über die blasenzellen bei *Bonnemaisonia, Trailliella, Antithamnion*. Z. Bot., 23, 217-226 (1930).

13. Sauvageau, C.: Sur quelques algues floridees renfermant de l´iode a l´etat libre. Bull. Stat. Biol. Arachon., 22, 5-45 (1925).

14. Chemin, M. E.: Sur l´etate de l´iode chez quelques floridees. Rev. de Bot., 40, 129-145 (1928).

15. Lami, M. R.: Liberation de l´iode des ioduques de *Bonnemaisonia asparagoides* sous l´action des rayons ultraviolets. C. R. Acad. Sc. Paris, 191, 863-865 (1930).

16. Wolk, C. P.: Role of bromine in the formation of the refractile inclusions of the vesicle cells of the Bonnemaisoniaceae (Rhodophyta). Planta, 78, 371-378 (1968).

17. Stosch, H. A. von: Jodbedarf bei meersalgen. Naturwissen, 49, 42-43 (1964).

18. Stosch, H. A. von: Wirkungen von jod und arsenit auf meersalgen a kultur. In: Proceedings of the Fourth International Seaweed Symposium (ed. D. D. Virville and J. Feldman), Pergamon Press, Oxford, 467 pp. (1964).

19. Freedman, R. B.: Applications of the chemical reactions of proteins in studies of their structure and function. Quart. Rev., 25, 431-454 (1971).

20. Fenical, W.: Polyhaloketones from the red seaweed *Asparagopsis taxiformis*. Tetrahedron Lett., 4463-4466 (1974).

21. McConnell, O. J. and Fenical, W.: Halogen chemistry of the red alga *Asparagopsis*. Phytochem., 16, 367-374 (1977).

22. Galbally, I. E.: Man-made carbon tetrachloride in the atmosphere. Science, 193, 573-576 (1976).

23. Morris, D. R. and Hager, L. P.: Chloroperoxidase, isolation and properties of the crystalline glycoprotein. J. Biol. Chem., 241, 1763 (1966).

24. Burreson, B. J., Moore, R. E. and Roller, P.: Haloforms in the essential oil of the alga *Asparagopsis taxiformis* (Rhodophyta). Tetrahedron Lett.(7) 473 (1975).

25. Burreson, B. J., Moore, R. E. and Roller, P.: Volatile halogen compounds in the alga *Asparagopsis taxiformis*. J. Agri. and Food Chem., 24, 856 (1976).

26. Woolard, F. X., Moore, R. E. and Roller, P.: Halogenated acetamides, but-3-en-2-ols, and isopropanols from *Asparagopsis taxiformis* (Delile) Trev. Tetrahedron, 32, 2843 (1976).

27. Moore, R. E., Woolard, F. X., and Roller P.: Halogenated acetamides, acetic acids, acrylic acids, but-3-en-2-ols and isopropanols from the red alga *Asparagopsis taxiformis*. Presented at the regional meeting of the American Chemical Society, San Francisco, Calif. (1977).

28. Wells, R. J.: Presented at IUPAC Conference in Dunedin, New Zealand, August (1976).

29. McConnell, O. J. and Fenical, W.: Unpublished results. We wish to thank Dr. Marianne Pedersén, Institute of Physiological Botany, University of Uppsala, for providing samples of *Falkenbergia rufolanosa*. (1977).

30. McConnell, O. J. and Fenical, W.: Polyhalogenated 1-octen-3-ones, antibacterial metabolites from the red seaweed *Bonnemaisonia asparagoides*. Tetrahedron Lett., 1851-1854 (1977).

31. McConnell, O. J. and Fenical, W.: Halogenated metabolites--including Favorsky rearrangement products--from the red seaweed *Bonnemaisonia nootkana*. Tetrahedron Lett., 4159-4162 (1977).

32. McConnell, O. J., Young, D. N. and Fenical, W.: Unpublished results. (1978).

33. Siuda, J. F., Van Blaricom, G. R., Shaw, P. D., Johnson, R. D., White, R. H., Hager, L. P. and Rhinehart, Jr., K. L.: 1-Iodo-3,3-dibromo-2-heptanone, 1,1,3,3-tetrabromo-2-heptanone, and related compounds from the red alga *Bonnemaisonia hamifera*. J. Am. Chem. Soc., 97, 937-938 (1975).

34. Kazlauskas, R., Murphy, P. T., Quinn, R. J. and Wells, R. J.: A new class of halogenated lactones from the red alga *Delisea fimbriata* (Bonnemaisoniaceae). Tetrahedron Lett., 37-40 (1977).

35. Pettus Jr., J. A., Wing, R. M. and Sims, J. J.: Marine natural products XII. Isolation of a family of multi-halogenated gamma-methylene lactones from the red seaweed *Delisea fimbriata*. Tetrahedron Lett., 41-44 (1977).

36. Rose, A. F. and Sims, J. J.: Marine natural products XIII. Isolation and synthesis of some halogenated ketones from the red seaweed *Delisea fimbriata*. Tetrahedron Lett., 1847-1850 (1977).

37. Keen, N. T., Sims, J. J., Erwin, D. C., Rice, E. and Partridge, J. E.: 6a-Hydroxy-phaseolin: An antifungal chemical induced in soybean hypocotyls by *Phytophthora megasperma* var. sojae. Phytopathology, 61, 1084 (1971).

38. Fenical, W., McConnell, O. J. and Stone, A.: Antibiotics and antiseptic compounds from red seaweeds of the family Bonnemaisoniaceae. Proceedings of the IX International Seaweed Symposium, Santa Barbara, Calif., August (1977).

39. Antibiotic bioassays at the Roche Institute involved incubating bacteria at 37°C in a liquid medium of the following composition: glucose (4 g/l), NH_4Cl (2 g/l), NH_4NO_3 (0.3 g/l), Na_2SO_4 (0.6 g/l), K_2HPO_4 (1 g/l), KH_2PO_4 (0.3 g/l), $MgSO_4 \cdot 7H_2O$ (0.03 g/l), ferric ammonium citrate (0.03 g/l) and oxoid trypticase soy broth (6 g/l). Yeasts and fungi were also grown in liquid medium containing glucose (20 g/l), inositol (0.01 g/l) and Difco bacto-yeast nitrogen base (6.7 g/l).

USE OF BROMOPEROXIDASE, AN ALGAL ENZYME, IN THE PREPARATION OF RADIOBROMINATED PROTEINS

Karen D. McElvany, Linda C. Knight, Michael J. Welch
Mallinckrodt Institute of Radiology, Washington University School of Medicine, St. Louis, Missouri, U.S.A.

Jerome F. Siuda
University of Pittsburgh, Pittsburgh, Pennsylvania, U.S.A.

Richard F. Theiler, Lowell P. Hager
University of Illinois, Urbana, Illinois, U.S.A.

CONSIDERATIONS IN THE CHOICE OF A PROTEIN LABEL

The use of radioactively labeled proteins has become an extremely useful tool for studying the fate of native unlabeled proteins in a biological system, and for radioimmunoassay. Radioisotopes of iodine (especially ^{131}I and ^{125}I) have been extensively used as protein labels because of their availability and the relative ease of attaching iodine to proteins to produce labeled proteins of high specific activity. For use in nuclear medicine, however, radioiodinated proteins suffer from several disadvantages: first, the high radiation dose to a patient receiving proteins labeled with ^{125}I or ^{131}I for an in vivo nuclear medicine study; second, the rapid rate of deiodination of iodine-labeled proteins due to the weak bond between carbon and iodine, resulting in a large amount of free radioiodide not associated with the protein it was supposed to trace; and third, any free radioiodide in vivo quickly concentrates in the thyroid, delivering a high radiation dose to that gland.

The chemical problems associated with iodine can be overcome by using a radioisotope of bromine as the label for proteins. Table 1 compares the carbon-halogen bond strengths for

Marine Algae in Pharmaceutical Science
edited by H. A. Hoppe, T. Levring, Y. Tanaka

various organohalogens.

Compound	D_{C-Br} (kcals/mole)	D_{C-I} (kcals/mole)
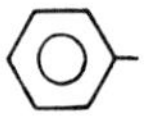 X	71 ± 2	61 ± 1.8
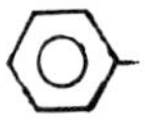 CH_2-X	50.5	38
CH≡C-CH_2-X	58 ± 3	45.7 ± 3.2
CH_2= CH-CH_2-X	47.7 ± 2.5	34.1 ± 1.5
CH_3-X	66.4 ± 2	52.6 ± 2.3

Table 1. Bond dissociation energies of some halogenated compounds (ref 1).

In every example, carbon-bromine bonds are stronger than carbon-iodine bonds in analogous compounds. To a first approximation (1), the rate of hydrolysis is proportional to

$$e^{-D/RT} \quad (D = \text{bond strength})$$

so brominated compounds should hydrolyze (dehalogenate) more slowly than the iodinated analogs.

A suitable isotope of bromine for medical use would be bromine-77. Its 2.33 day half-life is better suited to most nuclear medicine studies than are the longer lived ^{125}I or ^{131}I which are commonly used (see Table 2).

	^{77}Br	^{125}I	^{131}I
Half-life	2.33 days	60 days	8 days
Decay Mode	>99% EC <1% β^+	100% EC	100% β^-
Decay Energies	γ 242 (30%) γ 300 (6%) γ 520 (24%) γ 580 (7%)	γ 25 (100%)	γ 364 (main) β^- 608 (87%) 330 (9%)

Table 2. Physical characteristics of Bromine-77, Iodine-125, and Iodine-131 (ref 2).

Because of the longer half-lives of the two iodine isotopes, radiation exposure may be prolonged. Bromine-77, which emits medium-energy gamma radiation and very little particle emission, delivers only about one-tenth the radiation dose delivered by the same amount of ^{125}I or ^{131}I when distributed over the whole body (3). However, because radioiodide in vivo concentrates in the thyroid and radiobromide does not, the avoidance of an acute radiation dose to the thyroid by using ^{77}Br is even more significant.

FACTORS IN THE CHOICE OF A HALOGEN LABELING TECHNIQUE

Carrier-free radioiodine may be attached to proteins by several methods (4-6), all of which involve oxidation of iodide to a higher oxidation state so that it may electrophilically substitute into the protein. Carrier-free bromination is more difficult to accomplish because of the lower oxidation potential of the bromide (1); hence, it is difficult to find an oxidizing agent which will oxidize bromide without doing oxidative damage to the protein. A halogenating enzyme would be one solution; however, lactoperoxidase (LP), an enzyme commonly used for catalyzing radioiodination (5), does not catalyze bromination. Previously, the only reported brominating enzyme was chloroperoxidase (CPO) (7), which works quite well to attach carrier-free radiobromide to a variety of substrates. However, chloroperoxidase works best at pH 2.8 and does not function at all as a halogenating enzyme above pH 4.5. Such low pH conditions may damage many proteins, and also affect the site of attachment of the label (3, 8).

Sites of Halogenation

Under mild conditions, three amino acid residues are most susceptible to halogenation in proteins: tyrosine, histidine and cysteine (9). Tyrosine and histidine may both be dihalogenated, and the halogen atoms attach covalently to carbon

atoms: (X = Br, I)

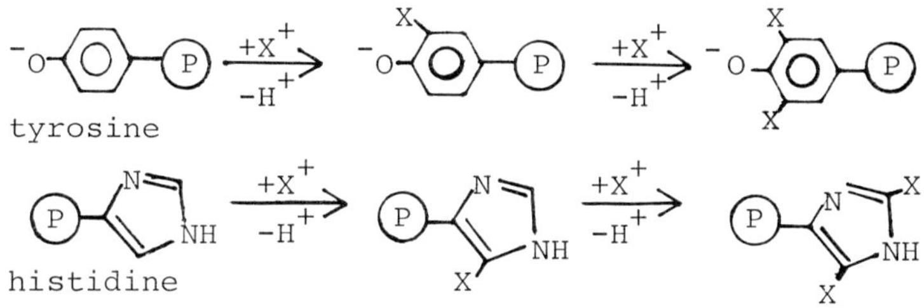

In the case of cysteinyl residues, however, a sulfur-halogen bond is formed which is unstable with respect to loss of the halogen by hydrolysis:

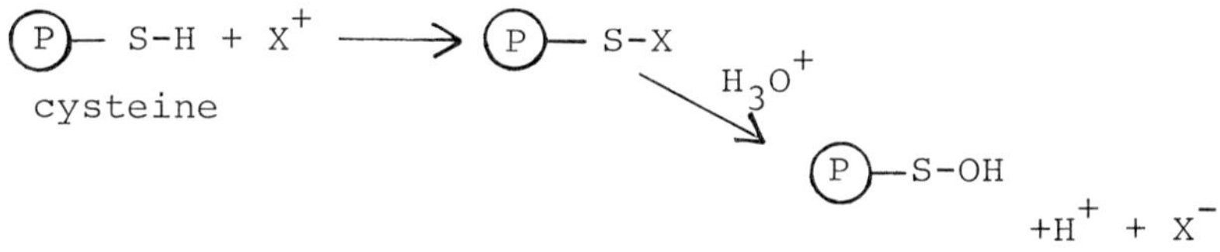

As might be expected, carbon-halogen and sulfur-halogen bonds hydrolyze at different rates. Previous experiments have shown that monoiodohistidine hydrolyzes about 3 times as fast as monoiodotyrosine, and iodocysteine hydrolyzes about 10 times as fast as monoiodotyrosine (10). Obviously, for the most stable label, proteins should be halogenated primarily at tyrosyl residues, and labeling of cysteinyl residues should be avoided if possible.

The effect of pH on the sites of iodination and bromination was previously studied in our laboratory (8). Human serum albumin was labeled with radioiodide by several methods under a variety of pH conditions, or with radiobromide using the chloroperoxidase method at pH's below 4.5. Each sample of labeled protein was purified by gel filtration to remove free radiohalide, then treated with a proteolytic enzyme (Pronase) at pH 7 to gently cleave peptide bonds but leave carbon-halogen bonds intact (10). The resulting mixture of labeled and unlabeled amino acids was analyzed by liquid chromatography using a flow-through scintillation detector to detect labeled compounds as they were eluted. The products of proteolysis of the labeled protein were identified by chromatographing standard compounds. A typical elution pattern is shown in Figure 1.

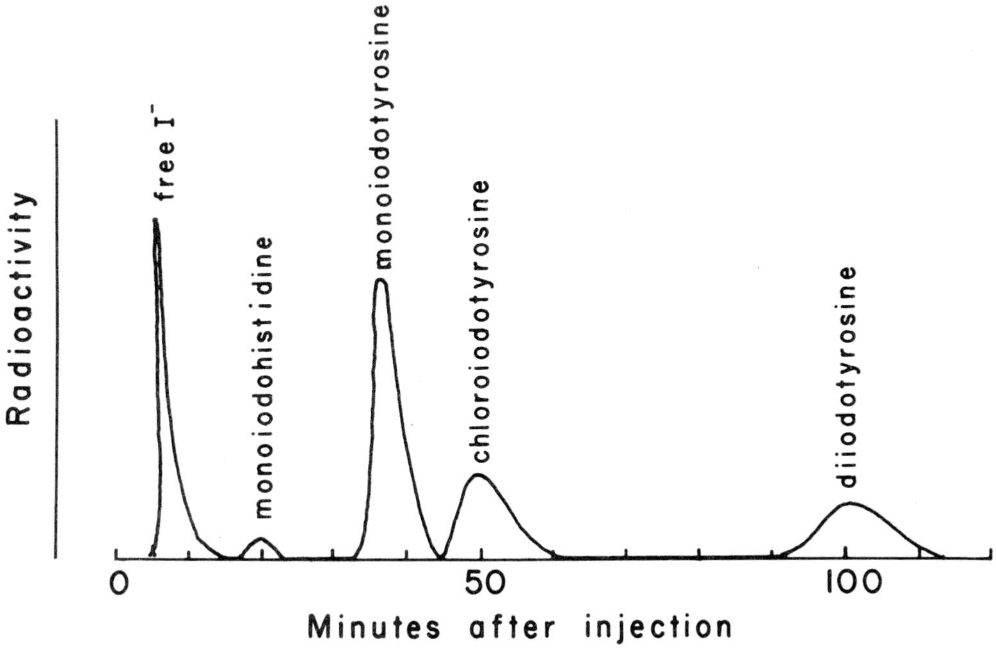

Figure 1: Typical elution pattern resulting from liquid chromatography of radioiodinated albumin after treatment with Pronase (ref 3).

Percentages of total radioactivity associated with each peak were calculated. The percents of the total eluted radioactivity which were associated with monoiodotyrosine are plotted for each pH condition in Figure 2 (for the iodinated albumin experiments). It is apparent that a pH of 7 is necessary for labeling predominately at tyrosyl residues. At other pH's, the label attached primarily to cysteinyl residues, as indicated by the appearance of free iodide after proteolysis (S-I bonds are easily hydrolyzed in a denatured protein

(9) and the existence of S-I bonds was confirmed by adding cysteine to intact labeled protein and observing free iodide as a result (8) see Figure 3.

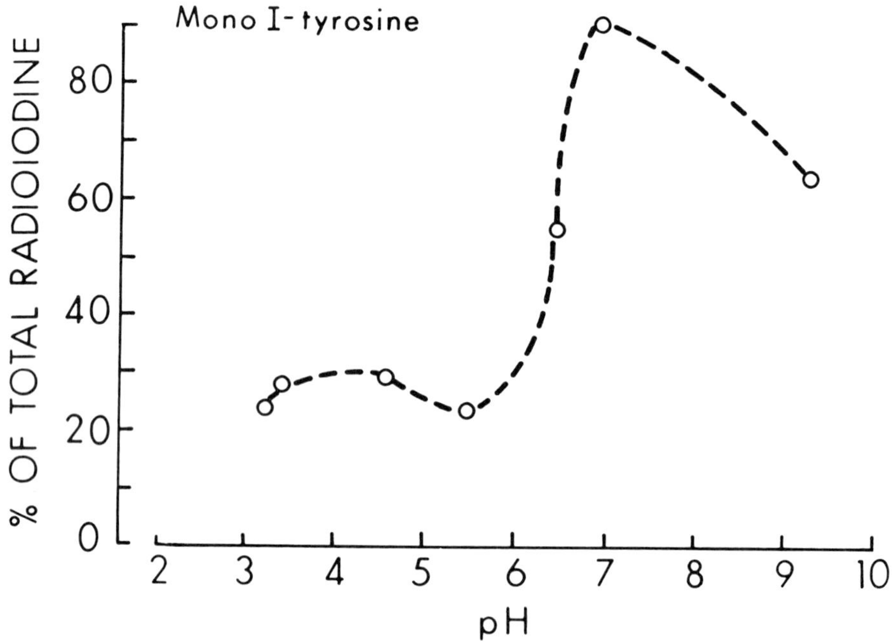

Figure 2: Percentages of total radioiodine associated with monoiodotyrosine peak vs pH of labeling with lactoperoxidase (ref 3).

We were unable to test the sites of bromination at pH 7 with the chloroperoxidase method, but we assume that the labeling pH dependence should be similar to that for iodination. When we compared the in vitro hydrolysis rates of brominated and iodinated proteins, we obtained the following results (8):

Label	^{77}Br	^{131}I	^{131}I
Method	CPO	LP	CPO
Time Elapsed (days)	% Bound	% Bound	% Bound
0	100.0	100	100.0
1	97.0	94	96.7
2	97.5	91	70.5
3	98.0	89	29.4
4	89.5	87	11.4

Table 3. In vitro hydrolysis rates of radioiodinated and radiobrominated albumin.

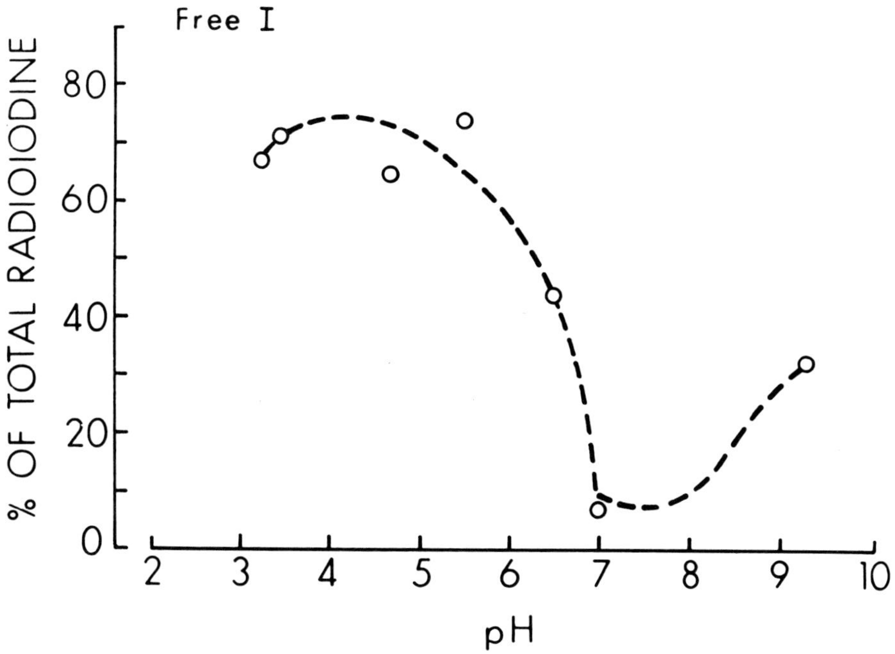

Figure 3: Percentages of total radioiodine eluted as free iodide vs pH of labeling with lactoperoxidase.

In spite of the fact that brominated proteins were expected to hydrolyze more slowly than iodinated proteins, these results indicate that brominated proteins prepared at pH 2.8 dehalogenate at about the same rate as iodinated proteins prepared at pH 7. Iodinated proteins prepared at pH 2.8 hydrolyze much more rapidly, presumably because of the large percentage of labeling at cysteinyl moieties. These results suggest that if bromination could be carried out at pH 7, predominantly tyrosyl groups would be brominated and the labeled protein would be more stable to hydrolysis.

In summary, a useful method for producing radiobrominated proteins for nuclear medicine should be carried out at neutral pH in order to insure optimum stability of the label. In addition, carrier-free radiobromide must be used to produce labeled proteins of high specific activity and to prevent loss of biological activity resulting from excessive bromination on any one molecule. This requirement eliminates traditional chemical bromination methods. Even in carrier-free systems, however, harsh oxidizing conditions must be avoided so that the protein is not denatured. Haloperoxidases in general work under conditions which are mild enough for labeling proteins.

BROMINATING ENZYMES

A new brominating enzyme has recently been discovered which will work at a higher pH than chloroperoxidase, the only previously known brominating enzyme. The new enzyme, called bromoperoxidase (BPO), is derived from algae rather than from a fungus, as is chloroperoxidase. Bromoperoxidase appears to catalyze only bromination and iodination, and its pH optimum is 6.5. Whereas chloroperoxidase is completely inactive above pH 4.5, bromoperoxidase works quite well at pH 7. Table 4 summarizes the properties of the two brominating enzymes.

	Chloroperoxidase (CPO)	Bromoperoxidase (BPO)
Source	Fungal	Algal
Usable Halides	Cl^-, Br^-, I^-	Br^-, I^-
pH Optimum	2.8 (Inactive above pH 4.5)	6.5 (Can be used at pH 7.0)

Table 4. Comparison of brominating enzymes.

Bromoperoxidase

The new brominating enzyme, bromoperoxidase, is extracted from the red algae Bonnemaisonia hamifera. Extracts of B. hamifera have been shown to contain large quantities of brominated derivatives of 2-heptanone (11); thus, the algae apparently use the enzyme to produce brominated ketones. Brominating activity appears to be in the form of a heme-containing glycoprotein. A 25-fold purification of the enzyme, which we use to catalyze radiobromination of proteins, is still quite rich in carbohydrates. The enzyme extract has an assayed brominating activity of 2.9 μmoles per minute per 100 μl enzyme extract, based on the bromination of monochlorodimedon at pH 5.5, as monitored by a decrease in absorbance at 278 nm.

The enzyme extract is obtained by first homogenizing and sonicating frozen batches of algae, and then isolating the supernatant fraction by centrifugation. A series of repetitive extractions with two buffer systems (TRIS-HCl and phosphate-citrate) yields an enzyme fraction purified 5-10 fold over the crude homogenate. Twenty-five fold purification of the enzyme extract can be achieved by subjecting the partially purified fraction to ion-exchange chromatography on DEAE cellulose.

Bromoperoxidase, like other peroxidases, has been found to be a heme-protein. Difference spectra between the oxidized and dithionite reduced forms of the enzyme show absorbance maxima identical to those for crystalline cytochrome b_5 (12), suggesting the presence of a low spin heme environment in the partially purified enzyme preparations. In addition, bromoperoxidase is sensitive to cyanide and azide ions, both of which are known to be inhibitors of heme-proteins.

Production of Bromine-77

The bromine-77 used to label proteins by the bromoperoxidase method was produced in the Washington University 52 inch cyclotron. Natural arsenic pentoxide was bombarded with 26 MeV alpha particles to produce ^{77}Br according to the reaction $^{75}As\ (\alpha,2n)\ ^{77}Br$ (13, 14). Bromine-77 so formed is purified and collected without adding any stable bromine, as previously described (7).

Bromine-77 should be easy to produce by spallation reactions in high-energy accelerators (15), which may eventually make ^{77}Br available to institutions without their own cyclotrons. Many materials can be used as targets for spallation production of bromine-77; the following reaction scheme is one example of such a spallation method for bromine-77 production.

$$\text{NaBr}\quad {}^{79}\text{Br(p,3n)}\quad {}^{77}\text{Kr} \xrightarrow[\text{gas}]{\text{sweep out}} {}^{77}\text{Kr}$$

freeze into trap (16)

let decay $\downarrow$ $t_{1/2} = 1.2$ hr

^{77}Br

rinse out with dil. NaOH

The intermediate krypton parent can be easily purified by freezing into a trap; it then decays to bromine-77.

PREPARATION AND PROPERTIES OF PROTEINS RADIOBROMINATED USING BROMOPEROXIDASE

Bromoperoxidase and bromine-77 have been used to label human serum albumin and canine fibrinogen as model compounds. The reaction involves mixing the protein to be labeled with bromoperoxidase and carrier-free radiobromide, all in pH 7 buffer, and lastly adding a small amount of hydrogen peroxide to activate the enzyme. The reaction mixture, with a volume less than 0.25 ml, is incubated at 37°C until labeling is essentially completed. Labeling efficiencies are determined by TCA precipitation: 1 mg carrier protein is added to the labeled protein and precipitated with 10% trichloroacetic acid. The precipitate is washed with 10% TCA, and the percentage of radiolabel attached to the protein is determined. Labeling efficiencies of more than 50% have been obtained for bromination of proteins using bromoperoxidase. The radiolabeled protein is purified of remaining free bromide by gel permeation chromatography (Sephadex G-10) or, in the case of fibrinogen, by ammonium sulfate precipitation.

This radiobromination technique seems to be gentle enough for even very fragile proteins, such as fibrinogen, which is of clinical interest but is easily denatured.

Optimization of Conditions for Radiobromination of Proteins Using Bromoperoxidase

Each of the essential reaction conditions was varied to obtain the conditions giving the highest labeling yields. Figure 4 shows the effect of the concentration of the protein to be labeled, in this case human serum albumin, on the labeling yield. Maximum labeling efficiencies are obtained when 25 µg of HSA are labeled.

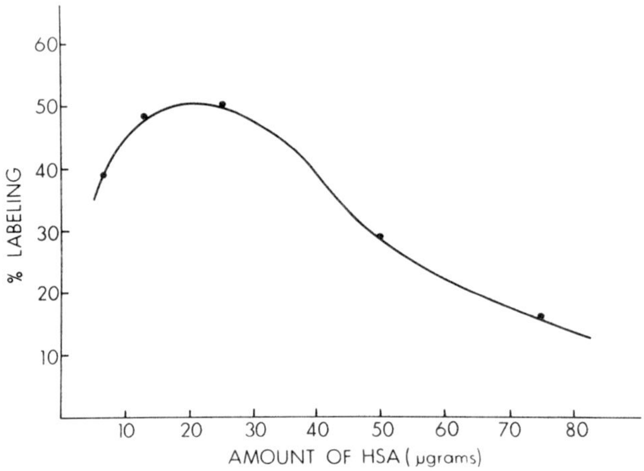

Figure 4: Effect of albumin concentration on BPO bromination of human serum albumin (HSA) at pH 7.

The amount of bromoperoxidase used to catalyze radiobromination of proteins also affects the labeling yield. As shown in Figure 5, optimum labeling efficiencies occur when 20-25 μl of partially purified bromoperoxidase extract is used.

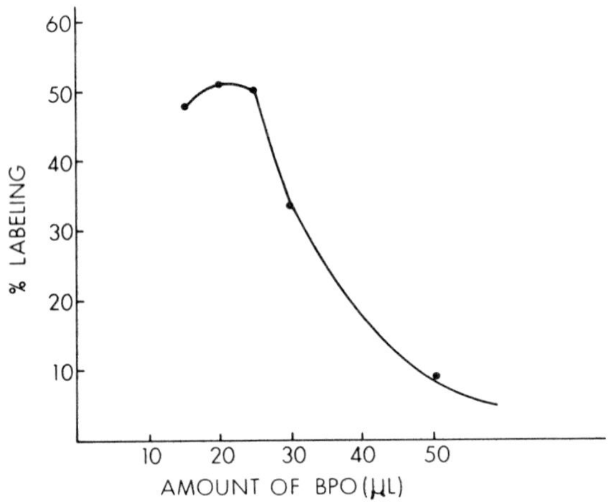

Figure 5: Effect of bromoperoxidase concentration on BPO bromination of albumin at pH 7.

Hydrogen peroxide concentration has a pronounced effect on the radiobromination labeling efficiency with bromoperoxidase (Figure 6). This is in agreement with observations that other peroxidases are sensitive to the relative concentrations of halide ion acceptor and hydrogen peroxide (17).

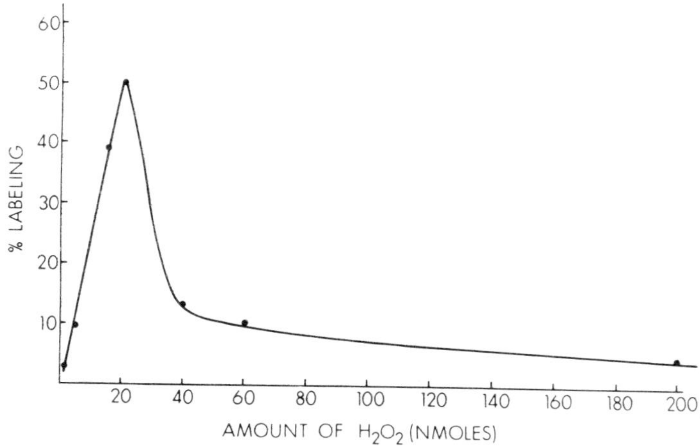

Figure 6: Effect of hydrogen peroxide concentration on BPO bromination of albumin at pH 7.

Bromoperoxidase-catalyzed radiobromination of proteins displays a pH optimum near 6.5 (Figure 7). This pH optimum is very close to pH 7, at which the most stably labeled proteins can be prepared. Bromoperoxidase-catalyzed radiobromination labeling efficiencies decrease sharply around the pH optimum of 6.5; however, satisfactory labeling yields can still be achieved at pH 7.

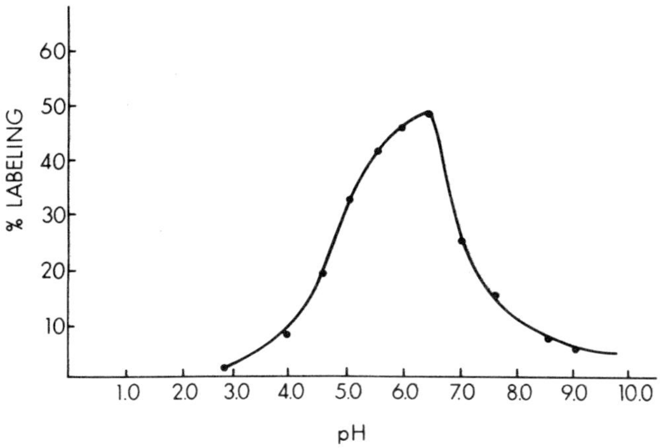

Figure 7: pH dependence of BPO bromination of albumin.

Thus, optimum conditions for bromoperoxidase-catalyzed radiobromination of proteins at neutral pH involves mixing 25 μg of protein (serum albumin or fibrinogen), 25 μl partially purified bromoperoxidase extract, 150-200 μl carrier-free ^{77}Br ($Na^{77}Br$) in pH 7 phosphate buffer (0.1M), and 20 nmole H_2O_2. The labeling mixture is incubated at 37°C for 30-35 minutes, at which time the reaction is complete (Figure 8).

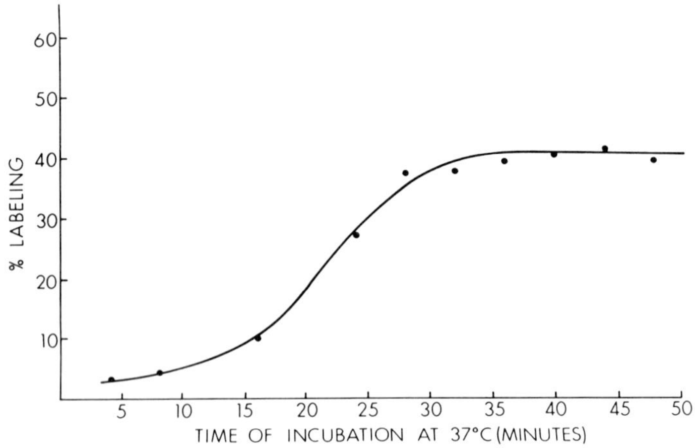

Figure 8: Time dependence of BPO bromination of albumin at pH 7.

In Vitro Hydrolysis of Proteins Radiobrominated with the BPO Method

Human serum albumin radiobrominated using the bromoperoxidase method was subjected to in vitro hydrolysis at pH 7. Radiolabeled albumin was purified of remaining free bromide by gel permeation chromatography (Sephadex G-10). The labeled albumin was kept in 0.1M phosphate buffer, pH 7, at 37°C for the duration of the experiment. Aliquots were removed each day and labeling yields determined by TCA precipitation. As predicted, proteins brominated at pH 7 were more stable toward hydrolysis of the radiolabel than any previously prepared radiohalogenated proteins. Radiobrominated human serum albumin prepared in this way was found to be completely stable to in vitro hydrolysis for 10 days. Table 5 compares in vitro hydrolysis results for human serum albumin brominated at pH 7 by the bromoperoxidase technique, to hydrolysis of albumin labeled with radioiodide at pH 7 using the lactoperoxidase technique.

Label	^{77}Br	^{131}I
Method Used	Bromoperoxidase pH7	Lactoperoxidase (8) pH7
Time Elapsed (days)	% Bound	% Bound
0	100	100
1	100	94
2	100	91
3	100	89
4	100	87
7	100	83
10	100	--

Table 5. In vitro hydrolysis of brominated and iodinated human serum albumin

Sites of Bromination With The BPO Method

The sites of attachment of bromine-77 to proteins labeled at pH 7 using bromoperoxidase were determined as previously described. Only tyrosyl resides are labeled in significant amounts: monobromo- and dibromotyrosine account for 97% of the labeled products of proteolysis; the remaining 3% is monobromohistidine. This confirms the prediction that stable carbon-halogen bonds (as in labeled tyrosyl residues) predominate when labeling occurs at neutral pH.

POTENTIAL USES OF RADIOBROMINATED COMPOUNDS

Bromine-77 may be used to label a variety of proteins for in vivo protein turnover studies (18). If fibrinogen is labeled, it may be used for renal transplant rejection studies (19), especially in patients who have already been studied with radioiodine in another procedure. Radiobrominated fibrinogen can also be used in place of radioiodinated fibrinogen for thrombus localization (19) in cases where imaging will not be carried out.

If proteins are labeled with another isotope of bromine, bromine-76, which is a positron-emitter, 3-dimensional organ images can be obtained using computer-assisted positron tomography (21).

Finally, it may be possible to use bromoperoxidase to attach isotopes of bromine to other types of compounds besides proteins to produce clinically useful radiopharmaceuticals for tumor localization, and for a variety of purposes in nuclear medicine.

Acknowledgement
This work was supported by S.C.O.R. Thrombosis Grant No. HL14147.

REFERENCES

1. Vedeneyev, V.I., Gurvich, L.V., Kondrat'yev, V.N., et al: Bond energies, ionization potentials and electron affinities. Edward Arnold, London, 1966.
2. Lederer, C., Hollander, J.M., Perlman, I.: Table of isotopes, 6th edition. John Wiley and Sons, New York, 1968.
3. Knight, L.C.: Radiohalogenated proteins: Preparation and properties. Ph.D. Thesis, Washington University, 1976, appendix.
4. Hunter, W.M., Greenwood, F.C.: Preparation of I-131 labelled human growth hormone of high specific activity. Nature 19: 495-496 (1962).
5. Marchalonis, J.J.: An enzymatic method for trace iodination of immunoglobulins and other proteins. Biochem. J. 113: 299-305 (1969).
6. Rosa, U., Scasselati, G.S., Pennisi, F.: Labelling of human fibrinogen with I-131 by electrolytic iodination. Biochim. Biophys. Acta 86: 519-526 (1964).
7. Knight, L., Krohn, K.A., Welch, M.J., et al: ^{77}Br: A new protein label. In Radiopharmaceuticals, Subramanian, G., et al, editors, Society of Nuclear Medicine, New York, 1975, Chapter 15.
8. Knight, L.C., Welch, M.J.: unpublished data.
9. Means, G.E., Feeney, R.E.: Chemical modification of proteins. Holden-Day, San Francisco, 1971, Chapter 9.
10. Krohn, K.A., Knight, L.C., Harwig, J.R., et al: Differences in the sites of iodination of proteins following four methods of radioiodination. Biochim. Biophys. Acta 490: 497-505 (1977).
11. Siuda, J.F., Van Blaricom, G.R., Shaw, P.D., et al: 1-Iodo-33-dibromo-2-heptanone, 1,1,3,3-Tetrabromo-2-heptanone, and related compounds from the red alga Bonnemaisonia hamifera. J. Am. Chem. Soc. 97: 937-938 (1975).
12. Deeb, S.S., Hager, L.P.: Crystalline cytochrome b. from escherichia coli. J. Biol. Chem. 239: 1024 (1964).

13. Sarantites, D.G., Erdal, B.R.: Anomalous $5/2^{+}$ states in Se-77 from the decay of Br-77. Phys. Rev. 177: 1631 (1969).

14. Helus, F.: Preparation of carrier-free bromine-77 for medical use. Radiochem. Radioanal. Lett. 3: 45 (1970).

15. Richards, P., Lebowitz, E., Stang, L.G.: The Brookhaven LINAC isotope producer (BLIP). In Radiopharmaceuticals and Labelled Compounds, IAEA-SM-171138, Vol. 1. Vienna, IAEA, 1173, pp 326-341.

16. Blue, J.W., Benjamin, P.P.: Production of ^{77}Br by the reaction ^{76}Se (α, 3n)^{77}Kr. J. Nucl. Med. 12: 416-417 (1971).

17. Brown, F.S.: Studies on the mechanism and active site of chloroperoxidase. Ph.D. thesis, University of Illinois, 1967.

18. Collen, D., Tytgat, G.N., Claeys, H., et al: Fibrinogen turnover in physiological conditions in humans. Br. J. Haematol. 22: 681-700 (1972).

19. Winston, M.A., Weiss, E.R., Blahd, W.H., et al: Use of ^{131}I-fibrinogen in detection of renal transplant rejection. Invest. Urol. 9: 119-123 (1971).

20. Kakkar, V.V.: The diagnosis of deep vein thrombosis using the ^{125}I fibrinogen test. Arch. Surg. 104: 152-159 (1972).

21. Ter-Pogossian, M.M., Phelps, M.E., Hoffman, E.J., et al: A positron-emission transaxial tomography for nuclear imaging (PETT). Radiology 114: 89-98 (1975).

PARTIAL CHARACTERIZATION OF A SPECIFIC ANTIBIOTIC, ANTIFUNGAL SUBSTANCE ISOLATED FROM THE MARINE DIATOM *CHAETOCEROS LAUDERI* RALFS CC

D. Pesando, M. Gnassia-Barelli

Laboratoire de Physique et Chimie Marines. BP n° 8. La Darse.
06230 Villefranche-sur-Mer

E. Gueho

Laboratoire de Cryptogamie. UER des Sciences Pharmaceutiques.
8, avenue Rockfeller. 69373 Lyon Cedex

Abstract

Aqueous extraction, ultrafiltration, electrophoresis, cold ehtanol precipitation and fractionation through A. E. A. cellulose and Sephadex G 200 chromatography are required for the isolation of a high molecular weight, acid polysaccharide from cells of a marine diatom *Chaetoceros lauderi* Ralfs CC. which have antibiotic and antifungal activity.

Introduction

The search for drugs from the sea is relatively recent. Many marine flora and fauna contain substances with antiviral, antimicrobial, tumor inhibitory, anticoagulant, cardioactive or neurophysiological properties.

Recent books "The Chemistry of Marine Material Products" by Scheuer (1973), "Compounds from marine organisms" by Baker and Murphy (1976),

Marine Algae in Pharmaceutical Science
edited by H. A. Hoppe, T. Levring, Y. Tanaka

"Marine Pharmacology" by Baslow (1977) and a review paper by Ruggieri (1976) describe some of the active compounds from marine organisms structurally defined.

Among them, most investigations were carried out on the antibacterial activity of extracts from marine unicellular algae, whereas very few took into consideration their fungicidal activity.

Burkholder et al. (1960) had tested 131 specimens of marine algae from the bay of Puerto-Rico against pathogenic bacteria, marine bacteria and the yeast Candida albicans, and they found that the dinoflagellate Goniaulax tamarensis showed noticeable activity against Candida albicans.

From a marine dinoflagellate, Goniodoma sp., collected in Puerto-Rico, Goniodomin (formula $C_{43}H_{58}O_{11}$) has been isolated by Sharma et al. (1968). Goniodomin was active against Cryptococcus sp., Trichophyton sp. and other fungi.

The antimicrobial activity of a marine diatom Asterionella notata Grunow has been studied by Gauthier (1969). The author reported that an aqueous extract from this diatom inhibited the growth of several bacteria and of some fungi as Candida albicans, Penicillium and Aspergillus species.

A systematic sreening for antifungal activity of planktonic algae strains in culture had been made (Pesando et al., in press). Aqueous extracts of 48 planktonic algae were examined and 5 were found to contain significative antifungal activity against pathogenic fungi of man. The active algal species were: Asterionella notata Grunow, Chaetoceros lauderi Ralfs CC., Chaetoceros pseudocurvisetus Mangin, Chaetoceros socialis Lauder and Fragilaria pinnata Ehrenberg.

Among these strains, we paid a particular attention to a marine diatom: Chaetoceros lauderi Ralfs CC., a cold neritic species which is sometimes taken to be oceanic as it often drifts to the off-sea. This diatom was previously studied by Aubert et al. (1966) (1972) for antibacterial activity[+], and by us for antifungal activity (Gueho et al., 1977).

This diatom has a very wide antibacterial spectrum towards aerobes, anaerobes and salmonella, and is expecially active against dermatophytes and some moulds.

This report deals with the identification of the antibacterial and antifungal substance isolated from the marine diatom Chaetoceros lauderi.

Material and methods

Diatom Culturing Techniques

A strain of Chaetoceros lauderi has been isolated from a sea-water sample taken in 1976 in the bay of Villefranche-sur-Mer, France. This species has been identified by P. Nival (Station Marine de Villefranche-sur-Mer).

The Chaetoceros lauderi cultures are carried out under axenic conditions in aged, sterilized (115° C), nutrilites enriched (ES Provasoli, 1966) sea-water. Batch cultures of diatom are grown without air bubbling in flasks containing 5000 ml ES medium. The cultures are maintained at 16° C under continuous light provided by fluorescent tubes Silvania Grolux and Mazda (irradiation at about 95 Wm^{-2}).

Cell extract preparation

After 12 days culturing, the cells are separated by filtration with Sartorius membranes (8 μ pore size) and deep-frozen. The cells are mixed

---

[+] The antibacterial activity attributed to Chaetoceros teres is now attributed to Chaetoceros lauderi after the examination of the spores.

with distilled water to obtain a 5 % wet weigh/volume mixture. The mixture is homogeneized at room temperature with an Ultraturax (20,000 rpm during 5 minutes). The suspended homogenate is then centrifugated (10,000 rpm during 5 minutes) and the supernatant is filtered through Sartorius membrane (0.2 μ pore size).

Test organisms

Staphylococcus epidermitis (C. I. P. 53 124) is the most sensitive of the bacteria tested, it is chosen as test bacteria. The tests are made using the antibiogram technic (Chabbert, 1963), on Trypticase Soy agar at pH 6 using agar plates (11 cm Ø).

The dermatophyte Microsporum gypseum and the mould Sporothrix schenckii are chosen as test fungi. The antifungal tests are similar to those described by Holt (1975) and Gueho et al. (1977).

Among these test-organisms, Staphylococcus epidermitis is used to follow the active product along the various steps of purification.

Ultrafiltration

A Diaflo apparatus is used (Berland et al., 1972). Filtering is conducted in the dark at 4° C with UM_{10} Diaflo membranes (11 cm Ø) which serve as barrier to the diffusive transport of molecules above 10,000 MW. The two phases are tested against Staphylococcus epidermitis. Antibiotic properties of extract is persistent in the above 10,000 MW fraction.

Electrophoresis on High MW Fraction

After concentration of > 10,000 MW fraction, electrophoresis is applied to 0.5 mg of sample for 20 mn, on a cellogel strip in a Tris, HCI buffer (0.02 M at pH 8) under 200 m. volts. The active product is deposited at the cathode and after migration, the strip is cut in two length-wise; one half is stained for proteins with Ponceau Red, the other half is plated over agar inoculated with Staphylococcus epidermitis. After 18 hours incubation at 37° C, the active spot can be located on the cellogel strip.

Anion exchange cellulose (A.E.A).

The active fraction (> 10,000 MW) is concentrated in presence of 0.005 M buffer solution of Veronal at pH 8 by ultrafiltration over UM10 Diaflo membrane and then is passed through an anion exchange cellulose column ("Cellex Δ" Bio-Rad) (Medcalf, 1975). The A.E.A. column is rinsed with buffer solution of Veronal (0.05 M, pH 8), then elution is carried out with continuous NaCl gradient from 0.2 M to 1 M in 0.05 M Veronal buffer (250 ml); 8 ml fractions are obtained and tested for their antibacterial activity.

The elution curve is obtained by measuring the absorption of fractions at 280 nm with an Unicam Sp 200 spectrophotometer and by measuring the quantity of polysaccharide, as saccharose equivalent, according to Morris technic with Anthrone reagent.

Ethanol precipitation

Antibiotic fractions obtained after treatment with A.E.A. cellulose are blended, ultrafiltrated and concentrated over Diaflo UM10 membrane with distilled water to discare NaCl; then, the sample is cold precipitated with 3 volumes ethanol (24 hours at 4° C) and centrifugated (12,000 rpm during 30 mn) (Allan et al., 1972 - Mithal et al., 1972). The precipitate and the supernatant are then tested against Staphylococcus epidermitis.

Purification by Sephadex G 200 Chromatography

The precipitate is passed through a Sephadex G 200 gel column (Mithal et al., 1972), of 1 cm diameter and 80 cm height. 5 ml fractions are obtained and tested for antibacterial activity, after elution with distilled water.

Chemical tests for Polysaccharide characterization

U. V. absorption: The U. V. absorption curve of the active fraction (220 to 300 nm) is obtained with an Unicam Sp 200 spectrophotometer.

Anthrone test (Morris, 1948): This reagent, in presence of hydrolysed polysaccharides, gives a blue or blue-green color. The blue-green complex shows maximum absorption at 625nm.

Reaction of hexuronic acids with carbazole and sulforic acid (Dische, 1962): Carbazole reagent, in presence of hexuronic acids gives a purple color, stable about 1 hour, and that begins to fade slowly. The purple complex shows maximum absorption at 535 nm.

Biological tests

Influence of pH and T° on biological activity:

- 1 ml fractions of the product solution are acidified or alcalinized from pH 0 to pH 14 with drops of dilute HCl or NaOH, then all neutralized to pH 7 and tested against *Staphylococcus epidermitis*.
- 10 ml solution of active product (200 to 300 µg/ml) is progressively heated from 20° C to 90° C (by 5° C steps) and aliquots of solution are tested for each temperature after cooling to 20° C.

Action of chymotrypsin and catalase on antibacterial activity: To determine if chymotrypsin could cause a reduction of antibacterial activity, the active product is treated with solution of 10^{-3} M and 10^{-5} M chymotrypsin (30 mn at 30° C) and these solutions are tested for antibacterial activity. The active product in presence or not of catalase is tested against *Staphylococcus epidermitis*, on agar plate or in deep agar.

Antifungal tests

The purified active product is tested against fungi which were previously inhibited by the aqueous extract of *Chaetoceros* lauderi. The I. M. C. (Inhibiting Minimum Concentration) is determinated with the test-fungi: *Microsporum gypseum* and *Sporothrix schenckii*.

Results

Ultrafiltration

After ultrafiltration on Diaflo UM10 membrane filter (cut off 10,000), the activity remains in the >10,000 MW fraction; it is suggested that the active product is a high MW molecule.

Electrophoresis

After electrophoretic migration, the reaction with Ponceau Red and the antibacterial test show that the active product migrates farther than a protein evidenced with Ponceau Red (Fig. 1), and that it contains anionic charges.

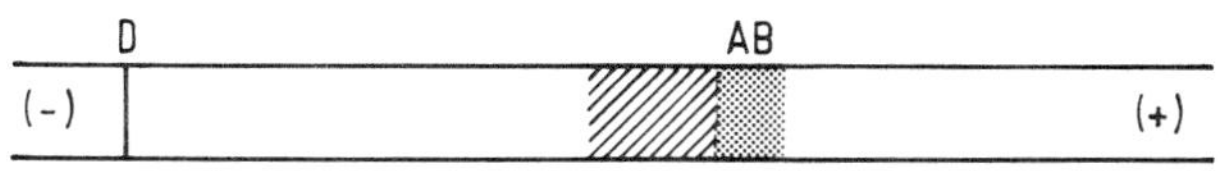

Fig. 1 - Electrophoresis of the dialysis residue in Tris HCl buffer at pH 8 on polyacrylamide gel

A. E. A. Cellulose

Since the active substance can migrate on cellogel in electrophoresis, logically it can be retained on A. E. A. cellulose column. Indeed no activity is found before the NaCl gradient is performed. The elution curve obtained (Fig. 2) shows that the active fractions corresponding to a 0.5 M NaCl gradient do not absorb in the U. V. range 240 to 300 nm, whereas they react positively with Antrone.

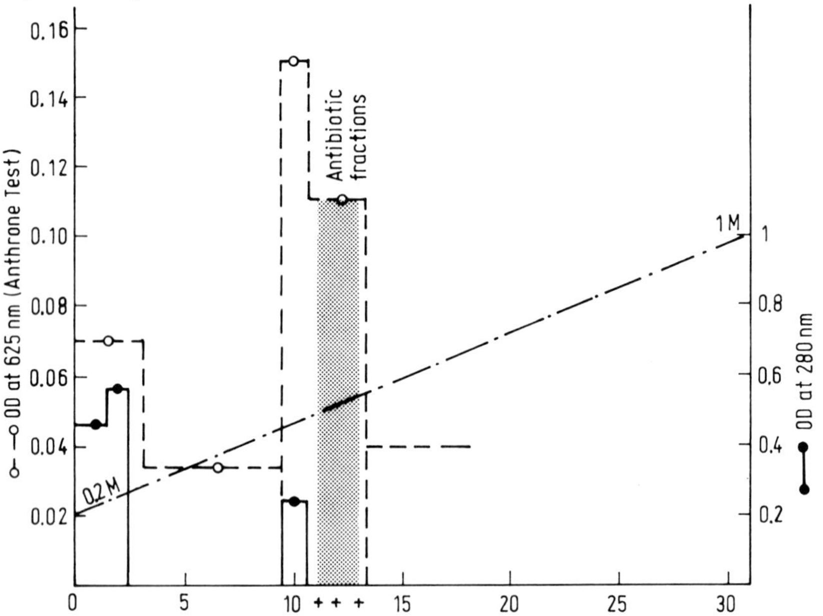

Fig. 2 - Elution curve on A. E. A. cellulose in Veronal buffer 0.05 M with continuous NaCl gradient from 0.2 M to 1 M of the dialysis residue

Ethanol precipitation

The antibacterial activity is recovered in the precipitate after cold ethanol precipitation; 200 μg of product give a 8 mm lysis on agar plate inoculated with <u>Staphylococcus epidermitis</u>.

Sephadex G 200

Antibacterial activity is persistent in the dead volume of the column, that indicates the antibacterial activity is associated with a very high molecular weight material (≃200,000 daltons).

Chemical tests

After purification on Sephadex G 200, the active product does not absorb in U. V. (240-300 nm range) and reacts positively with Anthrone and carbazole-sulfuric acid reagents.

Biological tests

Antibacterial activity is maintained from pH 2.5 to pH 11.5 (Fig. 3) and from -30° C to +80° C (Fig. 4).

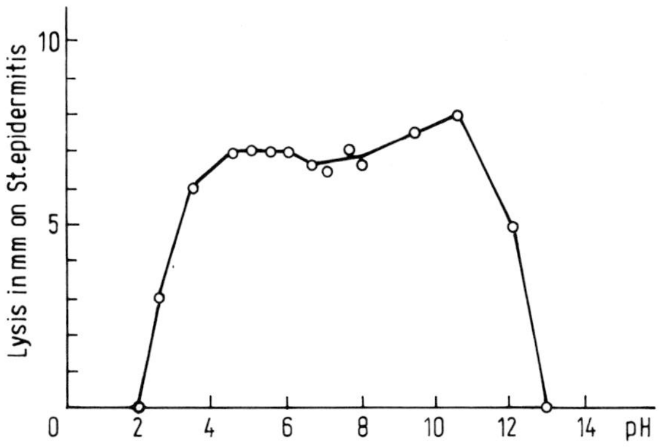

Fig. 3 - Influence of pH on biological activity

Chymotrypsin does not reduce antibacterial activity but catalase has an inhibiting action on it.

The purified antibacterial active substance is also responsible of the antifungal activity of the diatom extract.

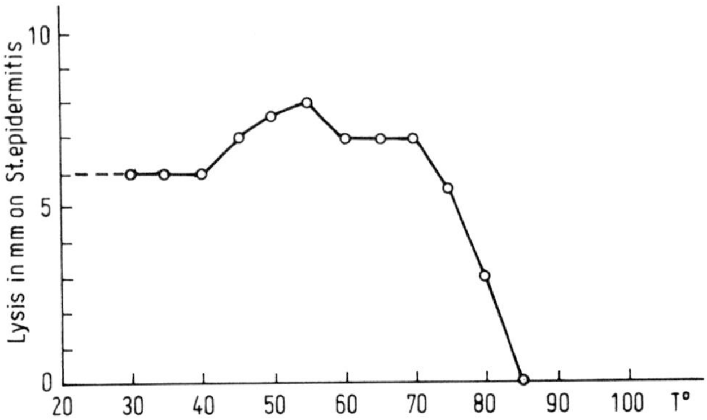

Fig. 4 - Influence of T^{O} on biological activity

We can note that 50 µg of purified product give a 5 mm lysis on agar plate inoculated with Staphylococcus epidermis.

The minimum concentration which inhibits the test fungi (Microsporum gypseum and Sporothrix schenckii) is about 150 - 200 µg/ml.

Discussion

With regards to these first results, it is suggested that the antibacterial and antifungal substance isolated from the marine diatom Chaetoceros lauderi Ralfs CC is a high molecular weight anionic polysaccharide. It is interesting to note that the optimum temperature for antibacterial activity is about 50° C - 55° C, perhaps in relation with a best solubility of the polymer in warm water.

The antifungal activity (I. M. C. 150-200 µg/ml) may be compared with those of classic antifungal products if we take into consideration the high molecular weight of the substance.

The inhibitory action of catalase seems to be an indication of its acting mainly on the respiration of bacteria. Steemann-Nielsen (1955) reported that extracts of Chlorella sp. can inhibit the bacterial respiration. Gauthier (1976) has studied the modification of bacterial respiration by a macromolecular polyanionic antibiotic produced by a marine Alteromonas.

Polysaccharides of biological interest have been isolated from marine algae. Among them, antiherpes virus polysaccharide has been found in extracts of Forlowia mollis and Constantinea simplex by Ehresmann et al. (1977).

Algae contain many polysaccharides both structural and metabolic. Some observations can be pointed out: the culture medium of Chaetoceros lauderi is not active against bacteria and fungi; the release of active substance is increased by grinding the cells in aqueous extract and the molecular weight of the active substance from Chaetoceros lauderi is determined to be greater than 200,000 daltons by Sephadex chromatography.

So it appears that all these properties are more compatible with those of a structural rather than a metabolic component.

Acknowledgements

This research was sponsored partly by I.N.S.E.R.M. (Institut National de la Santé et de la Recherche Médicale) - F.R.A. 20 - under contract 77.1.138.8.A and partly by C.N.R.S. (Centre National de la Recherche Scientifique) - E.R.A. 278 - under contract A.T.P. 1977.

We thank M. Rinaudo and J. Defaye (C.E.R.M.A.V. - C.N.R.S. Grenoble) for helpful criticism during the course of this work.

References

1. Allan, G. G., Lewin, J., Johnson, P. G.: Diatom polymers. IV Diatom polysaccharides. Botanica Marina 15, 102-108 (1972).

2. Aubert, M., Gauthier, M. J.: Origine et nature des substances antibiotiques présentes dans le milieu marin. VIIe partie: Note sur l'activité antibactérienne d'une Diatomée marine: Chaetoceros teres cleve. Rev. Intern. Océanogr. Méd. 4, 33-38 (1966).

3. Aubert, M., Gambarotta, J. P.: Etude de l'action antibactérienne d'espèces phytoplanctoniques marines vis-à-vis de germes anaérobies. Rev. Intern. Océanogr. Méd. 25, 39-48 (1972).

4. Baker, J. T., Murphy, V.: Compounds from marine organisms; in Handbook of Marine Science 1 (CRC Press, Cleveland, Ohio, 1976).

5. Baslow, M. H.: Marine pharmacology (Robert E. Krieger Publishing Company, Huntington, New York, 1977).

6. Berland, B. R., Bonin, D. J., Cornu, A. L.: The antibacterial substances of the marine alga Stichochrysis immobilis (Chrysophyta). J. Phycol. 8, 383-392 (1972).

7. Burkholder, P. R., Burkholder, L. M., Almadovar, L.: Antibiotic activity of some marine algae of Puerto Rico. Botanica Marina 2, 149-156 (1960).

8. Chabbert, Y. A.: L'antibiogramme. Coll. Techniques de base. Ed. Tourelle. Saint-Mandé. 1 vol. (1963).

9. Dische, Z.: Color reactions of Hexuronic acids; in Methods in carbohydrate chemistry. Vol. I: Analysis and preparation of sugars, 497-498 (Academic Press, N. Y. and London) (1962).

10. Ehresmann, D. W., Deig, E. F., Hatch, M. T., Disalvo, L. H., Vedros, N. A.: Antiviral substances from California marine algae. J. Phycol. 13, 37-40 (1977).

11. Gauthier, M. J.: Activité antibactérienne d'une Diatomée marine: Asterionella notata (Grun). Rev. Intern. Océanogr. Méd. 25-26, 103-165 (1969).

12. Gauthier, M. J.: Modification of bacterial respiration by a macromolecular polyanionic antibiotic produced by a marine Alteromonas. Antimicrobial agents and chemotherapy. Journal of Bacteriology 9, n° 3, 361-366 (1976).

13. Gueho, E., Pesando, D., Barelli, M.: Propriétés antifongiques d'une Diatomée marine Chaetoceros lauderi Ralfs C. C. Mycopathologia 60, 105-107 (1977).

14. Holt, R. J.: Laboratory tests of antifungal drugs. J. Clin. Path. 28, 767-774 (1975).

15. Medcalf, D. G., Scott, J. R., Brannon, J. H., Hemerick, G. A., Cunningham, R. L., Chessen, J. H., Shah, J.: Some structural features and viscometric properties of the extracellular polysaccharide from Porphyridium cruentum . Carbohydrate Research 44, 87-96 (1975).

16. Mithal, B. M., Hasgiwal, P. C.: Fractionation of polysaccharides. The Indian Journal of Pharmacy 34, 3, 51-57 (1972).

17. Morris, D. L.: Quantitative determination of carbohydrates with Dreywood's anthrone reagent. Science 107, 254-255 (1948).

18. Pesando, D., Gueho, E., Gnassia-Barelli, M.: Antifungal properties of some marine planktonic algae (in press).

19. Provasoli, L.: Media and prospects for cultivation of marine algae. A. Watanabe and A. Hahari (eds). Cultures and collection of algae. Jap. Conf. Hakane. Sept. 1966. Jap. Soc. Plant Physiol., 63-75 (1966).

20. Ruggieri, G. D.: Marine organisms with novel chemical constituents are excellent sources of new drugs. Science 194, 29 October, 491-497 (1976).

21. Scheuer, P. J.: Chemistry of natural products. Academic Press, New-York (1973).

22. Sharma, G. M., Michaels, L., Burkholder, P. R.: Goniodomin, a new antibiotic from a dinoflagellate. Journal of antibiotics 11, 659-664 (1966).

23. Steemann-Nielsen, E.: The production of antibiotics by planktonic algae and its effects upon bacterial activities in the sea. Papers in Marine Biol. and Oceanogr., London, 281-286 (1955).

ANTIFUNGAL PROPERTIES OF SOME MARINE PLANKTONIC ALGAE

D. Pesando, M. Gnassia-Barelli

Laboratoire de Physique et Chimie Marines. BP n° 8. La Darse.
06230 Villefranche-sur-Mer

E. Gueho

Laboratoire de Cryptogamie. UER des Sciences Pharmaceutiques.
8, avenue Rockfeller. 69373 Lyon Cedex

Abstract

Extracts of ca. 50 marine planktonic algae strains in culture were studied with regard to their antifungal activity; five diatoms were found to inhibit the growth of characteristic pathogenic fungi.

Introduction

Marine algae as a source of antifungal substances have not yet been extensively studied. In addition to the antifungal activity reported by Martinez-Nadal (1966) marine algae have also been found to be active against the fungi *Trichophyton mentagrophytes* and *Trichophyton rubrum* (Mauter et al., 1953), *Candida albicans* (Olesen et al., 1964) and a number of other fungal organisms (Welch, 1962). Welch tested aqueous homogenates of 22 marine algae *Rhodophyta*, *Chlorophyta* and *Phaeophyta* against 6 assay fungi and found that these preparations affected growth, fruiting and color of the test organisms. Few papers have been concerned with products which inhibit the growth of pathogenic fungi. Among them, an isozonarol hydroquinone active against phytophathological fungi and a cycloendesmol

Marine Algae in Pharmaceutical Science
edited by H. A. Hoppe, T. Levring, Y. Tanaka

active against Candida albicans have been isolated respectively from the brown seaweed Dictyopteris zonarioides and from the marine alga Chondria opposicloda Dawson by Fenical et al. (1973) (1974). Antifungal activity to Candida albicans has also been found in extracts from some marine algae by Sims et al. (1975). Extracts from some brown algae were found to contain active substances that inhibit the growth of some moulds isolated from leather material (Khaleafa et al., 1975).

The influence of the green alga Chlorella vulgaris extracts on the growth of some fungi has been studied by Matusiak and Krynicka (1975).

Most investigations were carried out on the antibacterial activity of extracts from marine unicellular algae, whereas very few took into consideration their fungicidal activity.

131 specimens of marine algae from Puerto-Rico were tested against indicatory micro-organisms (pathogenic bacteria, marine bacteria and Candida albicans) by Burkholder et al. (1960). The authors noted that the dinoflagellate Goniaulax tamarensis showed noticeable activity against Candida albicans; among the other species showing such an activity, they noted Chondria littoralis, Falkenbergia hillebrandii, Lyngbya majuscula and the genus Wrangelia.

From a marine dinoflagellate Goniodoma sp., collected in Puerto-Rico, Goniodomin, fungicidal substance, has been isolated by Sharma et al. (1968).

The antimicrobial activity of a marine diatom Asterionella notata has been studied by Gauthier (1969); an aqueous extract from this diatom inhibited the growth of several bacteria and of some fungi as Candida albicans, Penicillium and Aspergillus genus.

In a preliminary paper (Gueho et al., 1977), we reported a significative antifungal activity, especially against dermatophytes, in extracts from the marine diatom Chaetoceros lauderi.

In the present paper, a systematic screening for antifungal activity of planktonic algae strains in culture (ca. 50 species) was related. The tests were carried out on 6 indicator species of fungal diseases.

Material and Methods

Strains

Several planktonic algae strains tested were obtained from the "Laboratorio per lo Studio della Contaminazione radioattiva del Mare", CNEN EURATOM, Fiascherino, Italia (Dr. Berland) and from the "Station Marine d'Endoume", Marseille, France (Dr. Berland). The other strains were isolated off Villefranche Bay (identification by P. Nival, Station Marine de Villefranche-sur-Mer, France). Dinoflagellate were obtained from the Université de Nice (Dr. J. M. Pincemin).

Culture conditions

Sterilized sea-water and Provasoli ES solution (1966) were used as culture medium in bacteria free conditions.

Batch cultures of algae were grown without air bubbling in flasks containing 5000 ml ES medium. The cultures were maintained at 16° C under continuous light provided by fluorescent tubes Silvania Gro-Lux and Mazda (irradiation at about 95 W. m^{-2}).

After 12 days, phytoplankton cultures were filtered over Sartorius membrane filters (from 0.2 to 8 µ pore size) according to the size of plankton, and cells were deep-frozen.

Algal extracts

Each deep-frozen algal sample was mixed with distilled water to obtain a 5 % mixture wet weight/volume. The mixture was homogenized at room temperature with an Ultraturax (20,000 rpm during 5 minutes). The suspended homogenate was then centrifuged (10,000 rpm during 5 minutes) and the supernatant was filtered through Sartorius membrane filter (0,2 μ pore size) to obtain a sterile solution.

Antifungal tests

The test organisms: Aspergillus fumigatus (mould), Candida albicans (yeast), Epidermophyton floccosum, Microsporum gypseum, Trichophyton mentagrophytes (Dermatophytes) and Sporothrix schenckii (mould), indicators of fungal diseases of man were obtained from the "Laboratoire de Mycologie de l'Institut Pasteur (Paris)" and the "Laboratoire de Cryptogamie, UER des Sciences Pharmaceutiques (Lyon)".

The antifungal tests were similar to those described by Holt (1975). Fungi strains were grown on nutrient medium with 2 % glucose, 1 % peptone, 0.2 % yeast extract, 2 % agar-agar, or malt-agar with 4 % malt-flour, 2 % agar-agar.

When the cultures reach their optimal growth (from 48 to 192 hours depending on the species), suspensions of spores or mycelium were made under sterile conditions into 1 ml distilled water containing an anionic surfactant with high dispersant power (0.01 % Triton X100).

In the case of very filamentous fungi, the suspension was homogenized by using a Dounce homogenizer to a final inoculum of approximatively 1.10^5 cells/ml.

Antibiosis was researched on liquid medium because the methods using discs or diffusion were very difficult to interpret in the case of filamentous fungi.

Then, in 5 ml tubes, 1 ml liquid nutrient medium prepared with a double concentration of glucose, peptone, and yeast extract was mixed with the same volume of planktonic extract and 0.05 ml of fungal suspension. In control tubes, the cell extract was substituted by 1 ml sterile distilled water. Cultures were maintained at 25° C until a complete growth was obtained in the control tubes (from 24 to 192 hours).

Results and Discussion

Aqueous extract of 48 planktonic algae were examined and 5 were found to contain significative antifungal activity (Table 1). The active algal species were Asterionella notata Grunow, Chaetoceros lauderi Ralfs C.C., Chaetoceros pseudocurvisetus Mangin, Chaetoceros socialis Lauder and Fragilaria pinnata Ehrenberg.

The extract of Asterionella notata inhibited the growth of all the fungi tested. With 48 other fungi (yeasts, moulds and dermatophytes) also tested, in presence of the same extract, no growth was observed. Gauthier (1969) had observed from the same diatom an antibacterial activity on a very large spectrum of bacteria.

It is significant that the inhibition of the growth of bacteria and fungi was unselective and so it is suggested that a toxic phenomenon is involved in this action.

The growth of the three dermatophytes and of the two moulds tested was inhibited by the extract of Chaetoceros lauderi. This inhibition is function of the concentration of the suspension of fungi in the case of Aspergillus fumigatus. This diatom had also been tested against a broad spectrum of fungi (Gueho et al., 1977) and a significant activity against all the dermatophytes tested was observed. The antibacterial acitivty was previously known (Aubert et al., 1966) (1972)[+].

[+] The antibacterial activity attributed to Chaetoceros teres is now attributed to Chaetoceros lauderi after the examination of the spores.

Table 1 - Growth of fungi in presence of algal extract

	Candida albicans (yeast)	Sporothrix schenckii(mould)	Trichophyton mentagrophytes (Dermatophyte)	Microsporum gypseum(Dermatophyte)	Epidermophyton floccosum (Dermatophyte)	Aspergillus fumigatus (mould)
BACILLARIOPHYCEAE						
Asterionella japonica Cleve	+++	+++	+++	++	-	+++
Asterionella notata Grunow	-	-	-	-	-	-
Chaetoceros affinis Lauder	+++	+++	+++	+++	+++	+++
Chaetoceros brevis Schütt	+++	+++	+++	+++	A	+++
Chaetoceros danicus Cleve	+++	+++	+++	+++	+++	+++
Chaetoceros protuberans (var. Nivali ?)	+++	+++	+++	+++	++	+++
Chaetoceros lauderi Ralfs	+++	-	-	-	-	±
Chaetoceros lorenzianus Grunow	+++	+++	+++	+++	+++	+++
Chaetoceros pseudo curvisetus Mangin	+++	++	+	-	-	+++
Chaetoceros socialis Lauder	+++	++	+	-	-	+++
Chaetoceros sp.	+++	+++	+++	+++	++	+++
Fragilaria pinnata Ehrenberg	+	+	-	-	-	+++
Gyrosigma fasciola (Ehrenberg) Cleve	+++	+++	++	+++	-	+++
Lithodesmium undulatum Ehrenberg	+++	+++	+++	+++	++	+++

Navicula Incerta Grunow	+++	+++	+++	+++	+++	+++
Nitzschia ascicularis W. Sm.	+++	+++	+++	+++	+++	+++
Nitzschia seriata Cleve	+++	+++	+++	+++	+++	+++
Phaeodyctylum tricornutum Bohlin	+++	+++	+++	+++	+++	+++
Skeletonema costatum (Greville) Cleve	+++	+++	+++	+++	-	+++
Thalassiosira sp.	+++	+++	++	++	-	+++
DINOPHYCEAE						
Amphidinium sp.	+++	+++	+++	+++	+++	+++
Coolia monotis Meunier	+++	+++	+++	+++	+++	+++
Exuviella mariae labouriae Parke and Ballantine	+++	+++	+++	+++	+++	+++
Exuviella sp.	+++	+++	+++	+++	+++	+++
Glenodinium sp.	+++	+++	+++	+++	+++	+++
Gonyaulax sp.	+++	+++	+++	+++	+++	+++
Peridium Trochoideum (Stein) Lemmerman	+++	+++	+++	+++	+++	+++
Prorocentrum micans Ehrenberg	+++	+++	+++	+++	+++	+++
HAPTOPHYCEAE						
Coccolithus huxleyi (Lohmann) Kamptner	+++	+++	+++	+++	+++	+++
Coccolithosphora sp.	+++	+++	+++	+++	+++	+++
Cricosphaera elongata Parke and Green	+++	+++	+++	+++	+++	+++
Dicrateria sp.	+++	+++	+++	+++	+++	+++
Isochrysis galbana Parke	+++	+++	+++	+++	+++	+++

	Candida albicans (yeast)	Sporothrix schenckii(mould)	Tricophyton mentagrophytes (Dermatophyte)	Microsporum gypseum(Dermatophyte)	Epidermophyton floccosum (Dermatophyte)	Aspergillus fumigatus (mould)
Ochrosphaera neapolitana Schussnig	+++	+++	+++	+++	+++	+++
Monochrysis lutheri Droop	+++	+++	+++	+++	+++	+++
Pavlova pinguis Green	+++	+++	+++	+++	+++	+++
Prymnesium parvum Carter						
CHLOROPHYCEAE						
Dunaliella sp.	+++	+++	+++	+++	+++	+++
Dunaliella primolecta Butcher	+++	+++	+++	+++	+++	+++
Dunaliella salina (Dunal) Theod.	+++	+++	+++	+++	+++	+++
CRYPTOPHYCEAE						
Chroomonas fragarioides Butcher	+++	+++	+++	+++	+++	+++
PRASYNOPHYCAEA						
Platymonas suecica Kylin	+++	+++	+++	+++	+++	+++
Platimonas sp.	+++	+++	A	+++	A	+++
Tetraselmis maculata Butcher	+++	+++	+++	+++	+++	+++

RHODOPHYCEAE						
Porphyridium sp. Naëgeli	+++	+++	+++	+++	A	+++
XANTOPHYCEAE						
Heterothrix sp. Pashner	+++	+++	+++	+++	+++	+++
CYANOPHYCEAE						
Cyanophycea sp.	+++	+++	+++	+++	+	+++
Microcystis sp. Kützing	+++	+++	+++	+++	+++	+++

+++ normal growth

++ delayed growth

\+ weakly growth

\- no growth after 15 days incubation

A - not tested

We have isolated from this diatom an acid polysaccharide which is responsible of both antibacterial and antifungal activities (Pesando et al., in press).

The growth of dermatophytes: Trichophyton mentagrophytes, Microsporum gypseum and Epidermophyton floccosum, was delayed or inhibited by the extracts of Chaetoceros pseudocurvisetus and Chaetoceros socialis.

The extract of Fragilaria pinnata inhibited the growth of dermatophytes and delayed the growth of Candida albicans and Sporothrix schenckii. Berland et al. (1972) had isolated from this diatom two antibacterial substances active against Xanthomonas sp., Flavobacterium sp. and Achromobacter sp.

With regards to these results, it is significant that all the antifungal algae are members of the class of Bacillariophyceae. Three of the active diatoms are Chaetoceros species, and Asterionella notata, Fragilaria pinnata and Chaetoceros lauderi were previously known for their antibacterial activity.

The ability of marine planktonic algae to possess antifungal substances has not been extensively explored. This investigation represents a partial account of survey of marine planktonic algae for antifungal properties.

Acknowledgements

This research was sponsored partly by I.N.S.E.R.M. (Institut National de la Santé et de la Recherche Médicale) - F.R.A. 20 - under contract 77.1.138.8.A and partly by C.N.R.S. (Centre National de la Recherche Scientifique) - E.R.A. 278 - under contract A.T.P. 1977.

References

1. Aubert, M., Gauthier, M. J.: Origine et nature des substances antibiotiques présentes dans le milieu marin. VIIe partie: Note sur l'activité antibactérienne d'une Diatomée marine: Chaetoceros teres cleve. Rev. Intern. Océanogr. Méd. 4, 33-38 (1966).

2. Aubert, M., Gambarotta, J. P.: Etude de l'action antibactérienne d'espèces phytoplanctoniques marines vis-à-vis de germes anaérobies. Rev. Intern. Océanogr. Méd. 25, 39-48 (1972).

3. Berland, B. R., Bonin, D. J., Cornu, A. L.: The antibacterial substances of the marine alga Stichochrysis immobilis (Chrysophyta). J. Phycol. 8, 383-392 (1972).

4. Burkholder, P. R., Burkholder, L. M., Almadovar, L.: Antibiotic activity of some marine algae of Puerto-Rico. Botanica marina 2, 149-156 (1960).

5. Gauthier, M. J.: Activité antibactérienne d'une Diatomée marine. Asterionella notata (Grun). Rev. Intern. Océanogr. Méd. 25-26, 103-165 (1969).

6. Gueho, E. Pesando, D., Barelli, M.: Propriétés antifongiques d'une Diatomée marine Chaetoceros lauderi Ralfs C. C. Mycopathologia 60, 2, 105-107 (1977).

7. Fenical, W. F., Sims, J. J., Squatrito, D., Wing, R. N., Radlick, P.: Zonarol and isozonarol. Fungitoxic hyroquinones from the brown seaweed Dictyopteris zonarioides. J. Org. Chem. 38, 2383 (1973).

8. Fenical, N.: Cycloendesmol, an antibiotic cyclopropane containing sesquiterpene from the marine alga, Chondria opposicloda Dawson J. J. Tetrahedron Lett. 13, 1137-1140 (1974).

9. Holt, R. J.: Laboratory tests of antifungal drugs. J. Clin. Path. 28, 767-774 (1975).

10. Khaleafa, A. F., Kharboush, A. M., Metwalli, A., Mohsen, A. F., Serwi, A.: Antibiotic fungicidal action from extracts of some seaweeds. Botanica marina 18, 163-165 (1975).

11. Martinez-Nadal, N. G., Casillas-Chapel, C. M., Rodriguez, L. V., Rodriguez-Perazza, J. R., Torres-Vera, L.: Antibiotic properties of marine algae. III. Cympolia barbata. Botanica marina 9, 21-26 (1966).

12. Mautner, H. C., Gardner, G. M., Pratt, R.: Antibiotic activity of seaweed extracts. II. Rhodomela larix. J. Amer. Pharm. Ass. Sci. Ed. 42, 294-296 (1953).

13. Olesen, P.E., Maretzki, A., Almondovar, L. A.: An investigation of antimicrobial substances from marine algae. Botanica marina 6, 224-232 (1964).

14. Pesando, D., Gnassia-Barelli, M., Gueho, E.: Partial characterization of a specific antifungal substance isolated from a marine diatom Chaetoceros lauderi (Ralfs C.C.) (in press).

15. Provasoli, L.: Media and prospects for cultuvation of marine algae, Watanabe, A., Hahari, A. (eds.). Cultures and collection of algae. Jap. Conf. Hakane Sept. 1966. Jap. Soc. Plant Physiol. 63-75 (1966).

16. Sharma, G. M., Michaels, L., Burkholder, P. R.: Goniodomin, a new antibiotic from a dinoflagellaté. Journal of antibiotics 11, 659-664 (1966).

17. Sims, J. J., Donnel, M. S., Leary, J. L.: Antimicrobial agents from marine algae. Antimicrobial agents chemoth. 7, 320-321 (1975).

18. Welch, A. M.: Preliminary survey of fungistatic properties of marine algae. J. Bacteriol 83, 97-99 (1962).

Fatty Acids and Lipids of Marine Algae and the Control of their Biosynthesis by Environmental Factors.

P.Pohl and F.Zurheide
Institut für Pharmazeutische Biologie, Universität Kiel,
Grasweg 9, 2300 Kiel, Germany

1. Introduction

Fatty acids and lipids are well-known constituents of living cells. Fast and exact analyses of these compounds, however, have only been achieved since the introduction of chromatographic methods such as thin layer and gas chromatography, i.e. after 1950-1960. Since that time, fatty acids and lipids have been investigated in a multitude of species from all groups of the plant kingdom. As for marine algae, our knowledge is still rather poor. A good number of papers, however, has been published from which some general conclusions can be drawn.

There are significant differences in the compositions of fatty acids: a) between the individual classes of marine algae, b) between marine and freshwater algae, and c) between algae (in general) and terrestrial plants. Therefore it was necessary to make some comparisons with the fatty acids in the other groups of the plant kingdom, too.

In general, the main function of fatty acids and lipids in plants seems to be their participation in cell membranes. Furthermore, as storage products, they serve as a source of energy and metabolic precursors. In many plants they are components of the outer protecting layer.

In addition, there is one group of physiologically active fatty acids which is of particular importance for animals. These are the "essential fatty acids". All of them are unsaturated fatty acids with a special arrangement of double

Marine Algae in Pharmaceutical Science
edited by H. A. Hoppe, T. Levring, Y. Tanaka

bonds. They cannot be synthesized by animals and therefore have to be taken up from vegetable food. The most important essential fatty acids are of the linoleic acid type (see fig.1) such as linoleic acid (9,12-18:2)*) and arachidonic acid (5,8,11,14-20:4). Arachidonic acid has more recently become known as precursor of the prostaglandins in animal tissues (1).

When animals are fed without essential fatty acids over a longer period of time they develop an essential fatty acid deficiency (EFAD) syndrome characterized by the increase of unusual unsaturated fatty acids which normally are present only in relatively small amounts, such as 7,10,13-20:3, 5,8,11-20:3, 9-16:1 (palmitoleic acid), and 9-18:1 (oleic acid)(2,3). EFAD-rats develop dermal symptoms such as scaly skin with impaired barrier properties and a high rate of transepidermal water loss (4-6). The main physiological effects of essential fatty acids on EFAD-rats are promotion of growth, prevention or cure of dermal symptoms, and decrease of water consumption (7). Possibly prostaglandins are related to these effects (8). In man, EFAD normally does not occur because the food contains sufficient amounts of essential fatty acids. EFAD was observed, however, when patients (3,9-11) and newborn infants (2,12) had to be kept on a fat-free diet. Even cutaneous manifestations such as scalp dermatitis and hair depigmentation have been observed which could be reversed by daily topical application of linoleic acid to the patient's skin (3).

There is still another reason why essential fatty acids today are in the focus of research although this topic cannot be discussed here in detail: Essential fatty acids have an effect on hyperlipidemia by lowering the levels of plasma lipids (cholesterol and triglycerides) (13-15). Hyperlipide-

*) 9,12-18:2 is a straight chain fatty acid with 18 carbon atoms and two double bonds in the positions 9/10 and 12/13. The carboxylic group has carbon atom no.1.

mia, however, is frequently related to the development of atherosclerosis (16-18) and coronary heart disease (15).

Recently, lipids have gained attention in the fields of pharmacy and medicine, too. Phospholipids such as phosphatidylcholine and phosphatidylethanolamine become spontaneously organized in water into stable structures mostly with a lamellar state (19). These lamellae are made up of bimolecular layers (bilayers) in which the polar groups of the lipids are directed towards the water and the unpolar carbon chains stick together in the interior of the lamellae. Smaller bilayers are called vesicles. Liposomes are closed vesicles with a bilayer structure (19,20) and are able to fuse with cells (21,22).

In such vesicles, drugs can be entrapped and then be administered. Thus the phospholipids can be used as carriers for exogenous and endogenous biologically active agents as for example enzymes and antibiotics (23-27).

In this connection it is interesting that probably a new function of certain brain lipids has been discovered recently: Abood et al. (28) showed that phosphatidylserine (which is the major acidic brain lipid) and phosphatidylinositol enhance the binding of opiates to a rat brain membrane fraction. Hoss and Smiley (29) demonstrated that aqueous suspensions of the above two phospholipids immobilize (bind) opiates. Probably there is an opiate receptor in the brain in the form of a protein phospholipid membrane complex (28,30).

2. Classification of fatty acids and lipids of marine algae

Fatty acids

Most of the fatty acids synthesized by marine algae are straight chain saturated and unsaturated fatty acids with even-numbered carbon chains. Fatty acids with an uneven number of carbon atoms, and branched chain iso and anteiso fatty acids are present quite frequently, but only in small proportions.

Marine algae do not seem to synthesize fatty acids with "unusual" structures, such as acetylenic, hydroxy, epoxy, oxo, cyclopropanoic, and cyclopropenoic acids. These compounds have been found in bacteria, fungi, and terrestrial plants, respectively (31-33).

a) Saturated fatty acids

Marine algae synthesize the same saturated fatty acids commonly found in other plants, too, namely lauric acid (12:0), myristic acid (14:0), palmitic acid (16:0), and stearic acid (18:0). Arachic acid (20:0), behenoic acid (22:0), and fatty acids with still longer carbon chains usually are not present or only in trace amounts. Often one can find small quantities of uneven-numbered fatty acids such as the 13:0, 15:0, 17:0, and 19:0 fatty acids.

b) Unsaturated fatty acids

Among all groups of the plant kingdom, marine algae have the greatest variety of cis-unsaturated fatty acids. Normally, all of the double bonds are in the cis-configuration (34). There is, however, one important exception to this: In photosynthesizing algae (as well as in all other photosynthesizing eukaryotic organisms) a trans 3-16:1 fatty acid is found in only one lipid, phosphatidyl glycerol. The algal fatty acids usually do not occur as free fatty acids (FFA) but in a bound form, mostly in lipids. The total number of unsaturated fatty acids in algae is about 50. At first sight this variety may be confusing. There are, however, certain pathways in the biosynthesis of plant (and algal) fatty acids which lead to the formation of several biosynthetically related "types" of unsaturated fatty acids. These are often called "fatty acid families".

Within each "fatty acid family" the compounds are structurally related in that they have the same "methyl end" (which refers to the part of the carbon chain from the last double bond to the terminal methyl group). Fatty acids with a C_9 methyl end like 7-16:1 (see fig.1) are also called ω9-fatty acids, those with a C_6 methyl end (linoleic and arachidonic acids) ω6 fatty acids etc. The four major types of fatty acids are presented in fig.1. As mentioned before, the linoleic acid and also α-linolenic acid types are essential fatty acids which cannot be synthesized by animals. According to Schlenk (7), some odd numbered fatty acids (for example

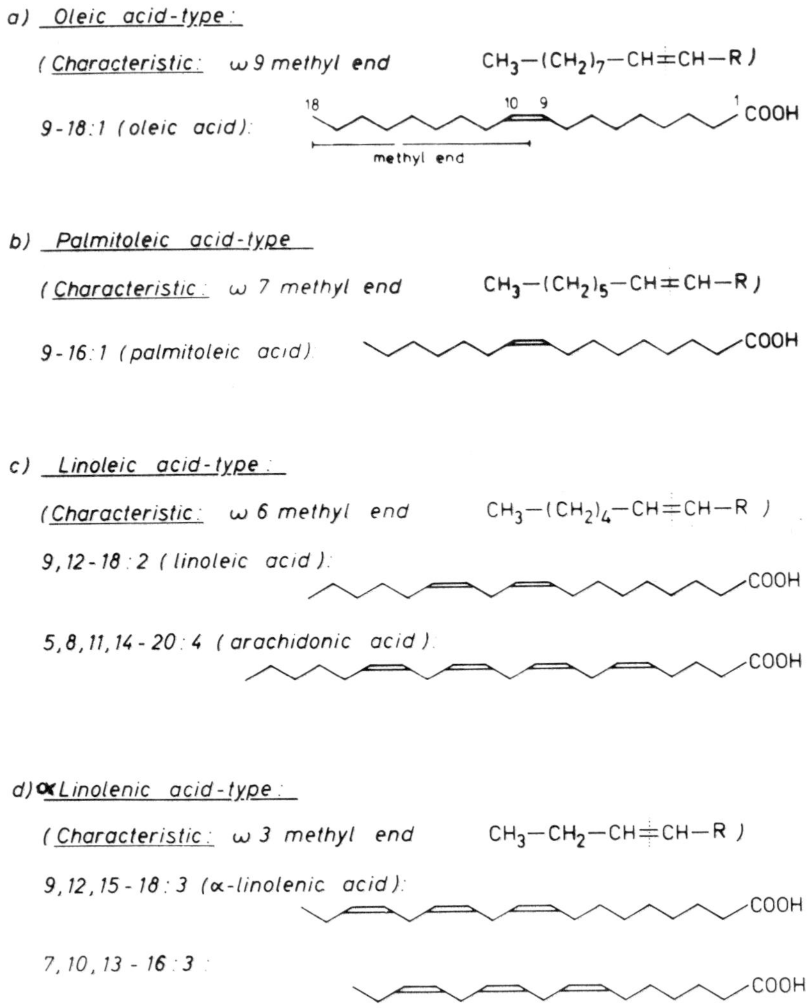

Fig.1: Structures of four different types of unsaturated fatty acids

9,12-17:2 and 6,9,12-17:3) have a similar effect as linoleic acid (9,12-18:2).

Lipids

Lipids (from "lipos " (gr.) = fat) are a group of compounds which are characterized by their good solubility in lipophilic organic solvents such as petrolether, ether, chloroform, and solvent mixtures (for example chloroform-methanol), respectively. In contrary to other constituents of the cell (proteins, saccharides, nucleic acids), they are not soluble in water. In earlier publications lipids were sometimes subdivided into fats and lipoids (fat-like substances). This subdivision and the term "lipoids", however, are no longer current today.

The majority of lipids are acyl lipids, i.e. they contain fatty acids bound to hydrophilic moieties such as alcohols (mainly glycerol), sugars, and bases.

If one extracts plant material with lipophilic organic solvents one will usually obtain other highly lipophilic compounds such as pigments (chlorophylls, carotenoids), sterols, and sterol esters. These are sometimes regarded as lipids, sometimes not. They will not be discussed in this paper.

The lipids can be subdivided according to their polarity which depends on the unpolar lipophilic carbon chains (fatty acids, long chain alcohols) and the hydrophilic moieties (phosphate and carboxylic groups, alcohols, sugars, bases etc.) in each lipid.

a) Unpolar (neutral) lipids

There are only relatively few investigations on the quantitative composition of algal lipids. The main unpolar lipids of algae are triglycerides, hydrocarbons, and free fatty acids (FFA). The latter are of a medium polarity because of their carboxylic groups. Wax esters belong to the unpolar lipids, too. Although they play an important role in the marine food chain, they do not seem to be synthesized in most algae to any extent.

Triglycerides: These are the major unpolar lipids in most marine algae and probably serve as storage products (35-38). They are often accompanied by minor proportions of mono- and diglycerides.

Hydrocarbons: These compounds belong to the unsaponifiable matter. Saturated and unsaturated hydrocarbons (1-6 double bonds) have been detected in a wide number of marine benthic and planktonic algae during the last 10 years. In red, brown, and green algae the content is 0.001-0.48 % of the dry weight (39). Brown algae seem to synthesize mainly saturated C_{15} (n-pentadecane), and red algae saturated C_{17} (n-heptadecane) hydrocarbons whereas two green algae (Enteromorpha compressa and Ulva lactuca) have a 17:1 (n-heptadecene) and a C_{17} cyclopropane hydrocarbon (39,40). The most unsaturated seems to be a 3,6,9,12,15,18-21:6 hydrocarbon found in numerous algae (Chrysophyceae, Bacillariophyceae, Dinophyceae, Phaeophyceae, Chlorophyceae, and Cryptophyceae) (41-43).

Hydrocarbons have also been isolated from many freshwater algae (Cyanophyceae, Chlorophyceae, Chrysophyceae). The normal range is from C_{15} - C_{19} in most species. Some of the algae show a bimodal distribution with maxima at C_{17} and C_{27} (44-48).

Chemotaxonomic considerations, however, should not be made mainly "in view of the observed gross differences in hydrocarbon chemistry at different growth stages" (39).

Wax esters: These are esters of fatty acids with long chain alcohols. There are only very few reports about the formation of this lipid class in algae. Saturated wax esters have been found in a marine cryptomonad, Chroomonas salina, grown photoheterotrophically on glycerol (49). The wax esters consisted of 12:0, 14:0, and 16:0 fatty acids while the main alcohols were 14:0 and 16:0. Wax esters of a similar composition have also been found in a unicellular freshwater alga,

Euglena gracilis (Euglenophyceae) (50-53). The capability to synthesize these lipids appears to be very little spread among microscopic and macroscopic marine algae.

It should be mentioned here that wax esters seem to play an important role in the marine food chain. The initial stage of this chain are phytoplankton algae which, however, do not make wax esters (54,55). The phytoplankters then serve as food for herbivorous copepods (crustaceans,zooplankton) which synthesize wax esters in large amounts (54-59). The copepods then are the principal food at some stage in the life of many species of fish (herrings, sardines, anchovies) (56,60).

Free fatty acids (FFA): To the authors' knowledge, there are at present no reports on the formation of FFA in marine algae. Current investigations (38) (see fig.7) show that in the red alga, Phycodrys sinuosa, the biosynthesis of FFA varies according to the season. In July/August the proportion of FFA is 50-70 % of the total lipids. However, in various brown algae (Fucus serratus, Fucus spiralis, Fucus vesiculosus, and Chorda tomentosa), and in the green alga, Enteromorpha linza, there were only trace amounts of FFA.

b) Polar lipids

The polar lipids of marine algae are all acyl lipids, i.e. they contain fatty acids and are saponifiable. According to their polar (hydrophilic) groups they can be subdivided into phospholipids and glycolipids. Both of them are synthesized by marine algae as well as by freshwater algae and terrestrial plants and mostly are associated with the cell membranes.

Phospholipids:The six major phospholipids of algae (and of other plants, too) are listed in fig.2. All of them contain glycerol esterified to fatty acids via the hydroxy groups 1 and 2. The remaining hydroxyl is bound to various polar groups (choline, ethanolamine, serine, inositol etc.) via phosphoric acid. The most simple phospholipid is phosphatidic acid (PA) which is present as a minor component.

phosphatidic acid

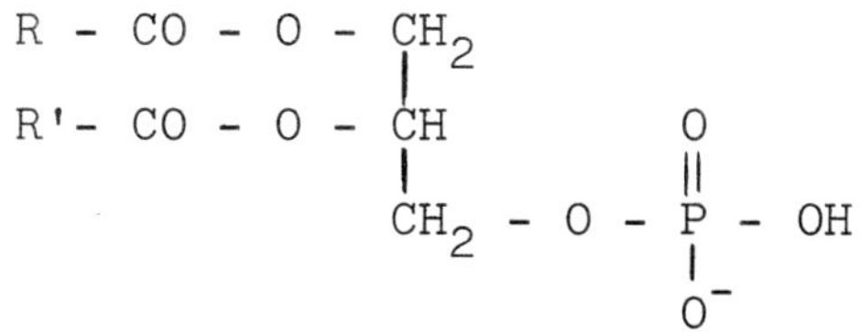

Usually, the predominant phospholipids of algae are phosphatidylcholine (PC) (also known as lecithin) and phosphatidylethanolamine (PE). One of the less abundant phospholipids,

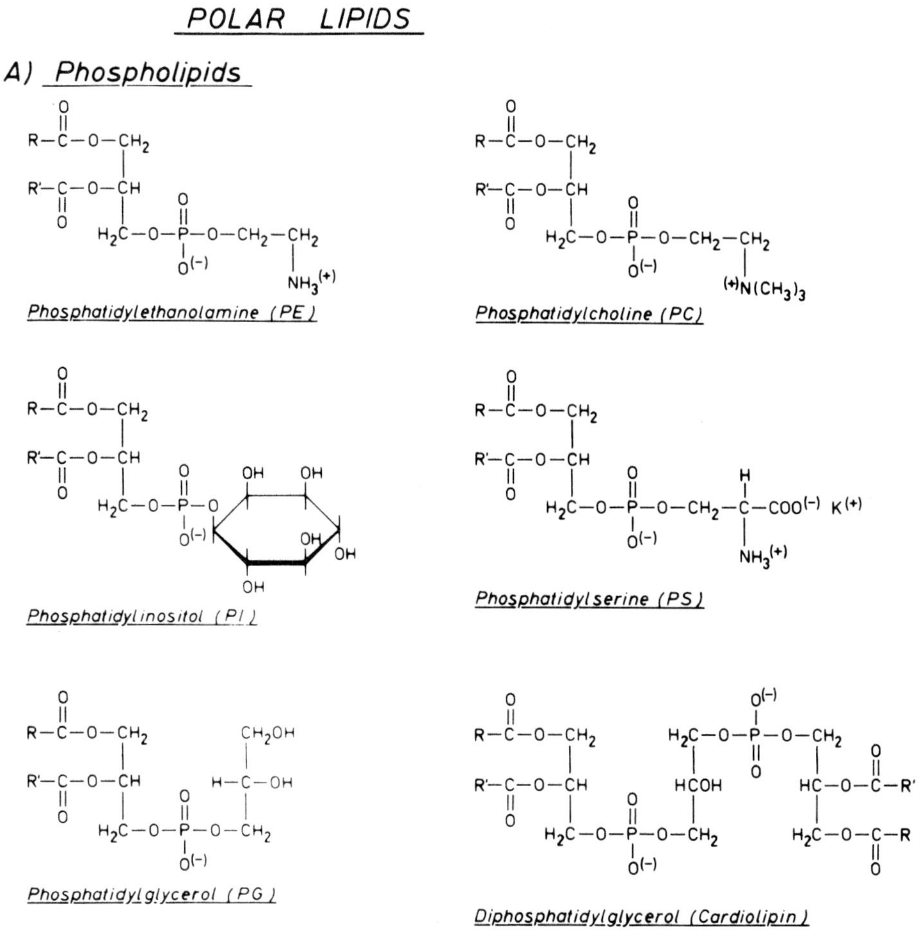

Fig.2: Polar lipids (A: phospholipids)

phosphatidylglycerol (PG) is a characteristic component of chloroplasts. It is the only lipid which contains a trans 3-16:1 fatty acid. There may also be minor quantities of "lyso" phospholipids with only one fatty acid bound to the glycerol.

Glycolipids: There are three major glycolipids which seem to

POLAR LIPIDS

B) Glycolipids

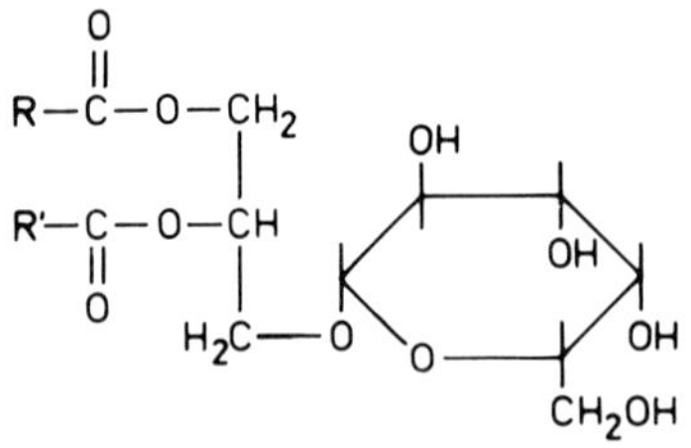

Monogalactosyldiglyceride (MGDG)

(1,2-diacyl-[β-D-galactopyranosyl-(1'→3)]-sn-glycerol)

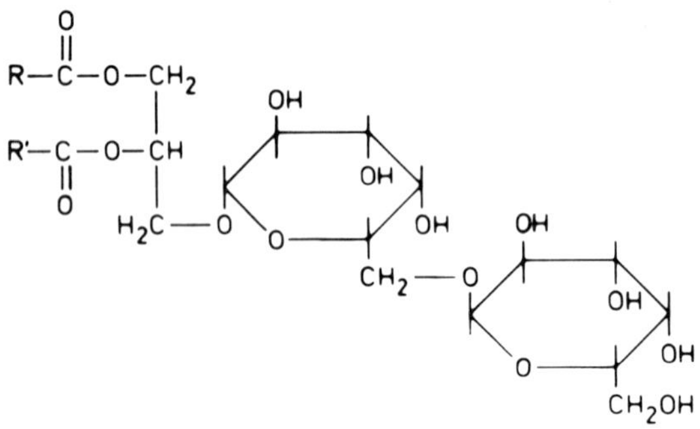

Digalactosyldiglyceride (DGDG)

(1,2-diacyl-[α-D-galactopyranosyl-(1''→6')-β-D-galactopyranosyl-(1'→3)]-sn-glycerol)

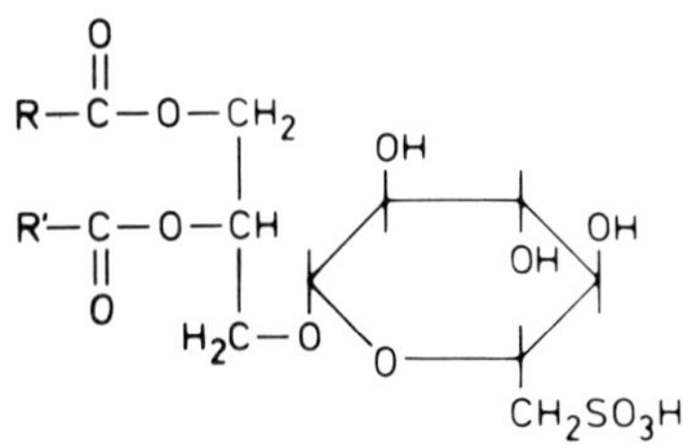

Sulphoquinovosyldiglyceride (Sulpholipid, SL)

(1,2-diacyl-[6-sulpho-α-D-quinovopyranosyl-(1'→3)]-sn-glycerol)

Fig.3: Polar lipids (B: glycolipids)

be produced by most, if not all photosynthesizing marine and freshwater algae as well as by terrestrial plants. They are shown in fig.3.

As in the phospholipids, the carbon skeleton of the glycolipids is made up by glycerol esterified to two fatty acids. The third hydroxy group is attached to one or two molecules of galactose leading to the formal formation of monogalactosyl diglyceride (MGDG) and digalactosyl diglyceride (DGDG), respectively. In the third glycolipid (sulpholipid, SL), the sugar is 6-sulpho-D-quinovose (D-quinovose = 6-desoxy-D-glucose).

c) Novel glycolipids and chlorosulpholipids: Recently, in brown algae, Pham Quang and Laur (61-63) detected a number of complex acyl lipids containing glucose, galactose, glycosyl sulphates, and glycosyl phosphates. These lipids will be discussed later (see Phaeophyceae).

Another novel class of lipids are chlorosulpholipids. They are chloro derivatives of N-docosane-1,14-diol and of N-tetracosane-1,15-diol disulphates and have been found in freshwater algae of the Chrysophyceae (64-66), Xanthophyceae (67, 68), Chlorophyceae (68), and Cyanophyceae (68).

For further information about topics concerning marine and freshwater algae, the reader is referred to the excellent book on "Algal Physiology and Biochemistry" by Stewart (69).

3. Occurrence of fatty acids and lipids in marine algae

Abbreviations: The following abbreviations will be used in the next chapters: MDGD (monogalactosyl diglyceride), DGDG (digalactosyl diglyceride), SL (sulphoquinovosyl diglyceride), PG (phosphatidyl glycerol), PC (phosphatidylcholine), PS (phosphatidylserine), PA (phosphatidic acid).

The algae were classified botanically according to Fott (70). Prior to regarding the data compiled in the following chapters, some general remarks seem to be necessary. As will be shown later, the formation of fatty acids and lipids in algae (and probably in terrestrial plants, too) is controlled by environmental factors. Primarily, these factors are the growth conditions (light, temperature etc.), and the composition of the nutrient media (especially the nitrogen content). There are probably other factors, too, such as the growthphase and the age of the algae.

As yet, environmental factors have only been studied in some microscopic freshwater algae and in few marine algae. This is a field of increasing actuality especially with regard to the influence of nitrogen-rich waste waters on marine algae growing in polluted areas close to the shore (see chapter 6).

So, for correct comparisons, one should investigate algae which have been grown in defined media under controlled growth conditions. With most marine algae, this is of course impossible, and one has to rely on field analyses (temperature, pH, composition of seawater etc.). The technique of "mariculture" (culture of marine algae) may be a prospective method for the future. So far, however, it has only been carried out with some carrageenan producing red algae (71).

Despite the above reservations, the data in the following chapters will show that there is considerable correspondence with regard to the fatty acids and lipids within each class of algae.

Chrysophyceae

Fatty acids (see table 1): The fatty acids of two of the Chrysophyceae listed in table 1 (Monochrysis lutheri and Syracosphaera carterae) have also been analyzed by Williams (72) with similar results. In general, the Chrysophyceae are characterized by a great variety of fatty acids. They produce saturated and unsaturated C_{16}, C_{18} C_{20}, and C_{22} fatty acids, the concentrations of which vary widely among the species. No fatty acid seems to prevail throughout all of the algae.

Lipids: There are no reports on lipids in marine Chrysophyceae. One freshwater alga of this class, Ochromonas danica,

Table 1: Fatty acids of marine Chrysophyceae (phytoplankton)

Fatty acids (% of total fatty acids)	Positions of double bonds	Monochrysis lutheri (74/73)	Syracosphaera carterae (73/72)	Pseudopedinella sp. (74)	Prymnesium parvum (74)	Isochrysis galbana (74)	Dicrateria inornata (74)	Coccolithus huxleyi (74)	Cricosphaera carterae (74)	Cricosphaera elongata (74)
12:0		0.6/ 0.2	0.1/ 0.5	0.6	-	0.1	0.4	0.4	0.3	0.2
14:0		9.2/11.4	0.6/ 1.4	4.9	5.6	0.6	1.3	6.4	8.6	6.6
14:1		- / 0.2	0.2/ tr	-	-	-	-	-	-	-
15:0		0.6/ 0.2	0.1/ -	0.4	-	0.3	0.3	0.4	0.2	1.3
16:0		10.1/15.1	21.8/49.1	7.9	15.7	15.6	15.9	16.9	8.8	9.5
16:1	all isomers	20.2/25.4	3.0/ 9.9	23.6	10.1	5.1	7.9	27.8	21.4	20.8
16:2	all isomers	7.1/ 1.9	3.8/ 6.1	7.4	2.3	2.9	1.7	3.8	9.8	7.6
16:3	all isomers	14.8/ 0.6	0.8/ 0.3	16.7	0.9	2.1	2.1	7.1	14.8	13.8
16:4	all isomers	1.5/ 0.3	0.6/ -	1.3	0.5	0.2	0.4	1.2	0.4	1.1
17:0		0.4/ -	0.1/ -	0.4	-	0.3	0.4	-	0.1	-
18:0		0.4/ -	0.7/ 3.3	-	-	-	0.1	0.7	0.4	-
18:1	9	5.7/ 0.8	7.3/ 7.9	2.6	24.9	14.6	16.6	10.0	3.3	1.8
18:2	9,12	1.6/ 0.7	9.5/ 8.8	1.9	17.8	10.7	5.0	2.1	2.8	2.3
18:3	6,9,12	- / 0.1	- / -	-	0.4	0.8	0.4	-	-	0.2
18:3	9,12,15	- / 0.2	8.0/11.6	-	11.2	14.3	12.8	1.3	-	0.7
18:4	6,9,12,15	0.6/ 4.0	24.1/ -	-	2.0	17.2	20.3	1.1	1.7	0.9
20:1	all isomers	- / 0.6	0.5/ -	0.5	-	-	-	0.2	-	0.3
20:3	8,11,14	1.7/ 0.1	- / -	0.7	1.2	0.5	1.4	0.4	-	0.2
20:4	all isomers	0.5/ 0.4	0.1/ -	1.3	1.4	0.4	0.6	0.5	3.2	2.0
20:5	5,8,11,14,17	18.9/16.3	3.9/ -	26.5	4.2	3.1	7.8	17.1	19.8	27.9
22:5	4,7,10,13,16	0.5/ 1.2	1.1/ -	-	0.2	-	-	0.1	-	-
22:5	7,10,13,16,19	3.3/ 0.4	0.1/ -	1.1	0.7	0.3	3.4	0.1	1.8	0.3
22:6	4,7,10,13,16,19	- /13.1	8.6/ -	-	-	-	-	-	-	-

Table 2:Fatty acids of marine Xanthophyceae (phytoplankton)

Fatty acids (% of total fatty acids)	Positions of double bonds	Olisthodiscus sp. (73/74)	Undescribed flagellate (74)	Monodus subterraneus (freshwater)(75)
12:0		0.5/ 0.6	0.4	-
14:0		7.4/ 9.1	6.4	2
14:1	5	0.4/ -	-	-
15:0		0.4/ 0.3	0.2	-
16:0		21.8/ 6.0	12.3	24
16:1	all isomers	19.9/ 0.4	18.1	24
16:2	6,9	0.2/ 0.5	0.6	-
16:2	9,12	2.0/ 0.5	3.4	-
16:3	all isomers	0.3/ 2.5	12.3	-
16:4	all isomers	0.2/ 0.9	1.2	-
18:0		0.2/ -	0.2	1
18:1	all isomers	1.1/ 7.3	0.6	9
18:2	9,12	0.9/ 6.8	0.8	4
18:3	6,9,12	0.3/ 0.4	-	tr
18:3	9,12,15	6.8/ 6.3	0.2	tr
18:4	6,9,12,15	7.9/29.3	0.2	-
20:3	all isomers	1.7/ -	-	1
20:4	all isomers	2.2/ 2.1	5.8	5
20:5	5,8,11,14,17	21.8/ 1.8	31.4	29
22:5	7,10,13,16,19	0.1/13.2	2.5	-
22:6	4,7,10,13,16,19	3.0/ -	-	-

has been analyzed (75). The lipids of this organism were composed of MGDG, DGDG, SL, PG, PC, PE, PI, cardiolipin, and triglycerides. Furthermore, a number of chlorosulpholipids (chloro derivatives of docosane-1,14-diol) has been detected in this alga (64-66). In several species of the Chrysophyceae (by some authors classified as "Haptophyceae") a highly unsaturated straight chain hydrocarbon (3,6,9,12,15,18-21:6) has been detected (43,76). In Syracosphaera carterae the main hydrocarbon was a C_{17} n-alkane (40).

Xanthophyceae

There are only few reports on fatty acids and lipids of algae from this class (see table 2). Two marine species (Olisthodiscus sp. and an unidentified flagellate) primarily synthesized the 16:0, 16:1, and 20:5 fatty acids (73,74). In Monodus subterraneus (freshwater alga) the fatty acid composition was quite similar (75). The lipids of Monodus s. were MGDG, DGDG, SL, PG, PE, PI, and triglycerides. Of these, the two galactolipids had high concentrations of the 20:5 acid.

In three freshwater species of the Xanthophyceae, Tribonema aequale (7), Botrydium granulatum, and Monodus subterraneus (68) chlorosulpholipids have been detected recently. These lipids were docosane-1,14-diol and several chlorinated derivatives. The hydrocarbons of Tribonema aequale mainly contained short chain alkanes and alkenes (14:0, 15:0, 15:1, 15:2) (76).

Bacillariophyceae (diatoms)

Of the Bacillariophyceae listed in table 3, three species (Cyclotella cryptica, Phaeodactylum tricornutum, and Skeletonema costatum) have been investigated by Kates (35), Chuecas(74), and coworkers,too. Diatoms play an important role in the marine food chain. They serve as a primary source of food for zooplankton copepods (42,54,55) but also for fish: According to De Silva and Wijeyaratne (77), the diet of young grey mullet (Mugil cephalus) consisted of 18 genera of dia-

Table 3: Fatty acids of marine Bacillariophyceae (diatoms, phytoplankton)

Fatty acids (% of total fatty acids)	Positions of double bonds	Nitzschia closterium (35)	Nitzschia angularis (35)	Nitzschia thermalis (35)	Cyclotella cryptica (73)	Phaedactylum tricornutum (73)	Skeletonema costatum (73)	Ditylum brightwelli (74)	Bidulphia sinensis (74)	Chaetoceros septentrionale (74)	Lauderia borealis (74)	Asterionella japonica (74)	Thalassiosira fluviatilis (73)	Navicula incerta (37) *)
12:0		-	-	-	0.1	0.1	0.2	0.9	-	0.6	-	0.5	0.1	-
14:0		5.5	5.3	2.2	7.2	7.8	32.7	7.8	11.4	6.3	7.3	6.3	7.9	5.5
14:1	5	0.4	0.5	0.3	0.1	0.2	1.0	-	-	-	-	-	0.1	tr
15:0		0.4	1.3	0.5	1.3	0.2	1.4	0.7	0.4	0.5	0.5	0.4	1.0	-
16:0		25.4	23.6	30.4	27.8	26.6	6.9	13.4	14.1	11.0	11.7	11.3	23.2	30.1
16:1	all isomers	24.9	30.5	45.5	42.7	46.9	22.4	33.0	27.6	21.2	21.3	21.7	44.8	38.3
16:2	all isomers	1.8	2.6	2.3	2.6	2.6	7.4	2.7	5.4	2.9	2.9	2.8	2.8	4.4
16:3	all isomers	3.7	3.6	2.4	4.1	2.9	2.3	9.7	4.0	8.7	12.1	9.4	6.5	2.7
16:4	all isomers	-	-	-	0.2	0.3	6.6	0.4	9.4	1.2	0.9	1.2	0.1	-
18:0		0.6	1.7	0.8	0.3	0.5	0.1	0.2	0.7	0.1	-	0.2	0.3	tr
18:1	all isomers	2.6	3.2	1.2	0.5	1.1	0.3	6.3	2.5	3.4	1.8	2.9	0.4	2.7
18:2	9,12	2.5	2.2	0.8	0.3	1.0	1.1	2.1	1.1	0.5	1.1	0.8	0.4	tr
18:3	6,9,12+9,12,15	-	1.6	1.3	0.1	0.6	0.6	1.0	tr	0.5	0.3	0.2	0.2	tr
18:4	6,9,12,15	-	-	-	2.6	0.1	2.2	0.2	2.8	0.1	-	0.4	0.2	-
20:4	5,8,11,14	9.4*)	-	3.7*)	-	0.2	-	-	0.4	0.5	-	6.5	0.5	-
20:4	8,11,14,17				0.1	-	-	1.4	-	1.2	1.3	4.3	0.1	-
20:5	5,8,11,14,17	17.2	21.1	6.9	7.5	8.6	13.8	10.9	15.1	20.8	30.3	19.8	8.0	16.4
22:5	all isomers	-	-	-	0.1	-	0.4	1.2	1.5	3.7	1.1	3.2	0.4	-
22:6	4,7,10,13,16,19	-	-	-	0.9	0.8	1.7	-	-	-	-	-	2.2	-

*) Positions of double bonds not reported

toms, together with green and blue-green algae and other food groups.

Fatty acids (see table 3): The major fatty acids are the 16:0, 16:1, and 20:5, which account for more than 50 % of the total fatty acids in most of the diatoms. A low content of C_{18} fatty acids (especially of 18:3) seems to be characteristic for this class. In the non-photosynthetic diatom, Nitzschia alba, the main fatty acids were 14:0, 16:0, 18:1, and 20:5 (78).

Lipids: The lipids of several diatoms have been analyzed by Kates (35) and Opute (37). The main storage lipids were triglycerides. Furthermore, MGDG, DGDG, SL, PG, PC, and PI have been found. The galactolipids contained high proportions of the 16:3 and 20:5 fatty acids. Recently, the total lipids of Nitzschia alba (marine, non-photosynthetic) were analyzed quantitatively by Kates, Volcani, and coworkers (78,79). They reported the presence of four sulpholipids (phosphatidylsulphocholine, lysophosphatidylsulphocholine, deoxyceramide sulphonic acid, sterol sulphate).

In many diatoms, a considerable portion of the total lipids consists of hydrocarbons (40-43). Among, these, a unique polyunsaturated hydrocarbon (3,6,9,12,15,18-21:6) often prevails (42,43,76).

There are data on some freshwater diatoms, too: Navicula pelliculosa (35), Navicula muralis and Nitzschia palea (37) appear to synthesize less 16:0 and more 16:3 than the marine species. No major differences, however, were observed in the lipid compositions.

Dinophyceae

Fatty acids (see table 4): The largest proportion of the fatty acids in most of the Dinophyceae is made up of one saturated (16:0) and three polyunsaturated fatty acids (18:4, 20:0, 22:6), which is a rather unique composition. Moreover, a very unusual octadecapentaenoic acid (3,6,9,12,15-18:5) was discovered recently in natural and laboratory-cultured Dino-

Table 4: Fatty acids of marine Dinophyceae (phytoplankton)

Fatty acids (% of total fatty acids)	Positions of double bonds	Amphidinium carteri (36/73)	Exuviella sp. (36)	Glenodinium sp. (36)	Gonyaulax catanella (36)	Gonyaulax polyedra (83)	Gonyaulax tamarensis (36)	Gymnodium nelsoni (36)	Peridinium triquetum (36)	Peridinium trochoideum (36/74)	Prorocentrum micans (74)	Cryptecodinium cohnii (heterotrophic) (82)
12:0		-/ 0.2	-	-	-	-	-	-	-	-/ -	0.5	8
14:0		2/ 3.3	3	6	13	2	11	6	7	14/17	13.6	19
14:1		-/ -	-	1	-	-	1	-	-	-/ -	-	<1
16:0		12/36	26	23	30	36	28	35	26	27/36	9.8	20
16:1	all isomers	1/ 0.5	2	4	3	1	4	5	5	8/ 4	22.7	1
16:2	all isomers	2/ 0.2	1	-	-	-	2	-	-	2/ 1	-	-
16:3	all isomers	-/ 0.1	-	-	-	-	-	-	-	-/ 3	11.6	-
18:0		2/ 5.7	10	3	4	-	3	1	2	4/ -	-	1
18:1	all isomers	2/ 7.5	5	5	7	3	5	16	2	9/ 7	6.4	14
18:2	9,12	1/ 0.1	9	5	3	2	4	1	9	3/ -	-	-
18:3	6,9,12	-/ -	-	-	-	-	-	-	-	-/ -	2.4	-
18:3	9,12,15	3/ 0.1	2	6	6	3	5	1	3	2/ 3	6.0	-
18:4	6,9,12,15	19/10,1	8	23	14	14	12	4	13	9/ 7	-	-
20:0		1/ 2.2	1	-	-	-	-	-	-	-/ -	-	-
20:2	8,11	-/ -	-	-	-	-	-	-	-	-/ 2	-	-
20:4	8,11,14,17	-/ -	-	-	-	-	-	-	-		-	-
20:5	5,8,11,14,17	20/ 7,4	4	2	1	14	3	5	2	1/13	12.4	-
22:5	7,10,13,16,19	-/ 0.6	-	1	-	-	1	-	2	1/ 3	-	<1
22:6	4,7,10,13,16,19	24/25.4	12	19	12	23	18	20	19	15/ -	-	30

flagellates (80-82). The major fatty acids of another species, Crypthecodinium cohnii (heterotrophic) were 18:1 and 22:6 (84).

<u>Lipids</u>: Information about lipids in Dinophyceae is very scarce. Harrington and coworkers (36) analyzed the lipids of two organisms: Glenodinium sp. (photoauxotroph) had MGDG, DGDG, PC, and triglycerides. The lipids of Gyrodinium cohnii (heterotroph) consisted primarily of triglycerides and PC.

The hydrocarbons of numerous species have been investigated by Lee and Loeblich (43) and Blumer et al. (76). A 3,6,9,12, 15,18-21:6 was the major hydrocarbon in the photosynthetic Dinophyceae.

<u>Phaeophyceae</u>

<u>Fatty acids (see tables 5 and 6)</u>:
The fatty acids of four brown algae from the Mediterranean (Dictyopteris polypodioides, Styptocaulon scoparium, Taonia atomaria, Undaria pinnatifida) which are not listed in tables 5 and 6 have been investigated by Pohl et al. with similar results (85). The Phaeophyceae generally have higher levels of both unsaturated C_{18} (18:3,18:4) and C_{20} (20:4, 20:5) fatty acids. The proportion of unsaturated C_{16} fatty acids usually is very low.

In some of the brown algae (especially Ascophyllum nodosum), Jamieson and Reid (86) found a 5,11,14-20:3 which is of an unusual structure. Similar Δ 5-fatty acids have also been found in lower (88) and higher plants (89).

<u>Lipids</u>: There are only few reports with a detailed analysis of the total lipids. Jamieson and Reid (86) only examined MGDG, DGDG, and "polar lipids". According to Laur and Pham Quang (90-93) and to our own investigations (38), the main lipids of several species of Fucus are MGDG, DGDG, SL, PG, PE, and triglycerides with smaller amounts of cardiolipin, PI, PS, and free fatty acids. PC was also detected (61,92). Radunz (94,132) isolated MGDG, DGDG, and SL from Fucus vesiculosus, too.

Table 5: A) Fatty acids of Phaeophyceae (marine, macroscopic)

Fatty acids (% of total fatty acids)	Positions of double bonds	Ectocarpus confervoides (86)	Leathesia difformis (86)	Chordaria flagelliformis (86)	Desmarestia aculeata (86)	Dictyosiphon foeniculaceus (86)	Punctaria plantaginea (86)	Chorda filum (86)	Chorda tomentosa (38)	Laminaria saccharina (86)	Laminaria digitata (86)	Alaria esculenta (86)
12:0		tr	0.2	0.2	0.1	0.1	tr	tr	tr	tr	tr	tr
14:0		2.4	5.4	5.8	4.1	4.5	3.6	3.5	3.3	4.0	4.0	4.8
16:0		11.0	23.2	17.8	12.3	12.4	16.2	27.6	25.1	10.2	11.9	10.1
16:1	all isomers	3.0	3.0	1.4	2.1	4.7	1.2	1.4	1.7	3.0	1.9	2.4
16:2	all isomers	0.2	0.3	0.2	0.4	1.1	0.1	tr	tr	0.1	tr	0.1
16:3	all isomers	0.3	0.5	0.1	0.3	0.2	0.1	0.1	tr	0.1	0.1	0.1
16:4	all isomers	0.4	tr	0.2	0.2	0.7	0.1	0.1	tr	0.1	tr	0.1
18:0		0.4	1.3	2.1	0.4	1.2	1.1	1.2	0.5	0.3	0.4	0.5
18:1	9	5.9	10.8	15.2	6.9	7.2	13.5	17.8	14.4	8.1	12.7	6.5
18:2	9,12	5.4	6.8	11.9	6.3	7.8	6.9	13.7	10.5	3.9	2.8	4.2
18:3	6,9,12	1.2	0.4	1.2	1.7	1.6	1.0	0.6	1.9	0.6	0.5	0.9
18:3	9,12,15	10.1	8.9	7.4	10.2	11.2	7.3	6.9	7.1	9.9	6.7	10.6
18:4	6,9,12,15	26.0	8.9	12.0	16.4	14.5	14.2	6.3	7.5	20.3	19.3	27.9
20:0+20:1		0.1	0.3	0.2	0.2	0.5	1.1	0.7	tr	0.1	0.4	0.7
20:2	11,14	0.2	0.3	0.2	0.1	0.1	0.1	0.1	tr	0.2	0.1	tr
20:3	11,14,17	0.4	0.4	0.6	0.4	0.9	0.6	0.7	0.5	0.2	0.4	0.4
20:4	5,8,11,14	6.7	9.1	12.4	19.1	14.5	8.1	8.5	10.3	9.2	12.6	11.9
20:4	8,11,14,17	0.7	0.4	0.6	0.5	0.7	0.8	0.5	1.1	0.7	1.1	0.4
20:5	5,8,11,14,17	24.2	19.3	10.4	18.6	14.2	22.8	9.8	14.2	28.4	24.8	18.4
$C_{22}+C_{24}$ *)		1.2	0.5	0.1	0.3	1.6	1.1	0.1	0.8	0.5	0.2	tr

*) 22:0 + 22:1 + 22:4 + 22:5 + 22:6 + 24:0 + 24:1

Table 6: B) Fatty acids of Phaeophyceae (marine, macroscopic)

Fatty acids (% of total fatty acids)	Positions of double bonds	Fucus spiralis (86/38)	Fucus vesiculosus (86/38)	Fucus vesiculosus var. evesiculosus (86)	Fucus serratus (86/38)	Fucus platycarpus (87)	Ascophyllum nodosum (86)	Pelvetia canaliculata (86)	Halydris siliquosa (86)
12:0		0.1/ tr	0.1/ -	0.1	0.1/ -	-	0.1	0.1	0.1
14:0		9.1/10.0	8.8/ 7.7	10.0	11.7/ 6.5	11.6	7.1	6.8	7.8
16:0		9.4/12.1	11.2/11.1	12.4	15.6/13.0	24.5	11.0	9.3	14.6
16:1	all isomers	1.3/ 0.9	1.2/ 0.8	2.2	1.7/ 1.3	2.0	2.3	1.4	1.1
16:2	all isomers	0.3/ 0.3	0.3/ 0.3	0.1	0.5/ 0.4	0.2	0.5	0.3	0.2
16:3	all isomers	0.4/ tr	0.3/ 0.2	0.3	0.1/ -	0.4	0.4	0.3	0.2
16:4	all isomers	0.4/ tr	0.7/ tr	0.5	Tr / -	0.3	0.3	0.4	0.2
18:0		1.1/ 1.1	0.2/ 1.2	0.3	0.4/ 0.6	0.8	0.4	2.2	0.5
18:1	9	38.2/30.2	17.1/30.3	12.4	27.4/36.3	16.2	34.4	38.8	12.0
18:2	9,12	8.1/ 8.6	9.1/10.0	8.0	9.6/13.1	7.6	7.9	10.1	6.1
18:3	6,9,12	0.7/ 2.5	0.5/ 1.8	0.3	0.6/ 2.3	-	0.3	1.1	0.6
18:3	9,12,15	5.0/ 5.9	7.8/ 6.9	10.2	6.1/ 3.9	7.1	4.0	3.2	9.9
18:4	6,9,12,15	2.2/ 4.8	7.8/ 4.7	11.0	4.3/ 4.1	7.0	4.3	1.3	12.6
20:0+20:1		0.5/ 0.6	0.9/ 1.1	0.4	0.9/ 0.3	tr	0.5	0.4	0.3
20:2	11,14	0.8/ tr	0.7/ tr	0.5	0.1/ tr	1.5	2.5	0.6	0.2
20:3	11,14,17	1.3/ 1.8	1.2/ 1.0	0.6	1.2/ 1.2	1.1	5.8	2.3	0.5
20:4	5,8,11,14	13.0/11.6	13.3/12.1	17.0	12.4/10.0	10.9	9.6	13.6	22.8
20:4	8,11,14,17	0.2/ 0.7	0.5/ 0.5	0.4	0.5/ 0.4	-	0.3	0.1	1.1
20:5	5,8,11,14,17	6.9/ 7.4	15.7/ 8.5	19.5	5.8/ 5.4	7.8	5.2	5.2	9.0
$C_{22}+C_{24}$ *)		0.9/ 1.2	2.5/ 1.5	2.6	0.9/ 0.8	tr	2.9	2.4	0.2

*) 22:0 + 22:1 + 22:4 + 22:5 + 22:6 + 24:0 + 24:1

MGDG seems to be the most unsaturated lipid having the highest proportion of 18:4, and 20:5 fatty acids. The proportion of these compounds in DGDG is somewhat lower (86,94). The other lipids (especially SL and triglycerides) are much less unsaturated (mostly 16:0 and 18:1).

Recently, Pham Quang and Laur(61-63) detected a series of novel glycolipids in Fucus vesiculosus, Fucus serratus, and Pelvetia canaliculata, containing glucose, galactose, glycosyl sulphates, and glycosyl phosphates. The predominant (semideveloped) lipid structures are presented below:

18:3 - O - CH_2
|
CH - O - 18:3
|
CH_2- O - Glucose (3 H_2O)

Some novel glycolipids in brown algae (61 - 63)

18:3 - CH_2
|
CH - O - 18:3
|
CH_2- O - Glucose - Galactose(3H_2O)

16:0 - O - CH_2
|
CH - O - 18:3
|
CH_2- O - 6S, Glucose (5H_2O)

The authors also reported the occurrence of four new sulpholipids (sulphate esters of C_{18}, C_{20}, and C_{30} mono - and dialcohols) in the same three Phaeophyceae (62,63).

Brown algae synthesize minor quantities of hydrocarbons (0.001-0.08 % of the dry weight) with C_{13}-C_{29} carbon chains (39,40). The main hydrocarbon is a C_{15} alkane which in some brown algae accounts for more than 90 % of the total hydrocarbons. In addition, there are smaller proportions of unsaturated uneven-numbered hydrocarbons with 15-21 carbon atoms and 1-6 double bonds (39).

Rhodophyceae

One red alga already represented in table 7 (Laurencia pinnatifida) and Rhodymenia palmata have also been analyzed by Chuecas (74).

Fatty acids (see tables 7 and 8): The red algae have one rather common feature, the formation of high amounts of polyunsaturated C_{20} fatty acids (20:4 and 20:5). Mostly 20:5 is much more abundant, in some cases it can be the contrary (see Phycodrys sinuosa). The unsaturated C_{18} fatty acids are present in lower quantities, and the polyunsaturated C_{16} fatty acids (16:2, 16:3, 16:4) are not present or in very small proportions.

Lipids: Our knowledge about the lipids in red algae is still poor. There is one report about MGDG, DGDG, and "polar lipids" in three Rhodophyceae (86). A detailed analysis of the total lipids of Phycodrys sinuosa (38) showed that the main lipids in this alga were MGDG, DGDG, SL, PG, PC, triglycerides, and free fatty acids. In the Rhodophyceae, the fatty acids with the highest degree of unsaturation (18:4, 20:5) tend to accumulate in the MGDG; DGDG is slightly less unsaturated (20:4, 20:5) (38,86). The unpolar (neutral) lipids of Phycodrys sinuosa (triglycerides, free fatty acids) were characterized by the 16:0 and 20:4 fatty acids (38).

The fatty acids and lipids of the unicellular marine red alga, Porphyridium cruentum (73,75,85), and of the freshwater red alga, Batrachospermum moniliforme (94,132) resemble those of marine macroscopic species.

The hydrocarbons of several red algae (0.003-0.05% of the dry weight) mainly consist of saturated C_{17} alkanes (96-99% of the total hydrocarbons), together with trace amounts of alkenes (39,40,76). In Porphyra leucosticta, a 19:5 alkene accounted for 62% of the total hydrocarbons (39).

Table 7: A) Fatty acids of Rhodophyceae (marine, macroscopic)

Fatty acids (% of total fatty acids)	Positions of double bonds	Gracilaria confervoides (85)	Sebdenia monardiana (85)	Laurencia obtusa (85/86)	Laurencia pinnatifida (86)	Rhodomela subfusca (87/86)	Ceramium rubrum (86)	Plocamium coccineum (87)	Dumontia crispata (86)	Schizymenia dubyi (86)
12:0		1.9	0.9	tr / 0.1	0.3	- / 0.1	0.1	0.1	0.1	0.1
14:0		8.1	6.0	6.7/ 1.5	4.3	3.9/ 2.3	2.9	10.4	2.2	1.4
16:0		18.2	26.2	54.4/18.0	20.9	29.1/17.8	23.3	26.6	17.0	22.0
16:1	all isomers	3.4	1.3	4.0/ 7.0	4.0	4.9/ 8.5	5.7	6.7	7.0	2.9
16:2	all isomers	tr	tr	- / 0.8	0.3	0.2/ 1.6	0.2	0.1	0.5	tr
16:3	all isomers	-	-	- / 1.1	1.0	0.2/ 0.9	0.4	0.1	0.4	0.3
16:4	all isomers	-	-	- / 1.8	0.9	tr / 1.0	0.4	0.2	0.6	0.4
18:0		1.0	1.8	3.7/ 0.2	3.7	0.3/ 1.3	1.2	0.9	0.3	0.5
18:1	9	16.1	5.3	11.7/ 4.7	7.9	14.9/11.1	11.1	5.6	4.4	4.6
18:2	9,12	1.7	22.1	4.3/ 2.0	4.9	1.4/ 4.3	4.2	3.0	1.9	1.9
18:3	6,9,12	-	-	- / 0.5	0.5	- / 0.3	0.5	-	0.5	0.6
18:3	9,12,15	0.8	-	0.9/ 2.7	3.3	1.1/ 1.7	2.5	0.4	3.3	2.6
18:4	6,9,12,15	0.5	-	- / 3.0	3.5	0.8/ 2.0	2.0	0.9	6.5	3.0
20:0+20:1		1.3	tr	tr / 0.8	0.3	- / 0.6	0.3	0.2	0.2	1.3
20:2	11,14	-	-	- / 0.4	0.4	- / 0.4	0.3	-	0.1	0.3
20:3	8,11,14	-	-	- / 0.4	0.5	0.5/ 0.5	1.0	6.6	0.3	0.3
20:4	5,8,11,14	45.5	28.1	5.3/ 9.0	6.3	14.0/ 9.1	9.1	11.5	2.5	7.5
20:4	8,11,14,17	-	-	- / 0.6	0.4	- / 0.1	1.0	-	0.4	0.4
20:5	5,8,11,14,17	0.4	7.6	8.9/42.5	35.4	24.1/35.8	32.4	21.6	51.3	49.8
$C_{22}+C_{24}$ *)		0.5	-	- / 2.8	0.9	0.9/ 0.4	1.0	3.3	0.5	0.1

*) 22:0 + 22:1 + 22:4 + 22:5 + 22:6 + 24:0 + 24:1

Table 8: B) Fatty acids of Rhodophyceae (marine, macroscopic)

Fatty acids (% of total fatty acids)	Positions of double bonds	Corallina officinalis (86)	Cryptoleura ramosum (86)	Odonthalia dentata (86)	Chondrus crispus (86)	Chyloclydia kaliformis (86)	Phycodrys sinuosa (38)	Delessaria sanguinea (38)	Cystoclonium purpurescens (38)	Ahnfeltia plicata (38)	Gigartina stellata (86)
12:0		0.1	0.2	0.1	0.2	0.3	0.8	1.1	0.5	0.3	0.1
14:0		1.7	7.1	3.0	2.1	4.3	4.9	3.4	4.3	1.4	3.4
16:0		23.5	19.9	26.6	17.6	22.3	21.5	35.2	30.1	18.3	20.0
16:1	all isomers	3.3	2.6	2.3	4.6	3.8	5.4	5.2	7.4	3.2	3.7
16:2	all isomers	0.8	0.1	0.2	2.4	0.6	1.5	4.2	tr	tr	0.2
16:3	all isomers	1.0	tr	0.2	1.0	0.8	0.5	tr	1.0	tr	0.4
16:4	all isomers	0.8	1.0	0.1	3.3	0.3	tr	4.3	tr	0.4	0.5
18:0		1.1	0.2	1.1	1.0	0.3	2.9	1.7	0.6	2.6	2.3
18:1	9	4.6	4.0	11.3	11.2	6.0	5.0	6.6	5.6	5.7	12.5
18:2	9,12	2.2	1.9	1.0	3.5	0.8	0.6	0.9	1.8	1.7	2.9
18:3	6,9,12	0.5	0.3	0.7	0.5	0.3	0.3	0.7	3.6	0.7	0.7
18:3	9,12,15	0.6	1.5	0.3	5.8	0.2	0.5	4.0	1.6	0.9	2.9
18:4	6,9,12,15	0.8	2.5	0.3	5.6	0.7	0.5	3.7	1.7	0.5	2.7
20:0+20:1		0.4	0.2	0.1	0.4	0.2	0.4	tr	tr	0.6	0.4
20:2	11,14	0.2	0.1	0.2	0.4	0.3	-	0.3	tr	0.1	0.2
20:3	8,11,14	0.5	0.8	1.0	0.7	0.2	0.2	0.7	0.4	4.6	0.5
20:4	5,8,11,14	5.5	3.8	13.2	12.5	5.7	43.9	15.5	14.4	22.0	17.7
20:4	8,11,14,17	0.4	0.4	0.1	0.4	tr	6.0	0.6	0.3	0.5	0.3
20:5	5,8,11,14,17	51.6	50.9	37.8	26.4	49.1	2.1	8.9	24.1	11.9	27.6
$C_{22}+C_{24}$ *)		0.5	2.4	0.4	0.4	3.6	tr	1.9	0.8	2.4	0.8

*) 22:0 + 22:1 + 22:4 + 22:5 + 22:6 + 24:0 + 24:1

Marine macroscopic Chlorophyceae

Many macroscopic marine green algae have as yet been analyzed for their fatty acids. There are, however, only few reports about the lipids.

Fatty acids (see table 9): In comparison to the Rhodophyceae and Phaeophyceae, the Chlorophyceae primarily synthesize C_{16} and C_{18} fatty acids, with larger proportions of the 16:3, 16:4, and 18:4. C_{20} and C_{22} fatty acid are usually much less abundant (mainly 20:4, 20:5, sometimes 22:4-22:6). In total, this class is characterized by a wide variety of saturated and unsaturated fatty acids.

Lipids: Jamieson and Reid (86) separated the total lipids of 5 marine algae into MGDG, DGDG, and "polar lipids". MGDG was the lipid with the highest concentration of polyunsaturated C_{16} and C_{18} fatty acids.

A detailed study of Enteromorpha linza (Baltic Sea) (38) showed that the lipids were composed of MGDG, DGDG, SL, PG, PC, PE, PI, PS, and unpolar lipids. The latter comprised triglycerides with small proportions of free fatty acids, hydrocarbons, wax esters, mono- and diglycerides.

Several marine green algae have been analyzed for their hydrocarbons (39,40,43). Youngblood et al. (39) reported the occurrence of a 17:0 alkane in Codium fragile (89 % of the total hydrocarbons), of a 17:1 alkene in Enteromorpha compressa (96 %) and Ulva lactuca (88 %), and of polyunsaturated alkenes (21:5 (22 %); 21:6 (60 %)) in Spongomorpha arcta. Low proportions of a C_{17} alkylcyclopropane were detected in Enteromorpha compressa and Ulva lactuca. In these four green algae, the contents of hydrocarbons ranged from 0.024 to 0.48 % of the dry weight.

Marine microscopic Chlorophyceae and Prasinophyceae

Some of the species mentioned in table 10 have formerly been regarded as Chlorophyceae and are now classified with a new class, the Prasinophyceae (95-97).

Table 9: Fatty acids of marine macroscopic Chlorophyceae

Fatty acids (% of total fatty acids)	Positions of double bonds	Ulva lactuca (86)	Codium fragile (87)	Enteromorpha compressa (86/87)	Enteromorpha intestinalis (86)	Enteromorpha linza (38)	Cladophora rupestris (86) *)	Cladophora albida (86)	Halimeda Tuna (85)	Valonia utricularia(85)	Codium elongatum (85)	Ulva fasciata (85)
12:0		0.2	0.3	0.1/ -	0.1	0.1	0.1	0.1	0.6	tr	2.4	1.3
14:0		0.4	0.8	1.0/ 0.6	0.8	0.9	4.9	6.4	10.1	3.4	4.4	0.8
16:0		17.9	28.1	26.6/21.6	12.5	21.3	16.3	18.2	18.3	37.3	10.9	19.2
16:1	all isomers	2.1	1.6	2.3/ 2.2	2.6	3.3	14.1	16.3	6.4	3.4	10.2	3.9
16:2	all isomers	0.3	0.9	1.3/ 1.1	0.8	0.3	3.2	1.0	1.7	1.2	3.0	1.3
16:3	all isomers	1.3	12.2	3.0/ 2.2	2.9	3.3	2.1	0.5	3.0	4.9	1.6	-
16:4	all isomers	17.8	-	7.4/14.9	15.7	14.8	16.0	7.2	-	-	-	3.0
18:0		0.1	0.9	0.4/ -	0.7	0.8	0.3	1.2	2.0	3.2	2.0	0.5
18:1	9	9.3	10.8	14.6/ 7.9	9.4	1.7	9.4	9.1	26.3	18.4	18.3	20.7
18:2	9,12	1.6	5.5	11.7/ 4.7	5.6	4.5	15.1	3.1	12.9	7.5	9.7	7.5
18:3	6,9,12	0.4	-	1.0/ -	0.6	0.8	0.4	0.6	-	-	-	-
18:3	9,12,15	16.7	27.2	14.2/25.8	20.8	21.4	2.3	14.0	5.2	9.8	16.4	13.5
18:4	6,9,12,15	24.2	1.5	6.4/ 8.6	16.9	12.5	2.8	4.0	1.9	tr	4.9	21.2
20:0+20:1		0.1	0.5	0.5/ -	0.5	tr	0.3	0.5	0.8	tr	6.2	tr
20:2	11,14	0.2	-	0.2/ -	0.2	-	0.3	0.1	-	-	-	-
20:3	8,11,14	0.2	-	1.0/ 2.1	0.4	tr	0.2	0.2	2.0	-	-	-
20:4	5,8,11,14	0.5	3.0	2.0/ 3.6	1.4	0.4	2.5	1.8	2.4	5.8	4.8	1.4
20:4	8,11,14,17	0.7		0.8/	1.0	0.5	0.5	0.2	-	-	-	-
20:5	5,8,11,14,17	2.1	1.9	2.0/ 2.2	2.3	1.0	6.3	12.0	3.6	1.5	3.6	0.4
22:0+22:1		0.6	3.2	0.7/ 1.7	0.7	0.8	0.2	0.3	0.5	tr	0.7	0.9
22:4	7,10,13,16	0.5	-	0.5/ -	0.8	tr	0.1	0.2	-	-	-	-
22:5	7,10,13,16,19	2.2	-	1.0/ -	2.6	2.8	1.3	1.9	-	-	-	3.4
22:6	4,7,10,13,16,19	0.4	-	0.1/ -	tr	-	0.7	0.3	1.5	-	-	-

*) Mean values

Fatty acids of marine microscopic Chlorophyceae and Prasinophyceae (phytoplankton)

Table 10:		Chlorophyceae				Prasinophyceae			
Fatty acids (% of total fatty acids	Positions of double bonds	Dunaliella primolecta (74)	Dunaliella tertiolecta (73)	Dunaliella salina (72)	Chlamydomonas sp. (74)	Heteromastix rotunda (74)	Tetraselmis (Platymonas) sp. (73)	Platymonas tetrathele (85)	Halosphaera viridis (97)
12:0		0.8	0.9/ 0.2	0.2	0.8/ 0.9	0.2	0.1	0.4	0.12
14:0		4.7	5.9/ 0.5	0.4	1.3/ 4.5	9.2	0.8	2.0	4.95
15:0		1.7	0.7/ 1.3	-	0.3/ 1.3	0.5	0.1	-	0.02
16:0		11.3	12.9/17.6	41.3	16.8/15.1	10.6	19.7	11.2	17.63
16:1	all isomers	9.8	9.5/ 3.4	14.8	6.8/ 2.4	15.7	2.9	4.0	1.25
16:2	all isomers	8.3	3.0/ 2.2	0.4	0.8/ 6.6	2.9	0.5	2.4	0.13
16:3	all isomers	7.4	4.9/ 4.3	-	0.3/ 6.6	3.1	1.8	3.5	0.07
16:4	all isomers	5.9	7.1/16.5	-	0.5/ 0.3	2.0	14.7	14.0	3.42
18:0		0.1	- / 0.4	-	0.2/ -	-	0.3	9.2	0.07
18:1	all isomers	5.8	7.7/ 3.1	11.2	7.1/16.7	2.3	14.8	12.9	44.75
18:2	9,12	5.8	5.8/ 5.1	7.8	28.4/10.1	2.8	3.3	12.4	7.38
18:3	6,9,12	2.1	0.8/ 5.8	-	0.6/ 3.1	-	0.6	-	0.91
18:3	9,12,15	10.4	7.7/37.4	18.5	0.2/28.4	3.7	18.9	16.0	7.28
18:4	6,9,12,15	6.7	8.3/ 1.2	-	18.6/ 0.8	9.3	7.5	7.6	6.55
20:0+20:1		0.3	- / 0.1	-	15.8/ 1.0	-	1.6	5.0	0.09
20:2	8,11	0.8	1.0/ -	-	- / 0.5	-	-	-	-
20:3	8,11,14	1.3	1.9/ 0.1	-	- / -	-	-	0.1	-
20:4	5,8,11,14	0.4	0.4/ -	-	- / -	0.6	1.3	2.0	- ?
20:4	8,11,14,17	1.7	3.5/ -	-	- / -	0.2	-	-	0.39
20:5	5,8,11,14,17	9.7	10.2/ -	-	- / -	27.9	7.5	3.8	tr
22:5	all isomers	3.9	6.2/ -	-	0.3/ -	6.8	-	-	0.43
22:6	4,7,10,13,16,19	-	- / -	-	- / -	-	1.4?	-	3.83

At present, there are only reports on the fatty acids (see table 10). As to these compounds, the marine microscopic resemble the marine macroscopic green algae. They contain many saturated and unsaturated fatty acids of all chain lengths and all degrees of unsaturation. Relatively high proportions of unsaturated C_{16} and C_{18} fatty acids are usual. There is, however, no fatty acid which predominates in all of the marine microscopic Chlorophyceae.

Freshwater Chlorophyceae

The fatty acids and lipids of freshwater Chlorophyceae are not the topic of this paper. Some general characteristics, however, should be mentioned. The freshwater green algae synthesize the same major lipids which are known from the marine species. There is, however, a remarkable difference with regard to the unsaturated fatty acids: Almost all of the freshwater Chlorophyceae are incapable of synthesizing the 18:4 and the polyunsaturated C_{20} and C_{22} fatty acids which are commonly found in the seawater species. The most unsaturated fatty acid of the freshwater Chlorophyceae normally is α-linolenic acid (9,12,15-18:3). Herein they resemble the terrestrial plants (31). The following freshwater green algae have so far been analyzed: Chlorella vulgaris (85,98,99), Chlorella pyrenoidosa (87,100,101), Chlorella fusca (99), Scenedesmus obliquus (85,87), Scenedesmus quadricauda (101), Chlamydomonas reinhardi (102,1 3), Halochlorella rubescens (99), Bractecoccus minor (99), Tetraedron sp. (101), Coelastrum microsporum (101), Botrycoccus braunii (104), Ulothrix aequalis (105), several filamentous green algae (106), colourless Chlorophyceae (107).
Several papers (43,45-48) deal with hydrocarbons in freshwater green algae.

Euglenophyceae

There are two species of this class which have so far been analyzed, Euglena gracilis and Astasia longa. Both of them are freshwater organisms. Euglena gracilis has frequently been used for studies in the field of lipids because it can be grown heterotrophically in the dark and photoauxotrophically in the light. In the dark, it makes large amounts of wax esters and phospholipids (PC,PE). Upon illumination,MGDG DGDG, SL, PG (chloroplast lipids) are the major lipids synthesized (50-53, 102,108-111). The fatty acids are composed of saturated and unsaturated C_{16}, C_{18}, C_{20}, and C_{22} fatty acids (111-113). Astasia longa produces the 14:0, 16:0, 16:1, 18:0,18:1, 18:2, and large amounts of C_{20}, C_{22}, and C_{24} polyunsaturated fatty acids (102).

Cryptophyceae

Fatty acids (see table 11): The marine species of the Cryptophyceae contain a wide range of saturated and unsaturated C_{16}, C_{18}, C_{20}, and C_{22} fatty acids with 5-6 major compounds. These are the 16:0, 16:1, 18:3, 18:4, 20:1, and 20:5 fatty acids. Perhaps the most striking feature is the high content of 20:1 which is unique among all other algae.

Lipids: There is only one rep ort on lipids in Cryptophyceae by Beach et al.(114): The lipids of Cryptomonas sp. were composed of MGDG, DGDG, SL, PG, and PC. No PE and PI were detected. MGDG contained high proportions of 18:4 (75%).

Three freshwater Cryptophyceae have also been analyzed:Cryptomonas ovata (114), Cryptomonas ovata var.palustris (115), and Chilomonas paramecium (non-photosynthetic)(114,116).The former exhibited a fatty acid composition similar to that of the marine species. Chilomonas paramecium contained no 18:4, but more 9,12,15-18:3 (27%).
The hydrocarbons of two Cryptomonas species (43,76) and of Rhodomonas lens (43) were rich in the 21:6 alkene.

Cyanophyceae

These organisms will not be treated here since there are only reports on freshwater algae. The fatty acid compositions are rather simple (mostly 16:0, 16:1, 18:0, 18:1, 18:2, 18:3; some contain 18:4). The main lipids are MGDG, DGDG, SL, and

Table 11: Fatty acids of marine Cryptophyceae (phytoplankton)

Fatty acids (% of total fatty acids)	Positions of double bonds	*Hemiselmis rufescens* (74)	*Hemiselmis brunescens* (74)	*Hemiselmis virescens* (114)	*Cryptomonas appendiculata* (74)	*Cryptomonas maculata* (74)	*Cryptomonas* sp. (114)	*Chroomonas* sp. (114)	*Rhodomonas lens* (114)
12:0		0.5	0.6	-	-	0.8	-	-	-
13:0		0.3	0.9	-	-	-	-	-	-
14:0		0.9	1.2	8	1.7	4.5	6	2	18
15:0		0.5	0.6	-	0.3	1.0	-	-	-
16:0		21.3	12.5	23	16.1	14.7	4	16	13
16:1	all isomers	10.0	3.4	7	11.8	6.5	2	2	6
16:2	all isomers	2.9	3.1	-	1.4	2.7			-
16:3	6,9,12	0.3	0.2	-	1.5	1.1	2	1	-
16:4	6,9,12,15	1.0	0.1	-	0.4	-			-
17:0		0.7	0.4	-	0.5	0.3	-	-	-
18:0		-	0.1	-	0.4	0.3	-	1	1
18:1	all isomers	2.3	2.3	7	2.5	3.5	5	5	10
18:2	9,12	1.3	0.1	3	4.2	-	-	3	2
18:3	6,9,12	2.0	0.2	-	-	-	-	1	-
18:3	9,12,15	6.8	8.4	22	12.0	6.3	7	23	16
18:4	6,9,12,15	17.4	30.5	16	13.3	15.6	44	23	13
20:0		1.8	0.2	-	0.9	-	-	-	-
20:1	all isomers	14.4	18.0	-	10.3	16.7	-	-	-
20:4	5,8,11,14	-	0.2	-	2.7	0.7	-	-	-
20:4	8,11,14,17	2.6	0.3	-	5.8	1.0	-	-	-
20:5	5,8,11,14,17	8.2	13.9	7	9.8	17.2	16	14	13
22:4	4,10,13,16	0.9	-	2	-	-	-	1	1
22:5	7,10,13,16,19	3.4	1.3	-	2.5	3.4	-	-	-
22:6	4,7,10,13,16,19	-	-	2	-	-	10	6	5

PG. Phospholipids seem to be present in very small amounts or absent. Many Cyanophyceae make hydrocarbons. The reader is referred to reports and reviews on fatty acids and lipids in these algae (69, 117-125).

4. Plant fatty acids and evolution

During the course of evolution, the algae have developed from the Bacteriophyta via intermediate stages (phytoflagellates). The Bacteriophyta have a very simple fatty acid composition, namely saturated and mono-unsaturated short chain fatty acids (mostly 16:0, 16:1, 18:0, 18:1). As was shown above, the marine algae synthesize polyenoic C_{16} - C_{22} fatty acids and indeed have the greatest variety of saturated and cis-unsaturated fatty acids among all groups of the plant kingdom. So, during the course of evolution from the Bacteriophyta to the marine algae, a considerable development of fatty acid biosynthesis took place.

From the marine (green) algae the terrestrial plants developed. These, however, have a strikingly different fatty acid composition: With only few exceptions, they lack the ability to synthesize unsaturated C_{20} and C_{22} and 18:4 fatty acids (31-33). The terrestrial plants produce fatty acids with relatively short carbon chains and a lower number of double bonds as compared to the marine algae, namely the 12:0, 14:0, 16:0, 16:1, 16:2, 16:3, 18:0, 18:1, 18:2, and 18:3 fatty acids. Only few terrestrial plants are able to synthesize the 18:4 or unsaturated C_{20} and C_{22} fatty acids (31-33).

The evolutionary transition from the marine green algae to the terrestrial plants probably proceeded in two stages: During the first stage, the Chlorophyceae adapted themselves from seawater to freshwater (via brackish water; for example in the mouths of rivers). In the second stage, the freshwater Chlorophyceae developed into the terrestrial plants. With regard to the fatty acids, the freshwater Chlorophyceae resemble the terrestrial plants: They almost entirely lack the ability to synthesize unsaturated C_{20} and C_{22} and 18:4 fatty acids.

It is not known whether this ability was at first acquired by the marine Chlorophyceae and then lost during the evolutionary transition to the terrestrial plants or whether the terrestrial plants developed from marine Chlorophyceae which did not yet possess this ability.

5. Control of fatty acid and lipid biosynthesis by environmental factors

From the different fatty acid compositions in marine algae and in terrestrial plants , one is tempted to presume that environmental (ecological) factors have an influence on fatty acid biosynthesis in plants. Indeed, several factors are known.

Some important factors seem to be the light and the light intensity. Investigations carried out by various authors (98, 108-111) showed that the formation of polyunsaturated C_{16} and C_{18} fatty acids (mainly 16:2, 16:3, 16:4, 18:2,18:3) as well as of MGDG, DGDG, SL, and PG in Euglena gracilis and Chlorella vulgaris is enhanced upon illumination. The above fatty acids and lipids are typical constituents of the chloroplasts. Fig.4 shows the fatty acid composition of Euglena gracilis grown in the dark and in light.

Another factor is the temperature. Lower temperatures favour the formation of polyunsaturated C_{18} fatty acids (18:2,18:3) in the oils of terrestrial plants (126). Similar results have been obtained with algae: According to Ackman (73), low temperatures (10°) increase the formation of polyunsaturated C_{18} fatty acids in Monochrysis lutheri as compared to 20°. In Cyanidium caldarium (thermophilic alga with unclear classification (127)), the ratio of unsaturated to saturated fatty acids and the formation of glycolipids decreased with increased temperature (20° to 55°)(128,129). Adams et al. (130) however, reported an increase of glycolipids by 90% in 45° as compared to 20° grown cells. Possibly there is a temperature optimum for unsaturated fatty acid

Fatty acids (% of total fatty acids)	Growth in the dark	Subsequent growth in light
12:0	tr	tr
14:0	9,2	3,4
16:0	27,6	7,5
16:1	9,1	4,9
16:2	1,1	25,8
16:3	tr	4,4
16:4	tr	14,4
18:0	1,2	2,4
18:1	24,1	1,9
18:2	1,9	8,5
18:3	1,8	22,5
18:4	3,4	tr
20:0	tr	tr
20:4	9,2	1,3
20:5	1,5	2,8
22:6	0,3	tr

Fig.4:

Fatty acids of Euglena gracilis grown in the dark and subsequently in the light (111)

biosynthesis between 20° and 55° in this alga. According to our own investigations (38), the marine macroscopic green alga, Enteromorpha linza, had a temperature optimum for the biosynthesis of polyunsaturated fatty acids (16:4, 18:3, 18:4, 20:5) and galactolipids at 15°, whereas at 10° and 20° the formation of these compounds was lower.

According to Constantopoulos (131) the biosynthesis of fatty acids and lipids in Euglena gracilis is also controlled by the concentration of manganese in the nutrient medium. At low levels of Mn^{++} (9×10^{-2} µmoles) Euglena gracilis synthesized only small amounts of chlorophylls, of polyunsaturated C_{16} and C_{18} fatty acids, and of galactolipids.

An important factor seems to be the content of nitrogen (NO_3^-, NH_4^+ etc) in the nutrient medium. The biosynthesis of polyunsaturated C_{16} and C_{18} fatty acids in the freshwater green algae, Bracteacoccus minor and Chlorella vulgaris, depends on the levels of KNO_3 (see fig.5) in the medium (99). At low concentrations of N, saturated and mono-unsaturated

fatty acids (% of total fatty acids)	Bracteacoccus minor				Chlorella vulgaris		
	KNO_3	0,0015 %	0,003 %	0,01 %	0,0003 %	0,0015 %	0,01 %
	µmoles N/l	149	297	990	30	149	990
12:0		tr	tr	tr	tr	tr	tr
14:0		0,5	0,4	0,5	1,1	1,0	0,8
16:0		15,0	14,4	15,7	28,8	20,6	13,8
16:1		1,0	4,1	3,6	0,8	1,7	2,9
16:2		4,1	6,8	7,8	0,4	3,4	5,8
16:3		1,2	1,4	0,7	0,9	6,2	8,9
16:4		1,9	6,1	6,3	-	-	-
18:0		1,0	2,0	2,6	4,8	1,0	tr
18:1		44,4	13,5	8,7	30,0	8,6	1,1
18:2		18,5	22,2	20,9	14,2	17,9	14,9
18:3		12,1	29,1	33,2	19,0	39,6	51,6

Fig.5: Fatty acid compositions of Bracteacoccus minor and Chlorella vulgaris grown at varying concentrations of KNO_3

fatty acids (16:0,18:1) prevailed, whereas at high nitrogen concentrations the 16:2, 16:3, 16:4, 18:2, and 18:3 fatty acids were synthesized in larger proportions. The fatty acids of Euglena gracilis were influenced in a similar way (111). The formation of lipids in Euglena gracilis was controlled by the nitrogen levels in the nutrient medium,too (see fig.6). At low levels of NH_4Cl (0.001%), the alga mainly produced unpolar lipids (wax esters), whereas at high levels (0.01%) the chloroplast lipids (MGDG, DGDG, SL, PG) predominated.

6. Formation of fatty acids and lipids in marine algae in relation to environmental factors

The above influence of nitrogen on fatty acid and lipid biosynthesis in algae may be of considerable importance because nitrogen is one of the main pollutants in agricultural waste waters which contaminate rivers and even oceans as for example the Baltic Sea and the Mediterranean. So there is the question whether environmental factors have an influence not only on fatty acid and lipid biosynthesis in microscopic

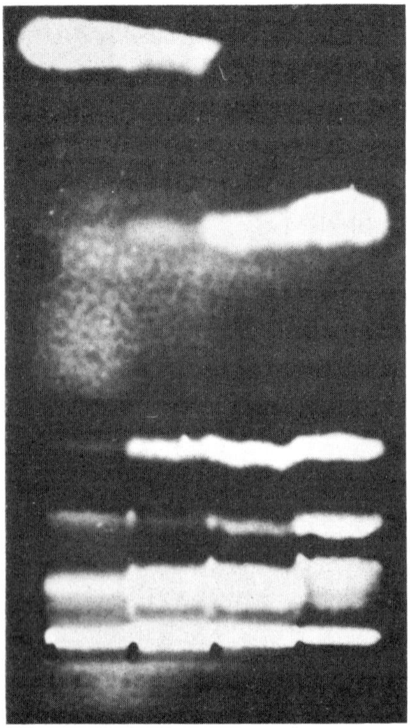

A B C D

Unpolar lipids
(wax esters)

MGDG

DGDG

SL

PE,PG

PC

Fig.6: Thinlayer chromatogram of the total lipids of Euglena gracilis grown in the light at varying concentrations of NH_4Cl in the nutrient medium (111)

A: 0.001 % NH_4Cl

B: 0.002 % "

C: 0.005 % "

D: 0.01 % "

freshwater algae but in marine macroscopic algae as well. An answer to this question would provide some arguments to the discussion whether and to what degree the metabolism of marine algae is influenced by pollution.

For this purpose the formation of fatty acids and lipids was studied in marine red, brown, and green algae from the Baltic Sea (38). The results obtained with the red benthic alga, Phycodrys sinuosa (macroscopic alga) are summarized in fig. 7a - 7d.

Phycodrys sinuosa was collected from a location (Strander Grasberg) in the inner bay of Kiel. The alga had two main growth periods: From the middle of February to the end of of June, and from the middle of September to October. The analysis of the seawater at this location (fig.7b) showed that

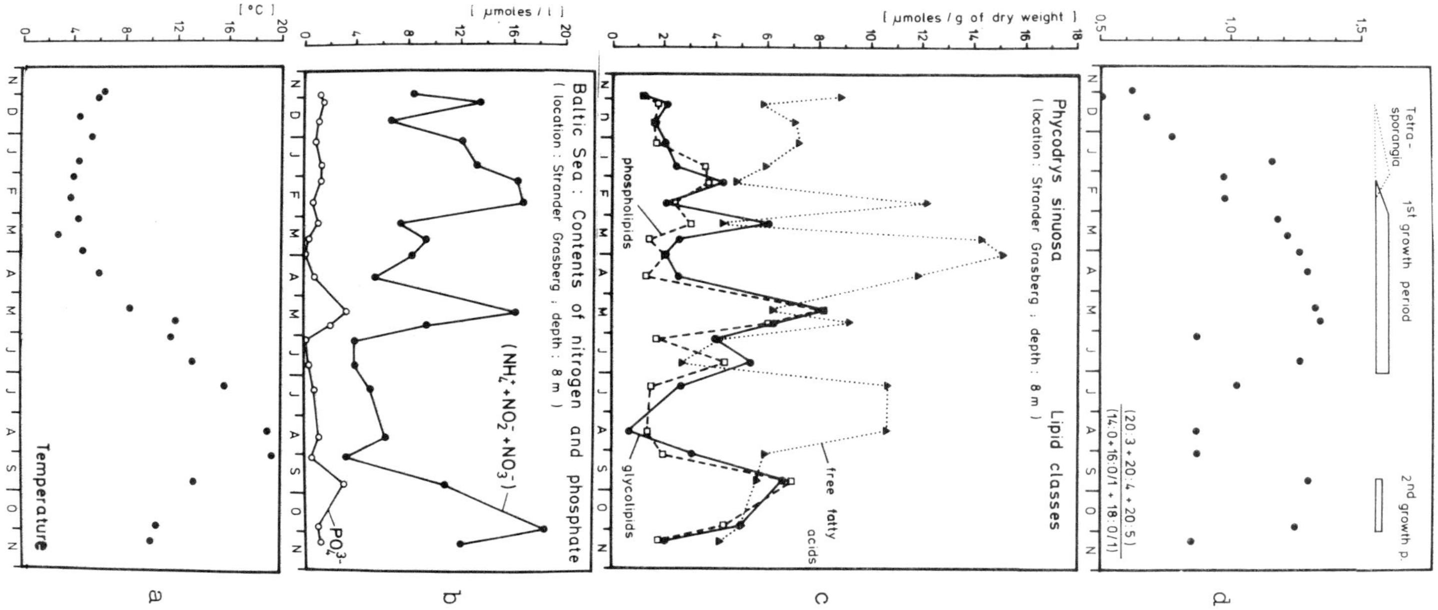

Fig.7: Phycodrys sinuosa (Rhodophyceae, Baltic Sea) (38)

a: Temperature of the seawater
b: Contents of total nitrogen ($NH_4^+ + NO_3^- + NO_2^-$) in the seawater
c: Formation of lipid classes : Unpolar lipids (free fatty acids), phospholipids, and glycolipids
d: Ratio of the major polyunsaturated fatty acids (20:3 + 20:4 + 20:5) to the saturated and mono-unsaturated fatty acids (14:0 + 16:0 + 16:1 + 18:0 + 18:1)

in the cold season (October-February) the total nitrogen content ($NH_4^+ + NO_3^- + NO_2^-$) was high (15-20 μmoles N/l). The nitrogen content then decreased during the warm season (March-August) due to nitrogen consumption by algae. However, fig.7b shows that there was another maximum of nitrogen in May. This was due to pollution by waste waters. Phosphate had a maximum at this time,too.
From the data obtained with Phycodrys sinuosa it is obvious that the formation of fatty acids and lipids varied throughout the year (fig.7c and 7d): When there were high levels of nitrogen and phosphate in the seawater (May and September/ October), the alga mainly produced polar lipids (phospho- and glycolipids). When the contents of nitrogen and phosphate decreased (June-August) the growth of the alga was poor , and the main lipids were unpolar (free fatty acids). Fig.7d shows the ratio of the major unsaturated fatty acids of Phycodrys sinuosa (20:3 + 20:4 + 2o:5) to the saturated and mono-unsaturated fatty acids. This curve had maximum values in April/May and in September/October.

From these data it appears probable that environmental factors have an influence on the biosynthesis of fatty acids and lipids in Phycodrys sinuosa. Our observations with brown and green marine algae of the Baltic Sea were of a similar kind. Among the various factors the nitrogen content seems to be one of the most important ones. With regard to the influx of nitrogen-rich waste waters into the sea it would be desirable to study the influence of this and other factors on the metabolism of algae over longer periods of time.

References

1 Pike,J.E.: Prostaglandins. Science 225,84-92(1971)

2 Friedman,Z., Danon,A., Stahlmann,M.T., Oates,J.A.: Rapid onset of essential fatty acid deficiency in the newborn. Pediatrics 58,640-649(1976)

3 Skolnik,P., Eaglstein,W.H., Ziboh,V.A.: Human essential fatty acid deficiency: Treatment by topical application of linoleic acid. Arch.Dermatol. 113,939-941(1977)

4 Privett,O.S., Phillips,F., Fukazawa,T., Kaltenbach,C.C., Sprecher,H.W.: Studies on the relationship of the synthesis of prostaglandins to the biological activity of essential fatty acids. Biochim.Biophysica Acta 280,348-355(1972)

5 Hartop,P.J., Prottey,C.: Changes in transepidermal water loss and the composition of epidermal lecithin after applications of pure fatty acid triglycerides to the skin of essential fatty acid-deficient rats. Br.J.Dermatol. 95,255-264(1976)

6 Prottey,C.: Investigations of functions of essential fatty acids in the skin. Br.J.Dermatol. 97,29-38(1977)

7 Schlenk,H.: Odd numbered and new essential fatty acids. Fed.Proc.31,1430-1435(1972)

8 Denko,C.W.: Modification of adjuvant inflammation in rats deficient in essential fatty acids. Agents Actions 6,636-641(1976)

9 Connor,W.E.: Pathogenesis and frequency of essential fatty acid deficiency during total parenteral nutrition. Ann. Intern.Med. 83,895-896(1975)

10 O'Neill,J.A., Caldwell,M.D., Meng,H.C.: Essential fatty acid deficiency in surgical patients. Ann.Surg. 185,535-542(1977)

11 Faulkner,W.J., Flint,L.M.,Jr.: Essential fatty acid deficiency associated with total parenteral nutrition. Surg. Gynecol.Obstet.144,665-667(1977)

12 Friedman,Z., Shochat,S.J., Maisels,M.J., Marks,K.H., Lamberth,E.L.,Jr.: Correction of essential fatty acid deficiency in newborn infants by cutaneous application of sunflower-seed oil. Pediatrics 58,650-654(1976)

13 Spritz,N., Mishkel,M.A.: Effects of dietary fats on plasma lipids and lipoproteins: An hypothesis for the lipid-lowering effect of unsaturated fatty acids. J.Clin.Invest. 48,78-86(1969)

14 Grundy,S.M.: Effects of polyunsaturated fats on lipid metabolism in patients with hypertriglyceridemia. J.Clin. Invest. 55,269-282(1975)

15 Ahrens,E.H.,Jr.: The management of hyperlipidemia: Whether, rather than how. Ann.Intern.Med. 85,87-93(1976)

16 Ross,R., Harker,L.: Hyperlipidemia and atherosclerosis. Science 193,1094-1100(1976)

17 Butkus,A., Tan,E., Koletzky,S.: Tissue lipid distribution in genetically obese and spontaneously hypertensive rat. Part.I. Plasma, aortic intima-media and erythrocyte lipids. Artery 2,53-69(1976)

18 Bondjers,G., Bjorkerud,S.: Arterial repair and atherosclerosis after mechanical injury:VI. Cholesterol elimination in vitro from experimental atherosclerotic lesions. Exp.Mol.Pathol. 26,341-349(1977)

19 Gregoriadis,G., Siliprandi,N., Turchetto,E.: Possible implications in the use of exogenous phospholipids. Life Sci. 20,1773-1786(1977)

20 Huang,C.: Studies on phosphatidylcholine vesicles. Formation and physical characteristics. Biochemistry 8,344-352(1969)

21 Papahadjopoulos,D., Mayhew,E., Poste,G., Smith,S.: Incorporation of lipid vesicles by mammalian cells provides a potential method for modifying cell behaviour. Nature 252,163-166(1974)

22 Pagano,R.E., Huang,L., Wey.C.: Interaction of phospholipid vesicles with cultured mammalian cells. Nature 252, 166-167(1974)

23 Gregoriadis,G., Leathwood,P.D., Ryman,B.E.: Enzyme entrapment in liposomes. FEBS Letters 14,95-99(1971)

24 Gregoriadis,G., Ryman,B.E.: Fate of protein-containing liposomes injected into rats. An approach to the treatment of storage diseases. Eur.J.Biochem.24,485-491(1972)

25 Gregoriadis,G.: Drug entrapment in liposomes. FEBS Letters 36,292-296(1973)

26 Gregoriadis,G., Wills,E.J., Swain,C.P., Tavill,A.S.: Drug-carrier potential of liposomes in cancer chemotherapy. The Lancet (June),1313-1316(1974)

27 Poste,G., Papahadjopoulos,D.: Drug-containing lipid vesicles render drug-resistant tumour cells sensitive to actinomycin D. Nature 261,699-701(1976)

28 Abood,L.G., Salem,N.,Jr., MacNeil,M., Bloom,L., Abood, M.E.: Enhancement of opiate binding by various molecular forms of phosphatidylserine and inhibition by other unsaturated lipids. Biochim.Biophysica Acta 468,51-62(1977)

29 Hoss,W., Smiley,C.: Immobilization of opiates by membrane lipids. Molec.Pharmacol.13,251-258(1976)

30 Abood,L.G., Hoss,W.: Stereospecific morphine adsorption to phosphatidyl serine and other membranous components of brain. Eur.J.Pharmacol. 32,66-75(1975)

31 Pohl,P., Wagner,H.: Fettsäuren im Pflanzen- und Tierreich (eine Übersicht)I: Gesättigte und cis-ungesättigte Fettsäuren. Fette, Seifen, Anstrichmittel 74,424-435 (1972)

32 Pohl,P., Wagner,H.: Fettsäuren im Pflanzen- und Tierreich (eine Übersicht) II: Trans-ungesättigte, Alkin-, Hydroxy-, Epoxy-, Oxo-, Cyclopropan- und Cyclopropen-Fettsäuren. Fette, Seifen, Anstrichmittel 74,541-550 (1972)

33 Hitchcock,C.: Structure and distribution of plant acyl lipids. in: Recent Advances in the Chemistry and Biochemistry of Plant Lipids, ed.Galliard,T., Mercer,E.I., Academic Press, London, New York, San Francisco, 1975, (pp.1-19)

34 Ackman,R.G.: Structural homogeneity in unsaturated fatty acids of marine lipids. A review. J.Fish.Res.Bd.Canada 21,247-253(1964)

35 Kates,M., Volcani,B.E.: Lipid components of diatoms. Biochim.Biophysica Acta 116,264-278(1966)

36 Harrington,G.W., Beach,D.H., Dunham,J.E., Holz,G.G.,Jr.: The polyunsaturated fatty acids of marine dinoflagellates. J.Protozool. 17,213-219(1970)

37 Opute,F.I.: Lipid and fatty acid composition of diatoms. J.Exp.Bot. 25,823-835(1974)

38 Zurheide,F., Pohl,P.: Fatty acids and lipids of marine algae and the control of their biosynthesis by environmental factors. Publications in preparation.

39 Youngblood,W.W., Blumer,M., Guillard,R.L., Fiore,F.: Saturated and unsaturated hydrocarbons in marine benthic algae. Marine Biol. 8,190-201(1971)

40 Clark,R.C.,Jr., Blumer,M.: Distribution of n-paraffins in marine organisms and sediment. Limnol.Oceanogr.12, 79-87(1967)

41 Lee,R.F., Nevenzel,J.C., Paffenhöfer,G.-A., Benson,A.A., Patton,S., Kavanagh,T.E.: A unique hydrocarbon from a diatom (Skeletonema costatum). Biochim.Biophysica Acta 202,386-388(1970)

42 Blumer,M., Mullin,M.M., Guillard,R.R.L.: A polyunsaturated hydrocarbon (3,6,9,12,15,18-heneicosahexaene) in the marine food web. Marine Biol. 6,226-235(1970)

43 Lee,R.F., Loeblich III,A.R.: Distribution of 21:6 hydrocarbon and its relationship to 22:6 fatty acid in algae. Phytochemistry 10,593-602(1971)

44 Oro,J., Tornabene,T.G., Nooner,D.W., Gelpi,E.: Aliphatic hydrocarbons and fatty acids of some marine and freshwater microorganisms. J.Bacteriol. 93,1811-1818(1967)

45 Han,C., McCarthy,E.D., Van Hoeven,W., Calvin,M., Bradley, W.H.: Organic geochemical studies,II. A preliminary report on the distribution of aliphatic hydrocarbons in algae, in bacteria, and in a recent lake sediment. Proc.Natl.Ac.Sci.US 59,29-33(1968)

46 Maxwell,J.R., Douglas,A.G., Eglinton,G., McCormick,A.: The botryococcenes - hydrocarbons of novel structure from the alga Botryococcus braunii Kützing. Phytochemistry 7,2157-2171(1968)

47 Gelpi,E., Schneider,H., Mann,J., Oro,J.: Hydrocarbons of geochemical significance in microscopic algae. Phytochemistry 9,603-612(1970)

48 Knights,B.A., Brown,A.C., Conway,E., Middleditch,B.S.: Hydrocarbons from the green form of the freshwater alga Botryococcus braunii. Phytochemistry 9,1317-1324(1970)

49 Antia,N.J., Lee,R.F., Nevenzel,J.C., Cheng,J.Y.: Wax ester production by the marine cryptomonad Chroomonas salina grown photoheterotrophically on glycerol. J.Protozool. 21,768-771(1974)

50 Rosenberg,A.: A comparison of lipid patterns in photosynthesizing and nonphotosynthesizing cells of Euglena gracilis. Biochemistry 2,1148-1154(1963)

51 Guehler,P.F., Peterson,L., Tsuchija,H.M., Dodson,R.M.: Microbiological transformations. XIII. Composition of the wax formed by Euglena gracilis. Arch.Biochem.Biophys. 106,294-298(1964)

52 Rosenberg,A.: Euglena gracilis: A novel lipid energy reserve and arachidonic acid enrichment during fasting. Science 157,1189-1191(1967)

53 Kolattukudy,P.E.: Reduction of fatty acids to alcohols by cell-free preparations of Euglena gracilis. Biochemistry 9,1095-1102(1970)

54 Lee,R.F., Nevenzel,J.C., Paffenhöfer,G.-A., Benson,A.A.: The metabolism of wax esters and other lipids by the marine copepod, Calanus helgolandicus. J.Lipid Res. 11, 237-240(1970)

55 Lee,R.F., Nevenzel,J.C., Paffenhöfer,G.-A.: Importance of wax esters and other lipids in the marine food chain: phytoplankton and copepods. Marine Biol. 9,99-108(1971)

56 Lee,R.F., Nevenzel,J.C., Paffenhöfer,G.-A.: Wax esters in marine copepods. Science 167,1508-1510(1970)

57 Nevenzel,J.C.: Occurrence, function and biosynthesis of wax esters in marine organisms. Lipids 5,308-319(1970)

58 Lee,R.F., Nevenzel,J.C., Paffenhöfer,G.-A.: The presence of wax esters in marine planktonic copepods. Naturwiss. 59,406-411(1972)

59 Cowey,C.B., Sargent,J.R.: Lipid nutrition in fish. Comp. Biochem.Physiol. 57 B,269-273(1977)

60 Kayama,M., Nevenzel,J.C.: Wax ester biosynthesis by midwater marine animals. Marine Biol. 24,279-285(1974)

61 Pham Quang L., Laur,M.H.: Teneur, composition et répartition cytologique des lipides polaires soufrés et phosphorés de Pelvetia canaliculata (L.) Decn. et Thur., Fucus vesiculosus (L.) et Fucus serratus (L.). Phycologia 15,367-375(1976)

62 Pham Quang,L., Laur,M.H.: Structures, teneurs et compositions des esters sulfuriques, sulfoniques, phosphoriques des glycosyldiglycérides de trois fucacées. Biochimie 58,1367-1380(1976)

63 Pham Quang,L., Laur,M.H.: Les alcohols aliphatiques sulfates: nouveaux lipides polaires isolés de diverses fucacées. Biochimie 58,1381-1396(1976)

64 Haines,T.H., Pousada,M., Stern,B., Mayers,G.L.: Microbial sulpholipids: (R)-13-Chloro-1-(R)-14-docosanediol disulphate and polychlorosulpholipids in Ochromonas danica. Biochem.J. 113,565-566(1969)

65 Mayers,G.L., Pousada,M., Haines,T.H.: Microbial sulpholipids .III. The disulfate of (+)-1,14-docosanediol in Ochromonas danica. Biochemistry 8,2981-2986(1969)

66 Elovson,J., Vagelos,P.R.: A new class of lipids: Chlorosulpholipids. Proc.Natl.Ac.Sci.US 62,957-963(1969)

67 Mercer,E.I., Davies,C.L.: Chlorosulpholipids of Tribonema aequale. Phytochemistry 13,1607-1610(1974)

68 Mercer,E.I., Davies,C.L.: Chlorosulpholipids in algae. Phytochemistry 14,1545-1548(1975)

69 Stewart,W.D.P.: Algal Physiology and Biochemistry, Botanical Monographs Volume 1o, Blackwell Scientific Publications, Oxford, London, Edinburgh, Melbourne, 1974

70 Fott,B.: Algenkunde. Gustav-Fischer-Verlag, Stuttgart, 1971

71 Dawes,C.J., LaClaire,J.W., Moon,R.E.: Culture studies on Eucheuma nudum J.Agardh, a carrageenan producing red alga from Florida. Aquaculture 7,1-9(1976)

72 Williams,P.M.: Fatty acids derived from lipids of marine origin. J.Fish.Res.Bd.Canada 22,1107-1120(1965)

73 Ackman,R.G., Tocher,C.S.: Marine phytoplankter fatty acids. J.Fish.Res.Bd.Canada 25,1603-1620(1968)

74 Chuecas,L., Riley,J.P.: Component fatty acids of the total lipids of some marine phytoplankton. J.mar.biol.Ass. UK 49,97-116(1969)

75 Nichols,B.W., Appleby,R.S.: The distribution and biosynthesis of arachidonic acid in algae. Phytochemistry 8, 1907-1915(1969)

76 Blumer,M., Guillard,R.L., Chase,T.: Hydrocarbons of marine phytoplankton. Marine Biol. 8,183-189(1971)

77 De Silva,S.S., Wijeyaratne,M.J.S.: Studies on the biology of young grey mullet, mugil cephalus L.II. Food and feeding. Aquaculture 12,157-167(1977)

78 Anderson,R., Livermore,B.P., Kates,M., Volcani,B.E.: The lipid composition of the non-photosynthetic diatom,Nitzschia alba. Biochim.Biophysica Acta 528,77-88(1978)

79 Anderson,R., Kates,M., Volcani,B.E.: Identification of the sulpholipids in the non-photosynthetic diatom Nitzschia alba. Biochim.Biophysica Acta 528,89-106(1978)

80 Joseph, J.D.: Identification of 3,6,9,12,15-octadecapentaenoic acid in laboratory-cultured photosynthetic dinoflagellates. Lipids 10,395-403(1975)

81 Mayzaud,P., Eaton,C.A., Ackman,R.G.: The occurrence and distribution of octadecapentaenoic acid in a natural plankton population. A possible food chain index. Lipids 11,858-862(1976)

82 Harrington,G.W., Holz,G.G.,Jr.: The monoenoic and docosahexaenoic fatty acids of a heterotrophic dinoflagellate Biochim.Biophysica Acta 164,137-139(1968)

83 Patton,S., Fuller,G., Loeblich III,A.R., Benson,A.A.: Fatty acids of the "red tide" organism, Gonyaulax polyedra. Biochim.Biophysica Acta 116,577-579(1966)

84 Beach,D.H., Harrington,G.W., Gellerman,J.L., Schlenk,H., Holz,G.G.,Jr.: Biosynthesis of oleic acid and docosahexaenoic acid by a heterotrophic marine dinoflagellate, Crypthecodinium cohnii. Biochim.Biophysica Acta 369, 16-24(1974)

85 Pohl,P., Wagner,H., Passig,T.: Inhaltsstoffe von Algen - II. Über die unterschiedliche Fettsäurezusammensetzung von Salz- und Süßwasseralgen. Phytochemistry 7,1565-1572 (1968)

86 Jamieson,G.R., Reid,E.H.: The component fatty acids of some marine algal lipids. Phytochemistry 11,1423-1432 (1972)

87 Klenk,E., Knipprath,W., Eberhagen,D., Koof,H.P.: Über die ungesättigten Fettsäuren der Fettstoffe von Süßwasser- und Meeresalgen. Hoppe-Seyler's Z.Physiol.Chem. 334,44-59(1963)

88 Jamieson,G.R., Reid,E.H.: The leaf lipids of some conifer species. Phytochemistry 11,269-275(1972)

89 Smith,C.R.,Jr., Kleiman,R., Wolff,I.A.: Caltha palustris L. seed oil. A source of four fatty acids with cis-5 unsaturation. Lipids 3,37-42(1967)

90 Laur,M.H.: Les lipides des algues. Actes du 95e Congrès National des Sociétés Savantes, Reims 1970 , Paris,Bibliothèque Nationale (1975)

91 Laur,M.H., Pham Quang,L., Heim,R.: Algologie. Sur les lipides neutres de trois Fucacées des cotes francaises: Fucus serratus L., Fucus vesiculosus L., et Pelvetia canaliculata (l.) Decn.et Thur.: Analyse qualitative et quantitative des différents composants. C.R.Acad.Sc. Paris, 271,1752-1755(1970)

92 Pham Quang,L., Laur,M.H.: Les lipides polaires de: Pelvetia canaliculata (L.) Decn.et Thur., Fucus vesiculosus L. et Fucus serratus L. Biochimie 56,925-935(1974)

93 Pham Quang,L., Laur,M.H.: Variations de la composition et de la teneur en lipides polaires des trois Fucacées communes des cotes bretonnes: Pelvetia canaliculata (L.) Decn.et Thur., Fucus vesiculosus L., Fucus serratus L. au cours de leur cycle annuel. Botanica Marina XVIII, 105-113(1975)

94 Radunz,A.: Die Fettsäuren der Galaktolipide aus der Rotalge Batrachospermum moniliforme und der Braunalge Fucus vesiculosus. Hoppe-Seyler's Z.Physiol.Chem. 349,1091-1094(1968)

95 Parke,M., Manton,I.: Preliminary observations on the fine structure of Prasinocladus marinus. J.mar.biol.Ass.UK 45,525-536(1965)

96 Craigie,J.S., McLachlan,J., Majak,W., Ackman,R.G., Tocher,C.S.: Photosynthesis in algae.II. Green algae with special reference to Dunaliella spp. and Tetraselmis spp. Can.J.Bot. 44,1247-1254(1966)

97 Ackman,R.G., Addison,R.F., Hooper,S.N.: Halosphaera viridis: Fatty acid composition and taxonomical relationship. J.Fish.Res.Bd.Canada 27,251-255(1970)

98 Nichols,B.W.: Light induced changes in the lipids of Chlorella vulgaris. Biochim.Biophysica Acta 106,274-279 (1965)

99 Pohl,P., Passig,T., Wagner,H.: Über den Einfluß von anorganischem Stickstoff-Gehalt in der Nährlösung auf die Fettsäure-Biosynthese in Grünalgen. Phytochemistry 10, 1505-1513(1971)

100 Schlenk,H., Mangold,H.K., Gellerman,J.L., Link,W.E., Morrissette,R.A., Holman,R.T., Hayes,H.: Comparative analytical studies of fatty acids of the alga Chlorella pyrenoidosa. J.Amer.Oil Chemists' Soc. 37,547-552(1960)

101 Schneider,H., Gelpi,E., Bennett,E.O., Oro,J.: Fatty acids of geochemical significance in microscopic algae. Phytochemistry 9,613-617(1970)

102 Erwin,J., Bloch,K.: Polyunsaturated fatty acids in some photosynthetic microorganisms. Biochem.Z. 338,496-511 (1963)

103 Eichenberger,W.: Lipids of Chlamydomonas reinhardi under different growth conditions.Phytochem.15,459-463(1976)

104 Douglas,A.G., Douraghi-Zadeh,K., Eglinton,G.: The fatty acids of the alga Botryococcus braunii. Phytochemistry 8,285-293(1969)

105 Jamieson,G.R., Reid,E.H.: The fatty acid composition of Ulothrix aequalis lipids. Phytochemistry 15,795-796 (1976)

106 Attavian,B.N., Floyd,G.L., Fairbrothers,D.E.: Fatty acids of filamentous green algae. Biochem.Syst.Ecol. 5,65-69(1976)

107 Glasl,H., Pohl,P.: Fettsäuren und Lipide in farblosen Algen. Z.Naturforsch. 29c,399-406(1974)

108 Rosenberg,A., Pecker,M.: Lipid alterations in Euglena gracilis cells during light-induced greening. Biochemistry 3,254-258(1964)

109 Rosenberg,A., Gouaux,J.: Quantitative and compositional changes in monogalactosyl and digalactosyl diglycerides during light-induced formation of chloroplasts in Euglena gracilis. J.Lipid Res. 8,80-83(1967)

110 Constantopoulos,G., Bloch,K.: Effect of light intensity on the lipid composition of Euglena gracilis. J.biol. Chemistry 242,3538-3542(1967)

111 Pohl,P., Wagner,H.: Control of fatty acid and lipid biosynthesis in Euglena gracilis by ammonia, light and DCMU. Z.Naturforsch. 27b,53-61(1972)

112 Korn,E.D.: The fatty acids of Euglena gracilis. J.Lipid Res. 5,352-362(1964)

113 Korn,E.D.: The polyunsaturated 20-carbon and 22-carbon fatty acids of Euglena. Biochem.Biophys.Res.Commun. 14, 1-6(1964)

114 Beach,D.H., Harrington,G.W., Holz,G.G.,Jr.: The polyunsaturated fatty acids of marine and freshwater cryptomonads. J.Protozool. 17,501-510(1970)

115 Collins,R.P., Kalnins,K.: The fatty acids of Cryptomonas ovata var.palustris. Phyton 26,47-50(1969)

116 Erwin,J., Hulanicka,D., Bloch,K.: Comparative aspects of unsaturated fatty acid synthesis. Comp.Biochem.Physiol. 12,191-207(1964)

117 Nichols,B.W., Harris,R.V., James,A.T.: The lipid metabolism of blue-green algae. Biochem.Biophys.Res.Commun. 20,256-262(1965)

118 Han,J., McCarthy,E.D., Calvin,M., Benn,M.H.: Hydrocarbon constituents of the blue-green algae Nostoc muscorum, Anacystis nidulans, Phormidium luridum and Chlorogloea fritschii. J.Chem.Soc.C,2785-2791(1968)

119 Nichols,B.W., Wood,J.B.: The occurrence and biosynthesis of gamma-linolenic acid in a blue-green alga, Spirulina platensis. Lipids 3,46-50(1968)

120 Nichols,B.W.: Fatty acid metabolism in the chloroplast lipids of green and blue-green algae. Lipids 3,354-360 (1968)

121 Holm-Hansen,O.: Ecology, Physiology, and Biochemistry of blue-green algae. Ann.Rev.Microbiol. 22,47-70(1968)

122 Holton,R.W., Blecker,H.H., Stevens,T.S.: Fatty acids in blue-green algae: Possible relation to phylogenetic position. Science 160,545-547(1968)

123 Winters,K., Parker,P.L., Van Baalen,C.: Hydrocarbons of blue-green algae: Geochemical significance. Science 163, 467-468(1969)

124 Kenyon,C.N., Stanier,R.Y.: Possible evolutionary significance of polyunsaturated fatty acids in blue-green algae. Nature 227,1164-1166(1970)

125 Paoletti,C., Pushparaj,B., Florenzano,G., Capella,P., Lercker,G.: Unsaponifiable matter of green and blue-green algal lipids as a factor of biochemical differentiation of their biomasses:I. Total unsaponifiable and hydrocarbon fraction. Lipids 11,258-265(1976)

126 Hilditch,T.P., Williams,P.N.: Chemical constitution of the natural fats. Halsted Press, 1964

127 Allen,C.F., Good,P., Holton,R.W.: Lipid composition of Cyanidium. Plant Physiol. 46,748-751(1970)

128 Kleinschmidt,M.G., McMahon,V.A.: Effect of growth temperature on the lipid composition of Cyanidium caldarium.I. Class separation of lipids. Plant Physiol. 46, 286-289(1970)

129 Kleinschmidt,M.G., McMahon,V.A.: Effect of growth temperature on the lipid composition of Cyanidium caldarium.II. Glycolipid and phospholipid components. Plant Physiol. 46,290-293(1970)

130 Adams,B.L., McMahon,V.A., Seckbach,J.: Fatty acids in the thermophilic alga, Cyanidium caldarium. Biochem. Biophys.Res.Commun. 42,359-365(1971)

131 Constantopoulos,G.: Lipid metabolism of manganese-deficient algae:I. Effect of manganese deficiency on the greening and the lipid composition of Euglena gracilis. Plant Physiol. 45,76-80(1970)

132 Radunz,A.: Über das Sulfochinovosyl-diacylglycerin aus höheren Pflanzen, Algen und Purpur-Bakterien. Hoppe-Seyler's Z.Physiol.Chem. 350,411-417(1969)

ALGAL POLYSACCHARIDES: THEIR POTENTIAL USE TO PREVENT CHRONIC METAL POISONING

Yukio Tanaka
St. Mary's Hospital and McGill University, Montreal, Quebec H3T 1M5, Canada
Jerry F. Stara
The United States Environmental Protection Agency, Cincinnati, Ohio 45237, U.S.A.

Water- and air-pollution have become serious problems in public health in recent years. Although attempts are being made to avert further deterioration of our environment, it is equally important to protect the public from the dangers of already polluted environment and from possible accidental contamination of the environment in the future. This article summarizes the results obtained by our joint effort in the past decade in search of specific in vivo binders of radioactive strontium, cadmium, and lead. The details of our investigation are found in our published and forthcoming reports cited herein.

1. GENERAL PRINCIPLE

Natural, ecological concentration processes by aquatic orgamisms have been known to biologists for many years (1). Marine algae, for example, are known to concentrate certain metals in their body from seawater. Using the Polikarpov terminology (1), the ability of algae, and of other marine organisms in general, to concentrate a certain metal can be expressed in terms of concentration factors K:

$$K = C / C'$$

where C and C' are the concentrations of the metal in the organism and in seawater, respectively. Strontium, for example, is concentrated in brown algae as shown by their high concentration factors (K = 7 - 60), as compared with those of other organisms such as red algae (K = 0.05 - 8.0), crustaceans (K = 0.1 - 8.0), and molluscs (K = 0.4 - 10). Other examples

Marine Algae in Pharmaceutical Science
edited by H. A. Hoppe, T. Levring, Y. Tanaka

include concentration of cesium in green algae (K = ~20,000) and of plutonium in red algae (K = ~1,000) (1).

The high concentration factors imply that these organisms contain substance(s) which can bind certain metals in a highly specific manner through ion-exchange reactions between their body and seawater. If these substances are isolated from the organisms, they can still undergo the same specific ion-exchange reactions in vitro. Furthermore, these reactions may be duplicated in the intestinal tract in order to bind some toxic metals, thus preventing the absorption into the body. These substances must satisfy certain prerequisites to be used in human beings; they must be non-toxic and non-absorbable, and should maintain a high degree of specificity for toxic metals, so that they do not affect the body balance of essential metals.

2. ALGINATES AS BINDERS OF RADIOACTIVE STRONTIUM

As mentioned earlier, strontium is known to be concentrated in brown algae, but NOT in red algae to any significant degree (1). Acidic polysaccharides in brown algae are likely to be responsible for this phenomenon. Between the two types of polysaccharides in brown algae, i.e. alginate (a polyuronate) and fucoidan (a partially sulfated fucan), the former was thought to be the substance which binds strontium specifically in their body. If fucoidan is responsible for it, red algae containing carrageenan (a partially sulfated galactan) should have high concentration factors for strontium as well. Alginate and its derivatives were therefore investigated for their in vivo reactions with radioactive strontium.

Alginates were first isolated by Stanford (2) in the hope that seaweeds would be better utilized than just burning them for the recovery of iodine. His hope was partially fulfilled and alginates found their way into various areas of applications with their unique physical properties. Their chemical properties are now utilized as described in this article.

The chemistry of alginates has drawn the attention of many investigators for many years. Alginic acid was first believed to be a pure 1,4-linked poly-β-D-mannuronic acid, until in 1953 Fischer and Dörfel (3) isolated L-guluronic acid from the hydrolyzates of alginic acid. They also found that alginate preparations obtained from various species of Phaeophyceae contained different amounts of guluronic acid. Their finding prompted further investigations on the homogeneity of alginic acid. McDowell (4), for example, was able to prepare fractions with different proportions of mannuronic and guluronic acid by precipitation with manganous salts. Thus, the original concept on the structure of alginic acid had to be modified and possibilities now existed that alginic acid was either (i) a mixture of polysaccharides, in which some molecules consisted of mannuronic acid and others of guluronic acid or (ii) a pure or impure mixed polysaccharide which consisted of these two acids. The second alternative was favoured since a disaccharide with covalently linked L-gulose and D-mannose was isolated from an acid hydrolyzate of reduced alginic acid (5).

The next major development in the structure of alginic acid came in 1966, when Haug, Larsen, and Smidsrød (6) fractionated partially hydrolyzed alginic acid into "polymannuronic" and "polyguluronic acid,"each of which, contained a small amount of the other uronic acid. From this finding, they concluded that the two different building units, i.e. polymannuronic and polyguluronic acid chains, were held together by connecting chains of mixed uronic acids to form alginic acid.

The ion-exchange properties of alginates have been studied in great detail by several workers. Divalent cations, except magnesium, form an insoluble gel with algiantes and it is possible to form a uniform gel by releasing a cation slowly into an alginate solution, and the gel formed is birefringent, i.e. the polymer chains are oriented during the precipitation process. Thiele and Andersen (7) found that the gel formed had a smaller volume than the original solution, the shrinkage being parallel with the size of birefringence. If other conditions were identical, the shrinkage and birefringence were dependent on the metal ions used and followed in the order of ionotropic series:

$$Pb > Cu > Cd > Ba > Sr \approx Ca > Co \approx Ni > Zn > Mn$$

Alginates undergo ion-exchange reactions (Alg = alginate and M = metal):

$$\underset{\text{gel}}{M(Alg)_2} + \underset{\text{solution}}{2Na^+} \rightleftharpoons \underset{\text{gel}}{2Na(Alg)} + \underset{\text{solution}}{M^{++}}$$

and an equilibrium:

$$(M_{gel})(Na^+_{sol})^2 / (Na_{gel})^2(M^{++}_{sol}) = K$$

is established; the selectivity coefficient K is obtained when the concentrations in the gel are expressed as equivalent fractions and those in solutions as normalities. Even in earlier days when the structure of alginates was believed to be a pure polymannuronate, attempts were made to understand the rections involving metal ions and alginate. However, only more recent studies by Haug (8) took into consideration the fact that alginate samples are different in their chemical composition, i.e. the ratio of mannuronate to guluronate units (M/G ratio). He found that there were great differences in selectivity coefficients depending on the M/G ratios, and there was also a wide difference in the selectivity coefficients of different ions but they generally followed the same order for most of the metals as the ionotropic series of Thiele. The results obtained by Thiele using alginate isolated from Laminaria digitata were comparable to the selectivity coefficients for the same alginate studied by Haug. It was also found that strontium had a high affinity for alginates rich in guluronate and stood well ahead of calcium in the series.

The ion-exchange reactions of alginates described above were studied from a purely chemical point of view. The application of alginates to prevent the intestinal absorption of radioactive strontium using their ion-exchange properties was first reported in 1964 (9), when a marked reduction of bone uptake of radioactivity in rats was demonstrated by addition of sodium alginate to the diet contaminated with radioactive strontium. When a sufficient amount of long-life strontium-90, one of the most hazardous air pollutants at the time, is accumulated in the bone, tumors and other pathological changes develop. If alginate is given with the diet contaminated with the radioisotope, however, an ion-exchange reaction takes place in the intestine, forming an insoluble strontium alginate gel which is eventually excreted with the feces, thus causing no significant absorption of the isotope.

The in vivo action of alginate with radiostrontium may be expressed chemically as follows:

$$
\begin{array}{llllll}
Sr^{++}(\text{diet}) & & & & & \\
\downarrow & & & & & \\
Sr^{++}(\text{intestine}) & + \; 2Na(Alg) & = & 2Na^{+} & + & Sr(Alg)_2 \\
\uparrow\downarrow & \quad\text{soluble} & & & & \text{insoluble} \\
Sr^{++}(\text{blood}) & & & & & \downarrow \\
\uparrow\downarrow & & & & & \\
Sr^{++}(\text{bone}) & & & & & \text{excreted}
\end{array}
$$

The equation is, of course, over-simplified; sodium alginate will be acidified to alginic acid in the stomach and converted to a mixed salt with the cations present in the intestine. Strontium entering the intestine is preferentially bound by alginate because of its high affinity toward algiante. The specificity of this in vivo reaction was demonstrated by the fact that both sodium and calcium salts of an alginic acid showed a comparable degree of strontium-binding capacities; even the chemically closest element, calcium, was replaced by strontium (10).

This finding was quickly confirmed by other workers using different animal species and human volunteers (11). Furthermore, Stara (12) demonstrated in cats that the body-burden of radioactive strontium can be reduced by keeping them on an alginate diet. Here, the radiostrontium already deposited in the bone is slowly released into the blood and re-secreted into the intestine, where it is bound by alginate and can not be re-absorbed into the body. Thus, alginate can be used as an effective preventive as well as therapeutic measure against radiostrontium poisoning.

The chemical composition of alginate, i.e. M/G ratio, differs widely depending on the taxonomy of the species and on such environmental factors as season, temperature, and tide. Since mannuronic and guluronic acid are stereochemically different only in the configuration of the carboxyl group, their reactivity with strontium can be expected to be different; hence, the strontium-binding properties of alginates are likely to be dependent on the relative proportion of these uronic acid units, i.e. M/G ratio.

Sodium alginate preparations obtained from over 100 seaweed samples were therefore examined chemically for their M/G ratios and biologically for their in vivo strontium-binding properties. In general, guluronate-rich alginates were found to be more effective than mannuronate-rich alginates in binding strontium, as shown in Table I, where a number of alginate preparations obtained from samples collected in North America are listed (13). However, there were many exceptions, and no direct relationship between the chemical composition and the in vivo binding properties was observed.

Most of these experiments were performed in rats using the ligated intestinal segment technique, which includes injection of sodium alginate solutions into the intestinal segments. Sodium alginate gives a viscous aqueous solution, and the high viscosity made it difficult to inject the solution. It was further assumed that strontium can approach its reaction sites (carboxylate groups) more easily in less viscous media, increasing strontium-binding properties of alginates.

Partial hydrolysis is known to reduce the viscosity of alginates. Several alginate samples with different M/G ratios and strontium-binding properties were therefore partially degraded under various conditions. In all degradation products obtained, the M/G ratios were found to be lower, i.e. the guluronate contents were higher, than those of their respective parent alginates. In vivo assay of these products showed that all degradation products, except the methanolysis products, were higher in their selective strontium-binding properties, while their calcium-binding capacities remained low. The low strontium-binding properties of the methanolysis products are probably due to partial esterification of the carboxyl groups during methanolysis. An example of these degradation studies is shown in Table II (14).

As shown in these studies, partial hydrolysis of alginate resulted in increased guluronate contents as well as increased strontium-binding capacities. An immediate conclusion from these findings is that the guluronate units can bind strontium more strongly or effectively than the mannuronate units. The failure to obtain a correlation between the chemical composition and the strontium-binding properties among undegraded alginates (13) was probably due to factors other than the chemical compo-

Table I. Inhibition of intestinal absorption of radioactive strontium and calcium by sodium alginate isolated from North American seaweeds

Sodium alginate from:	Mannuronate / Guluronate	% Inhibition of: Sr-89	Ca-45
A) Highly effective samples			
Macrocystis pyrifera	2.37	80	45
Egregia menziensi	2.13	74	47
Nereocystis luetkeana	1.78	69	30
Alaria marginata	2.06	60	15
Laminaria digitata	1.99	60	30
B) Moderately effective samples			
Ascophyllum nodosum	2.77	58	46
Nereocystis luetkeana	2.88	54	32
Pelvetia spp.	3.45	50	1
Nereocystis luetkeana, blades	2.30	44	1
C) Least effective samples			
Hedeophyllum sessile	6.28	26	1
Alaria marginata	6.48	3	34

Table II. Inhibition of intestinal absorption of radioactive strontium and calcium by acid degradation products of commercial sodium alginate

Conditions of degradation:	Mannuronate / Guluronate	% Inhibition of: Sr-89	Ca-45
1N H_2SO_4, 100°, 6 hours	2.01	61	34
1N H_2SO_4, 100°, 12 hours	2.21	65	43
2N H_2SO_4, 100°, 6 hours	2.00	71	30
1M $(COOH)_2$, 100°, 6 hours	1.55	77	34
1N HCl-methanol, under reflux, 6 hrs	2.03	55	27
Undegraded sodium alginate	2.54	46	33

sition.

In order to demonstrate the difference in the reactivity of mannuronate and guluronate units more clearly, the following degradation study was undertaken. Alginic acid isolated from Laminaria hyperborea with an M/G ratio of ca. 1.0 was partially degraded with 1M oxalic acid at 100° for 20 hours. The insoluble part of the degradation products was dissolved in sodium carbonate solution, which was then dialyzed against water, concentrated in vacuo, and diluted with ethanol to yield partially degraded sodium alginate. It had, as expected, a lower M/G ratio (ca. 0.5) and a dilute solution of this substance was acidified to pH 2.8. A fraction with a low M/G ratio (0.27) precipitated, while another fraction with a high M/G ratio (11.1) remained in the solution. These two fractions were converted into their sodium salt for assay. They were judged to be similar in their chain length based on their viscosity. The larger fraction (polyguluronic acid, M/G = 0.27) was further degraded under the same conditions, and the product was found to be further lower in its M/G ratio (0.14) and viscosity than sodium polyguluronate, indicating that the chain length was shortened (Fig. 1).

The parent alginate and its degradation products obtained were tested for their ability to inhibit the intestinal absorption of radioactive calcium and strontium, using the ligated intestinal segment technique in rats. The results are shown in Table III. Sodium polyguluronate was found to be the most effective in binding strontium and its corresponding polymannuronate was only as effective as the parent alginate. Another pair of polyguluronate and polymannuronate obtained from Laminaria digitata showed a similar difference in the reactivity between the two. However, degraded sodium polyguluronate was found to be less effective in binding strontium than sodium polyguluronate, although the former had a higher proportion of guluronate units (15).

It may be concluded from the above results that the guluronate units of alginate have a stronger affinity toward strontium than the mannuronate units; however, other factors such as chain length also play a role in determining the strontium-binding capacity. A certain chain length ap-

Fig. 1. Partial hydrolysis and fractionation of alginic acid

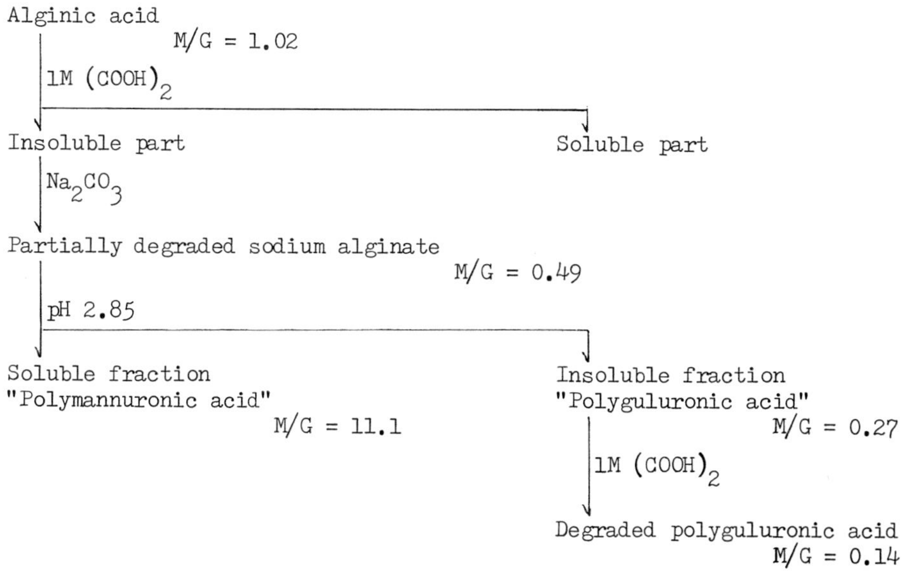

Table III. Inhibition of intestinal absorption of radioactive strontium and calcium by sodium alginate and its degradation products

Inhibitor :	Mannuronate / Guluronate	% Inhibition of : Sr-89	Ca-45
Undegraded sodium alginate	1.02	63	36
Degraded sodium alginate	0.49	77	35
Sodium polymannuronate	11.1	64	32
Sodium polyguluronate	0.27	85	30
Degraded sodium polyguluronate	0.14	78	32

pears to be essentail for the maximum gel formation with strontium and hence the highest capacity of in vivo binding.

All the fractions obtained in this study were found to be low in inhibition of radiocalcium absorption. This is an important feature of the action of alginate with strontium and alginates can be administered without affecting the body balance of calcium to any significant extent. Perhaps, calcium alginate may be preferred to sodium alginate when alginate is to be administered for a long period of time, as no excessive sodium is absorbed.

In order to determine the optimum chain length for the maximum inhibition of strontium absorption, sodium polymannuronate, polyguluronate, and degraded polyguluronate obtained in the above study were fractionated into fractions with different molecular weight ranges, using a membrane filtration technique. All the fractions obtained were assayed in rats for their strontium-binding properties. In all three substances, the fraction with a molecular weight of ca. 30,000 was found to be the most effective. It can be interpreted from the results that the maximum gel formation is achieved between polyuronates with a degree of polymerization of ca. 150 and strontium, and polyguluroantes with this chain length should be the most effective in vivo binders of radiostrontium (16).

Degradation of alginates can also be achieved by certain enzymes. Sodium alginate isolated from Laminaria hyperborea was partially degraded with a crude enzyme obtained from the hepatopancreas of abalone (17). The degradation products were fractionated by precipitation with ethanol as shown in Fig. 2, and the fractions obtained were assayed in rats for their radiostrontium-binding capacities. The results of the bioassay and the M/G ratios of the fractions, as well as those of their parent alginate are shown in Table IV. As expected, the fraction with the lowest M/G ratio was found to be the most effective. Perhaps, it would be the best way to use a purified mannuronate lyase for partial degradation of alginates in order to prepare guluronate-rich strontium binders. Highly specific degradation of mannuronate units in alginates using the specific enzyme would certainly be superior to random chemical processes.

Fig. 2. Partial enzymic degradation and fractionation of sodium alginate

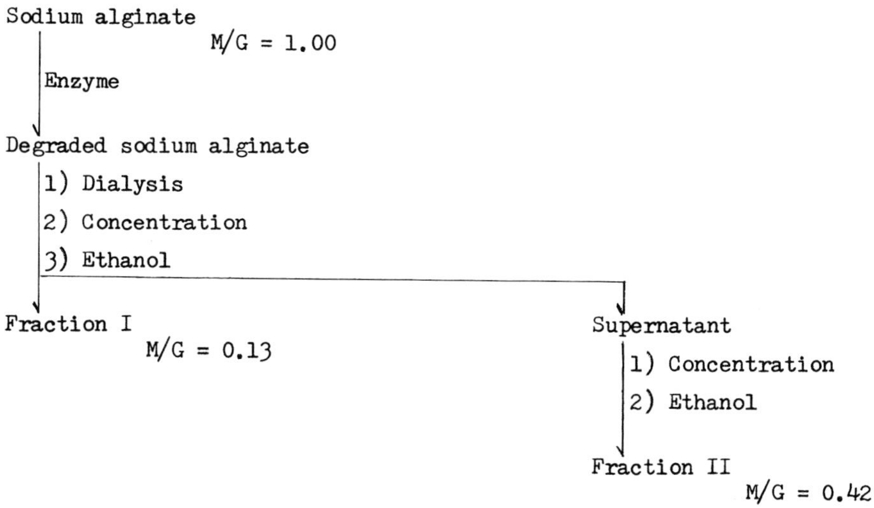

Table IV. Inhibition of intestinal absorption of radioactive strontium by sodium alginate and its enzymic degradation products

Inhibitor :	Mannuronate / Guluronate	% Inhibition of Sr-89
Undegraded sodium alginate	1.00	55
Degraded sodium alginate		
Fraction I	0.13	81
Fraction II	0.42	69

The nature of the unique chemical reactions of alginates with divalent cations to form a gel was investigated by Schweiger (18), who demonstrated that partially acetylated alginates did not precipitate with divalent cations such as calcium. This finding led him to suggest that the free hydroxyl groups of alginates participate in the gel formation. One would therefore expect that the in vivo binding of strontium by alginates could also be reduced by blocking the free hydroxyl groups. Indeed, partially acetylated and methylated alginates were found to be much less effective than their parent alginate in preventing the absorption of radioactive strontium (16). The presence of free hydroxyl groups is therefore essential for the in vivo reaction of alginates with strontium.

Another feature in the structure of alginate is the six-membered ring. Would the ring structure also be essential for alginate to exhibit its effectiveness in binding strontium in vivo? In order to test the essentiality of the ring structure, sodium alginate isolated from Laminaria hyperborea was oxidized with periodate and the oxidation product was then reduced with sodium borohydride (Only the mannuronate unit is shown for illustration):

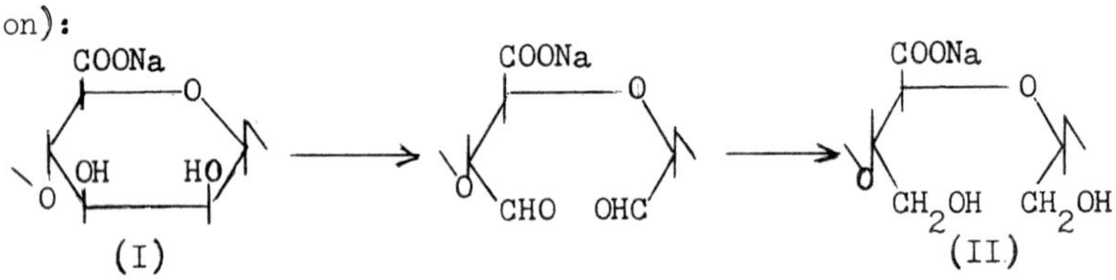

Because of slow over-oxidation, no clear end-point of the periodate oxidation was observed. However, the oxidation was terminated when 1 mole of periodate was consumed per hexuronate unit. The reduction which followed created necessary free hydroxyl groups in the same positions as in alginate, i.e. on carbon-2 and -3.

The oxidation-reduction product (II) was examined, along with its parent alginate (I), for its inhibitory action on the absorption of radioactive strontium and calcium, using the ligated intestinal segment technique in rats. The bone uptake of radioactivity in both groups of experimental animals was found to be only 39% for (I) and 34% for (II) as compared with that of control animals in radioactive strontium, while that of radioactive calcium for both (I) and (II) was 77% of the controls. It may be concluded from the results that the six-membered ring structure is not essential for the specific strontium-binding property of alginate, provided that free hydroxyl groups are available at the appropriate positions to allow them to participate in the gel formation reaction with strontium. Perhaps, the most significant fact in the results is that the low in vivo binding of calcium by alginate was not affected by ring-opening (19).

In summary, our findings as well as other's may be condensed as follows:
(i) Sodium and calcium alginates are effective in vivo binders of radioactive strontium, both in inhibition of its intestinal absorption and in reduction of its body burden.
(ii) Alginates which have high guluronate contents are more effective, and the maximum efficacy may be expected when the degree of polymerization is ca. 150.
(iii) Free hydroxyl groups on carbon-2 and -3 in the uronate units appear to be essential, but the covalent bond between the two carbon atoms do not seem to be necessary for the specific in vivo binding of strontium.

3. ALGINATES AS BINDERS OF OTHER TOXIC METALS

Thiele and Andersen (7) demonstrated, as confirmed by Haug and Smidsrød (8), that barium, the next higher element in the same group of the periodic table as strontium, is more preferentially bound by alginate in vitro than strontium or calcium. This finding immediately leads us to expect that radium, the highest element in the group of alkali earth metals, is most likely bound preferentially by alginate; this was confirmed by van der Borght et al. (20) using mice.

As the fall-out scare of atomic explosion faded out toward the end of 1960's and the early 1970's, we became aware of hazardous non-radioactive metal pollutants, particularly cadmium and lead. Cases of acute cadmium poisoning have been reported around the world, but of particular interest among the cadmium poisoning cases are those called "Ouch-ouch disease" in Japan. The epidemiological out-break was observed in Japan and the origin was traced to poisoning from cadmium, which contaminated water irrigating rice fields. The name "Ouch-ouch disease" stems from painful joints suffered by the victims.

The in vitro results obtained by Thiele (7) as well as by Haug (8) suggest that cadmium is another metal which may be bound in vivo by alginate. Sodium alginate from Laminaria hyperborea was therefore examined for its effectiveness in reducing the absorption of cadmium in the rat intestine. A lethal dose (8 mg cadmium per 100 g body weight as cadmium chloride) was given orally to control rats by a forced feeding technique. In addition to the same amount of cadmium, sodium alginate (57 mg/100 g body weight = 2 equivalent amount) was given to the experimental rats. The mortality rate and the renal contents of cadmium in both groups were as follows (21):

	Mortality	Renal Cd
Control	85%	9.8 µg/g
Experimental	7%	5.8 µg/g

The results indicate that alginate can indeed prevent the intestinal absorption of cadmium. Similar findings were also reported by Silva et al. (22) using radioactive cadmium in rats. Since cadmium is excreted in the feces, alginate may also be used to reduce the body burden of cadmium, as it did that of radioactive strontium.

The third metal which may be bound by alginate, judging from the in vitro experiments (7,8), is lead, another hazardous pollutant with a long history. Harrison et al. (23), however, failed to demonstrate a positive effect of alginate to reduce the absorption of radioactive lead in man, and Carr et al. (24) found that the effect of alginate on the absorption of radiolead in rats was small and dependent on dietary conditions.

4. CARRAGEENAN AND FUCOIDAN AS LEAD BINDERS

Since Harrison et al. (23) were not able to show any inhibitory effect on the intestinal absorption of lead in man, our search for an effective lead binder was extended beyond alginate (25). It is claimed that germanium, one of the elements in the same group of the periodic table as lead, is concentrated in both red and brown algae (1), which also contain partially sulfated polysaccharides, carrageenan and fucoidan, respectively. Preferential in vitro reaction of alginates with lead was shown by Thiele (7) and Haug (8), but for some unknown reasons, guluronate-rich alginates did not prevent the absorption of lead in vivo. If so, perhaps, these sulfated polysaccharides may be able to do so. Therefore, both carrageenan and fucoidan were investigated for their lead binding properties in vitro and in vivo.

Carrageenan used in our study was obtained commercially. Fucoidan was isolated from Ascophyllum nodosum by water extraction; the extracts were treated with lead acetate to remove alginate, and alginate-free extracts were made basic by addition of a barium hydroxide solution (5). The fucoidan-lead hydroxide complex which precipitated was treated with a dilute sulfuric acid and the freed fucoidan was precipitated with ethanol after neutralization. Crude fucoidan was then purified by precipitation with cetyl trimethyl ammonium bromide. The quarternary ammonium complex of fucoidan was decomposed with sodium chloride and fucoidan was isolated by precipitation with ethanol. It should be noted in this process that fucoidan did not precipitate with lead in acidic conditions, as in the stomach, but formed an insoluble precipitate in alkaline media, as in the intestine.

Our fucoidan preparations contained up to 90% of fucose, as examined by gas-chromatographic estimation, and ca. 9% of sulfur, which is equivalent to ca. 0.7 sulfate group per fucose unit. Carrageenan and fucoidan were examined for their lead binding capacities in rats by the ligated intestinal segment technique, by-passing the stomach acidity, and the amount of lead left unabsorbed (i.e. bound by the binders) in the segment was measured by the flameless atomic absorption technique. The following results were obtained (26, 27):

	Equivalent Ratio	% Inhibition
Carrageenan : Lead	1 : 1	41
Carrageenan : Lead	2 : 1	50
Fucoidan : Lead	1 : 1	38
Fucoidan : Lead	2 : 1	56
Fucoidan : Lead	3 : 1	75

The kidney and liver samples from the same animals as well as those of untreated animals were also examined for their lead contents. Evidence was obtained that these polysaccharides may also be able to bind lead which is re-secreted into the intestine.

Carrageenan or fucoidan may not be used in vivo as lead binders because of their digestibility in the intestine and/or instability in acidic conditions of the stomach. However, the knowledge obtained on these algal polysaccharides may be of great value, and appropriately sulfated, non-digestible polysaccharides may be prepared for practical use as effective lead binders. Further investigation on this matter is now in progress.

5. CONCLUDING COMMENTS

As briefly described in this article, ecological concentration processes are not only of purely scientific interest. They can provide us with effective in vivo binders of toxic metals or direct us toward new avenues to be explored. Industrial applications of algal polysaccharides may become practical as specific binders of toxic metals in industrial effluents either for their removal before discharge into our environment or for the recovery of rare elements from them. As our modern technology progresses, more exotic elements will be utilized and resulting potential pollution will present new problems. Some algal polysaccharides may be of great value in counteracting these future hazards. No doubt, algal polysaccharides and the knowledge we can gather therefrom will be immensely important to face the existing and future challenges of mankind.

REFERENCES

(1) Polikarpov, G.G.: "Radioecology of aquatic organisms," North-Holland, Amsterdam (1966).

(2) Stanford, E.C.C.: Chem. News 96, 254 (1883); quoted from Percival, E. and McDowell, R.H.: "Chemistry and enzymology of marine algal polysaccharides," Academic Press, London (1967).

(3) Fischer, F.G. and Dörfel, H.: Die papierchromatographische Trennung und Bestimmung der Uronsäuren, Hoppe-Seyler Z. physiol. Chem. 301, 224-35 (1955).

(4) McDowell, R.H.: A method for the fractionation of alginates, Chem. Ind. 1401-2 (1958).

(5) For a comprehensive review on the chemistry of alginic acid, see Percival, E. and McDowell, R.H.: "Chemistry and enzymology of marine algal polysaccharides," chapter 5, Academic Press, London (1967).

(6) Haug, A., Larsen, B., and Smidsrød, O.: A study of the constitution of alginic acid by partial acid hydrolysis," Acta Chem. Scand. 20, 183-90 (1966).

(7) Thiele, H. and Andersen, G.: Ionotrope Gele von Polyuronsäuren, Kolloidzeitsch. 140, 76-102 (1955).

(8) Haug, A. and Smidsrød, O.: The effect of divalent metals on the propties of alginate solutions, Acta Chem. Scand. 19, 341-51 (1965).

(9) Skoryna, S.C., Paul, T.M., and Waldron-Edward, D.: Studies on inhibition of intestinal absorption of radioactive strontium. I. Prevention of absorption from ligated intestinal segments, Canad. Med. Ass. J. 91, 285-8 (1964).

(10) Paul, T.M., Skoryna, S.C., and Waldron-Edward, D.: Studies on inhibition of intestinal absorption of radioactive strontium. V. The effect of administration of calcium alginate, Canad. Med. Ass. J. 95, 957-60 (1966).

(11) Stara, J.F.: Metabolism of internal emitters-repressive action of sodium alginate on absorption of radiostrontium in kittens, Abst. Symp. Nucl. Med., Omaha, Nebraska (1965).

Van der Borght, O., Colard, J., van Puymbroeck, S., and Kirchman, R.: Radiocontamination from milk in piglets (swine): influence of sodium alginate on the $^{85}Sr/^{134}Co$-ratio of the body burden and on the comparative ^{85}Sr-^{47}Ca absorption, Abst. Int. Symp. Radioecol. Concentra-

tion processes, Stockholm (1966).
Harrison, J., McNeil, K.G., and Janiga, A.: The effect of sodium alginate on the absorption of strontium and calcium in human subjects, Canad. Med. Ass. J. 95, 532-4 (1966).
Hesp, R. and Ramsbottom, B.: Effect of sodium alginate in inhibiting uptake of radiostrontium by the human body, Nature 208, 1341-2(1965).
Hodgkinson, A., Nordin, B.E.C., Hambleton, J., and Oxby, C.B.: Radiostrontium absorption in man: suppression by calcium and by sodium alginate, Canad. Med. Ass. J. 97, 1139-43 (1967).
Sutton, A.: Reduction of strontium absorption in man by the addition of alginate to the diet, Nature 216, 1005-7 (1967).
Patrick, G., Carr, T.E.F., and Humphreys, E.R.: Inhibition by alginates of strontium absorption studied in vitro and in vivo, Int. J. Radiat. Biol. 12, 427-34 (1967).

(12) Stara, J.F.: Repressive action of sodium alginate on absorption of radioactive strontium and calcium in cats, in "Diagnosis and Treatment of Deposited Radionuclides," Kornberg, H.A. and Norwood, W.D. (Eds.), Excerpta Medical Foundn. (1969).

(13) Tanaka, Y., Skoryna, S.C., and Waldron-Edward, D.: Studies on inhibition of intestinal absorption of radioactive strontium. VII. Relationship of biological activity to chemical composition of alginates obtained from North American seaweeds, Canad. Med. Ass. J. 99, 169-75 (1968).

(14) Tanaka, Y., Waldron-Edward, D. and Skoryna, S.C.: Studies on inhibition of intestinal absorption of radioactive strontium. VI. Alginate degradation products as potent in vivo sequestering agents of radioactive strontium, Canad. Med. Ass. J. 98, 1179-82 (1968).

(15) Tanaka, Y.: Application of metal-binding properties of marine algae in medicine, Proc. Food-Drugs from the Sea Symp., Marine Technology Society, Washington, D.C., p. 351-7 (1969).

(16) Tanaka, Y., Hurlburt, A.J., Angeloff, L., and Skoryna, S.C.: Application of algal polysaccharides as in vivo binders of metal pollutants, Proc. VII. International Seaweed Symp., Univ. Tokyo Press, Tokyo, p. 602-4 (1972).

(17) Skoryna, S.C., Hong, K.C., and Tanaka, Y.: The effects of enzymic degradation products of alginates on intestinal absorption of radio-

strontium, Proc. VII. International Seaweed Symp., Univ. Tokyo Press, Tokyo, p. 605-7 (1972).

(18) Schweiger, R.G.: Acetylation of alginic acid. I. Preparation and viscosities of algin acetates, J. Org. Chem. 27, 1786-9 (1962). Schweiger, R.G.: Acetylation of alginic acid. II. Reaction of algin acetates with calcium and other divalent ions, J. Org. Chem. 27, 1789-91 (1962).

(19) Tanaka, Y. and Stara, J.F.: To be published.

(20) Van der Borght, O., Van Puymbroeck, S., and Collard, J.: Intestinal absorption and body retention of 226-radium and 47-calcium in mice: Effect of sodium alginate, measured in vivo with a Ge(Li) detector, Health Phys. 21, 181-96 (1971).

(21) Skoryna, S.C., Tanaka, Y., Moore, Jr., W., and Stara, J.F.: Inhibition of intestinal absorption of excessive intakes of toxic metals, Trace Subst. Environ. Health, 6, 1-6 (1972).

(22) Silva, A.J., Fleshman, D.G., and Shore, B.: The effects of sodium alginate on the absorption and retention of several divalent cations. Health Phys. 19, 245-51 (1970).

(23) Harrison, G.F., Carr, T.E.F., Sutton, A., Humphreys, E.R., Rundo, J.: Effect of alginate on the absorption of lead in man, Nature 224, 1115-6 (1969).

(24) Carr, T.E.F., Nolan, J., and Durakovic, A.: Effect of alginate on the absorption and excretion of lead-203 in rats fed milk and normal diets. Nature 224, 1115 (1969).

(25) Our work on the effect of polygalacturonate is not described here, since it is beyond the scope of this article.

(26) Paskins-Hurlburt, A.J., Tanaka, Y., and Skoryna, S.C.: Carrageenan and the binding of lead. Bot. Marina 19, 59-60 (1976).

(27) Paskins-Hurlburt, A.J., Tanaka, Y., and Skoryna, S.C.: Isolation and metal binding properties of fucoidan. Bot. Marina 19, 327-8 (1976).

Polyhydroxyphenols in Some Brown Algae

Yasuhiko Tsuchiya
Kitasato University, School of Fisheries Sciences, Iwate Prefecture 022-01 Japan

It was reported by several workers[1] that brown algae contained tannin-like substances consisting of phloroglucinols. Recently Glombitza and co-workers[2] have made an elavolate study of them and found a variety of highly hydroxylated phenolic compounds referring to tannins constructed of phloroglucinol moieties linked by aryl-aryl or aryl-ether bonds or of a mixed type.

Those compounds were named phorotannins. On the other hand, Noguchi[3] found in 1943 that the aqueous extracts from Sargassum ringgoldianum were **as capable as quebracho tannins in converting hide into leather.** Oya and Fujikawa[4] described in their book that a roasted product of Chorda filum had been used as a substitute for coarse tea in Yamagata, north-eastern part of Honshu.

The present investigation was undertaken to make a semi-quantitative determination of polyhydroxyphenols [PHP] in some brown algae.

MATERIALS AND METHODS

Materials used in the experiments were 12 different species, of which 2 belong to Desmarestiales, 2 to Laminariales and the rest to Fucales, collected from Okirai Bay, Iwate Prefecture, from April to July, 1975.

The fresh materials were cut into small pieces and 30g of them were extracted with 1500ml of water in a steam bath for 60 minutes. Then they were filtered with a buchner funnel by means of suction. The filtrates and the washings were combined and concentrated to about 150ml under reduced pressure passing through nitrogen gas below 40°. Thereafter, they were transfered to a 200ml volumetric flask and diluted to the mark. In the case of Undaria pinnatifida, ethyl alcohol was added to

Marine Algae in Pharmaceutical Science
edited by H. A. Hoppe, T. Levring, Y. Tanaka

the viscous concentrated extracts for a final concentration of 70 per cent and the resulted precipitates of mucilagenous substance were removed by centrifugation. The supernatant was again concentrated to a small volume in vacuo and measured up 200ml in the same way as mentioned above. The quantity of PHP in the solution was semi-quantitatively determined as gallotannic acid by the slightly modified method,[5] shown in an Official and Tentative Methods of Analysis of the A.O.A.C., using the $KMnO_4$ solution diluted at a tenth of the concentration described. For assaying the phaeophyte tannins it is necessary to establish a more exact method, but the preliminary experiments showed that the values found by the tentative method were almost directly proportional to varying degree in concentration of the algal extracts and that the amount of tannic acid added to the extracts was recovered at a level of 88 to 95 per cent under the conditions employed.

RESULTS AND DISCUSSION

Amount of PHP in Brown Algae

Table 1 shows that PHP occurred in all of the samples examined and that the amount of it greatly varied from 0.54 to 54.51mg per g of dry matter, It is pointed out that the PHP content was

Table 1. Amount of PHP in brown algae (mg per g of dry matter)

Species	Date	Amount
Desmarestia ligulata	May 29	1.04
D. viridis	" "	6.39
Laminaria japonica f.membranacea	"	3.03
Costaria costata	" "	0.54
Undaria pinnatifida f.distans	" "	0.54
Hizikia fusiforme	" "	23.64
Sargassum Thunbergii	" 9	7.97
S. fulvellum	" 29	3.61
S. hemiphyllum	" "	26.81
S. micracanthum	June 3	32.47
S. horneri	" "	9.12
S. confusum	July 25	54.51

high in *Fucales*, especially predominant in *Hizikia fusiforme*, *Sargassum hemiphyllum*. *S. miracanthum* and *S. confusum* but low in *Desmarestiales* and *Laminariales*.

Species of *Sargassum* are unacceptable as food not only because of their lack of sweetness but also because of their astringency. This is probably due to the presence of a fair amount of PHP which actually tans the proteins. Suehiro and Searashi[6] reported that the extracts from *Undaria pinnatifida* completely failed to give colour reactions characteristic of tannins. On the contrary it proved otherwise as seen in the table. Further discussion of the subject will be found in the paragraph subsequent to next.

Distribution of PHP in the Different Portions of *Sargassum miracanthum*

It is conceivable that PHP is not uniformly distributed in the different portions of the seaweed. The next table, giving the milligram per gram of dry matter, illustrates the effect of organs upon PHP in *Sargassum micracanthum*. Contrary to expectation, there was little difference in amount between them, but it was slightly larger in air bladder and leaf-like portion than receptacle and stem-like portion. There is need of further investigation in connection with the season and habitat.

PHP Content in *Undaria pinnatifida*

Species of *Undaria* are known as one of the most important edible plants in Japan and are second to the laver in seaweed cultivation industry. It is also said that this alga is a very good food for convalescence after childbirth. In 1975 Japan produced 86,178 tons of cultivated species and 13,523 tons of wild ones.[7] Table 3 indicates that a small amount of PHP was

Table 2. Distribution of PHP in different portions of *Sargassum miracanthum*

Portion	Date	Amount (mg per g of dry matter)
Receptacle	July 10	25.84
Air bladder	"	33.89
Leaf-like portion	"	32.65
Stem-like portion	"	28.94

Table 3. PHP content in Undaria pinnatifida (mg per g of dry matter)

Source		Cultivated		Wild
Date		April 17	May 1	June 17
	Outgrowth	0.77	0.25	0.52
Portion	Midrib	0.13	1.33	1.23
	Sporophyll	1.59	0.58	1.58

always contained in various portions of Undaria pinnatififida harvested at different months. The alga reared in the cultivation ground of Okirai Bay was rapid in growth and large in size as compared with the wild one which grew on rocks in places where there was a strong current of water. However, they were subtly different from each other in quantity of PHP.

CONCLUSION

The studies described here dealt with the semi-quantitative examination of PHP in 12 different species of brown seaweeds. The content of it was highest in Sargassum confusum and least in Undaria pinnatifida and Costaria costata. Species of Sargassum that are unacceptable as food contained more PHP than others with the exception of Hazikia fusiforme and Desmarestia viridis. The fluctuation was found to be insignificant in different portions of Undaria pinnatifida, whereas in Sargassum miracanthum the amount of PHP was slightly larger in air bladder and leaf-like portion than receptacle and stem-like portion. Although the date available for discussion about the effect of habitat on PHP of Undaria pinnatifida were insufficient, it seemed that there was little difference between the cultivated alga and wild one.

ACKNOWLEDGMENT

The author wishes to thank Dr. Y.Sato for helpful advice and discussion and Mr. K.Ito for technical assistance.

REFERENCES

(1) Ogino, C., Taki, Y.: Studies on the tannin of brown alga, Sargassum ringgoldianum Harv. J.Tokyo Univ.Fisheries 43, 1-5(1957).

Haug, A., Larsen, B.: Phenolic compounds in brown algae I. Presence of reducing compounds in Ascophyllum nodosum. Acta Chem.Scand. 12, 650-657(1958).

Craigie, J. S., McLachlan, J.: Excretion of colored ultraviolet-absorbing substances [flavonols or tannins] by marine algae. Can.J.Botany 42, 23-33(1964).

(2) Glombitza, K.-W., Rösener, H.U., Vilter, H., Rauwald, W.: Antibiotica aus Algen. 8 Mitt. Phloroglucin aus Braunalgen. Planta Medica 24, 301-303(1973).

Glomzitza, K._W.: Highly hydroxylated phenols of the phaeophyceae. P. 191-204(1977). In marine Natural Products Chemistry, ed. Faulkner, D. J., Fenical, W. H. Plenum Press. New York and London.

(3) Noguchi, E.: The use of Sargassum ringgoldianum in tanning. Bull.Japan.Soc.Sci.Fisheries 12, 52-53(1943).

(4) Oya, T., Fujikawa, K.: Chemistry of Seaweeds. Kosei-sha Kosei-kaku, Tokyo, P. 249(1935).

(5) Horwitz, W., Chichilo, P., Reynolds, H.: Tannin (11)-Official Final Action. P. 240-241(1970). In Official Methods of Analysis of the Association of Official Analytical Chemists. Association of Official Analytical Chemists, Washington.

(6) Suehiro, Y., Searashi, T.: The utilizatin of the inedible seaweeds. Bull.Physiol.Sci.Research Inst, Tokyo Univ. 1, 44-50(1948).

(7) Iwate-ken Gyokyo Kyodokumiai Rengokai: Production of Undaria in Japan. Iwate Gyokyo Jiho 3, 9-10(1976).

SOME REMARKS ON ALGAL CAROTENOIDS AND THEIR INTERCONVERSION INTO ANIMAL CAROTENOIDS

Michael Wettern and Adolf Weber
Institute for General Botany, University of Hamburg
D-2000 Hamburg - 36, Jungiusstraße 6, F.R. Germany

Carotenoids are polyenic pigments, composed of isoprenoic units. They are commonly divided into two groups: the carotenes, consisting only of hydrocarbons, and the xanthophylls. These possess, in addition to the hydrocarbon skeleton, one or more functional groups, as for instance acetylenic allenic, glycosidic, epoxidic, ketonic, hydroxy, ester, or carboxylic group.

The carotenoids are widely distributed in nature. The highest structural diversity has been observed in organisms of the aquatic environment.

In algae the carotenoids are synthesized de novo and are reported to have several important functions, absorption of light for photosynthesis and phototaxis (Evstigneyev and Paramonova, 1976; Krinsky, 1971; Prézelin, 1976). They also probably protect cells against photodynamic destruction (Anderson and Krinsky, 1973; Asada, 1976; Mathews, 1964; Mc Millan et al., 1966) and may play a role in oxygen transport (Cox and Bendall, 1974; Yamamoto et al., 1962) and membrane stabilization (Krinsky, 1971).

To give a general impression of the distribution pattern and structural diversity of carotenoids in algae, table 1 was compiled. It should be kept in mind, that this is a more generalized survey and we intentionally omit a list of the carotenoids structure. For more details the following literature is recommended: Berger et al., 1977; Bjørnland and Aguilar-Martinez, 1976; Czeczuga, 1975, 1976a; Jeffrey et al., 1975; Johansen et al., 1974; Norgard et al.,1974a+b;

Marine Algae in Pharmaceutical Science
edited by H. A. Hoppe, T. Levring, Y. Tanaka

Tab. 1 Distribution pattern of carotenoids in algae

		Cyanophyceae	Chrysophyceae	Xanthophyceae	Bacillariophyceae	Phaeophyceae	Dinophyceae	Rhodophyceae	Chlorophyceae	Conjugatophyceae	Charophyceae	Euglenophyceae	Cryptophyceae	Chloromonadophyceae	colorless flagellates
a	myxoxanthophyll	+													
	4-keto-myxoxanthophyll	–													
	aphanizophyll	–													
	oscillaxanthin	–													
	myxol-2'-O-methyl methylpentoside	o													
	4-keto-myxol-2'-O-methyl-methylpentoside	o													
	oscillol-2,2'-di(O-methyl)-methylpentoside	o													
b	phytoene						o								
	phytofluene						o								
	lycopene	–							o		–				
c	γ-carotene	o	–				o		o		–				
d	α-carotene		+				o	–	+	o			+		
	β-carotene	+	+	+	+	+	+	+	+x	+	+	+	–	+	–
	ε-carotene				–	–							o		
e	crocoxanthin												+		
	monadoxanthin												–		
	diatoxanthin		–	+	+	o	–					+			

Tab. 1 ctd.

		Cyanophyceae	Chrysophyceae	Xanthophyceae	Bacillariophyceae	Phaeophyceae	Dinophyceae	Rhodophyceae	Chlorophyceae	Conjugatophyceae	Charophyceae	Euglenophyceae	Cryptophyceae	Chloromonadophyceae	colorless flagellates
e	diadinoxanthin		–	+	+	o	–					+		?	
	heteroxanthin			+								o		?	
	alloxanthin												+		
	manixanthin												–		
	pyrrhoxanthin						o								
	pyrrhoxanthinol						o								
f	β-carotene-epoxide		–												
	β-carotene-diepoxide			–											
	cryptoxanthin-epoxide		–	–					o			–			
	cryptoxanthin-diepoxide			–											
	antheraxanthin		–	o		–		o	+	+	+				
	violaxanthin		–	o		+		o	+	+	+				
	lutein-epoxide							o	o					?	
	mutatochrome	o													
	aurochrome	o													
	auroxanthin	o													
g	4-hydroxy-3',4'-diketo-β-carotene								×						
	3-hydroxy-3',4,4'-tri-keto-β-carotene								×						
	3-hydroxy-canthaxanthin								×						
	4-hydroxy-4'-keto-β-carotene	o													

Tab. 1 ctd.

	Cyanophyceae	Chrysophyceae	Xanthophyceae	Bacillariophyceae	Phaeophyceae	Dinophyceae	Rhodophyceae	Chlorophyceae	Conjugatophyceae	Charophyceae	Euglenophyceae	Cryptophyceae	Chloromonadophyceae	colorless flagellates
echinenone	+	–						–X			–			–
canthaxanthin	+							–X			–			–
3'-hydroxy-echinenone	–										–			
4'-hydroxy-echinenone								X						
[g]astaxanthin						o		X			+			
siphonaxanthin								–						
fritschiellaxanthin								X						
micronone								o						
phoenicopterone								o						–
neoxanthin	o	–	+	–	o		–	+	+	+	+			
desepoxy-neoxanthin		–												
dinoxanthin		o				–								
vaucheriaxanthin			+											
fucoxanthin		+	+	+	+	o	o							
[h]fucoxanthinol		–			o									
19'-hexanoyl-fucoxanthin		–												
peridinin						+								
peridiniol						o								
9'-hexanoyloxy-paracentrone-3-acetate		–												
cryptoxanthin	o	–					+	o			–			
[i]zeaxanthin	–	–	o		–		+	+	+	+	+			
caloxanthin	o													

Tab. 1 ctd.

		Cyanophyceae	Chrysophyceae	Xanthophyceae	Bacillariophyceae	Phaeophyceae	Dinophyceae	Rhodophyceae	Chlorophyceae	Conjugatophyceae	Charophyceae	Euglenophyceae	Cryptophyceae	Chloromonadophyceae	colorless flagellates
	nostoxanthin	o													
	β-carotene-2-ol								o						
	β-carotene-2,2'-ol								o						
	α-carotene-2-ol								o						
i	α-cryptoxanthin							o	o						
	lutein							–	+	+	+				
	loroxanthin								–						
	crustaxanthin								o						
	trihydroxy-α-carotene								–	–					
	trihydroxy-β-carotene								X						

pigment occurs : + ≙ frequently – ≙ occasionally
o ≙ only rarely X ≙ as secondary carotenoid

Distribution pattern of carotenoids in algae according to : Goodwin (1971,1974); Hager and Stransky (1970a+b); Jeffrey et al. (1975); Liaaen-Jensen (1977); Stransky and Hager (1970a-c); Thommen (1971); Weber (1969,1973,1975); Weber and Czygan (1972)

The carotenoids were roughly grouped according to chemical aspects :
a: glycosidic-, b: acyclic-, c: monocyclic-, d: bicyclic-, e: acetylenic-, f: epoxidic-, g: ketonic-, h: allenic-, i: hydroxy-carotenoids

Weber, 1969; Whittle and Casselton, 1975a+b. The table is based on the algal classification of Fott (1971).

From table 1 it can be concluded, that several carotenoids occur in many algal classes. These pigments are β-carotene, antheraxanthin, violaxanthin, neoxanthin, and zeaxanthin, although their quantity differs in the various classes[1]. On the other hand greater concentrations of some carotenoids occur only in a few classes, for instance crocoxanthin and alloxanthin in the Cryptophyceae, peridinin in the Dinophyceae, vaucheriaxanthin in some Xanthophyceae, and all quoted glycosidic carotenoids merely in the Cyanophyceae. Many of the listed carotenoids are restricted to one or few algal species, e.g., fritschiellaxanthin (3'-epi-α-doradexanthin) to Fritschiella tuberosa (Chaetophorales, Chlorophyceae) (Buchecker et al., 1978).

The ketonic xanthophylls echinenone, canthaxanthin, astaxanthin, and fritschiellaxanthin occur under favourable growth conditions only as minor components in green algae. Under insufficient developmental conditions, for example nitrogen deficiency, these and some other carotenoids are synthesized in greater amounts as secondary carotenoids, in addition to or instead of the primary carotenoids (Brown et al., 1967; Czygan, 1964, 1966b, 1968a+b; Donkin, 1976; Lichtenthaler and Verbeek, 1973; Tupa, 1974; Weber, 1975).

As the ability to synthesize secondary carotenoids is mostly restricted to the Chlorophyceae, Czygan (1970) and Deason et al. (1977) discussed the feasibility of using this as a taxonomic criterion. Hager and Stransky (1970a) reported that only those green algae which synthesize a tri-

---

1) These differences may be unimportant as the carotenoid content of algae is variable, depending on environmental conditions.

hydroxy-α-carotene are able to synthesize secondary carotenoids. Weber (1973, 1975), however showed, that although most of the desmids he examined contain this tri-hydroxy-α-carotenoid, these algae are not able to produce secondary pigments. In contrast, Fritschiella tuberosa is able to produce secondary pigments, although this alga lacks the tri-hydroxy-α-carotenoid (Weber, 1975).

It is well documented and generally believed that the carotenoids are located within the chloroplast membranes. Jeffrey et al. (1974) reported a unique carotenoid composition of chloroplast envelope membranes. The xanthophylls lutein and neoxanthin were detected by means of specific antisera in the outer surface of the thylakoid membranes (Radunz and Schmid, 1975; Radunz, 1976).

The secondary carotenoids, on the other hand, are not located within membranes. They are present as hydrophobic molecules in lipid globules either as plastoglobules inside the chloroplast as in Haematococcus pluvialis (Sprey, 1970) or as extrachloroplastic pigmented globules as in Protosiphon botryoides (Berkaloff, 1970) and Fritschiella tuberosa (Weber, 1975). In Protosiphon botryoides the carotenoids occur in the globules as carotenoproteins (Berkaloff, 1977). In Peridinium foliaceum Withers and Haxo (1975) identified β-carotene and γ-carotene in extraplastidic oil globules. In addition to these pigments the authors found chlorophyll c_1 and c_2 as well as fucoxanthin in this alga. The authors discussed the hypothesis that Peridinium foliaceum is a heterotrophic dinoflagellate hosting an algal symbiont of non-dinoflagellate origin, because γ-carotene and fucoxanthin are only present in some species of the Dinophyceae (Jeffrey et al., 1975). Presumably the extraplastidic carotenes belong to the apochlorotic host dinoflagellate and do not originate from the chloroplasts of the endosymbiont. Further evidence for a membranebound endosymbiont within the dinoflagellate <u>Peridinium foliaceum</u> was given by Jeffrey and Vesk (1976) in their study on the fine structure of this dinoflagellate.

Beside the occurrence of the carotenoids in algae the carotenoids of aquatic animals have attracted the interest of biochemists because of their diversity. As it is generally accepted that animals are not able to synthesize carotenoids de novo, these animal pigments must be of plant origin, partially modified by the animals.

In recent years several studies have been made on the distribution of carotenoids, <u>for example</u> in aves (Fox, 1974), tracheata (Czygan, 1972; Kayser, 1977a), annelida (Czeczuga, 1977b), pisces (Czeczuga, 1977c; Katayama et al. 1970; Khare et al. 1973; Webber, 1973), anthozoa (Czeczuga, 1973a), echinodermata (Czeczuga, 1977a; Gross et al., 1975), tunicata (Czeczuga, 1977d), mollusca (Czeczuga, 1976b), and crustaceae (Lee and Gilchrist, 1972; Tanaka et al., 1976).

The following presentation will be restricted to some examples as there is not enough space to discuss all the data concerning the biogenetical relation of algal and animal carotenoids.

The dependence of animal carotenoid content of the diet is well documented. Jensen and Sakshaug (1970a+b) found a good correlation between planktonic algae and the pigmentation of the mussel <u>Mytilus edulis.</u> These authors and Campbell (1969a) described a seasonal carotenoid variation in <u>Mytilus adulis</u>, which was controlled by the phytoplankton blooms and the sexual cycle of the mussel. Lack of food affected both carotenoid content and the maturation of the gonads.

Kayser (1977b) investigated the metabolism of the larvae of two notodontid moth species fed with (^{14}C)-β-carotene. He found a specific but different labelling. In one species the labelled precursor was incorporated to 3.8% into β-carotene-2-ol, in the other to 65% into β-carotene-3-ol (cryptoxanthin). It is noteworthy that the two moth species, belonging to the same family, have developed different ways of carotene hydroxylation. This is in agreement with the results of Czeczugas studies on carotenoids in fishes (1973b) where, independent of the type of food taken up by the animal, the divergent enzyme specificities play an essential part in the accumulation of carotenoids. This may result in either different carotenoids in species of the same family or the same carotenoids in species of different families.

The carotenoid metabolism in crustaceae has been studied for several years. Astaxanthin, alloxanthin, β-carotene, and canthaxanthin are often the most prominent pigments present, followed by some minor components (Campbell, 1969b; Czygan, 1966a; Katayama et al., 1973; Tanaka et al., 1976a). The keto-carotenoids are either ingested and stored or synthesized from β-carotene involving a series of hydroxylations and the insertion of ketonic groups. There is evidence, that in aquatic animals several metabolic pathways may be available for the production of astaxanthin (Katayama et al., 1973; Tanaka et al., 1976b). Katayama et al. (1970) in their study on goldfish carotenoid biosynthesis set up a possible biosynthetic route starting with lutein to astaxanthin via α-, and β-doradexanthin. Based on the studies of fritschiellaxanthin chirality (Buchecker et al., 1978), and that of lutein (Andrewes et al., 1974; Buchecker et al., 1972), astaxanthin (Andrewes et al., 1974b) and α-doradexanthin from goldfish (Buchecker et al., 1978) an epimerization at C-3' by animals enzyme would be required. This was confirmed by Buchecker et al. (1978) as they found 3'-epi-lutein in goldfish (c.f.fig.1, solid arrows). An alternative pathway may exist starting with lutein and its tranceformance into fritschiellaxanthin in the alga itself and after ingestion by animals to astaxanthin (see fig.1, dotted arrows).

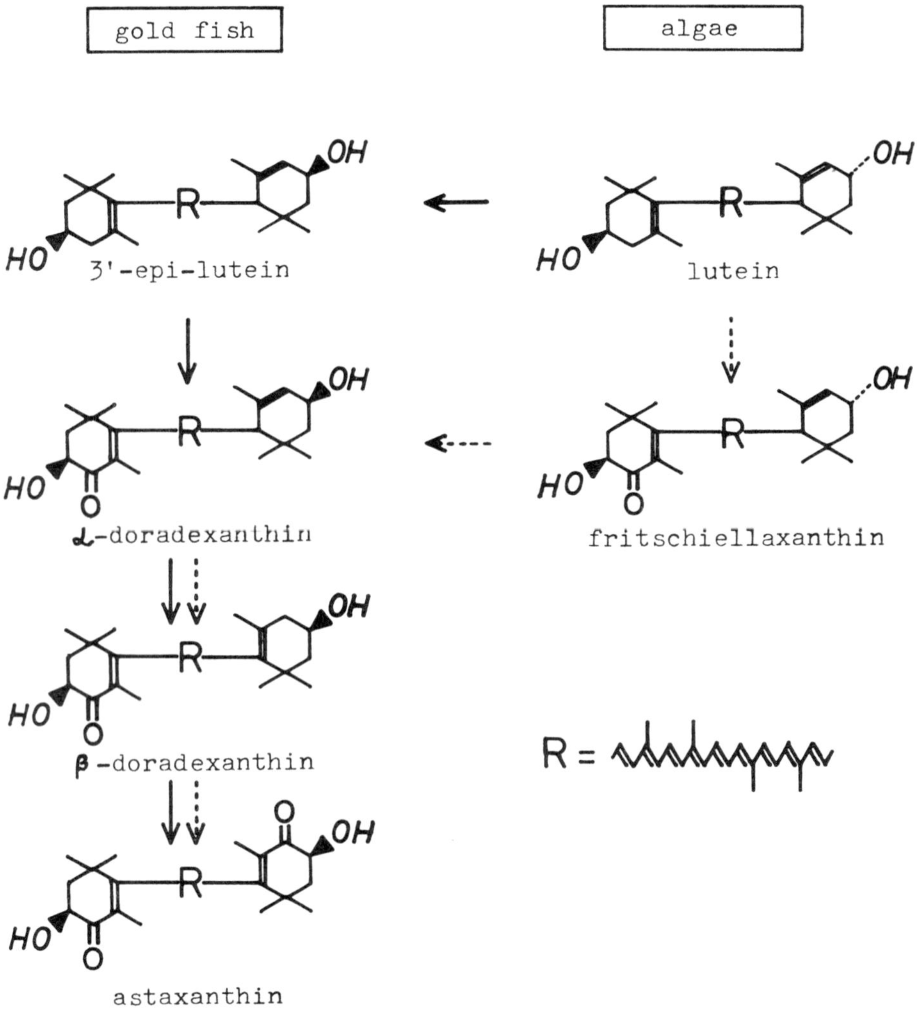

Fig. 1 Possible interconversion of algal lutein to gold fish astaxanthin

Andrewes and Starr (1976) on the other hand reported the occurrenceof astaxanthin in different chirality. This is explained with a different chirality of the β-end group of the precursors which is not affected during the following reactions leading to astaxanthin . The results of Buchecker et al. (1978) however showed a specific change in chirality by animals enzyme. We think it should be interesting to determine the chirality of astaxanthin from organisms which modify ingested (C^{14})-carotenoids to astaxanthin.

References

Anderson, S.M., and Krinsky, N.I.: Protective Action of Carotenoid Pigments Against Photodynamic Damage to Liposomes. Photochem. Photobiol. 18: 403-408 (1973).

Andrewes, A.G., Borch, G., and Liaaen-Jensen, S.: Carotenoids of Higher Plants. 7. On the Absolute Configuration of Lutein. Acta Chem. Scan. B 28: 139-140 (1974a).

Adrewes, A.G., Borch, G., Liaaen-Jensen, S., and Snatzke, G.: Animal Carotenoids.9. On the Absolute Configuration of Astaxanthin and Actinoerythrin. Acta Chem. Scan. B 28: 730-736 (1974b).

Andrewes, A.G., and Starr, M.P.: (3R, 3'R)-Astaxanthin from the YeastPhaffia rhodozyma. Phytochem. 15: 1009-1011 (1976).

Asada, K.: Photosynthesis and Photooxidative Damage. J. Agricult. Chem. Soc. Jap. 50: R115 (1976).

Berger, R., Liaaen-Jensen, S., Mc Alister, V., and Guillard, R.R.L.: Carotenoids of Prymnesiophyceae (Haptophyceae). Biochem. System. Ecol. 5: 71-75 (1977).

Berkaloff, C.: Essai d'isolement des globules pigmentés extraplastidaux de l'algue Protosiphon botryoides. C.R. Acad. Sc. Paris, Ser. D 271: 1518-1521 (1970).

Berkaloff, C.: Carotenoproteins in Extrachloroplastic Structures of the Green Alga Protosiphon botryoides. Plant Sci. Lett. 10: 45-48 (1977).

Bjørnland, T., and Aguilar-Martinez, M.: Carotenoids in Red Algae. Phytochem. 15: 291-296 (1976).

Brown, T.E., Richardson, F.L., and Vaughn, M.L.: Development of Red Pigmentation in Chlorococcum wimmeri (Chlorophyta, Chlorococcales). Phycologia 6: 167-184 (1967).

Buchecker, R., Hamm, P., and Eugster, C.H.: Synthese von (±)-cis- und (±)-trans-Methoxy α-jonon; ein weiterer experimenteller Beitrag zur Chiralität des natürlichen (+)-Xanthophylls (Luteins). Chimia 26: 134-136 (1972).

Buchecker, R., Eugster, C.H., Weber, A.: Absolute Konfiguration von α-Doradexanthin und von Fritschiellaxanthin, einem neuen Carotinoid aus Fritschiella tuberosa Iyeng.. Helv. Chim. Acta 61: to be published

Campbell, S.A.: Seasonal Cycles in the Carotenoid Content in Mytilus edulis. Marine Biol. 4: 227-232 (1969a).

Campbell, S.A.: Carotenoid Metabolism in the Commensal Crab Pinnotheres pisum. Comp. Biochem. Physiol. 30: 803-812 (1969b).

Cox. R.P., and Bendall, D.S.: The Functions of Plastoquinone and β-Carotene in Photosystem II of Chloroplast. Biochem. Biophys. Acta 347: 49-59 (1974).

Czeczuga, B.: Comparative Studies of Carotenoids in the Fauna of Gullmar Fjord (Bohuslän, Sweden) I Alyconium digitatum and Pennatula phosphora (Anthozoa). Marine Biol. 19: 206-209 (1973a).

Czeczuga, B.: Carotenoids in Fish II. Carotenoids and Vitamin A in some Fishes from the Coastal Region of the Black Sea. Hydrobiol. 41: 113-125 (1973b).

Czeczuga, B.: Carotenoids in three Algal Species from the Mediterranean Sea. Nova Hedwigia 26: 157-163 (1975).

Czeczuga, B.: Carotenoid Pigments in some Phytobenthos Species of Different Systematic Position in the Coastal Area of Ofotfjord/Norway. Nova Hedwigia 27: 223-229 (1976a).

Czeczuga, B.: Investigations of Carotenoids in some Faunal Elements of the Adriatic Sea IV. Molluscs. Hydrobiol. 51: 71-75 (1976b).

Czeczuga, B.: Investigations of Carotenoids in some Animals of Adriatic Sea V. Echinodermata. Hydrobiol. 54: 177-180 (1977a).

Czeczuga, B.: Astaxanthin, a Dominating Carotenoid in 4 Species of Leeches. Bull. L' Acad. Pol.Sci.Ser.Biol. 25: 85 (1977b).

Czeczuga, B.: Carotenoids in Fish 12. Coregonus peled (Gmel.) from Polish Waters. Acta Hydrobiol. 19: 183-190 (1977c).

Czeczuga, B.: Comparative Studies of Carotenoids in the Fauna of Gullmar Fjord (Bohuslän, Sweden) IV. Ascidiella aspersa (O.F.Müller) (Tunicata). Hydrobiol. 55: 77-79 (1977d).

Czygan, F.-Chr.: Canthaxanthin als Sekundär-Carotinoid einiger Grünalgen. Experentia 20: 573-574 (1964).

Czygan, F.-Chr.: Über den Stoffwechsel von Keto-Carotinoiden in niederen Krebsen. Z. Naturforsch. 21b: 801-805 (1966a)

Czygan, F.-Chr.: Echinenon als Sekundär-Carotinoid einiger Grünalgen. Z. Naturforsch. 21b: 197-198 (1966b).

Czygan, F.-Chr.: Sekundär-Carotinoide in Grünalgen I. Chemie, Vorkommen und Faktoren, welche die Bildung dieser Polyene beeinflussen. Arch. Microbiol. 61: 81-102 (1968a).

Czygan, F.-Chr.: Sekundär-Carotinoide in Grünalgen II. Untersuchungen zur Biogenese. Arch. Microbiol. 62: 209-236 (1968b).

Czygan, F.-Chr.: Untersuchungen über die Bedeutung der Biosynthese von Sekundär-Carotinoiden als Artmerkmal bei Grünalgen. Arch. Microbiol. 74: 77-81 (1970).

Czygan, F.-Chr.: Biogenetische Beziehungen zwischen den Keto-Carotinoiden des Kartoffelkäfers (Chrysomela decemlineata L.) und den Chloroplastenpigmenten der Kartoffelpflanze (Solanum tuberosum L.). Z. Pflanzenphysiol. 67: 33-42 (1972).

Deason, T.R., Czygan, F.-Chr., and Soeder, C.J.: Taxonomic Significance of Secondary Carotenoid Formation in Neospongiococcum (Chlorococcales, Chlorophyta). J.Phycol. 13: 176-180 (1977).

Donkin, P.: Ketocarotenoid Biosynthesis by Haematococcus lacustris. Phytochem. 15: 711-715 (1976).

Evstigneyev, V.B., and Paramonova, L.I.: The Role of Fucoxanthin in Photosynthesis, IUB 10. Internat. Congr. Biochem., Hamburg 25.-31.7.1976, Nr. 06-2-211.

Fott, B.: Algenkunde. G. Fischer Verlag Stuttgart 1971

Fox, D.L.: Feather Carotenoids of an Interspecific Hybrid Flamingo. Comp. Biochem. Physiol. B 48: 295-298 (1974).

Goodwin, T.W.: Algal Carotenoids in: Goodwin, T.W. (ed.) Aspects of Terpenoid Chemistry and Biochemistry, Acad. Press London + New York 1971, 315-356.

Goodwin, T.W.: Carotenoids and Biliproteins in: Stewart, W. D.P. (ed.) Algal Physiology and Biochemistry, Blackwell Sci. Publ. Oxford London Edinburgh Melbourne 1974, 176-205.

Gross, J., Carmon, M., Liftshitz, A., and Sklarz, B.: Carotenoids of the Invertebrates of the Red Sea (Eilat Shore). Carotenoids of the Crinoid Lamprometra klunzingeri (Echinodermata). Comp. Biochem. Physiol. B 52: 459-464 (1975).

Hager, A., and Stransky, H.: Das Carotinoidmuster und die Verbreitung des lichtinduzierten Xanthophyllcyclus in verschiedenen Algenklassen. III. Grünalgen. Arch. Microbiol. 72: 68-83 (1970a).

Hager, A., and Stransky, H.: Das Carotinoidmuster und die Verbreitung des lichtinduzierten Xanthophyllcyclus in verschiedenen Algenklassen. V. Einzelne Vertreter der Cryptophyceae, Euglenophyceae, Bacillariophyceae, Chrysophyceae und Phaeophyceae. Arch. Microbiol. 73: 77-89 (1970b).

Jeffrey, S.W., Douce, R., and Benson, A.A.: Carotenoid Transformations in the Chloroplast Envelope. Proc.Nat.

Acad.Sci. USA 71: 807-810 (1974).

Jeffrey, S.W., Sielicki, M., and Haxo, F.T.: Chloroplast Pigment Patterns in Dinoflagellates. J. Phycol. 11: 374-384 (1975).

Jeffrey, S.W., and Vesk, M.: Further Evidence for a Membrane-Bound Endosymbiont within the Dinoflagellate Peridinium foliaceum. J. Phycol. 12: 450-455 (1976).

Jensen, A., and Sakshaug, E.: Producer-Consumer Relationships in the Sea. I. Prelimenary Studies on Phytoplankton Density and Mytilus Pigmentation. J. exp. mar. Biol. Ecol. 5: 180-186 (1970a).

Jensen, A., and Sakshaug, E.: Producer-Consumer Relationships in the Sea. II. Correlation between Mytilus Pigmentation and the Density and Consumption of Phytoplanktonic Populations in Inshore Waters. J.exp. mar. Biol. Ecol. 5: 246-253 (1970b).

Johansen, J.E., Svec, W.A., Liaaen-Jensen, S., and Haxo, F. T.: Carotenoids of Dinophyceae. Phytochem. 13: 2261-2271 (1974).

Katayama, T., Yokoyama, H., and Chichester, C.O.: The Biosynthesis of Astaxanthin I. The Structure of α-Doradexanthin and β-Doradexanthin. Intern. J.Biochem. 1: 438-444 (1970).

Katayama, T., Kunisaki, Y., Shimaya, M., Simpson, K.L., and Chichester, C.O.: The Biosynthesis of Astaxanthin XIV. The Conversion of Labeled β-carotene-15,15-$^3H_2$ into Astaxanthin in the Crab, Portunus trituberculatus. Comp. Biochem. Physiol. B 46: 269-272 (1973).

Kayser, H.: Metabolites of β,β-carotene in the Stick Insect, Carusius morosus Br., Compounds with 2-one and 3,4-Didehydro-2-one Structure. Z. Naturforsch. 32c: 327-336 (1977a).

Kayser, H.: Conversion of (^{14}C)β-Carotene to its 2-Hydroxy and 3-Hydroxy Metabolites by two Moth Species.Comp. Biochem. Physiol. B 58: 177-181 (1977b).

Khare, A., Moss, G.P., Weedon, B.C.L., and Matthews, A.D.: Identification of Astaxanthin in Scottish Salmon . Comp. Biochem. Physiol. B 45: 971-973 (1973).

Krinsky, N.I.: Function in: Isler, O. (ed.) Carotenoids. Birkhäuser Verlag Basel + Stuttgart 1971, 669-716

Lee, W.L., and Gilchrist, B.M.: Pigmentation Color Change and the Ecology of the Marine Isopod Idotea resecata. J. exp. mar. Biol. Ecol. 10: 1-27 (1972).

Liaaen-Jensen, S.: Algal Carotenoids and Chemosystematics in: Faulkner, D.J. and Fenical, W.H. (eds.) Marine Natural Products Chemistry, Plenum Press New York + London 1977, 239-259.

Lichtenthaler, H.K., and Verbeek, L.: Die Hemmung der Carotinoidsynthese im Stickstoffmangel. Planta (Berl.) 112: 265-272 (1973).

Mathews, M.M.: Protective Effect of β-Carotene against Lethal Photosensitization by Haemetoporphyrin. Nature 203 : 1092 (1964)

Mc Millan, J.D., Maxwell, W.A., and Chichester, C.O.: Function of Carotenoids in Protection of Rhodotorula glutinis against Irradiation from a Gas Laser. Photochem. Photobiol. 5: 555-565 (1966).

Norgård, S., Svec, W.A., Liaaen-Jensen, S., Jensen, A., and Guillard, R.R.L.: Chloroplast Pigments and Algal Systematics. Biochem. System. Ecol. 2: 3-6 (1974a).

Norgård, S., Svec, W.A., Liaaen-Jensen, S., Jensen, A., and Guillard, R.R.L.: Algal Carotenoids and Chemotaxonomy. Biochem. System. Ecol. 2: 7-9 (1974b).

Prézelin, B.B.: The Role of Peridinin-Chlorophyll a - Proteins in the Photosynthetic Light Adaption of the Marine Dinoflagellate, Glenodinium sp. Planta (Berl.) 130: 225-233 (1976).

Radunz, A., and Schmid, G.H.: The Effect of Antibodies to Neoxanthin on Electron Transport on the Oxygen Evolving Side of Photosystem II and the Reactions of this Antiserum with Various Chloroplast Preparations. Z. Naturforsch. 30c: 622-627 (1975).

Radunz, A.: Investigations on the Localization and Function of Lipids and Pigments in the Thylakoid Membrane by Means of Specific Antisera. Abstract of the Symposium "Lipid And Lipidpolymers in Higher Plants", Karlsruhe, F.R. Germany 18.-21.7.1976.

Sprey, B.: Die Lokalisierung von Sekundär-Carotinoiden in Haematococcus pluvialis. Protoplasma 71: 235-250 (1970).

Stransky, H., and Hager, A.: Das Carotinoidmuster und die Verbreitung des lichtinduzierten Xanthophyllcyclus in verschiedenen Algenklassen II. Xanthophyceae. Arch. Microbiol. 71: 164-190 (1970a).

Stransky, H., and Hager, A.: Das Carotinoidmuster und die Verbreitung des lichtinduzierten Xanthophyllcyclus in verschiedenen Algenklassen IV. Cyanophyceae and Rhodophyceae. Arch. Microbiol. 72: 84-96 (1970b).

Stransky, H., and Hager, A.: Das Carotinoidmuster und die Verbreitung des lichtinduzierten Xanthophyllcyclus in verschiedenen Algenklassen VI. Chemosystematische Betrachtung. Arch. Microbiol. 73: 315-323 (1970c).

Tanaka, Y., Matsuguchi, H., Katayama, T., Simpson, K.L., and Chichester, C.O.: The Biosynthesis of Astaxanthin XVI. The Carotenoids in Crustaceae. Comp. Biochem. Physiol. B 54: 391-393 (1976a).

Tanaka, Y., Matsuguchi, H., Katayama, T., Simpson, K.L., and Chichester, C.O.: The Biosynthesis of Astaxanthin -XVIII. The Metabolism of the Carotenoids in the Prawn, Penaeus japonicus Bate. Bull. Jap. Soc. Sci. Fish. 42: 197-202 (1976b).

Thommen, H.: Metabolism in: Isler, O. (ed.) Carotenoids. Birkhäuser Verlag Basel + Stuttgart 1971, 637-668.

Tupa, D.D.: An Investigation of Certain Chaetophoralean Algae. Beihefte zur Nova Hedwigia 46: 1-155 (1974).

Webber, R., Barlow, G.W., and Brush, A.H.: Pigments of a

Color Polymorphism in a Cichlid Fish. Comp. Biochem. Physiol. B 44: 1127-1135 (1973).

Weber, A.: Über die Chloroplastenfarbstoffe einiger Conjugaten. Flora, Abt. A 160: 457-473 (1969).

Weber, A., and Czygan, F.-Chr.: Chlorophylle und Carotinoide der Chaetophirineae (Chlorophyceae, Ulotrichales) I. Siphonaxanthin in Microthamnion kuetzingianum Neagl.. Arch. Microbiol. 84: 243-253 (1972).

Weber, A.: Über die Chlorophylle und Carotinoide einiger Ulotrichales. Mitt. Staatsinst. Allg. Bot. Hamburg 14: 25-29 (1973).

Weber, A.: Chlorophylle und Carotinoide der Chaetophorineae (Chlorophyceae, Ulotrichales) II. Der Einfluß unterschiedlicher Stickstoffkonzentrationen auf die Pigmentgarnitur und die Morphogenese der Grünalge Fritschiella tuberosa Iyeng. Arch. Microbiol. 102: 45-52 (1975).

Whittle, S.J., and Casselton, P.J.: The Chloroplast Pigments of the Algal Classes Eustigmatophyceae and Xanthophyceae I. Eustigmatophyceae. Br. phycol. J. 10: 179-191 (1975a).

Whittle, S.J., and Casselton, P.J.: The Chloroplast Pigments of the Algal Classes Eustigmatophyceae and Xanthophyceae II. Xanthophyceae. Br. phycol. J. 10: 192-204 (1975b).

Withers, N., and Haxo, F.T.: Chlorophyll c_1 and c_2 Extraplastidic Carotenoids in the Dinoflagellate, Peridinium foliaceum. Plant Sci. Lett. 5: 7-15 (1975).

Yamamato, H.Y., Chichester, C.O., and Nakayama, T.O.: Biosynthetic Origin of Oxygen in the Leaf Xanthophyll. Arch. Biochem. Biophys. 96: 645-649 (1962).

The Distribution of Chemical Elements in Algae

Toshio Yamamoto, Yukiko Otsuka, Masaaki Okazaki
Ken-ichi Okamoto
Department of Chemistry, Kyoto University of Education
Fushimi-ku, Kyoto, 612, Japan

Thorough knowledge of the distribution of chemical elements in seaweeds is of vital importance to such areas of study as biogeochemistry of the ocean, algal physiology, and pharmaceutical application of algae. Vinogradov(1) compiled data from about 2300 reports on the elemental composition of marine organisms, and more recent reports on trace elements in marine organisms were reviewed by Goldberg(2). The biochemistry of trace elements was described by Bowen(3). Thus, a considerable amount of knowledge on the elemental composition of various marine organisms has been accumulated. However, few systematic studies on the same subject in seaweeds appeared in literature.

Because of differences in the treatment of materials and in analytical procedures, it is often difficult to compare the published results by various authors. The purpose of our study is therefore a systematic investigation on the contents of twenty elements in Japanese seaweeds.

SECTION I describes the analytical procedures employed for each element and the results thus obtained. A general regularity in the distribution of these elements among the species emerged from the results, and is discussed in SECTION II, along with some mathematical treatment of the results, indicating the characteristics of each algal species and the behavior of each element on the general distribution in the algae.

Marine Algae in Pharmaceutical Science
edited by H. A. Hoppe, T. Levring, Y. Tanaka

References

1) Vinogradov,A.P. :"The elementary chemical composition of marine organisms." Yale University, New Haven(1953).
2) Goldberg,E.D. :"Review of trace element concentrations in marine organisms." Puerto Rico Nuclear Center.
3) Bowen,H.J.M. :"Trace element in biochemistry." Academic Press, London and New York(1966).

Experimental

Most of the seaweeds used in the present study were collected on the shore of Tomo Island, Wakayama Prefecture, Japan. The details of collection sites and collection date of individual samples were described in our original articles cited in SECTION I. The samples were washed thoroughly with distilled water and dried. Before analysis, the samples were dried to constant weight at 105°C, and gradually ashed at 400-500°C in an electronic furnace. The ashed samples were analysed, and the mean values of total ash of each species are summarized in Table 1, where the specimens are arranged according to algal taxonomy. Species listed in the tables throughout this article should be referred to by the code number given in Table 1 for identification. An analytical method was selected for each element with the aim of achieving maximum accuracy. The analytical methods and results are summarized in SECTION I. Only mean values of each species are given in the following tables for each element. The values for individual samples of each species are described in our original articles cited under each element.

Table 1. List of samples

Code No.	Species	Total ash(%) (mean value)	Number of samples	Genus	Division
1	*Monostroma nitidum*	22.5	2	*Monostroma*	CHLOROPHYTA
2	*Ulva pertusa*	18.8	18	*Ulva*	
3	*Ulva conglobata*	15.9	1		
4	*Letterstedtia japonica*	10.3	1	*Letterstedtia*	
5	*Enteromorpha compressa*	22.6	7	*Enteromorpha*	
6	*Enteromorpha linza*	18.0	2		
7	*Cladophora wrightiana*	18.0	1	*Cladophora*	
63	*Cladophora densa*	19.7	2		
8	*Chaetomorpha spiralis*	9.10	1	*Chaetomorpha*	
9	*Chaetomorpha crassa*	11.3	3		
10	*Caulerpa okamurai*	18.2	1	*Caulerpa*	
11	*Codium pugniformis*	36.0	1	*Codium*	
12	*Codium fragile*	9.47	3		
64	*Spathoglossum pacificum*	23.0	2	*Dictyota*	PHAEOPHYTA
13	*Dictyopteris prolifera*	7.80	3	*Dictyopteris*	
14	*Padina arborescens*	14.2	8	*Padina*	
15	*Tinocladia crassa*	11.0	1	*Tinocladia*	
16	*Chordaria sp.*	11.4	1	*Chordaria*	
17	*Ishige okamurai*	13.1	3	*Ishige*	
18	*Ishige foliacea*	14.7	4		
19	*Desmarestia vilidis*	3.90	2	*Desmarestia*	

Table 1. (continued)

20	_Myelophycus caespitosus_	14.2	1	_Myelophycus_	PHAEOPHYTA
21	_Scytosiphon lomentaria_	23.3	5	_Scytosiphon_	
65	_Endarachne binghamiae_	18.8	2	_Endarachne_	
22	_Eisenia bicyclis_	14.9	9	_Eisenia_	
23	_Ecklonia cava_	12.7	1	_Ecklonia_	
24	_Undaria pinnatifida_	20.9	5	_Undaria_	
25	_Cystophyllum sisymbrioides_	18.3	2	_Cystophyllum_	
26	_Hizikia fusiforme_	20.6	12	_Hizikia_	
27	_Sargassum piluliferum_	12.3	3	_Sargassum_	
28	_Sargassum patens_	16.3	2		
29	_Sargassum horneri_	16.7	3		
30	_Sargassum serratifolium_	16.4	1		
31	_Sargassum tortile_	13.5	5		
32	_Sargassum giganteifolium_	12.2	2		
33	_Sargassum ringgoldianum_	14.4	11		
66	_Sargassum confusum_	23.6	2		
34	_Sargassum fulvellum_	14.6	6		
35	_Sargassum thunbergii_	20.7	23		
36	_Sargassum hemiphyllum_	16.9	7		
37	_Sargassum micracanthum_	13.7	1		
38	_Sargassum nigrifolium_	11.1	1		
39	_Sargassum tenuifolium_	15.2	1		
40	_Galaxaura fastigiata_	76.3	1	_Galaxaura_	RHODOPHYTA
41	_Gelidium amansii_	7.66	10	_Gelidium_	

Table 1. (continued)

42	*Acanthopeltis japonica*	12.1	5	*Acanthopeltis*	RHODOPHYTA
43	*Chondrococcus japonicus*	12.7	1	*Chondrococcus*	
44	*Rhodopeltis borealis*	70.2	1	*Rhodopeltis*	
45	*Grateloupia filicina*	21.3	4	*Grateloupia*	
46	*Grateloupia okamurai*	9.20	1		
47	*Grateloupia turuturu*	15.9	1		
48	*Carpopeltis rigida*	3.60	1	*Carpopeltis*	
49	*Carpopeltis flabellata*	13.8	5		
50	*Gloiosiphonia capillaris*	13.1	1	*Gloiosiphonia*	
51	*Gloiopeltis tenax*	11.5	2	*Gloiopeltis*	
67	*Gloiopeltis furcata*	20.8	1		
52	*Meristotheca papulosa*	16.4	1	*Meristotheca*	
53	*Plocamium telfairiae*	19.9	1	*Plocamium*	
54	*Hypnea saidana*	11.0	1	*Hypnea*	
55	*Hypnea sp.*	16.4	1		
56	*Hypnea charoides*	21.2	2		
57	*Sarcodia ceylanica*	10.0	1	*Sarcodia*	
68	*Gracilaria textorii*	15.2	1	*Gracilaria*	
58	*Ceratodictyon spongiosum*	20.6	1	*Ceratodictyon*	
59	*Gymnogongrus flabelliformis*	12.5	2	*Gymnogongrus*	
61	*Chondrus ocellatus*	18.4	2	*Chondrus*	
62	*Ceramium kondoi*	21.1	1	*Ceramium*	
0	*Polysiphonia urceolata*	13.7	1	*Polysiphonia*	

SECTION I (Method and Result)

Element : Boron

To the acetic acid solution of ashed sample, the spectrophotometric method(1) with curcumin was applied. For compensating the small salt error, the determination of boron content was carried out by adding the standard solution to each sample(2,3).

Table 2. Boron content in algae(μg/g of dried alga)

Code* No.	B content (mean value)	Number of samples	Code* No.	B content (mean value)	Number of samples
CHLOROPHYTA					
1	50.0	1	26	109	4
2	64.7	6	28	65.0	2
4	97.0	1	29	64.0	2
5	116	4	31	65.5	2
6	114	1	33	97.0	3
63	155	2	66	147	2
8	8.0	1	34	119	2
9	140	1	35	122	8
10	100	1	36	122	3
11	83.0	1	39	165	1
12	17.0	1	RHODOPHYTA		
PHAEOPHYTA			40	285	1
64	122	2	41	177	3
13	67.0	1	45	54.0	3
14	125	2	47	37.0	1
17	95.0	2	48	101	1
18	73.0	3	49	54.5	2
19	16.0	1	50	319	1
20	102	1	51	34.0	1
21	52.6	4	67	18.0	1
65	57.5	2	52	119	1
22	96.7	3	56	337	1
24	50.3	4	68	18.0	1
25	171	1	58	91.0	1

59	217	1	0	133	1
61	76.0	1			

* The species are shown in Table 1 according to this code No.

References

1) Greenhalge,R., Riley,J.P. : The development of a reproducible spectrometric curcumin method for determing boron, and its application to sea water. Analyst 87, 970-976(1962).
2) Yamamoto,T.,Yamaoka,T., Fujita,T., Isoda,C. : Boron content in seaweeds. Rec. Oceanogr. Wks. Japan 11, 7-13(1971).
3) Yamamoto,T., Yamaoka,T., Fujita,T., Isoda,C. : Boron content in marine plankton. Rec. Oceanogr. Wks. Japan 12, 13-21(1973).

Element : Sodium

Ashed sample was extracted with boiling water, acidified with acetic acid. Sodium in the solution was determined gravimetrically with uranyl magnesium acetate reagent(1,2).

Table 3. Sodium content in algae(μg/g of dried alga)

Code* No.	Na content (mean value)	Number of samples	Code* No.	Na content (mean value)	Number of samples
CHLOROPHYTA					
1	17900	1	22	14800	5
2	3600	2	26	16900	1
3	1000	1	27	4000	1
9	6500	1	29	3600	1
11	66600	1	31	2700	1
PHAEOPHYTA			32	4800	1
13	4700	1	33	7170	3
14	16100	1	34	7100	1
15	2300	1	35	8750	6
16	5400	1	37	6700	1
17	15300	1	RHODOPHYTA		
19	400	1	41	200	1
21	9900	1	42	1800	2
			48	1100	1
			49	10700	1

* The species are shown in Table 1 according to this code No.

References

1) Ishibashi,M., Yamamoto,T. : The content of ash, sodium and potassium in seaweeds. Rec. Oceanogr. Wks. Japan, Special No. 3, 109-115(1959).
2) Fujita,T. : Concentration of major chemical elements in marine plankton. Geochem. J. 4, 143-156(1971).

Element : Magnesium

After removing iron and aluminum as hydroxide from hydrochloric acid solution of ashed sample, calcium was separated as oxalate. Then, magnesium in the filtrate was determined gravimetrically as magnesium oxinate(1,2).

Table 4. Magnesium content in algae(μg/g of dried alga)

Code* No.	Mg content (mean value)	Number of samples	Code* No.	Mg content (mean value)	Number of samples
CHLOROPHYTA					
1	13000	2	29	16900	1
2	25800	12	30	16500	1
3	36500	1	31	8670	3
4	7200	1	32	13600	1
5	19800	1	33	10700	5
9	11300	2	34	12600	1
11	15400	1	35	9710	16
12	11100	1	36	8930	3
PHAEOPHYTA			37	11800	1
13	4800	2	RHODOPHYTA		
14	7970	3	40	12400	1
15	7900	1	41	5340	5
16	6700	1	42	2300	3
17	8400	1	44	9100	1
18	9050	2	47	5900	1
19	300	1	49	7800	1
21	10700	1	51	3300	1
22	9040	7	52	8400	1
25	9600	2	56	3400	1
26	9870	4	59	4200	1
27	6800	1	61	10800	1
28	9900	1	0	15700	1

* The species are shown in Table 1 according to this code No.

References

1) Ishibashi,M., Yamamoto,T. : The content of calcium, magnesium and phosphorus in seaweeds. Rec. Oceanogr. Wks. Japan, Special No. 4, 73-78(1960).
2) Fujita,T. : Concentration of major chemical elements in marine plankton. Geochem. J. 4, 143-156(1971).

Element : Aluminum

Ashed sample was dissolved in hydrochloric acid, aluminum in the silicic-excluded solution was determined by a spectrophotometric method combined with chloroform extraction of the oxinate(1,2).

Table 5. Aluminum content in algae(μg/g of dried alga)

Code* No.	Al content (mean value)	Number of samples	Code* No.	Al content (mean value)	Number of samples
CHLOROPHYTA					
1	1150	1	65	62	2
2	559	6	22	96	6
3	1010	1	23	57	1
5	688	3	24	226	14
6	630	2	26	203	9
7	765	1	27	663	3
63	1380	2	28	340	2
9	559	2	29	115	3
10	3320	1	30	110	1
11	717	1	31	81	3
12	1650	2	32	165	1
PHAEOPHYTA			33	96	8
64	1860	2	66	1350	2
13	633	2	34	102	2
14	767	3	35	1030	15
15	2370	1	36	338	3
16	590	1	37	438	1
17	98	2	38	81	1
18	140	1	RHODOPHYTA		
20	1430	1	41	142	3
21	1400	4	42	321	2

43	1080	1	53	1470	1
45	276	2	54	517	1
46	141	1	55	1250	1
49	241	4	56	329	1
50	60	1	57	151	1
67	65	1	68	143	1
52	731	1			

* The species are shown in Table 1 according to this code No.

References
1) Morii,F. : Chloroform extraction of oxinates of iron, aluminum and manganese, and their spectrophotometric determinations. Nippon Kagaku Zasshi 82, 1507-1509(1961).
2) Morii,F. : Aluminum and iron contents in seaweeds. Nippon Kagaku Zasshi 83, 77-81(1962).

Element : Silicon

After fusion of ashed sample with calcium carbonate, silicon in the leached solution was determined spectrophotometrically by the silicon molybdate method(1,2).

Table 6. Silicon content in algae(μg/g of dried alga)

Code* No.	Si content (mean value)	Number of samples	Code* No.	Si content (mean value)	Number of samples
CHLOROPHYTA			21	2480	1
2	5230	8	24	1120	9
5	21800	1	35	6820	9
PHAEOPHYTA			RHODOPHYTA		
18	13900	1	41	3270	4

* The species are shown in Table 1 according to this code No.

References
1) Watanabe,S. : A rapid analysis of nonmetals in plant tissues. Japan Anal. 7, 771-775(1958).
2) Fujita,T. : Concentration of major chemical elements in marine plankton. Geochem. J. 4, 143-156(1971).

Element : Phosphorus

From silicic-excluded hydrochloric acid solution of ashed sample, phosphorus was separated as ammonium phosphomolybdate, and determined gravimetrically by the magnesium pyrophosphate method(1,2).

Table 7. Phosphorus content in algae(μg/g of dried alga)

Code* No.	P content (mean value)	Number of samples	Code* No.	P content (mean value)	Number of samples
CHLOROPHYTA					
1	900	1	26	900	2
2	1450	2	27	700	1
3	800	1	29	1200	1
9	800	1	30	200	1
11	2800	1	31	700	1
12	1500	1	32	1100	1
PHAEOPHYTA			33	750	2
13	1800	1	34	900	1
14	980	1	35	1100	7
15	1100	1	37	500	1
16	700	1	RHODOPHYTA		
17	1700	1	41	900	2
19	900	1	42	1050	2
21	1800	1	49	1200	1
22	900	4	52	1300	1

* The species are shown in Table 1 according to this code No.

References

1) Ishibashi,M., Yamamoto,T. : The content of calcium, magnesium and phosphorus in seaweeds. Rec. Oceanogr. Wks. Japan, Special No. 4, 73-78(1960).

2) Fujita,T. : Concentration of major chemical elements in marine plankton. Geochem. J. 4, 143-156(1971).

Element : Potassium

After extracting ashed sample with boiling water, potassium in the solution was determined gravimetrically with sodium cobaltinitrite reagent(1,2).

Table 8. Potassium content in algae(μg/g of dried alga)

Code* No.	K content (mean value)	Number of samples	Code* No.	K content (mean value)	Number of samples
CHLOROPHYTA					
1	7600	1	26	34300	1
2	5100	2	27	3700	1
3	1700	1	29	6400	1
9	13200	1	31	5400	1
11	25800	1	32	8500	1
PHAEOPHYTA			33	17400	3
13	7000	1	34	13400	1
14	35700	1	35	20900	6
15	1400	1	37	12500	1
16	11200	1	RHODOPHYTA		
17	7600	1	41	300	1
19	700	1	42	1100	2
21	8700	1	48	1000	1
22	22100	5	49	6500	1

* The species are shown in Table 1 according to this code No.

References

1) Ishibashi,M., Yamamoto,T. : The content of ash, sodium and potassium in seaweeds. Rec. Oceanogr. Wks. Japan, Special No. 3, 109-115(1959).
2) Fujita,T. : Concentration of major chemical elements in marine plankton. Geochem. J. 4, 143-156(1971).

Element : Calcium

From the silicic-excluded hydrochloric acid solution of ashed sample, iron, aluminum and phosphorus were separated by precipitation with ammonium hydroxide, calcium in the filtrate was then determined gravimetrically by calcium oxalate method(1,2).

Table 9. Calcium content in algae(μg/g of dried alga)

Code* No.	Ca content (mean value)	Number of samples	Code* No.	Ca content (mean value)	Number of samples
CHLOROPHYTA					
1	13600	2	29	16400	1
2	8030	12	30	29200	1
3	8300	1	31	28800	3
4	7300	1	32	15700	1
5	11900	1	33	20800	5
9	10300	2	34	19800	1
11	8600	1	35	26200	16
12	11400	1	36	27600	3
PHAEOPHYTA			37	24100	1
13	15000	2	RHODOPHYTA		
14	19200	3	40	316000	1
15	6300	1	41	7000	4
16	11900	1	42	4770	3
17	14800	1	44	307000	1
18	12400	2	47	9000	1
19	8700	1	49	3800	1
21	31900	1	51	5100	1
22	15000	7	52	6200	1
25	48700	2	56	9600	1
26	17200	4	59	2800	1
27	19000	1	61	8300	1
28	16700	1	0	30300	1

* The species are shown in Table 1 according to this code No.

References

1) Ishibashi,M., Yamamoto,T. : The content of calcium, magnesium and phosphorus in seaweeds. Rec. Oceanogr. Wks. Japan, Special No. 4, 73-78(1960).
2) Fujita,T. : Concentration of major chemical elements in marine plankton. Geochem. J. 4, 143-156(1971).

Element : Titanium

From silicic-excluded hydrochloric acid solution of ashed sample, titanium was separated from most of the other elements by the cupferron chloroform extraction method. The extract was evaporated and ashed with nitric acid. The ash was dissolved in hydrochloric acid. Iron in the hydrochloric acid was separated from titanium by isopropyl ether extraction. Titanium in the hydrochloric acid solution was determined spectrophotometrically by hydrogen peroxide method(1,2).

Table 10. Titanium content in algae(μg/g of dried alga)

Code* No.	Ti content (mean value)	Number of samples	Code* No.	Ti content (mean value)	Number of samples
CHLOROPHYTA			26	89.4	1
1	23.6	1	30	14.5	1
2	39.3	5	31	7.7	1
3	33.1	1	33	5.7	2
4	47.5	1	34	16.5	1
5	113	2	35	36.9	8
8	14.4	1	36	16.8	1
9	8.7	1	RHODOPHYTA		
10	161	1	40	29.6	1
12	22.9	1	41	23.1	3
PHAEOPHYTA			42	40.7	1
14	82.3	2	47	13.5	1
15	178	1	49	19.3	2
17	4.5	1	51	8.4	1
18	27.6	2	58	6.2	1
22	5.8	4	59	2.4	1
25	6.5	1	0	53.3	1

* The species are shown in Table 1 according to this code No.

References

1) Sandell,E.B. : "Colorimetric determination of traces of metals". Interscience Publishers Inc.,New York,870-874(1959).
2) Yamamoto,T., Fujita,T., Ishibashi,M. : Vanadium and titanium contents in seaweeds. Rec. Oceanogr. Wks. Japan 10, 125-135 (1970).

Element : Vanadium

From silicic-excluded nitric acid solution of ashed sample, vanadium was coprecipitated with ferric hydroxide. The hydroxide was dissolved in sulfuric acid. The solution was extracted with N-benzoylphenyl-hydroxylamine reagent in the mixed solvent of chloroform-alcohol. Vanadium in the extract was determined spectrophotometrically(1,2).

Table 11. Vanadium content in algae(μg/g of dried alga)

Code* No.	V content (mean value)	Number of samples	Code* No.	V content (mean value)	Number of samples
CHLOROPHYTA			26	1.36	2
1	6.00	1	28	2.27	1
2	3.32	10	30	2.59	1
4	2.84	1	31	6.05	1
5	7.31	2	32	0.79	1
7	2.02	1	33	1.51	3
8	1.15	1	34	0.74	2
9	1.91	1	35	4.37	9
10	10.1	1	36	3.11	2
12	1.72	1	RHODOPHYTA		
PHAEOPHYTA			40	2.28	1
13	1.87	1	41	2.48	3
14	4.59	3	42	3.42	1
15	3.70	1	47	2.20	1
17	2.74	1	49	3.15	2
18	5.69	2	51	0.66	1
21	2.31	1	58	5.38	1
22	0.83	5	59	2.03	2
25	8.68	1	0	4.80	1

* The species are shown in Table 1 according to this code No.

References

1) Tomioka,T. : Extraction-absorptiometric determination of vanadium with N-benzoylphenyl-hydroxylamine. Japan Anal. 12 271-279(1963).

2) Yamamoto,T., Fujita,T., Ishibashi,M. : Vanadium and titanium contents in seaweeds. Rec. Oceanogr. Wks. Japan 10,125-135 (1970).

Element : Chromium

After fusing ashed sample with potassium pyrosulfate, the cake was dissolved in sulfuric acid. Chromium in the sulfuric acid solution was coprecipitated with aluminum hydroxide. Chromium in the hydroxide was determined spectrophotometrically by the diphenylcarbazide method(1,2).

Table 12. Chromium content in algae(μg/g of dried alga)

Code* No.	Cr content (mean value)	Number of samples	Code* No.	Cr content (mean value)	Number of samples
CHLOROPHYTA			22	0.74	1
1	2.07	1	26	0.63	1
2	0.95	3	32	0.52	1
3	1.21	1	33	0.66	2
9	1.82	1	35	1.54	3
11	2.53	1	36	2.83	1
PHAEOPHYTA			37	1.24	1
14	1.42	1	RHODOPHYTA		
17	0.89	1	41	0.50	1
18	1.18	1	58	1.42	1

* The species are shown in Table 1 according to this code No.

References

1) Yamamoto,T. : Quantitative determination of chromium in seaweeds. Nippon Kagaku Zasshi 81, 388-391(1960).
2) Yamamoto,T., Kadowaki,S., Carpenter,J.H. : Spectrophotometric determinations of chromium in marine substances. Geochem. J. 8, 123-133(1974).

Element : Manganese

From silicic-excluded sulfuric acid solution of ashed sample, manganese was determined spectrophotometrically. A modified method of oxidation was applied : a small amount of silver nitrate was added as a catalyser to oxidize manganese by sodium bismuthate(1,2,3).

Table 13. Manganese content in algae(µg/g of dried alga)

Code* No.	Mn content (mean value)	Number of samples	Code* No.	Mn content (mean value)	Number of samples
CHLOROPHYTA			26	29	7
1	55	1	27	38	2
2	97	3	29	14	1
3	97	1	30	4	1
7	87	1	31	20	1
9	151	2	32	13	1
11	52	1	33	13	5
12	299	1	34	28	1
PHAEOPHYTA			35	243	10
13	70	1	37	120	1
14	146	3	RHODOPHYTA		
15	227	1	41	43	3
16	122	1	42	60	4
17	7	1	43	270	1
18	31	1	48	44	1
19	9	1	49	28	1
20	54	1	50	76	1
21	78	1	52	337	1
22	8	5	57	48	1

* The species are shown in Table 1 according to this code No.

References

1) Ishibashi,M., Yamamoto,T. : Quantitative determination of manganese in seaweeds(I). Nippon Kagaku Zasshi 79, 1184-1187 (1958).
2) Ishibashi,M., Yamamoto,T. : Quantitative determination of manganese in seaweeds(II). Nippon Kagaku Zasshi 79, 1187-1190(1958).
3) Fujita,T., Yamamoto,T., Yamazi,I., Shigematsu,T. : The contents of ash, iron and manganese in marine plankton. Nippon Kagaku Zasshi 90, 680-686(1969).

Element : Iron

With silicic-excluded hydrochloric acid solution of ashed sample, iron was determined spectrophotometrically by O-phenanthroline method(1,3). In the simultaneous analysis of aluminum and iron contents in seaweeds, iron was determined by a spectrophotometric method combined with chloroform extraction of the oxinate(2).

Table 14. Iron content in algae(μg/g of dried alga)

Code* No.	Fe content (mean value)	Number of samples
CHLOROPHYTA		
1	900	2
2	764	15
3	609	1
4	847	1
5	1130	5
6	828	2
7	582	1
63	1610	2
8	371	1
9	356	3
10	3020	1
11	591	1
12	1330	3
PHAEOPHYTA		
64	1440	2
13	722	2
14	838	6
15	2140	1
16	584	1
17	89	3
18	399	3
19	402	2
20	1190	1
21	1490	5
65	93	2
22	98	8
23	85	1
24	193	14
25	256	1
26	158	10
27	541	3
28	256	2
29	112	3
30	92	1
31	112	4
32	105	2
33	110	10
66	885	2
34	161	5
35	709	21
36	406	5
37	367	1
38	75	1
39	81	1
RHODOPHYTA		
40	196	1
41	356	8
42	234	3
43	1090	1
45	130	2
46	185	1
47	168	1
49	299	5
50	58	1
51	160	1
67	67	1
52	465	1
53	1500	1
54	424	1

55	1010	1	58	1070	1
56	516	1	59	278	2
57	343	1	61	431	1
68	121	1	0	854	1

* The species are shown in Table 1 according to this code No.

References

1) Ishibashi,M., Yamamoto,T. : Iron content in seaweeds. Rec. Oceanogr. Wks. Japan, Special No. 4, 79-85(1960).
2) Morii,F. : Aluminum and iron contents in seaweeds. Nippon Kagaku Zasshi 83, 77-81(1962).
3) Fujita,T., Yamamoto,T., Yamazi,I., Shigematsu,T. : The contents of ash, iron and manganese in marine plankton. Nippon Kagaku Zasshi 90, 680-686(1969).

Element : Cobalt

From silicic-excluded nitric acid solution of ashed sample, cobalt was separated from most of other elements by 1-nitroso-2-naphtol chloroform extraction method. The extract was evaporated and ashed with nitric acid and perchloric acid. Cobalt in the ash was determined spectrophotometrically by the nitroso -R salt method(1,2,3).

Table 15. Cobalt content in algae(μg/g of dried alga)

Code* No.	Co content (mean value)	Number of samples	Code* No.	Co content (mean value)	Number of samples
CHLOROPHYTA			14	1.42	4
1	1.47	1	17	0.24	2
2	0.80	8	18	0.57	3
3	0.75	1	21	1.06	1
4	0.29	1	22	0.29	6
5	1.17	2	25	0.60	1
8	0.41	1	26	0.48	2
9	0.15	1	28	0.76	1
10	2.51	1	30	0.40	1
11	0.68	1	31	0.27	2
12	1.52	1	32	0.80	2
PHAEOPHYTA			33	0.32	3
13	0.53	1	34	0.56	3

35	1.70	10	47	0.26	1
36	1.23	2	49	0.63	2
39	0.55	1	51	0.53	1
RHODOPHYTA			58	0.72	1
40	0.43	1	59	0.75	2
41	0.34	3	61	0.65	1
42	0.40	1	0	0.52	1

* The species are shown in Table 1 according to this code No.

References

1) Saltzmann,B.E. : Microdetermination of cobalt in biological materials. Anal. Chem. 27, 284-287(1955).
2) Yamamoto,T., Fujita,T., Ishibashi,M. : On the inorganic constituents of seaweeds, determination of cobalt in seaweeds. Nippon Kagaku Zasshi 86, 49-53(1965).
3) Ishibashi,M., Yamamoto,T., Fujita,T. : Cobalt content in seaweeds. Rec. Oceanogr. Wks. Japan 7, 17-24(1964).

Element : Nickel

Silicic-excluded nitric solution of ashed sample was neutralized with ammoniacal citrate solution. Nickel in the solution was extracted with dimethyl-glyoxime-chloroform, and re-extracted with a dilute hydrochloric acid from the chloroform layer. Nickel in the hydrochloric acid layer was determined spectrophotometrically by the dimethyl-glyoxime-oxidizing agent method(1,2).

Table 16. Nickel content in algae(μg/g of dried alga)

Code* No.	Ni content (mean value)	Number of samples	Code* No.	Ni content (mean value)	Number of samples
CHLOROPHYTA			10	5.64	1
1	1.43	1	11	2.09	1
2	2.84	9	12	2.08	1
3	1.57	1	PHAEOPHYTA		
4	3.10	1	13	3.86	1
5	1.93	1	14	3.85	3
8	1.66	1	17	2.86	2
9	0.75	1	18	2.25	3

21	2.92	1	39	5.54	1
22	0.87	6	RHODOPHYTA		
25	1.04	1	40	4.31	1
26	4.05	2	41	1.22	3
28	4.10	1	42	1.32	1
30	1.10	1	47	0.23	1
31	0.84	2	49	1.42	2
32	0.54	2	51	0.71	1
33	1.28	4	58	5.04	1
34	4.24	3	59	1.67	2
35	5.01	9	61	1.53	1
36	3.52	2	0	3.68	1

* The species are shown in Table 1 according to this code No.

References

1) Sandell,E.B. : "Colorimetric determination of traces of metals". Interscience Publishers Inc., New York, 668-673 (1959).

2) Ishibashi,M., Yamamoto,T., Fujita,T. : Nickel content in seaweeds. Rec. Oceanogr. Wks. Japan 7, 25-32(1964).

Element : Copper

Copper in the hydrochloric acid solution of ashed sample was reduced with hydroxylamine hydrochloride. Tartaric acid was added to the solution as a masking agent. Copper in the solution was determined spectrophotometrically by the cuproine-isoamyl alcohol extraction method(1,2).

Table 17. Copper content in algae(μg/g of dried alga)

Code* No.	Cu content (mean value)	Number of samples	Code* No.	Cu content (mean value)	Number of samples
CHLOROPHYTA			12	19.6	1
2	13.1	4	PHAEOPHYTA		
3	21.1	1	13	17.8	1
7	13.2	1	14	9.4	3
9	11.7	2	15	12.7	1
10	18.1	1	17	9.9	1
11	27.7	1	18	7.4	1

19	9.2	1	37	13.3	1
20	12.6	1	RHODOPHYTA		
21	24.1	1	41	7.9	2
22	13.8	6	42	7.9	2
26	16.0	6	43	26.7	1
27	17.1	2	49	8.2	2
29	10.5	1	50	13.8	1
30	12.3	1	52	24.0	1
31	15.2	1	53	25.3	1
32	9.5	1	54	16.6	1
33	10.1	4	55	15.1	1
34	6.1	1	57	11.4	1
35	14.8	8			

* The species are shown in Table 1 according to this code No.

References

1) Host,J., Eeckout,J., Gills,J. : Spectrophotometric determination of copper with cuproine. Anal. Chim. Acta. 9, 263-274 (1953).
2) Ishibashi,M., Yamamoto,T., Morii,F. : Copper content in seaweeds. Rec. Oceanogr. Wks. Japan 6, 157-162(1962).

Element : Zinc

From silicic-excluded hydrochloric acid solution of ashed sample, zinc was extracted with a dithizone-carbon tetrachloride solution,and then re-extracted into an aqueous hydrochloric acid. After removing the hydrochloric acid by evaporation, the residue was dissolved in water, and after adding the supporting electrolyte, zinc in the solution was determined by the polarographic procedure(1,2,3).

Table 18. Zinc content in algae(μg/g of dried alga)

Code* No.	Zn content (mean value)	Number of samples	Code* No.	Zn content (mean value)	Number of samples
CHLOROPHYTA			4	214	1
1	213	2	5	245	2
2	138	11	7	80	1
3	62	1	8	153	1

9	154	3	35	323	12
11	332	1	36	106	2
12	164	2	37	53	1
PHAEOPHYTA			39	179	1
13	47	1	RHODOPHYTA		
14	137	6	40	230	1
15	43	1	41	155	5
17	109	2	42	172	2
18	170	3	43	124	1
19	39	2	47	128	1
20	65	1	49	165	3
21	189	1	50	205	1
22	104	6	51	83	1
25	340	1	52	139	1
26	79	3	53	64	1
27	40	2	54	37	1
28	232	1	55	128	1
30	18	1	57	84	1
31	170	1	58	83	1
32	63	2	59	119	2
33	60	6	61	169	1
34	83	2	0	194	1

* The species are shown in Table 1 according to this code No.

References

1) Ishibashi,M., Fujinaga,T., Morii,F., Kanchiku,Y., Kamiyama,F. : Determination of zinc, copper, lead, cadmium and nickel in seaweeds using dithizone extraction and polarographic method. Rec. Oceanogr. Wks. Japan 7, 33-36 (1964).

2) Ishibashi,M., Fujinaga,T., Yamamoto,T., Fujita,T., Watanabe,K. : The zinc contents of seaweeds. Nippon Kagaku Zasshi 86, 728-733(1965).

3) Fujita,T. : The zinc content in marine plankton. Rec. Oceanogr. Wks. Japan 11, 73-79(1972).

Element : Gallium

After fusing ashed sample with potassium pyrosulfate, the cake was dissolved in a dilute hydrochloric acid. Gallium in the solution was coprecipitated with aluminum hydroxide and was re-dissolved in hydrochloric acid. Gallium was then extracted with ethyl ether. After evaporating ether and dissolving the residue in a dilute hydrochloric acid, the 2-methyl-8-hydroxy-quinoliol chloroform extraction method was applied to the solution. Gallium in the chloroform layer was then determined fluorometrically(1,2).

Table 19. Gallium content in algae(μg/g of dried alga)

Code* No.	Ga content (mean value)	Number of samples	Code* No.	Ga content (mean value)	Number of samples
CHLOROPHYTA			29	0.03	1
2	0.17	6	30	0.04	1
5	0.20	1	31	0.03	1
6	0.07	1	33	0.05	5
11	0.20	1	34	0.64	1
12	0.17	2	35	0.22	5
PHAEOPHYTA			36	0.26	2
13	0.11	2	38	0.03	1
14	0.21	1	RHODOPHYTA		
18	0.08	1	41	0.07	3
21	0.62	1	42	0.08	1
22	0.05	5	46	0.06	1
23	0.02	1	49	0.15	3
24	0.16	1	50	0.11	1
26	0.11	3	52	0.18	1
27	0.27	2	62	0.26	1
28	0.08	1			

* The species are shown in Table 1 according to this code No.

References

1) Nishikawa,Y. : Determination of gallium in silicate rocks with 8-hydroxyquinaldine. Nippon Kagaku Zasshi 79, 236-238 (1958).
2) Yamamoto,T., Otsuka,Y., Uemura,K. : Gallium content in sea-weeds. J. Oceanogr. Soc. Japan 32, 182-186(1976).

Element : Strontium

With silicic-excluded hydrochloric acid solution of ashed sample, strontium was determined by atomic absoption spectrometry method using a standrad addition(1,2).

Table 20. Strontium content in algae(μg/g of dried alga)

Code* No.	Sr content (mean value)	Number of samples	Code* No.	Sr content (mean value)	Number of samples
CHLOROPHYTA			31	1980	2
1	180	1	33	1480	4
2	218	9	35	1650	12
4	170	1	36	2460	3
5	330	1	RHODOPHYTA		
9	230	1	40	9280	1
PHAEOPHYTA			41	90	3
13	1040	1	42	20	1
14	1310	2	44	11500	1
17	440	1	47	130	1
18	1170	2	51	30	1
22	1100	5	56	180	1
25	2720	2	59	90	1
26	1140	2	61	180	1
28	1740	1	0	2360	1
30	2360	1			

* The species are shown in Table 1 according to this code No.

References

1) Fujita,T., Shigematsu,T. : Atomic absorption spectrometric determination of strontium, calcium and magnesium in seaweeds. Japan Anal. 19, 893-899(1970).

2) Yamamoto,T., Fujita,T., Shigematsu,T. : Strontium content in seaweeds. Rec. Oceanogr. Wks. Japan 10, 29-38(1969).

Element : Molybdenum

From silicic-excluded hydrochloric acid solution of ashed sample, molybdenum was extracted with 2-benzoinoxime-chloroform. The chloroform extract was evaporated and the residue was ashed with nitric acid and perchloric acid. The ash was then dissolved in hydrochloric acid, and molybdenum in the solution was determined spectrophotometrically by the dithiol-buthyl acetate extraction method(1,2,3).

Table 21. Molybdenum content in algae(μg/g of dried alga)

Code* No.	Mo content (mean value)	Number of samples	Code* No.	Mo content (mean value)	Number of samples
CHLOROPHYTA			31	0.18	3
1	0.06	1	33	0.19	4
2	0.29	10	35	0.54	12
4	0.09	1	36	0.29	4
5	0.29	3	RHODOPHYTA		
8	0.11	1	40	0.64	1
9	0.11	1	41	0.27	3
12	0.09	1	42	0.25	1
PHAEOPHYTA			44	0.80	1
13	0.21	1	45	0.33	1
14	0.28	2	47	0.35	1
17	1.16	1	51	0.18	1
18	0.57	3	56	0.22	1
22	0.23	5	58	0.30	1
25	0.45	2	59	0.18	1
26	0.32	2	61	0.30	1
28	0.33	1	0	0.23	1
30	0.42	1			

* The species are shown in Table 1 according to this code No.

References

1) Ishibashi,M., Shigematsu,T., Nakagawa,Y. : Quantitative determination of tungsten and molybdenum in sea water. Bull. Inst. Chem. Res., Kyoto Univ. 32, 199-202(1954).
2) Clark,L.J., Axley,J.H. : Molybdenum determination in soils and rocks with dithiol. Anal. Chem. 27, 2000-2003(1955).
3) Yamamoto,T., Fujita,T., Shigematsu,T. : Molybdenum content in seaweeds. Rec. Oceanogr. Wks. Japan 9, 209-217(1968).

SECTION II (Discussion)

In the present phase of biogeochemical study of the sea, it has been important to find out the relations between the species of marine organisms and their contents of elements, and deduce a general law governing the concentrations of these elements in marine organisms and the environmental seawater. The following discussion is our attempt to approach the above-mentioned problem from the viewpoint of phycology(1, 2).

Data on the mean concentration of 19 elements in *Eisenia bicyclis*(code No.22) were composed in Table 22 from SECTION I (Table 2- Table 21). The values of oceanic residence time x of each element were obtained from the table given in literature(3). Concentration factor y for each element by *Eisenia bicyclis* was also calculated in the same table from the values of seaweed concentration and sea water concentration(3) of each element. From this table, it can be seen that the observed concentration factor y of elements varies inversely with the increase of residence time x of those elements in a general aspect.

The concept of oceanic residence time of an element was introduced by Barth(4) and now it is widely accepted as an indication of geochemical reactivity which elements exhibit in sea water. In environmental science, the importance of the concentration of trace elements to marine organisms from sea water has been appreciated for a long time. From the definition, it can be said that the concentration factor reflects biochemical reactivity. This means that elemental uptake into algae from the surrounding sea water is concerned with the relative reactivity of the elements in sea water. Therefore, it is presumed that residence time x and concentration factor y may be connected.

Table 22. Calculation of x and y values on <u>Eisenia bicyclis</u>(Code No.22)

	Residence time x* (year)	Sea water concentration (mg/liter)	Seaweed concentration (mean values) (µg/g of dried alga)	Concentration factor y**
Al	100	0.001	95.5	24000
Fe	200	0.003	98.0	8200
Mn	10000	0.002	8.40	1100
Ga	10000	0.00003	0.048	400
Ti	13000	0.001	5.78	1500
Si	18000	3		
Cu	20000	0.003	13.8	1200
Zn	20000	0.01	104	2600
Cr	20000	0.0005	0.74	370
V	80000	0.002	0.826	100
Ni	90000	0.007	0.873	31
Co	160000	0.0004	0.292	180
P	180000	0.09	9000	2500
Mo	700000	0.01	0.23	5.8
Ca	1000000	410	15000	9.2
Sr	4000000	8	1100	34
K	7000000	390	22100	14
Mg	12000000	1300	9040	1.7
B	18000000	4.5	96.7	5.4
Na	68000000	11000	14800	0.34

* Residence time in the ocean : $x = A / (dA/dt)$

A : the total amount of element in suspension or solution in the ocean

dA/dt : the amount introduced or precipitated per unit time

** Concentration factor : y = ppm in wet seaweed / ppm in sea water of the element concerned (assuming that the seaweeds contain 75% water)

The mutual relationship between residence time x and the cocentration factor y on Table 22 was graphically illustrated in Figure 1. Logarithum of concentration factor y was taken as the vertical axis and logarithum of residence time x as the horizontal axis. From the figure, it can be seen that there is a general linear relatioship between log x and log y. Consequently, the following formulas can be assumed.

[1] $y = ax^b$

[2] $\log y = \log a + b\log x$

In Figure 1, the values of a and b were calculated by method of the least squares from the formula [1], and the straight line was drawn by the linear equation [2]. The log a is the intercept with the ordinate and b is the slope of the straight line. Correlation coefficient γ between log x and log y was also calculated from the individual data. The calculated log a was 6.0, b was -0.76, and γ was -0.91 in Figure 1.

Figure 2 was based on the data of an individual sample of Eisenia bicyclis(No.1). The data was used from 13 elements. The calculated values of log a, b and γ of Figure 2 were close to those of Figure 1. Figure 3(No.1') was based on the data of the same sample(No.1) of Figure 2. The data was also used from 13 elements, but molybdenum and strontium were added to Figure 2 without aluminum and gallium. In sample No.1', the values of log a and -b were a little higher than those of the previous sample No.1, and the value of γ was lower.

Figure 4 was based on the data of an other sample of *Eisenia bicyclis*(No.2). The data was used from the same 13 elements as sample No.1 . The values of log a and -b on sample No.2 were a little higher than those for sample No.1 .

In these calculation on *Eisenia bicyclis*, it was seen that there was some variation in the values of log a, -b and γ, even in the same species from different samples. However, the variation was not so great. Thus, it may be possible to use these numerical values as an indication of the character of the distribution of chemical elements in each species of seaweed. In order to give a representative indication for one species, x values should be plotted within a wide range of residence time.

Similar correlations between log x and log y on other species of seaweed were graphically illustrated in Figure 5-21 from the related data showing in SECTION I. From these figures, it can be found that absolute values of γ were generally more than 9.0. Although there is such an excellent linear relationship in all species tested, the position of the straight line differs slightly for different species. The values of log a and -b show the difference for each species. Such a coexistence of generality and specificity throughout many species may be consistent in living matters.

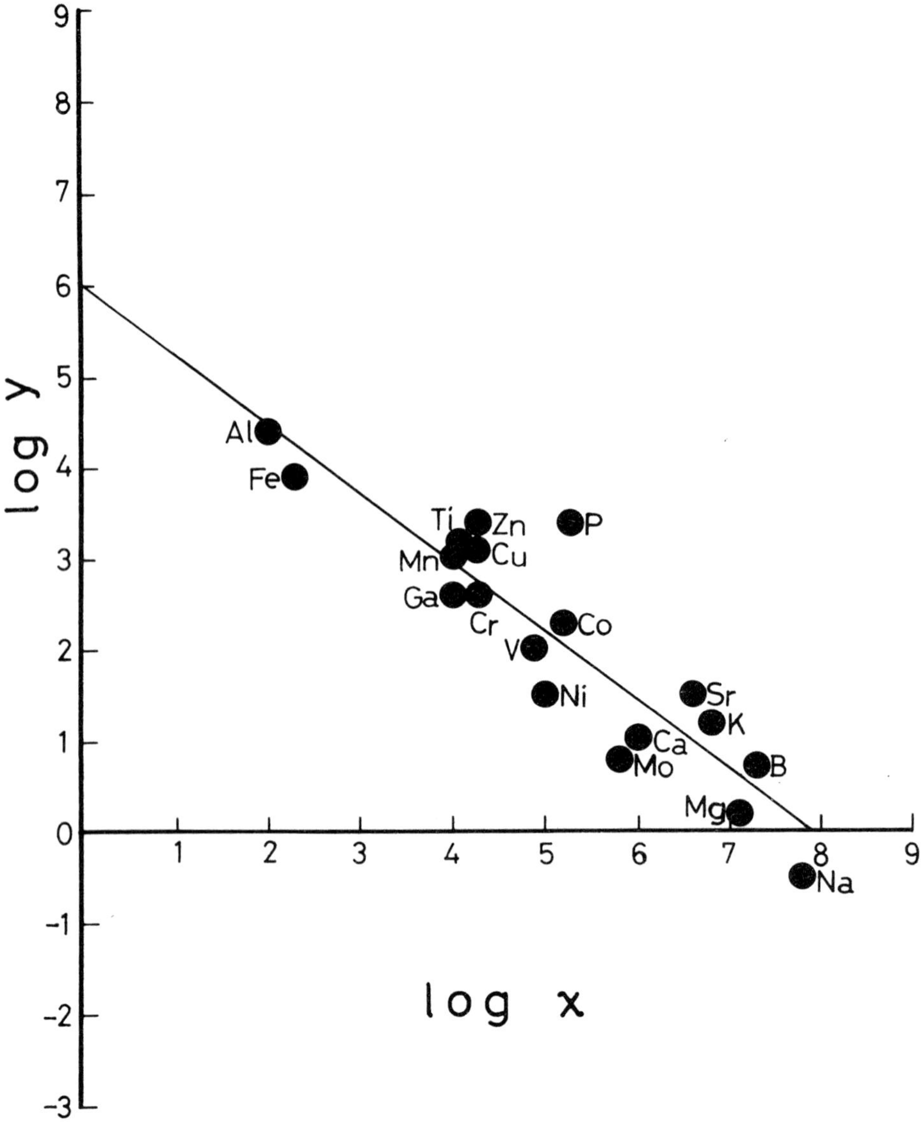

log a = 6.0 b = -0.76 γ = -0.91

EISENIA BICYCLIS

Fig.1 Code No.22

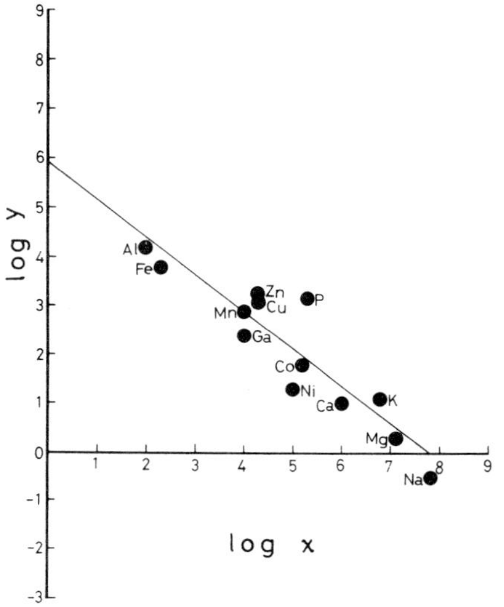

log a = 5.9 b = -0.75 γ = -0.92

EISENIA BICYCLIS (No.1)

Fig. 2

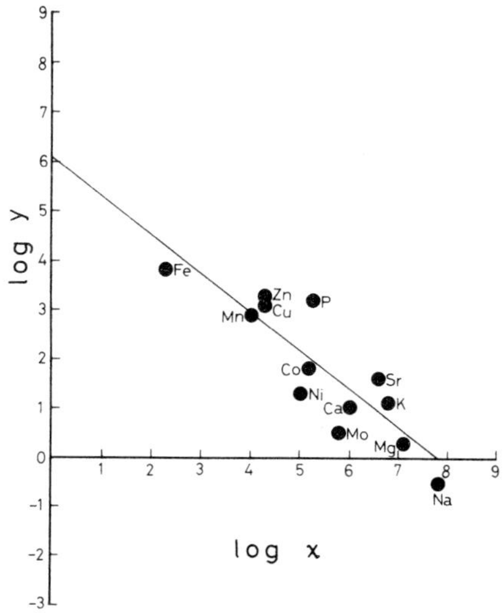

log a = 6.1 b = -0.78 γ = -0.88

EISENIA BICYCLIS (No.1')

Fig. 3

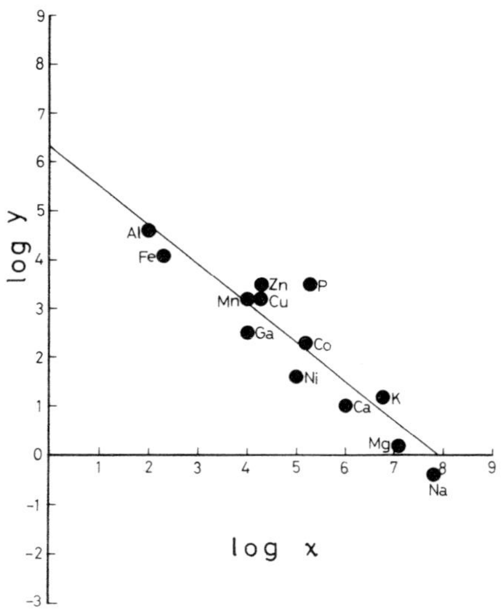

log a = 6.3 b = -0.80 γ = -0.92

EISENIA BICYCLIS (No.2)

Fig. 4

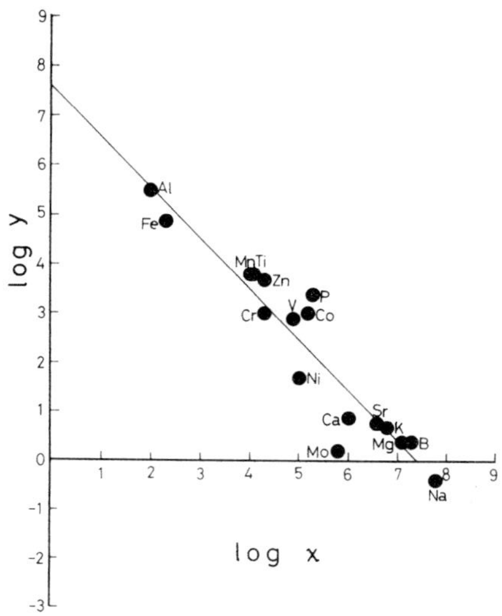

log a = 7.6 b = -1.03 γ = -0.94

MONOSTROMA NITIDUM

Fig. 5 Code No.1

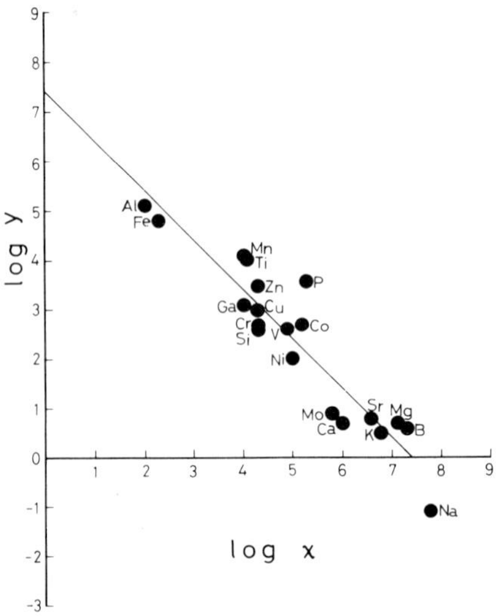

log a = 7.4 b = -1.00 γ = -0.94

ULVA PERTUSA

Fig. 6 Code No.2

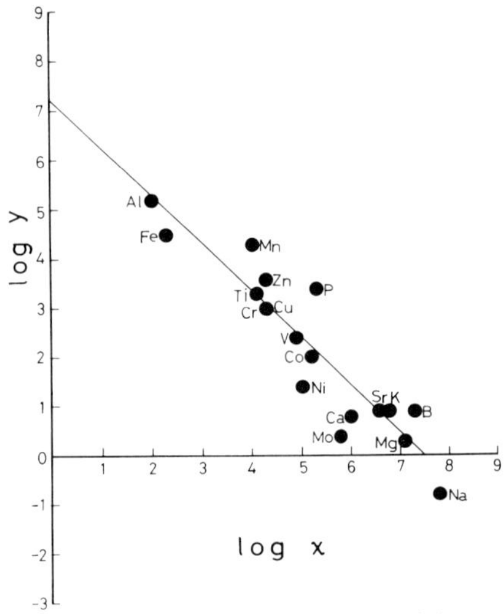

log a = 7.2 b = -0.96 γ = -0.93

CHAETOMORPHA CRASSA

Fig. 7 Code No.9

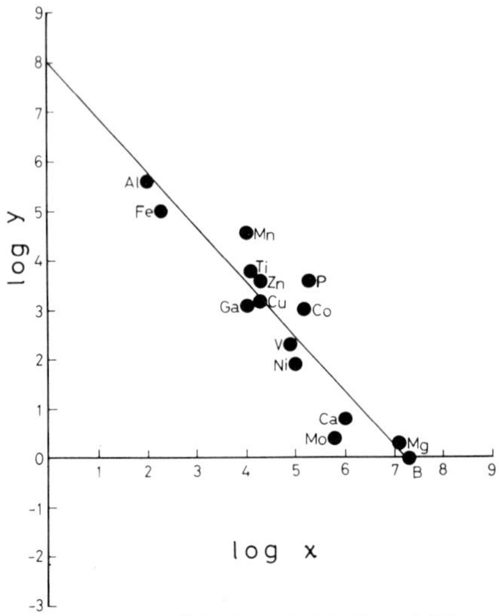

log a = 8.0 b = -1.11 γ = -0.92

CODIUM FRAGILE

Fig. 8 Code No.12

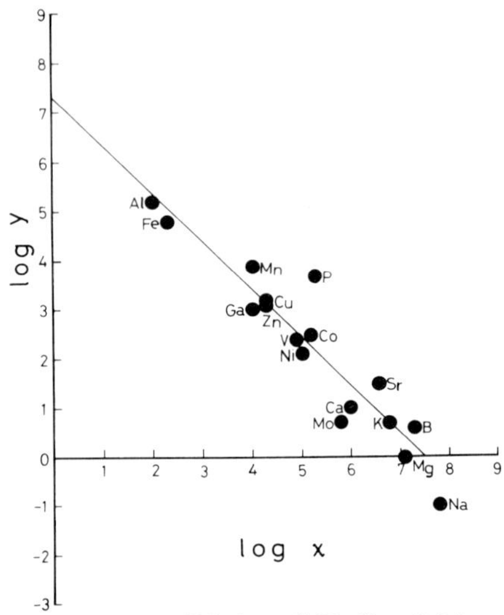

log a = 7.3 b = -0.98 γ = -0.94

DICTYOPTERIS PROLIFERA

Fig. 9 Code No.13

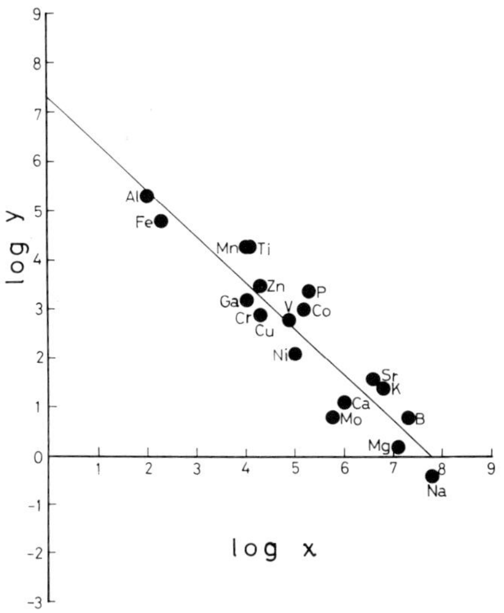

log a = 7.3 b = -0.94 γ = -0.94

PADINA ARBORESCENS

Fig. 10 Code No.14

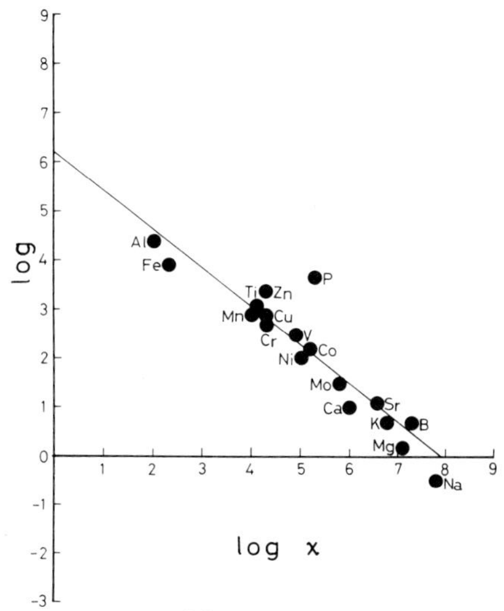

log a = 6.2 b = -0.79 γ = -0.93

ISHIGE OKAMURAI

Fig. 11 Code No.17

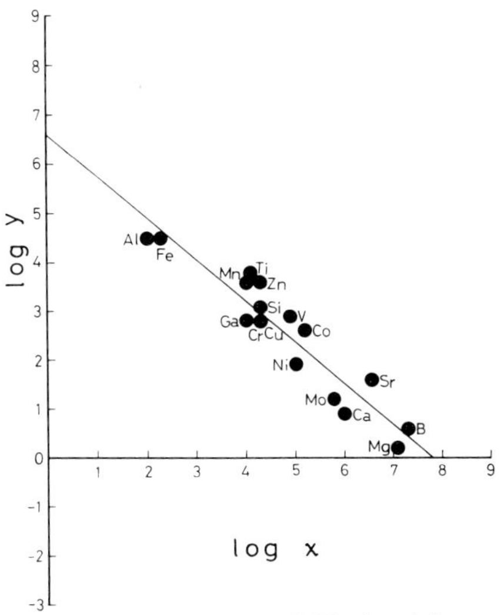

log a = 6.6 b = -0.85 γ = -0.94

ISHIGE FOLIACEA

Fig. 12 Code No.18

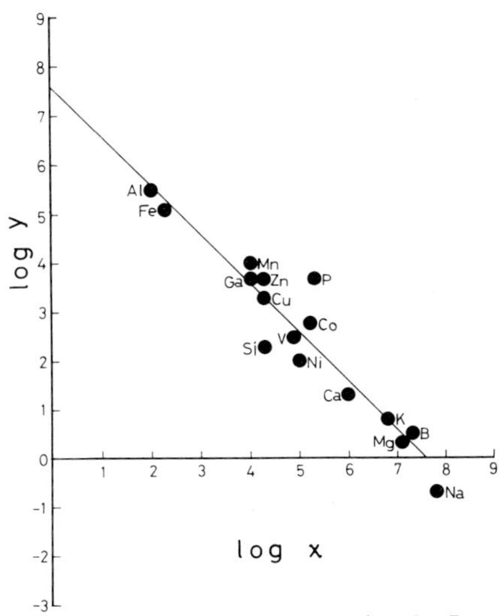

log a = 7.6 b = -1.01 γ = -0.95

SCYTOSIPHON LOMENTARIA

Fig. 13 Code No.21

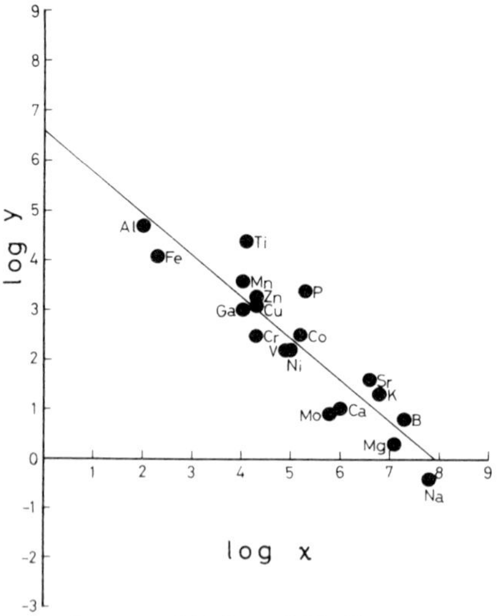

log a = 6.6 b = -0.83 γ = -0.92

HIZIKIA FUSIFORME

Fig. 14 Code No.26

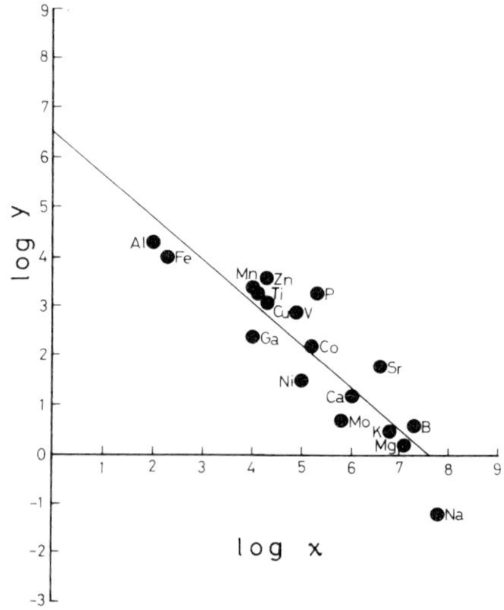

log a = 6.5 b = -0.85 γ = -0.90

SARGASSUM TORTILE

Fig. 15 Code No.31

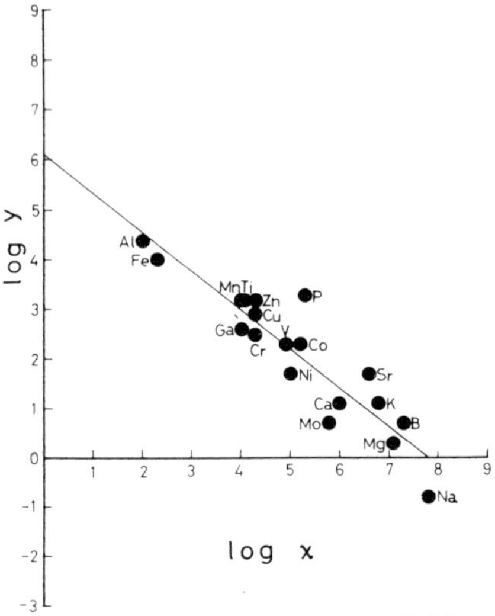

log a = 6.1 b = -0.78 γ = -0.92

SARGASSUM RINGGOLDIANUM

Fig. 16 Code No.33

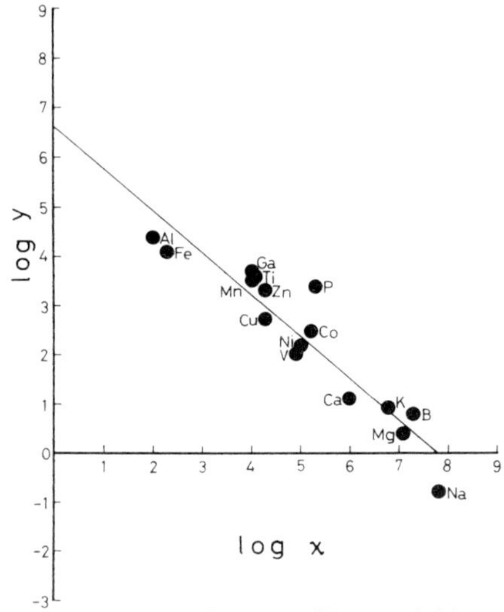

log a = 6.6 b = -0.85 γ = -0.94

SARGASSUM FULVELLUM

Fig. 17 Code No.34

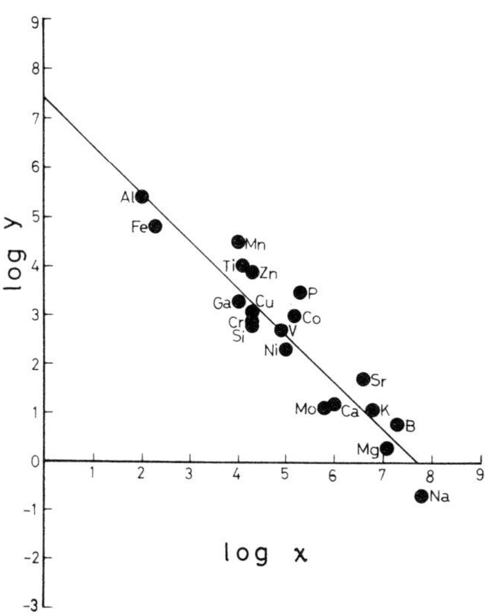

log a = 7.4 b = -0.96 γ = -0.94

SARGASSUM THUNBERGII

Fig. 18 Code No.35

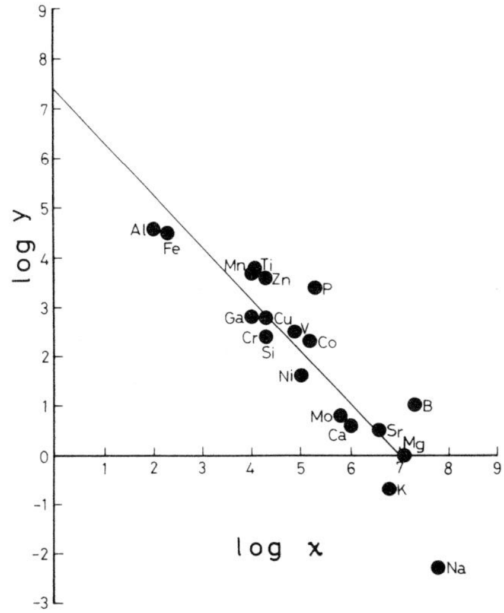

log a = 7.4 b = -1.05 γ = -0.91

GELIDIUM AMANSII

Fig. 19 Code No.41

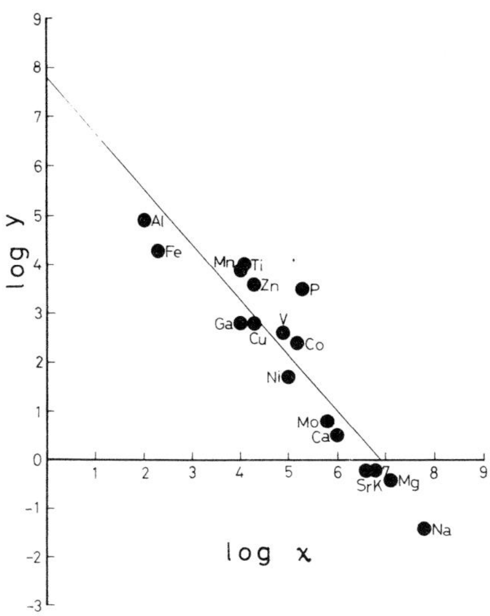

log a = 7.8 b = -1.13 γ = -0.94

ACANTHOPELTIS JAPONICA

Fig. 20 Code No.42

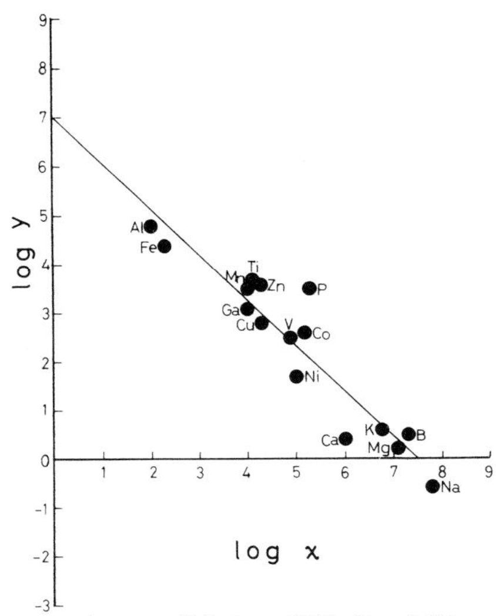

log a = 7.0 b = -0.93 γ = -0.94

CARPOPELTIS FLABELLATA

Fig. 21 Code No.49

Table 23. List of y/y_o value

Code* No.	Al	Fe	Mn	Ga	Ti	Si	Cu	Zn	Cr	V	Ni	Co	P	Mo	Ca	Sr	K	Mg	B	Na
1	0.7	0.4	2.0	-	2.3	-	-	3.2	0.6	1.9	0.1	4.7	14	0.1	0.3	0.8	1.2	1.1	1.8	1.1
2	0.5	0.5	4.6	0.5	4.8	0.3	0.8	2.6	0.4	1.3	0.3	3.1	28	0.2	0.2	1.0	0.9	2.3	2.5	0.2
9	0.8	0.3	9.2	-	1.4	-	0.9	3.7	0.9	0.9	0.1	0.7	18	0.1	0.3	1.1	2.3	1.0	5.2	0.4
12	0.6	0.4	9.4	0.3	1.9	-	0.9	2.2	-	0.5	0.2	5.1	26	0.1	0.3	-	-	1.4	0.9	-
13	0.7	0.5	3.6	0.4	-	-	1.2	1.0	-	0.7	0.5	2.1	35	0.1	0.3	4.8	1.2	0.4	2.4	0.3
14	0.7	0.5	4.9	0.5	7.1	-	0.4	1.8	0.4	1.1	0.3	3.3	10	0.1	0.2	3.1	3.0	0.3	2.2	0.4
17	0.6	0.3	0.8	-	1.2	-	1.3	4.2	0.7	1.6	0.5	1.2	41	0.7	0.3	1.4	0.8	0.4	1.8	0.3
18	0.4	0.7	2.4	0.4	5.2	1.2	0.7	4.7	0.6	2.5	0.3	2.3	-	0.3	0.2	3.5	-	0.4	1.4	-
21	0.9	0.6	2.5	1.3	-	0.1	1.0	2.4	-	0.6	0.2	2.8	24	-	0.5	-	1.1	0.7	1.5	0.4
22	0.8	0.4	1.1	0.4	1.9	-	2.1	4.8	0.7	0.5	0.2	1.6	25	0.2	0.3	3.7	2.3	0.4	1.8	0.3
26	0.6	0.3	2.0	0.5	16	-	1.3	2.0	0.3	0.5	0.5	1.7	15	0.1	0.3	2.9	2.8	0.4	1.7	0.3
31	0.3	0.3	2.1	0.2	2.0	-	1.9	6.4	-	3.7	0.2	1.5	19	0.1	0.7	8.3	0.7	0.6	1.8	0.1
33	0.7	0.5	1.7	0.4	1.8	-	1.5	2.7	0.6	1.0	0.3	1.8	21	0.1	0.5	5.2	1.9	0.5	1.9	0.2
34	0.3	0.3	2.0	3.1	3.0	-	0.5	2.2	-	0.3	0.6	2.1	17	-	0.3	-	1.3	0.6	2.2	0.2
35	0.8	0.4	7.6	0.5	3.0	0.3	0.6	3.9	0.4	1.0	0.4	3.8	12	0.2	0.3	4.0	1.8	0.4	2.2	0.2
41	0.2	0.3	3.9	0.4	5.5	0.4	1.0	5.9	0.4	2.0	0.3	2.9	38	0.4	0.4	1.1	0.1	1.3	19	0.1
42	0.2	0.1	4.1	0.4	7.5	-	0.8	5.1	-	2.4	0.3	3.1	42	0.4	0.3	0.3	0.6	0.7	-	0.5
49	0.4	0.3	1.8	0.6	3.2	-	0.7	4.1	-	1.4	0.2	2.7	25	-	0.1	-	1.0	0.6	1.7	0.5
Mean	0.6	0.4	3.7	0.7	4.2	0.4	1.0	3.5	0.5	1.3	0.3	2.6	24	0.2	0.3	3.0	1.4	0.7	3.1	0.3
S.D.**	0.2	0.1	2.6	0.7	3.5	0.4	0.5	1.5	0.2	0.9	0.1	1.1	10	0.2	0.1	2.1	0.8	0.5	4.2	0.2

* The species are shown in Table 1. according to this code No.

** Standard deviation

Although the position of the points for each element in relation to the straight line is variable among species of seaweed, there is still a limited range for each element. When the values of a and b could be obtained, a theoretical value of concentration factor y_0 for each element can be calculated from the assuming formula [1]. Table 23 shows a list of y/y_0 for each element in 18 species observed in the previous figures. Some elements such as phosphorus, titanium and zinc have high values, while other elements such as iron, chromium and aluminum have low values, but the standard deviation is limited to within a small range of the mean value of each element. So, the values of y/y_0 may be an indication of a characteristic of each element on the distribution in the algal body through various species of seaweed.

Both oceanic residence time and the concentration factor are essentially due to experimental results. Therefore, the proposed treatments of the regularity between residence time and concentration factor should improve with advance of experiments. This regularity may occur in other phyla of marine organisms. This may be a basis on which to consider the general distribution of elements in the biosphere.

References

1) Yamamoto, T., Ishibashi, M. : The content of trace elements in seaweeds. Proceedings of the seventh international seaweed symposium, University of Tokyo Press, 511-514(1971).

2) Yamamoto, T. : The relations between concentration factor in seaweeds and residence time of some elements in seawater. Rec. Oceanogr. Wks. Japan 11, 65-72(1972).

3) Goldberg, E.D., Broecker, W.S., Gross, M.G., Turekian, K.K. : Radioactivity in the marine environment. Publication ISBN 0-309-01865-x. Comittee on oceanography national academy of science, Washington D.C., 137-146(1971).

4) Barth, T.W. : Theoretical petrology. John Wiley, New York (1952).

MICROBIAL TRANSFORMATION OF MARINE STEROLS: FUCOSTEROL AND ISOFUCOSTEROL

H. W. Youngken, Jr., and F. M. Soliman
Department of Pharmacognosy, College of Pharmacy
University of Rhode Island, Kingston, Rhode Island 02881

Since 1937 when Mamoli and Vercellone (1) demonstrated the transformation of adrostane type steroids by yeast cells, numerous investigators have carried out similar steroid biotransformations for pharmaceutical and other purposes. The oxidation of pregnenolone and progesterone at C_{11} by microbiol transformation in the production process of adreno-cortical hormones are examples of commercial applications of this method.

The literature on biotransformation is replete with reviews on this subject, for example, the articles by Capek (2), Charney (3) and Iizuka (4). Included are descriptions of cell cultures and substrates, numerous enzymatic reactions which influence the transformations and the discussion of possible mechanisms by which the biotransformations occur.

Idler (5) reported the presence of 8 different sterols among 35 species of marine red algae and Meunier (6) identified 5 sterols among 14 red algae from widely separated geographical areas. In both studies cholesterol and desmosterol were the most commonly found sterols. Most all sterols thus far found in marine algae vary considerably in amounts in the sea weeds that produce them, even within the same species in different geographical areas of the oceans.

Marine Algae in Pharmaceutical Science
edited by H. A. Hoppe, T. Levring, Y. Tanaka

Goodwin (7), who reviewed the subject of sterols in algae, stated that "Although only relatively few algae have been examined thoroughly for sterols by modern methods, it is reasonable to conclude that these compounds are constant components of algae. The time is ripe for much careful survey work on pure cultures of the less common algae; a picture of some considerable phylogenetic significance could emerge." So also it may be said that biodegradation studies of sea weed constituents and some of their untested sterols, using laboratory models and microbiol transformation methods, may also contribute to this knowledge. Fucosterol commonly found in brown algae and isofucosterol similarly found in green marine algae have to our knowledge not been extensively used in such transformation studies.

Since cholesterol, a C_{27} type sterol widely found among marine organisms, has been successfully transformed by microbial organisms in studies by others (8,9,10), it serves a useful model for similar transformation experiments of other carbon compounds with different but similar structures.

Fucosterol and isofucosterol are marine sterols with structures similar to cholesterol, excepting for their C_{29} skeleton and the extra ethylidene group on the side chain. The C_{29} sterols, stigmasterol and β-sitosterol, closely relate structurally to fucosterol. These have been transformed into 17-keto steroids of the androstane and C_{21} pregnane-types by Mycobacterium species (11). Theoretically a pattern and the products of transformation for fucosterol and isofucosterol should be similar to those reported for cholesterol, stigmasterol and/or β-sitosterol. However, with the exception of a report by Nes (12), who reported the biotransformation of these C_{29} sterols by the protozoan, Tetrahymena pyriformis, there appears to be no other report in the literature regarding their use for such purpose or of other microbial organisms with which they were used as substrates.

Fucosterol has been reported as an antilipemic (hypocholesterolemic) agent in animal experiments and is extracted in good yields from marine weeds of the Phaeophyceae species. Thus far, no substantiated pharmacologic action has been reported for isofucosterol. It occurs as the major sterol in many of the green algae (13).

In view of the sparce information available regarding possible uses of fucosterol and isofucosterol as substrates in biotransformation studies, it was therefore of interest to us to subject these sterols to microbial action similar to that effected by substrates in which C_{19}, C_{21}, C_{27} and C_{29} sterols have been successfully used.

Species of *Nocardia*, *Pseudomonas* and *Saccharomyces* have been used with considerable success in biotransformation studies (2,3,4) using either mature organisms or induced cells. Sih and his co-workers (10,14,15) have used *Nocardia* species successfully in transforming cholesterol into a number of 17-keto steroids. Turfitt (8) and others similarly used *Proactinomyces* to convert the C_{29} sterol, stigmasterol, into its 22-diene-3-one and Merskeck (11) used a *Mycobacterium* species to transform stigmasterol and β-sitosterol into 17-keto anstrostane steroids and 20-keto steroids of the pregnane type.

Mention of some sterol biotransformations would not be complete without a general reference to the effects of chelating compounds and such mineral elements as cobalt and iron in their substrate mixture. Other investigators (16,17) have altered the products of transformation by using these adjuncts in the reaction substrate. For example, Nagasawa (16) prevented the total decomposition of cholesterol, β-sitosterol and stigmasterol by *Arthrobacter* using the chelating compound, 2,2'-dipyridyl. Investigators at the Noda Institute (37) similarly used the inhibitors hydroxyquinoline, phenanthroline and others to influence the transformation of β-sitosterol and stigmasterol by *Nocardia* and other microbial organisms. In the presence of cobalt chloride transformations yielded 17 keto C_{19} steroids (18). Thus the results of experiments reported in our laboratory described below are of some of these agents in the study of fucosterol and isofucosterol.

EXPERIMENTAL

Fucosterol was extracted from the brown sea weed, *Ascopyllum nodosum* after the method of Tsuda (19) and isofucosterol was similarly obtained from the sea lettuce, *Ulva lactuca*. The IR data for each purified compound conformed to that published by Gibbons (20) and for the reference standards.

Alcohol solutions of the sterols were used for each incubation experiment. The pH of the reaction substrates was kept at 6.05.

Preliminary experiments in our laboratories, during which the methods of other investigators were used for other sterols, failed to produce any discernible biotransformation of fucosterol or isofucosterol when type culture organisms of *Saccharomyces* and *Pseudomonas* species were employed. Commercial samples of yeast (Fleischmann$^{(R)}$) constituted the source for Saccharomyces for some experiments, also without results.

Cell cultures of *Nocardia restrictus* ATCC 14887 did, however, convert fucosterol and isofucosterol into their 3-keto structures, fucostenone and isofucostenone respectively (Fig. 1). Evidence was also obtained for the presence of androstadienediones and dehydrofucosteneone in the reaction mixture.

A typical transformation was performed by inoculation of a 24 hour culture of Nocardia restrictus ATCC 14887 into 60 ml of Difco nutrient broth in a 500 ml Erlenmeyer flask.* The cultures to which fucosterol and isofucosterol were added were allowed to grow at room temperature for 24-48 hours with continuous shaking. Alcoholic solutions of fucosterol and isofucosterol (each 20 mg in 0.5 ml of N,N-dimethyl formamide) were then added in the presence or absence of the chelating compound and fermentation was continued at room temperature with continuous shaking for from 6-252 hours. Each culture was then extracted with 4 x 20 ml of methylene chloride. The extract was washed with 5 ml of 2.5% aqueous solution of sodium bicarbonate and 3 x 5 ml of water, then dried with anhydrous sodium sulfate and evaporated to dryness under vacuum. The residue was spotted on preparative thin layer plates (0.5 mm thick) of Silica Gel HF254 and developed with benzene-isopropanol (9:1). The separated bands were marked under UV light (254 mm). The UV quenching bands were scraped off the plate and eluted with ether. The remaining constituent was located by spraying with water, then allowed to dry and treated as above. The purity of the isolated compounds was checked by T. L. C. using 0.25 mm thick plates developed with $CHCl_3$-ether (10.1) and benzene ethylacetate (9:1), and sprayed with the universal reagent (9). Gas

---

*For complete details of the experimental methods used in this study see Soliman, F. M. "Microbiological Transformation of Marine Sterols" Ph.D. Thesis, pp. 88; University of Rhode Island Library, Kingston, R. I. 02881.

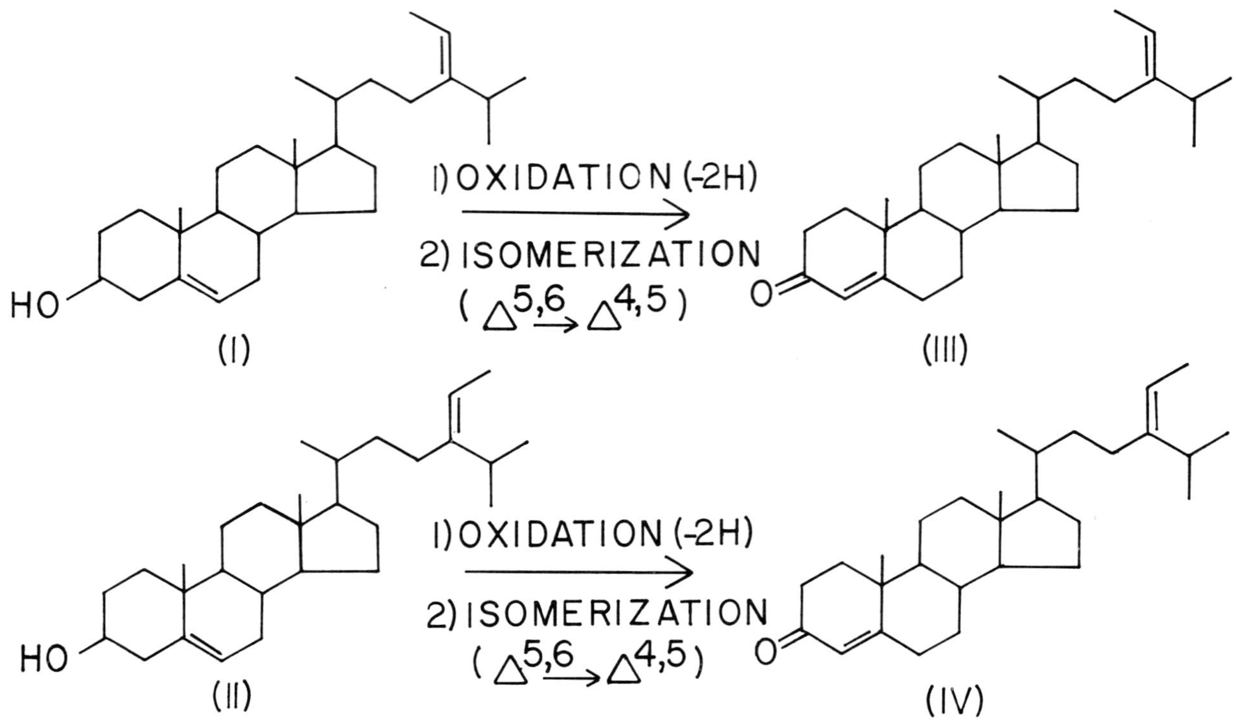

FIG. 1. STEROL TRANSFORMATION BY <u>NOCARDIA</u>: I FUCOSTEROL, II ISOFUCOSTEROL, III FUCOSTENONE, IV ISOFUCOSTENONE.

chromatography was also used after the method of Gibbons (20) in the identification work-up.

Chemical oxidation of fucosterol to fucostenone and isofucosterol to isofucostenone was performed for reference purposes by the methods of Bowers (21). Conventional U.V. absorbance was determined for each steroid product of the biotransformation.

The chelating compounds used in fucosterol biotransformation culture substrates were alcoholic solutions of 2,2-dipyridyl, 8-hydroxyquinoline, and disodium ethylenediamine tetraacetate (Na_2EDTA) at concentrations of 0.5-4.0 m Mole for a period of 48 and 78 hours. Concentrations of from 0.5 to 2.0 m Mole Fe_2SO_4 were added to 48 hour cultures of _N. restrictus_ with and without sterol and chelating compounds to measure effects of Fe^{++} ions on the biotransformation of fucosterol.

Fucostenone isolated from the fermentation experiments from hot methanol gave colorless crystalline clusters, m.p. 93-94°, U.V.: λmax 242 nm (ε=14,350), IR: νcm^{-1}: 1675 (3-ketone) 1620(4-3ne) and 825; which was identical to authentic fucostene prepared by Jones oxidation of fucosterol (26).

Isofucostenone similarly was isolated and yielded colorless cluster crystals, m.p. 109-110°, U.V.: λmax 242 nm, (ε=16,000) IR: νcm^{-1}: 1675 (α,β-unsaturated six member ketone); 1620 4,5-double bond; 812 (trans-ethylidene side chain).

Androsta^{-1},4-diene-3, 17-dione was isolated in the presence of chelating compounds as substrates as colorless prisms (kexane), m.p. 126°: U.V. λmax 242 nm (ε=16,000): νcm^{-1} 1740 (17-ketone); 1662, 1612 and 1602 ($\Delta^{1,4}$- 3-ketone). UV and IR spectra matched reference samples.

Other compounds isolated in trace amounts and identified by UV absorption, T.L.C. and G.C. comparisons with reference standards, were 1-dehydrofucostenone, androst-4-ene-3, 17-dione and 3β-hydroxy-9, 10-seco-1,3,5(10)-triene-9-10-dione. None of these latter compounds was, however, isolated in pure crystalline form.

RESULTS AND CONCLUSIONS

The transformation of fucosterol by *N. restrictus* to fucostenone and isofucosterol to isofucostenone (Fig. 1), probably involved enzymatic oxidation and isomerization of the double bond as has been shown by Talalay *et al.* (22) and Wang, *et al.* (23).

These investigators reported the oxidation of 3β-hydroxy steroids by a crystalline enzyme isolated from *Pseudomonas testosteroni*. Talalay (22) characterized the reaction as NAD-linked as:

$$H-\!\!>\!C\text{-}OH + NAD^+ \rightarrow\ >\!C{=}O + NADH + H^+.$$

Wang, Kawahara, and Talalay (23) have also shown the isomerization of the double bond of 3β-hydroxy steroids at carbons 5 and 6 to carbons 4 and 5 by a pure crystalline isomerase enzyme which they isolated from *P. testosteroni*. They postulated that isomerization involved the transfer of hydrogen from carbon-4 to carbon-6.

Enzymes involved in the degradation of steroids have been reported by others. Among these 9α-hydroxylase and dioxygenase were isolated from *Nocardia* sp. (24,25). No attempt was made in this study to isolated and study such enzyme systems in *Nocardia restrictus*.

The incubation of fucosterol over a period of from 6 to 252 hours at room temperature showed that the optimum period of transformation of fucosterol into fucostenone occurred within 24 hours, while the maximum accumulation of fucostenone was noticed in 76 hours (Fig. 2). This was followed by a gradual decrease in the concentration of fucostenone, suggesting that degradation of the product then occurred.

The biotransformation of fucosterol by N. restrictus in the presence of the three chelating compounds revealed trace amounts of another intermediate which showed similar physical and spectral characteristics of androsta-1,4-diene-3,17-dione.

Transformation of fucosterol by N. restrictus in the presence of the different chelating compounds at different concentrations for a period of 78 hours or less (Fig. 3,4,5) showed that the accumulation of fucostenone was not significantly reduced. The chelating compounds did not interfere with fucosterol transformation. The accumulation of androsta-1, 4-diene-3,17-dione was however variable.

Workers at Noda Institute (11) suggested that these chelating agents inhibit the 3β-hydroxylation in the degradative pathway of cholesterol. Sih and his co-workers (26) confirmed the result, and concluded that the inhibitors actually work directly on the 9β-hydroxylase itself. The results obtained here may indicate that androsta-1,4-diene-3,17-dione, once formed, is quickly hydroxylated at the 9β position and degraded; whereas the conversion of fucostenone to androsta-1,4-diene-3,17-dione is not significantly affected. Sih also reported that the chelating agents have little activity on enzymes affecting the sterol side chain.

An unknown intermediate was isolated under the influence of Na_2ETDA which showed λ_{max}^{EtOH} = 244 nm. Based on its TLC behavior as well as its ultraviolet spectrum, it is possible that this unknown intermediate is 1-dehydrofucostenone.

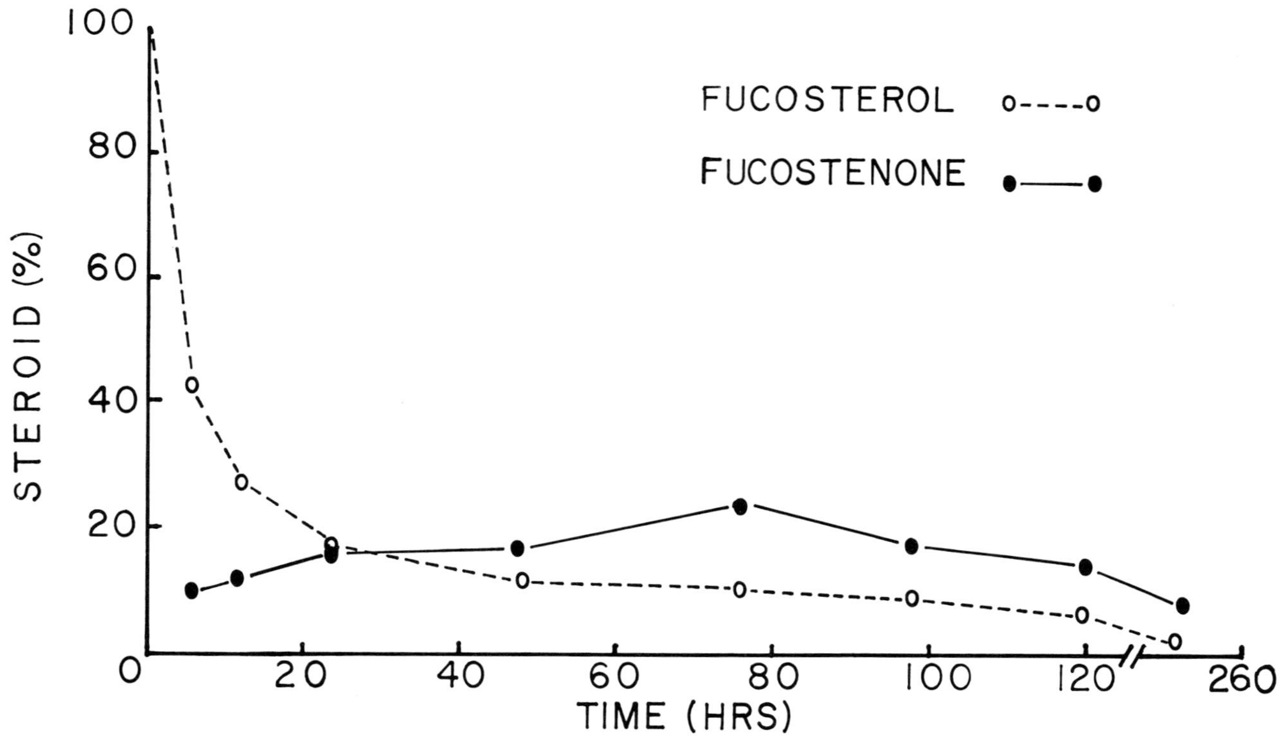

FIG. 2. FUCOSTEROL CONVERSION TO ISOFUCOSTEROL: TIME IN HOURS. _N_. _RESTRICTUS_.

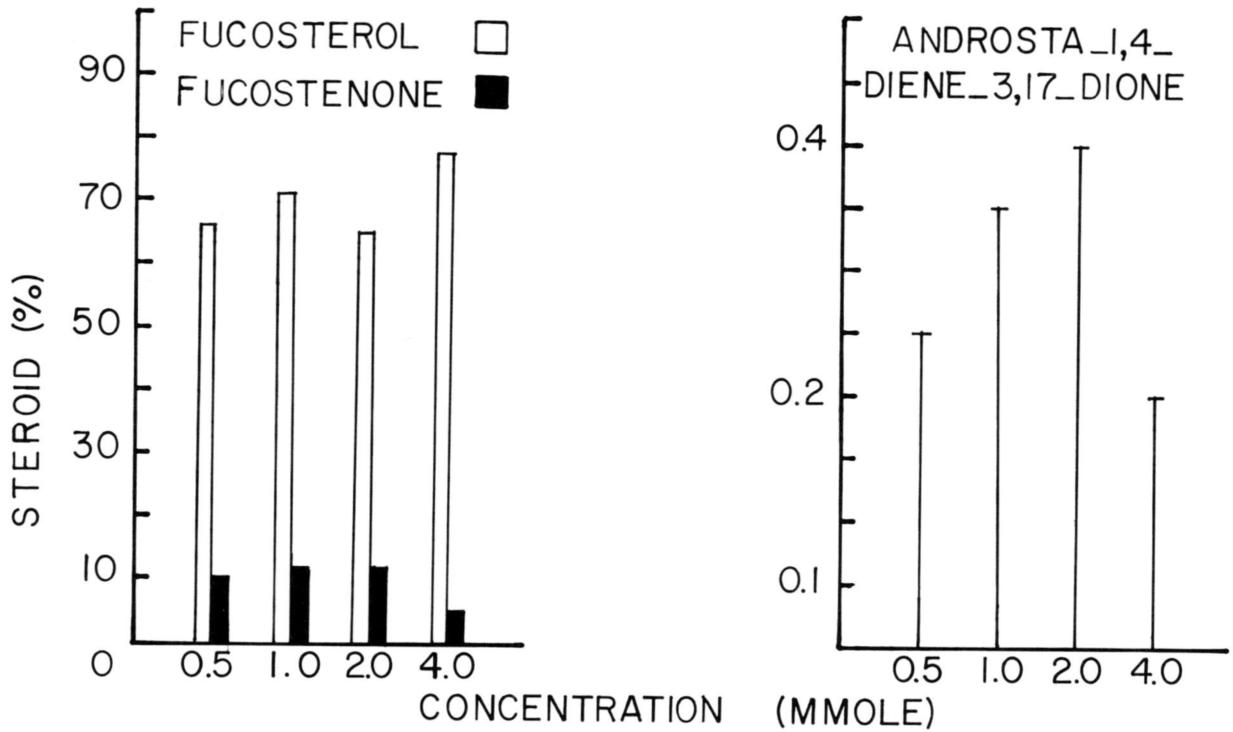

FIG. 3. EFFECT OF 2,2'-DIPYRIDYL IN N. RESTRICTUS/FUCOSTEROL SUBSTRATES - 78 HOURS.

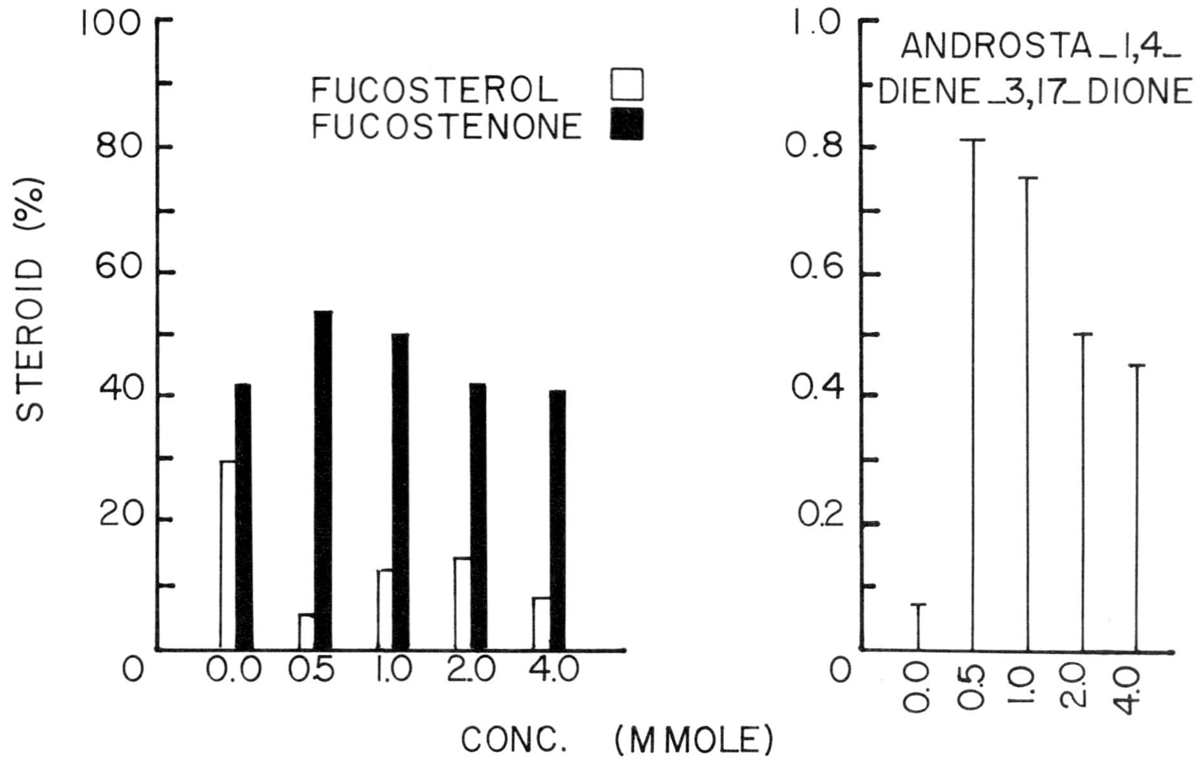

FIG. 4. EFFECT OF 8-HYDROXYQUINOLINE, NOCARDIA/FUCOSTEROL SUBSTRATES.

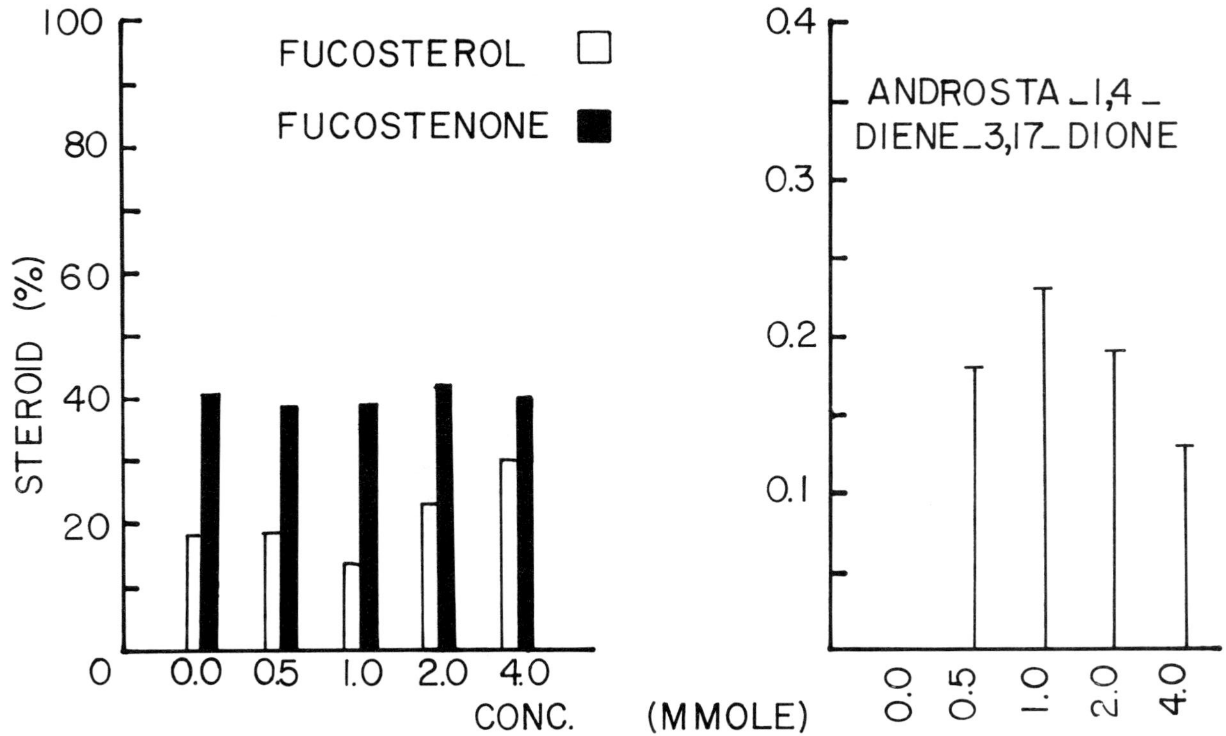

FIG. 5. EFFECT OF DISODIUM ETHYLENEDIAMINE TETRAACETATE (EDTA), <u>NOCARDIA</u>/ FUCOSTEROL SUBSTRATES.

A mixture of two components was isolated from cultures to which the chelating compound 2,2'-dipyridyl was added. This mixture showed maxima of 283 and 243 nm in UV spectrum. The result suggests the presence of a seco-phenolic intermediate together with androsta-4-ene-3,17-dione.

Based on these findings and reports by Sih (14,15,25) a pathway for the biotransformation of fucosterol by Nocardia restrictus (Fig. 6) is suggested as follows:
1. Fucosterol (I) was oxidized by an oxidoreductase system to a 3-keto compound, which in turn was isomerized to fucostenone (III) by an isomerase enzyme; 2. The fucostenone was transformed either to 1-dehydrofucostenone (VI), under the influence of a 1,2-dehydrogenase enzyme or to androst-4-ene-3,17-dione (VIII), through a side chain degradation. The androst-4-ene-3,17-dione (VIII) may be transformed into its 1-dehydro form by 1,2-dehydrogenase to give androsta-1,4-diene-3,17-dione (V). Androsta-1,4-diene-3,17-dione (V) can also be obtained as a result of the transformation of 1-dehydrofucostenone (VI). This involved a side chain degradation after possible hydrogenation of the ethylidene side chain. Successive oxidation of the side chain in a manner similar to fatty acids degradation could yield androsta-1,4-diene-3,17-dione (V).

Further degradation of androsta-1,4-diene-3,17-dione into CO_2 and water has been demonstrated by several investigators, noteably Schubert (28). This involves the enzymatic formation of a 9α-hydroxy intermediate which undergoes reverse aldol reaction resulting in 3β-hydroxy-9,10-seco-androsta-1,3,5(10)-triene-9,17 dione (VII) as shown by Wang (29). Further degradation of this product to CO_2 and H_2O has been reported by others (30).

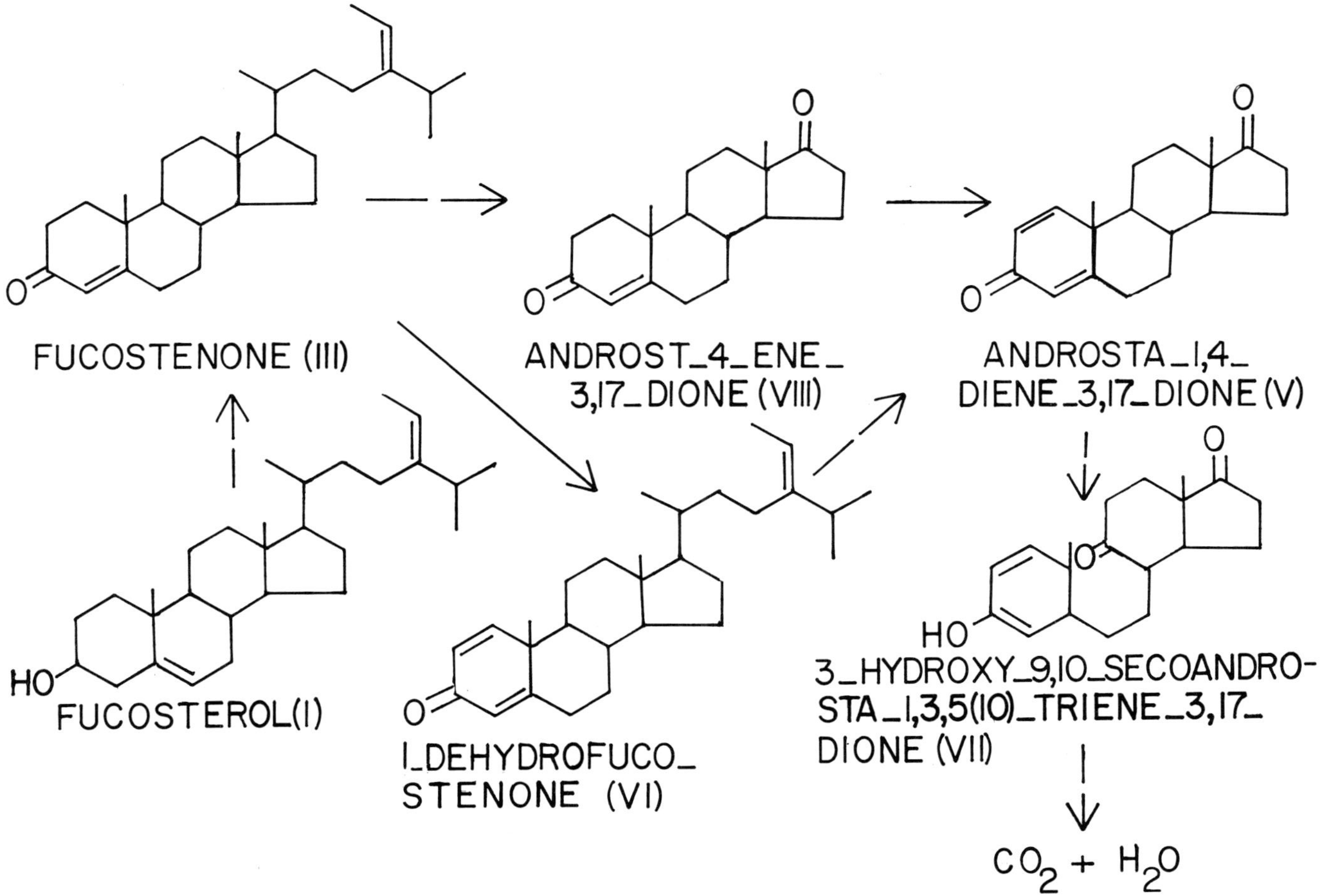

FIG. 6. DEGRADATION SCHEME, FUCOSTEROL TRANSFORMATION.

The transformation of fucosterol to fucostenone when used with 2 m Moles Fe^{++} ions (Fe_2SO_4) and 1 m Mole dipyridyl added to the substrate for _N. restrictus_ increased by about 30 percent (Fig. 7). The results were similar to those found when Fe^{++} ions and no 2,2'-dipyridyl was used. This could be due to the binding of ferrous ions with the chelating compound resulting in the removal of any inhibitory effect of the latter if such occurred, thus permitting the transformation to progress. Gibson (31) reported that the oxidation of 3,4-dihydroxy-9,10-seco-androsta-1,3,5(10)-triene-3,17-dione by cell extracts of _Nocardia restrictus_ was stimulated by the presence of ferrous ions. Our results are in agreement with these findings.

SUMMARY

A study was made of the biotransformation of the marine sterols, fucosterol and isofucosterol, when used as substrates for certain species of _Saccharomyces_, _Pseudomonas_ and _Nocardia_. The following results were observed:

1) No biotransformation products of these sterols were found when Fleischman's Yeast(R) (_Saccharomyces cerevisiae_) and _Pseudomonas aeruginosa_ were used.

2) Fucostenone and isofucostenone were obtained when the sterols were used as substrates for _Nocardia restrictus_ ATCC 14887.

3) Optimum transformation of fucosterol by _Nocardia_ was observed within a period of 24 hours, while maximum accumulation of fucostenone occurred in 76 hours.

4) There was an accumulation of androsta-1,4-diene-3,17-dione in trace amounts when the chelating compounds 2,2'-dipyridyl, 8-hydroxyquinoline, and disodium ethylenediamine

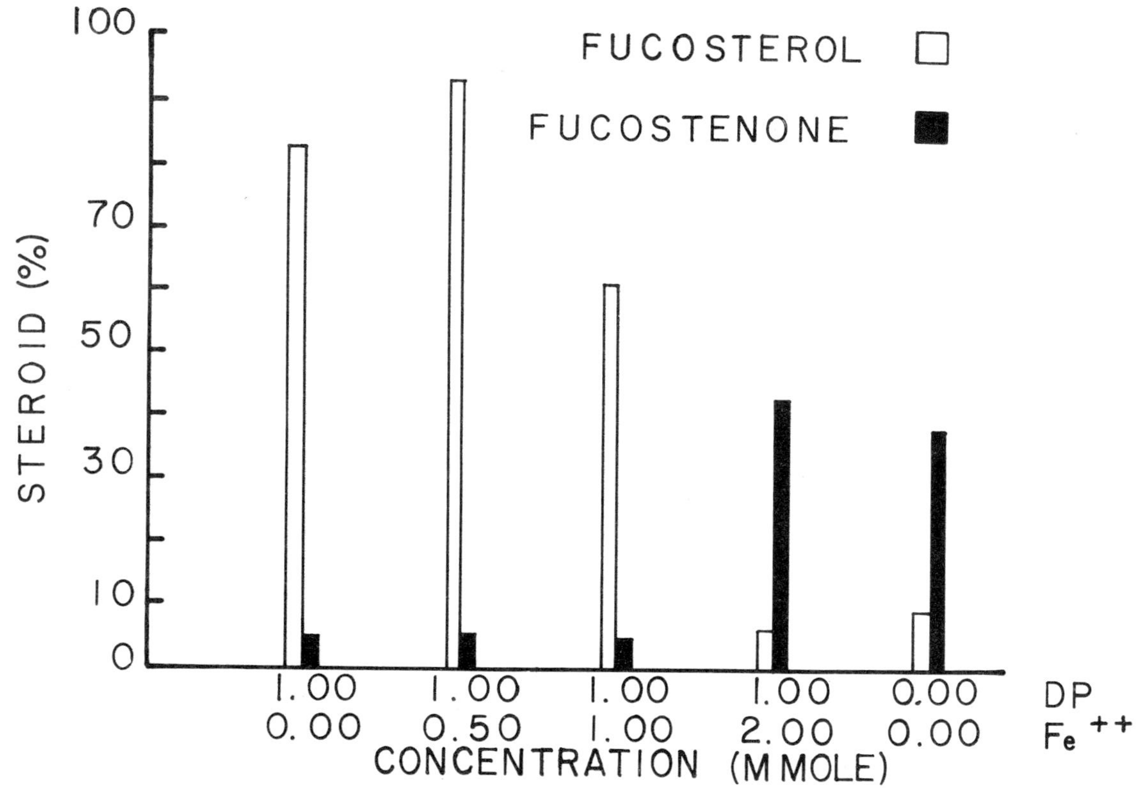

FIG. 7. EFFECT OF Fe^{++} AND 2,2'-DIPYRIDYL (DP), TRANSFORMATION OF FUCOSTEROL BY N. RESTRICTUS.

tetraacetate (EDTA) were added with fucosterol.

Other products of biotransformation are also suggested and some evidence is offered for the presence of 1-dehydro-fucostenone, androst-4-ene-3,17-dione and 3β-hydroxy-9,10-seco-androsta-1,3,5(10)-triene-9,17-dione as intermediate products in the cultures.

5) Fe^{++} ions were found to stimulate the biotransformation of fucosterol with intact cells of Nocardia restrictus in the presence of 2,2'-dipyridyl.

The authors acknowledge the assistant in this study of Professors Y. Shimizu and L. R. Worthen, Department of Pharmacognosy, University of Rhode Island. A portion of the project was supported by Grant No. 175-025 from the National Sea Grant Program, Department of Commerce, U.S.A.

REFERENCES

1. Mamoli, L., Vercellone, J. A.: Biochemische Unwandlung von Δ^5-androstendiol in isoandrostandiol and Δ^4-testosteron. Ber. 2079-2082 (1937).
2. Capek, A., Hanc, O., Tandra, M.: Microbiol transformation of steroids. Czechoslovak Acad. Sci. 140 (1966).
3. Charney, W., Herzog, H.: Microbial transformation of steroids - A Handbook. Academic Press, New York (1967).
4. Iizuka, H., Naito, A.: Microbial transformation of steroids and alkaloids. University Park Press, State College, Pa. (1967).
5. Idler, D. R., Saito, A., Wiseman, P.: Steroids in Red Algae (Rhodophyceae). Steroids 11, 465-473 (1968).
6. Meunier, H., Zelenski, S., Worthen, L.: Comparison of the sterol content of certain Rhodophyta. Food-Drugs from the Sea Proceedings, Youngken, H. ed., Marine Technology Society, Washington, D.C., 319-325 (1969).

7. Goodwin, T. W.: Sterols in "Algal Physiology and Biochem. istry." Stewart, W. D. P., ed., University of California Press, Berkeley and Los Angeles, 266-280 (1974).
8. Turfitt, G. E.: The microbiological degradation of steroids. Biochem. J. 42, 376-383 (1947).
9. Kramli, A., Horvath, J.: Microbial oxidation of sterols, Nature, 163, 219 (1949).
10. Sih, C. J., Wang, K. C., Tai, H. H.: Acid intermediate in the microbiological cleavage of the cholesterol side chain. J. A. C. S. 89, 1956-7 (1967).
11. Mersheck, W. J., Kraychy, S., Muir, R. D.: Microbiological degradation of sterols. Appl. Microbiol., 23, 72-77 (1972).
12. Nes, W. R. et al.: Conformational analysis of the enzyme substrate complex in the dehydrogenation of sterol by *Tetrahymena pyriformis*. J. Biol. Chem. 246, 561-568.
13. Petterson, G.: The distribution of sterols in algae. Lipids, 6, 120 (1971).
14. Sih, C. J., Wang, K. C., Tai, H. H.: Mechanisms of steroid oxidation by microorganisms. Biochem. 1, 796-807 (1968).
15. Sih, C. J., et al.: Mechanisms of steroid oxidation by microorganisms. Ibid. 808-818 (1968).
16. Nagasawa, M., Bae, M., Tamura, G., Arima, K.: Microbial transformation of sterols. Agr. Biol. Chem. 33, 1644-50 (1969).
17. Noda Institute for Scientific Research: Androst-4-ene-3,17-dione and androsta-1,4-diene-3,17-dione, Neth. Appl. 6, 502, 883 (1965) cited in C.A. 65, 6262g (1966).
18. Koninklijkl. Nederlandsche Gisten Spritusfabriek-N.V. Microbiological preparation of steroids. Neth. Appl. 67, 05, 450 (1968) cited in C.A. 70, 76473x (1969).
19. Tsuda, K.: Steroid studies XVI. Isolation of 22-dehydrocholesterol from Hypenea japonica, J.A.C.S. 82, 1442 (1960).

20. Gibbons, G. F.: The identification of 28-isofucosterol in the marine green algae Enteromorpha intestinalis and Ulva lactuca. Phytochem. 7, 983-988 (1968).

21. Bowers, A. et al.: The chemistry of triterpenes and related compounds III. J. Chem. Soc. 520 (1952).

22. Talalay, P.: Enzymic analysis of steroid hormones. In. Methods Biochem. Analysis 8, Glick, D., ed., Interscience Pub. Inc., New York, 119-143 (1960).

23. Wang, S. F., Kawahara, F. S., Talalay, P.: Absorption and fluorescence of spectra of enzyme steroid complexes. J. Biol. Chem. 283, 571 (1963).

24. Tai, H. H., Sih, C. J.: Nocardia restrictus I. Isolation of the enzyme and study of its physical and chemical properties. J. Biol. Chem. 245, 5062-5071 (1970).

25. Ibid.: II. Kinetic studies. J. Biol. Chem. 245, 5072-5078 (1970).

26. Sih, C. J., Rahim, A. M.: Mechanisms of steroid oxidation by microorganisms III. J. Pharm. Sci. 52, 1075-1080 (1963).

27. Sih, C. J.: Mechanisms of steroid oxidation by microorganisms. Biochim. Biophys. Acta. 62, 541-547 (1962).

28. Schubert, K. et al.: Abbau von Steroiden XII. J. Steroid Biochem. 2, 245-249 (1971).

29. Wang, K. C., Sih, C. J.: Mechanisms of steroid oxidation by microorganisms IV. Seco intermediates. Biochem. 2, 1238-43 (1963).

30. Sih, C. J., Weisenborn, F. L.: The use of metabolic inhibitors in microbial transformation of organic compounds. J. Am. Chem. Soc. 82, 2653-1654 (1960).

31. Gibson, D. T. et al.: Mechanisms of steroid oxidation by microorganisms IX. J. Biol. Chem. 241, 551-559 (1966).

Part III. Articles on Selected Algae and Algal Products

The Use of Alginates in Dentistry.

S. Ahing and M.Hébert
Department of Dentistry, St. Mary's Hospital
3830 Lacombe Avenue, Montreal, Quebec H3T 1M5, Canada

The major use of alginates in dentistry is for the recording of intraoral anatomical features in order to obtain replicas of a patient's mouth in dental stone or plaster of paris. This replica is called a model or cast and may be used for a variety of purposes depending on their accuracy; however the two most common reasons for which casts are made are for case records and the fabrication of prosthetic appliances.

Case records usually involve pre-and post-treatment records; however they are also useful in medico-legal situations such as Forensic Dentistry when intraoral anatomical features are used to identify cadavers in which facial features are obliterated by fire, decomposition or intentional mutilation. Models are useful in planning treatment, especially in surgery, orthodontics or prosthodontics, where mock jaw surgery can be performed, teeth moved into more functional positions, contours of teeth modified and missing teeth replaced on the models before similar procedures are done in the mouth. A variety of orthodontic and removable prosthetic appliances are fabricated on casts made from alginate impressions. Although alginates can be used for the fabrication of crowns and bridges the degree of accuracy required is usually more predictable with other types of impression materials. However with careful attention to handling the material the accuracy of the alginates is comparable to other more " accurate " materials such as the silicones and polysulphides (Hampson 1950, Hosoda 1961, Rapuano 1973)

Marine Algae in Pharmaceutical Science
edited by H. A. Hoppe, T. Levring, Y. Tanaka

Alginate material finds a variety of other uses in dentistry depending on the imagination of the clinician and his familarity with and manipulation of the material. On occasion alginates can be used to record anatomy in other parts of the body such as vertebrae and vaginal canals (Hebert 1977).

COMPOSITION OF ALGINATE IMPRESSION MATERIAL.

The exact composition of each commercial brand of alginate is proprietary. However as a group the alginates or irreversible hydrocolloids have the following composition as given in Table 1 (Asgar 1971, Walter 1971, DCNA 1971, Skinner and Phillips 1973, Combe 1975, Fainhoffer 1975).

Compound	Percentage by weight Average	Range
Salt of Alginic acid (Na, K, NH)	15	12-20
Calcium Sulphate	13	7-30
Sodium Phosphate	2.5	1-6
Filler	64.5	50-75
Fluoride	5	2-6
Lead Silicate	4.4	.008-17.5
Flavouring agents	Small	
pH indicator	.025	.005-.03

Table 1. Composition of alginate impression material.

ROLE OF THE COMPOUNDS IN ALGINATE IMPRESSION MATERIAL.

Salt of Alginic Acid

Alginic acid, an extract from seaweed, is insoluble in water but its sodium, potassium and ammonium salts are not. These soluble salts of alginic acid can be chemically converted to a gel state by calcium phosphate and the properties of the resulting gel, such as strength and elasticity, are modified by additives such as fillers and fluorides.

Calcium Sulphate

Calcium sulphate serves to release calcium ions which react with sodium alginate to form a calcium alginate gel. This reaction occurs almost instantaneously and in order to provide the dentist with more working time the reaction is retarded.

Sodium Phosphate

Calcium sulphate reacts preferentially with sodium phosphate, thereby preventing the release of free calcium ions to react with the sodium alginate(Skinner and Phillips 1973).In this way a longer working time is acheived. Consequently two types of alginates have been developed, namely Type 1 (Fast set) and Type 2 (Regular set). Type 1 alginate sets in 1 - 2 minutes and Type 2 sets in 2 - 4 minutes. The setting time can also be regulated by varying the temperature of the water used for mixing.

Fillers

The three previously mentioned compounds, sodium alginate, calcium sulphate and sodium phosphate, account for approximately 30% by weight of the impression material. A filler, usually diatomaceous earth, accounts for about 64% and improves dimensional stability by imparting bulk and strength to the impression material. It also produces a smooth, non-tacky gel surface by reducing syneresis (Buchan and Peggie 1966, Skinner and Phillips 1973) and also serves to increase cohesive forces within the material (Combe 1975). Other fillers may be zinc oxide, kieselguhr, barytes and whiting (Buchan and Peggie 1966).

Fluoride

The use of fluorides is primarily to ensure a hard, non-powdery stone surface in the cast when positive models of the impression are made. It does this by producing a firmer, more definite set of the alginate, thereby reducing surface syneresis. Fluoride also accelerates the set of gypsum products and increases the elastic properties of the alginate (Buchan and Peggie 1966, Skinner and Phillips 1973).

Lead Silicate

Lead silicate is usually most effective when used with a specific concentration of a silicofluoride. Under these conditions the setting time is decreased, the elasticity is increased, the permanent set deformation is decreased, the Brinell hardness increased and syneresis decreased - all desirable properties (Buchan and Peggie 1966). Lead silicate also has the ability to form lead alginate

pH indicator

When alginate is initially mixed it is alkaline (pH = 11) and when set it is neutral or nearly so. Consequently some manufacturers have added a pH indicator such as phenolphthalein so that the clinician can use colour changes to determine mixing, loading and setting times rather than a clock. Since the setting time may vary between different materials and between batches of the same material it is preferable to determine the development of optimum biophysical properties for each batch by a biochemical reaction rather than a standard average mixing time.

BIOPHYSICAL PROPERTIES OF ALGINATE IMPRESSION MATERIAL.

Properties controlled by the Manufacturer
1. Composition
2. Quality control
3. Shelf life
4. Elective additives
5. Economics
6. Biodegradability

As mentioned previously the exact composition of an alginate material is protected by patent laws. However the quality control of the individual materials used in its formulation are outlined in the USP or NF monologues. Quality control of the final impression material in North America is outlined in the American Dental Association's Guide to Dental Materials. Each country may follow its own guidelines such as Australia or pattern themselves after recommendations of International Organisations such as the International Dental Federation which set out guidelines for the standardisation of alginates in 1965. A flow sheet for the manufacture of an alginate is outlined in Table 2.

RAW MATERIALS					
Sodium alginate	Calcium sulphate	Sodium phosphate	Fillers	Fluoride	Other

↓ Sampling for Quality control

PROPORTIONED BY WEIGHT ACCORDING TO A FORMULA

↓

MIXED BY MECHANICAL TUMBLING

↓ Sampling for Quality control

SIFTING AND PACKAGING

↓

COMMERCIALLY AVAILABLE UNITS (1,5,20,100 lb units)

↓ Sampling for Quality control

DISTRIBUTING TO DENTAL SUPPLY HOUSES

Table 2. Steps in manufacturing dental alginates.

Manufacturers suggest their materials have a shelf life in excess of two years. However clinicians favour buying only enough alginate for one year at a time since storage conditions can affect shelf life. It is suggested that alginate be stored in a cool (50 - 80°F), dry enviroment and if possible alginate bought in bulk should be dated and weighed into individual packages to ensure the correct water: powder ratio.

A certain minimum of information should be provided on the tin of alginate. However this is not always the case. The following information is regarded as essential:-

1. Date of manufacture or a coded batch number.
2. Water:powder ratio by volume and weight respectively.
3. Recommended mixing times or colour changes.
4. Type 1 or Type 2 alginate
5. Addition of lead and fluorine compounds.

This information is of use to the clinician in making certain decisions. For example alginates containing fluoride do not usually have to be pretreated prior to pouring up the impression whereas alginates without fluorides should be fixed in 2% potassium sulphate or some other fixing solution prior to pouring the impression in order to ensure a hard, non-powdery stone surface.

The cost of various brands vary quite widely and the difference in cost per impression may range from 4 - 29 cents (Hawkins 1969, Walter 1971). Some other variables affecting the cost are the quantity purchased, available discounts and whether it is purchased from the manufacturer or a dental supply house. As far as is known the product is not biodegradable.

Properties controlled by the Dentist.

1. Water:Powder ratio
2. Setting time
3. Tray selection
4. Mixing material
5. Removal from the mouth
6. Storage of impression
7. Compatability with stone
8. Toxicity
9. The impression as a method of cross contamination
10. Recovery of used alginate

Although volumetric measurements of powders are inaccurate due to formation of voids the most common means of dispensing alginate powder is still by a scoop despite an average of 13% variability in weight measurements between scoops (Walter 1971). To overcome this problem pre-weighed packages of alginate can be purchased from the manufacturer at an increased cost per impression or the dentist can package his own. Distilled water has been suggested as preferable to tap water because minerals in the water may contaminate the mix and vary the setting time (Rudd et al 1969)

The most reliable way of altering the setting time without affecting the physical properties of the gel is by varying the temperature of water used for mixing. The amount of retardation or acceleration for each degree increase or decrease in temperature is different for each material. However all alginates will take a longer time to set if mixed with cold water and a shorter time with warm water. These relationships for three different alginates are shown in Table 3. (Harris 1969).

°C	Setting time (mins) Alginate A	Alginate B	Alginate C
3	5.05	3.30	9.30
10	9.10	2.30	6.20
20	3.15	2.10	3.35
30	2.40	1.40	1.30

Table 3 Variation of setting time with temperature.

When an impression is removed from the mouth two types of forces are acting at the junction of the impression material and the tray (Wilson and Smith 1963, Fusayama and Nakazato 1969). The sides of the alginate are stressed in shear while the other portions are stressed in tension (Figure 1).

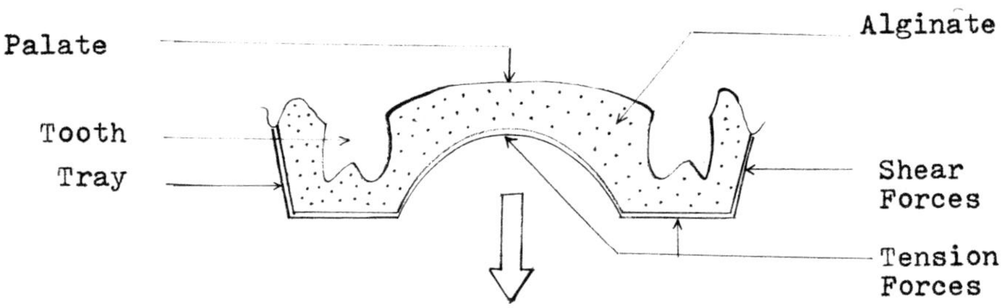

Figure 1 Stress forces in removing maxillary tray.

It has been found that perforated trays resist shearing forces best and plain stainless steel with an adhesive specific to each alginate resist tensional forces best.

At the present time commercially available trays are either perforated or plain stainless steel. Of the two the best choice is probably the perforated tray used with adhesive in the areas subjected to tensional forces. Plastic trays are available but warp easily and are therefore not accurate. The tray also must be selected to allow approximately 3-7 mm of impression material uniformly between the tray and the structure to be captured in the impression material. Excessively bulky areas of alginate may in fact shrink more in volume than thinner portions thereby leading to poorly fitting prosthetic appliances (Skinner and Phillips 1973).

All techniques of mixing alginates inherently incorporate air bubbles in the mix. The most common method of mixing is by using a rubber bowl and metallic spatula with a rotary motion rather than a whipping action. Once all the particles are wetted the mix is homogenised between the bowl and the spatula in a wiping motion. Mechanical spatulation under vacuum is available and is the method which shows the least amount of air bubbles in the mix. Another method for reducing air bubbles in the mix is the " bag technique " advocated by Diamond (1968). Bubbles, if excessive, contribute to the inaccuracy of the surface characteristics in the impression. Mixing time is also important in that over-spatulation will begin to break up the brush heap structure of the gel whereas under spatulating the mix will not ensure the incorporation of all the compounds evenly throughout the material. In both cases the mechanical properties of the gel will be affected.

Even when alginate is set it is not 100% elastic, in other words it will permanently deform if stressed. The extent of this deformation is approximately 4% or less when subjected to 10% compressive strain for 30-60 seconds 3 minutes after set.(Gilmore et al 1958,Anderson 1960, Wilson 1966, Bliss and Saggs 1967) It has also been found that the tear strength increased when the rate of loading increased or in other words when strain is applied for a short time. Thus when the tray is removed it should be done quickly with a snap (to increase rate of loading) and in the direction of the long axes of the structures within the impression without rotating,rocking or turning of the tray (in order to decrease compressive strain).

Due to imbibition, syneresis and the release of internal strains in the impression (created if the tray is moved during gelation) alginates should be poured within fifteen minutes after removal from the mouth (Combe 1975) If it is to be delayed it is best stored at 100% relative humidity in a humidor. It is felt that when impressions are inverted water lost through surface evaporation is replaced by gravitational flow of fluids through the bulk of the material (Civjan et al 1972).

Several articles have shown that not all dental stones are equally compatible with alginates hence the surface of the cast may show discrepancies and inaccuracies (Choy et al 1965, Morrow et al 1971). The choice is basically one of trial and error but guidelines can be found in the above two references.

Certain manufacturers add lead compounds to their formulas in order to affect the physical properties of the material. Apparently there are only three alginates on the world market without lead compounds (Benari 1977) and they are the alginates manufactured by International Dental Products, Unitek and Kerr. At the present time only one study has attempted to correlate the blood concentration of lead due to inadvertent inhalation of alginate powder (Mjor 1974) although several authors suggest a possible health hazard (Buchan and Peggie 1966, Smith 1969, Walter 1971). Mjor found blood levels of lead in dental personel using an alginate with high lead content to be 18 micrograms/ml and in those using a low lead content alginate to be 9.8 micrograms/ml. However studies at St. Mary's Hospital indicate these levels to be rather high.

The average blood lead level for dental personnel at St. Mary's Hospital was 0.087 micrograms/ml with a range of 0.02 - 0.12 micrograms/ml. Nevertheless,in view of the fact that lead is cumulative a good practice would be to wear face masks when handling alginates. There are no known published studies on blood lead levels in workers who manufacture the alginate.

A perusal of the literature could reveal no cases of cross contamination via alginate impression material. However the theoretical possibility of sterilisation of alginates without affecting the dimensional stability of the impression is ever present. Only one study is presently available to determine the effect of sterilising solutions on imbibition and syneresis in alginates and it was found that in attempting to sterilise against the Hepatitis B virus gluteraldehyde distorted the impression less than did hypochlorite and additionally did not corrode metallic trays (Trevelyan 1974).

Recovery techniques for alginates are non existant although alginates may be dissolved in .1N hydrochloric acid or 8% sodium bicarbonate (Smith 1969, Fusayama and Nakazato 1969). The significance of this is not known although this author speculates that polymerised alginate may be used as a binder or filler in the manufacture of bricks for use in the construction industry or possibly as a fuel source. Crushing or powdering of used alginate may find some use as a filtration medium for liquids, possibly as a sewage treatment.

Literature cited.

Asgar et al Restorative Dental Materials
CV Mosby 1971 St Louis.

Anderson Flow and Elasticity in Alginates
Dent. Progress 1, 63-70 (1960)

Benari Vice President, International Dental Products
Private Communication.

Buchan and Peggie Role of ingredients in alginate impression compounds.
J. Dent. Res. 45, 1120-1129 (1966)

Bliss and Saggs Note on alginate impression materials
New Zeland Dent. J. 63, 282-284 (1967)

Civjan et al Surface characteristics of alginate impressions.
J. Pros Dent 28, 373-378 (1972)

Choy et al The surface of gypsum cast in alginate impression.
Dent Pract. 16, 107-109 (1965)

Combe Notes on Dental Materials
Churchill Livingstone 1975 London

Diamond Bag mix, a new mixing method for alginates.
J. Pract. Ortho 11, 311 (1968)

Fainhoffer Scientific Aspects of Dental Materials
Butterworths 1975 London.

Fusayama and Nakazato The designs of stock trays and the retention of irreversible hydrocolloid material.
J. Pros. Dent. 21, 136-142 (1969)

Gilmore et al Effect of residual stress and water change on the deformation of hydrocolloid impression materials
J. Dent. Res. 37, 816-823 (1958)

Hebert Chief, Department of Dentistry, St. Mary's Hospital
Private Communication.

Hampson Use of irreversible hydrocolloids in Conservative Dentistry.
Br. Dent. Jour. 89, 171-175 (1950)

Hosoda and Fusayama Distortion of irreversible hydrocolloid and mercaptan rubber base impressions.
J. Pros. Dent. 11, 318-333 (1961)

Hawkins An evaluation of currently available alginate impression materials.
Aust. Ortho 2, 61-64 (1969)

Harris Water temperature and accuracy of alginate impressions.
J. Pros. Dent. 21, 613-617 (1969)

Mjor Blood lead analyses and alginate impression materials
Scand. J. Dent. Res. 82, 401-402 (1974)

Morrow et al Compatability of alginate impression materials with dental stones.
J. Pros. Dent. 25, 556-565 (1971)

Rapuano Irreversible hydrocolloid in fixed prosthodontics
J. Acad Gen Dent 21, 29-32 (1973)

Robbins and Angel Basic Pathology
WB Saunders 1976 Philadelphia.

Rudd et al Accurate alginate impressions
J. Pros Dent 22,294-300 (1969)

Smith Biologic Properties of Dental Materials
Dent Prac Dent Rec 20, 81-86 (1969)

Skinner and Phillips The Science of Dental Materials
WB Saunders Co. 1973 Philadelphia

Trevelyan The Prosthetic treatment of Hepatitis B antigen positive patients.
Br. Dent. Jour. 137, 63-64 (1974)

Wilson Some properties of alginate impression materials relevant to clinical practice.
Br. Dent. Jour. 117, 536-540 (1964)

Wilson and Smith The bonding of alginate impression material to impression tray.
Br. Dent. Jour. 115, 291-294 (1963)

Walter An assessment of alginate materials
Dent. Pract. Dent. Rec. 21, 377-384 (1971)

References.

Anderson Flow and Elasticity in alginates
Dent. Prog. 1,63-70 (1960)

Elborn and Wilson Temperature attained by impression materials in the mouth.
Br. Dent. Jour. 118, 80-82 (1965)

Einfeldt The alginate impression
Quintessence 4, 57-61 (1973)

Fairchild Versatile uses for alginate impression material.
J. Pros. Dent 31, 266-269 (1974)

Fung and Osborne The setting characteristics of two alginate materials in relation to temperature change.
Aust. Dent. Jour 17, 33-36 (1972)

Greener et al Material Science in Dentistry
Williams and Wilkins 1972 Baltimore

Heartwell et al Comparison of impressions made in perforated and non-perforated rimlock trays.
J. Pros. Dent. 27, 494-500 (1972)

Johnson Alginate Impression Materials
J. Can Dent Assoc. 94, 713-716 (1977)

Koski Comparative study of selected alginate materials and devices.
J. Am Dent. Assoc 94, 713-716 (1977)

Miller Syneresis in alginate impression materials
Br. Dent. Jour. 139, 425-430 (1975)

Macpherson et al Mechanical properties of hydrocolloid and rubber impression materials.
J. Dent Res. 46, 714-721 (1967)

McAdam Management of impression materials in restorative dentistry.
J. Ont. Dent 54, 18-23 (1977)

Smith and Nakamoto Alginate for duplicating procedure
J. Am. Dent. Assoc. 84, 588-590 (1972)

Sawyer Accuracy of casts produced from alginate and hydrocolloid impression materials.
J. Am Dent. Assoc 93, 806-808 (1976)

ENZYMIC AND N.M.R. SPECTROSCOPIC ANALYSES OF AGAR-TYPE POLYSACCHARIDES

S.S. BHATTACHARJEE AND W. YAPHE
Department of Microbiology and Immunology
McGill University, Montreal, Quebec H3A 2B4 (Canada)

INTRODUCTION

Agar represents a complex mixture of polysaccharides extracted from several species of the red algae known as agarophytes. The mixture previously considered to be composed of agarose and agaropectin has been fractionated by ion-exchange chromatography into essentially neutral agarose, charged agarose and non-gelling sulfated galactan (1-4). The relative portion of these components varies in agars obtained from different marine algae and in commercial agars (5). The polysaccharides, in general, are composed of alternating residues of 3-O-linked-β-D-galactopyranosyl and 4-O-linked-3,6-anhydro-α-L-galactopyranosyl residues (Fig. 2). The hydroxyl group of sugar residues may be substituted by O-methyl, sulfate and O-(1-carboxyethylidene) groups, and 3,6-anhydro-α-L-galactopyranosyl residue may be replaced by α-L-galactoyranosyl-6-sulfate residue.

Commercial agars of poorly defined chemical structure have been used for many years in pharmacological and other biomedical studies. Since the property of agar depends both on polysaccharide composition and macro-molecular structure, it is desirable in a biological study to use agars of known structural characteristics (6). In the past, agar-preparations were characterized by chemical and enzymic methods (5,7-12). Recently it has been shown that ^{1}H- and ^{13}C-nuclear magnetic resonance (n.m.r.) spectroscopy offers a convenient rapid physical method for the identification and characterization of agar-type polysaccharides (13-19). In view of these developements, we propose a simplified scheme for the rapid and definitive characterization of an agar-preparation.

Marine Algae in Pharmaceutical Science
edited by H. A. Hoppe, T. Levring, Y. Tanaka

^{1}H-N.M.R. SPECTROSCOPIC ANALYSIS

To obtain a reasonably well-resolved ^{1}H-n.m.r. spectrum of agar, a partial depolymerization of the polymer is necessary, and this can be achieved by either acid or enzymic hydrolysis. Izumi determined the main structural features of agar by analyzing partial acid hydrolysates with ^{1}H-n.m.r. spectroscopy (13). In his investigation, different polysaccharide fractions from Gelidium amansii, Gracilaria verrucosa and Gloiopeltis furcata were used. The ^{1}H-n.m.r. spectrum provided the molar ratio of galactose to 3,6-anhydrogalactose, and an estimate of 6-O-methyl, 6-sulfate and pyruvate groups based on the signals due to specific protons as illustrated in Fig. 1. The basis of this analytical approach is that the areas of the signal due to specific protons are proportional to their relative numbers present in the structure. The molar ratio of D-galactose to 3,6 anhydro-L-galactose was calculated from the area ratio of signals H-4 and H-5 (τ 5.55 - 5.65) in anhydrogalactose residue to those of all other protons except for the anomeric ones (τ 5.9 - 6.5). The signal (singlet) for 6-O-methyl group appears at τ 6.60, where the spectrum of agarobiose does not show any peak. An estimate of 6-O-methyl group can, therefore, be easily obtained from the area ratio of the peak at τ 6.60 to the signals of other protons. Signals at τ 5.79 and 5.76 have been attributed to H-6 of β-D-galactose-6-sulfate and α-L-galactose-6-sulfate, respectively. Total sulfate content was estimated from the area ratio of these peaks to others. The C-methyl group of pyruvic

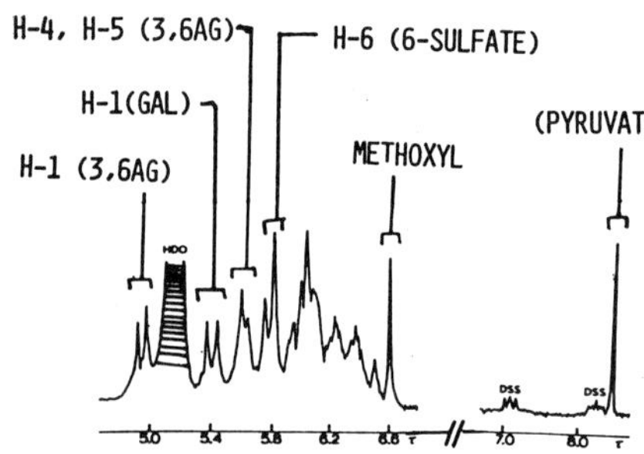

Fig. 1. ^{1}H-N.M.R. spectrum of partial acid hydrolysates of agar fractions in D_2O measured at 100 MHz; Gal, galactopyranose; 3,6 AG, 3,6 anhydrogalactopyranose. Adapted from Izumi, K. (13).

acid residue gives rise to a sharp singlet at τ8.44, and a quantitive analysis was attained in a similar way.

Agarobiose (Fig. 2a) is the major product formed under the conditions of partial acid hydrolysis with signals at τ 4.95 (doublet, J 6.0 Hz) due to H'-1 of the hydrated aldehyde group of 3,6-anhydrogalactose residue.[a] We have also found that the presence of higher oligosaccharides in the partial hydrolysate, for example, agarotetraose (Fig. 2b) is indicated by an additional signal at τ 4.85 (doublet, J 2.0 Hz) due to the anomeric hydrogen (H"-1) of internal 4-O-linked 3,6-anhydro-α-L-galactosyl residues. The ratio of the combined signal area for H'-1 and H"-1 to that of the remaining signals should also give an estimate of 3,6-anhydrogalactose residues.

The enzymic hydrolysates of agar, like the acid hydrolysate as described, can be used to provide ^{1}H-n.m.r. spectral data. The agaro-oligosaccharides (Fig. 2b) obtained by the hydrolysis of agarose with an

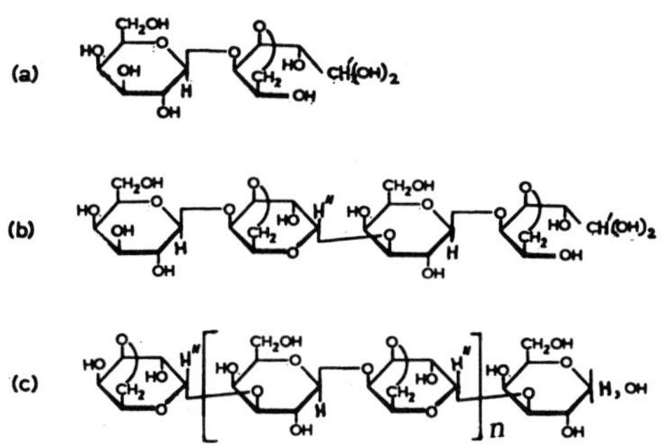

Fig. 2. Hydrolysis products of agarose; a) agarobiose; b) agarotetraose c) neoagaro-oligosaccharides.

---

[a] Partial hydrolysis was carried out by heating the polysaccharide either in 0.1 M DCl in D_2O or with Amberlite IR120 acid resin (after being exchanged with D_2O) in D_2O for 20 min. at 95°; chemical shift values were obtained form 220 MHz ^{1}H-n.m.r. spectra using 10% solution in D_2O. [S.S. Bhattachajee and W. Yaphe, unpublished results.]

α-agarase (20) gave a spectrum similar to that described above for a partial acid hydrolysate. An enzymic hydrolysate of agarose with β-agarase contains neoagaro-oligosaccharides (21) (Fig. 2c), and yields a ^{1}H-n.m.r. spectrum with signals (doublet) in the region of τ 4.85 (H"-1 of 3,6-anhydro-α-L-galactopyranose residue), τ 5.32 (H-1 of 3-O-linked-β D-galactopyranose residue), and τ 4.38 (H-1 of 3-O-linked-α-D-galactose at the reducing end); the spectrum does not contain any signal in the region of τ 5.75 and τ 6.60 where the signals due to 6-sulfate and 6-O-methyl groups normally appear. The spectra of O-(1-carboxyethylidene)-neoagaro-tetra and hexasaccharides obtained by the hydrolysis of charged agarose with β-agarase (7) contain a signal at τ 8.50 indicative of the pyruvate group. An advantage to using enzymic hydrolysates for ^{1}H-n.m.r. spectral analysis is the stability of 3,6-anhydrogalactose residue, 6-sulfate and pyruvate groups to enzymic hydrolysis, all of which are labile under acid condition.

^{1}H-n.m.r. spectroscopic analysis of a partial hydrolysate of agar provides quantitative data on the main chemical characteristics, that is, an estimate of 3,6-anhydrogalactose and galactose residues, and the degree of substitution (methoxyl and sulfate) at C-6 of galactose residues. A limitation of the ^{1}H-n.m.r. method seems to be that, on the basis of the present data, it does not provide a means to determine 2-and 4-sulfate or 2-O-methyl groups which probably exist in agar-type polysaccharides. In this respect, measurement of the ^{1}H-n.m.r. spectrum at a higher frequency, for example, at 220 MHz, or ^{13}C-n.m.r. spectroscopy as described in the following section, may be useful.

^{13}C-N.M.R. SPECTROSCOPIC ANALYSIS

We have reported[b)] that well-resolved ^{13}C-n.m.r. spectra of agars can be obtained without prior acid or enzymic degradation as required for the ^{1}H-n.m.r. analysis (14-18). The presence of 12 distinct signals in

---

[b)] In collaboration with G.K. Hamer; proton-decoupled ^{13}C-n.m.r. spectra were recorded at 22.63 MHz using solutions (30-50 mg/ml) in D_2O at 95°; other conditions of measurement as reported elsewhere (14, 15).

the ^{13}C-n.m.r. spectrum of agarose is indicative of a regular structure with the disaccharide repeating unit as shown in Fig. 3 (15,17). The signals at 103.4 and 99.2 p.p.m. have been attributed to the anomeric carbons (carbon -1) of β-D-galactopyranosyl and 3,6-anhydro-α-L-galactopyranosyl residues, respectively; the signal at 62.4 p.p.m. has been assigned to carbon-6 of the D-galactosyl residue; the other signals have been assigned to the remaining 9 carbon atoms in the disaccharide repeating unit (18). Identical spectra to agarose were obtained with a commercial agar (Difco Bacto agar) and agar from *Ahnfeltia plicata*. The agars from *Porphyra umbilicalis*, *Gracilaria bursapastoris* and *Gracilaria foliifera* yielded similar spectra with distinct signals at 103.4, 99.2 p.p.m. and new signals at 60.0, and 72-75 p.p.m. The signal at 60.0 p.p.m. has been attributed to the methoxyl group and two signals at 72-75 p.p.m. to the effect of substituents. The spectrum of the agar from *Gracilaria bursapastoris* contains a diminished signal at 62.4 p.p.m. and another new signal in the range of 68.9 p.p.m., possibly due to the presence of 6-sulfate group in galactosyl residue. Recently, Usov *et al* (19) have reported ^{13}C-n.m.r. spectral data (Table 1) obtained from an agar-preparation from *Odonthalia corymbyfera*, with a characteristic signal (signal b' Fig. 4) due to the presence of 4-sulfate group in the disaccharide repeating structure of agar.

In the past, due to the lack of a rapid and definitive analytical technique, seaweed gels, such as, furcellaran and phyllophoran have been

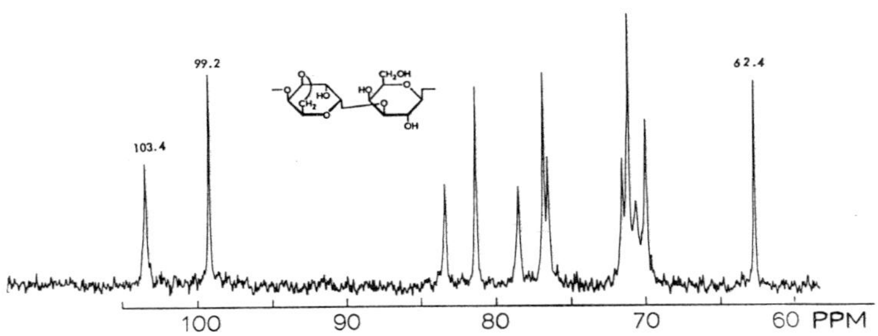

Fig. 3. Disaccharide repeating unit and ^{13}C-n.m.r. spectrum of agar (*Ahnfeltia plicata*).

occassionally designated as agar although structurally they are similar to carrageenan. Recent findings on ^{13}C-n.m.r. spectroscopic analysis of algal-polysaccharides suggest that this confusion in identification can be easily avoided. The carrageenans are a related group of polysaccharides present in red algae having structural features comparable to agar, with the 4-O-linked-3,6,-anhydro galactosyl residue in D-configuration instead of the L-configuration present in agar. The hydroxyl groups of sugar residues are substituted by sulfate group with varying degrees and sites of substitution. We have reported that different types of gel-forming carrageenans yield characteristic ^{13}C-n.m.r. spectra, each being different from the spectrum of agar (16,17). The well-resolved signal in the anomeric region related to the specific disaccharide repeating units (15), provides a means for the identification of the type of the polysaccharide (Fig.4, Table 1). Because of the difference in the configuration of 3,6-anhydrogalactosyl residue and variation in sulfate-distribution, the chemical shift of the anomeric carbon of 3,6-anhydrogalactosyl residue is characteristic of the polysaccharide-type (Table 1).

Hence ^{13}C-n.m.r. spectroscopy offers a novel physical method to differentiate agars from different types of carrageenans. In addition, this

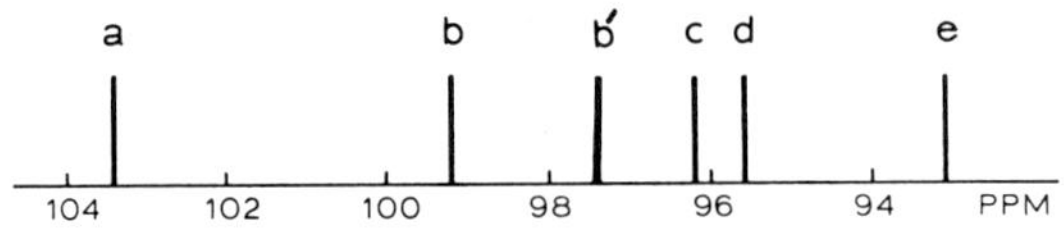

Fig.4. Characteristic signals (stick-diagram) in the anomeric region of ^{13}C-n.m.r. spectra for different agars and carrageenans (see Table 1). Chemical shift values (15,17); a, 103.4 - 103.6 p.p.m. (102.3 - 102.6 p.p.m.); b, 99.2 p.p.m. (98.2 - 98.3 p.p.m.) b', 97.6 p.p.m. (96.6 p.p.m.); c, 96.1 - 96.4 p.p.m. (95.1 p.p.m.); d, 95.6 p.-p.m. (94.6 p.p.m.); e, 93.1 p.p.m.; the chemical shift values in parenthesis are reported by Usov et al (19); the consistant difference of about 1.0 p.p.m. between two sets of data is attributed to the variation in methods for spectral measurement; the chemical shift of 97.6 p.p.m. for the signal b' is calculated from the reported value (96.6 p.p.m.) for comparison on the same scale.

technique can be extended to obtain information about the presence and location of methoxyl and sulfate groups. On the basis of the present data, it is not certain whether the ^{13}C-n.m.r. spectroscopy can be used to obtain quantitative data at the polysaccharide level, although we have shown that quantitative evaluation of ^{13}C-n.m.r. signals for agar-oligosaccharides is possible (14).

TABLE 1

Classification of different types of agars and carrageenans by characteristic ^{13}C-n.m.r. signals.

	Anomeric Signal*	
	β-Galactopyranosyl residue	3,6-anhydro-α-galactopyranosyl residue
Agar[1]	a	b
Agar sulfate[2]	a	b'
Furcellaran[3]	a	c d
Phyllopharan[4]	a	c d
kappa-Carrageenan[5]	a	c
iota-Carrageenan[6]	a	e
Hybrid-carrageenan[7] (*kappa*-*iota*)	a	c e

*The chemical shifts are given in Fig. 4;

1, *Ahnfeltia plicata*; *Gracilaria foliifera*; *Gracilaria bursapastoris*; *Porphyra umbilicalis*. 2, *Odonthalia corymbyfera*. 3, *Furcellaria fastigiata*. 4, *Phyllophora nervosa*. 5, *Eucheuma striatum*. 6, *Eucheuma spinosum*. 7, *Chondrus crispus* (traces of iota-type), *Ahnfeltia concinna*.

It is expected that future work involving a fuller interpretation of ^{13}C-n.m.r. spectra obtained from various types of algal-polysaccharides will widen the applicability of this n.m.r. technique.

ENZYMIC ANALYSIS

Structural analysis of the major components in agar has been obtained by enzymic hydrolysis with a β-agarase (7,8). This enzyme cleaves the $\beta(1 \rightarrow 4)$-linkage in agars from several marine algae yielding the neutral neoagarobiose series of oligosaccharides (Fig. 2c). The mode of action of the enzyme is altered by substitution of the hydroxyl group in neutral agarose with methoxyl, pyruvate and sulfate groups. Analysis of the end-products by thin-layer chromatography and electrophoresis provides information about the structure of methylated agarose, charged agarose and sulfated galactan (5,7). The enzymic cleavage of methylated agarose yields methylated di-and tetrasaccharides in addition to the neoagarbiose series of oligosaccharides (9). The pyruvated tetra and hexasaccharides are obtained from agarose containing O-(1-O-carboxyethylidene) substituents, whereas partially sulfated agarose yields high molecular weight sulfated oligosaccharides in addition to a small proportion of neutral oligosaccharides. It has been shown that the pyruvated oligosaccharides are free of sulfate indicating that the majority of pyruvate groups are present away from the sulfated segment of the galactan (7). Enzymic studies can thus provide information about the distribution of substituents along the macromolecule of agar, data which is not obtained by n.m.r. analysis.

SUMMARY

The enzymic and n.m.r. spectroscopic methods of analysis are complimentary, and represent definitive means for the characterization of agars. The n.m.r. spectroscopic methods are particularly attractive since a single spectrum can yield various analytical and structural information within a short time, and thus provides an additional approach to the conventional chemical methods of analysis of agars which have been reviewed recently by Craigie et al (12). Although n.m.r. spectrometers may not be easily accessible to biological research laboratories spectral service is becoming available on a commerical basis. Structural analysis of an agar-preparation for pharmacological and other biomedical studies can be facilitated by these advanced techniques, as outlined here:

1. A ^{13}C-n.m.r. spectrum is recorded to determine whether the preparation has the characteristic signals for agar or carrageenan-type polysaccharides. The spectrum can also provide information about the presence of methoxyl and sulfate groups, and in some cases, the location of the substituents.

2. The agar-preparation is fractionated by ion-exchange chromatography to determine the proportion of the three fractions: a) neutral agarose, b) charged agarose, and c) sulfated galactan (1).

3. Each of the above fractions is hydrolysed partially with acid or enzyme. The ^{1}H-n.m.r. spectrum is recorded to determine the ratio of 3,6-anhydrogalactose to galactose residues, and concentration of 6-*O*-methyl, pyruvate and sulfate groups.

4. For further structural evaluation, the neutral, charged and sulfated fractions are hydrolysed with β-agarase, confirming the presence of an agar-type structure. Analysis of the products of enzymic hydrolysis by thin-layer chromatography, electrophoresis and n.m.r. spectroscopy provides information on the type and distribution of substituents present in each of the major fractions of agar.

ACKNOWLEDGMENTS

We thank Drs. G.K. Hamer and N.M.K. Ng Ying Kin for valuable discussions, and the National Research Council of Canada for financial support.

REFERENCES

1. Duckworth, M., Yaphe, W.: The structure of agar. Part I. Fractionation of a complex mixture of polysaccharides. Carbohydr. Res., 16, 189-197 (1971).

2. Duckworth, M., Yaphe, W.: Preparation of agarose by fractionation from the spectrum of polysaccharides in agar. Anal. Biochem., 44, 636-641 (1971).

3. Izumi, K.: Chemical heterogeneity of the agar from *Gelidium amansii*. Carbohydr. Res. 17, 227-230 (1971).

4. Izumi, K.: Chemical heterogeneity of agar from *Gracilaria verrucosa*. J. Biochem. 72,135-140 (1972).

5. Duckworth, M., Hong, K.C., Yaphe, W.: The agar polysaccharides of Gracilaria species. Carbohydr. Res. 18, 1-9 (1971).

6. Duckworth, M., Yaphe, W.: The relationship between structures and biological properties of agars. In Proc. Seventh Int. Seaweed Symp. University of Tokyo press. 15-22 (1972).

7. Duckworth, M., Yaphe, W.: The structure of agar. Part II. The use of a bacterial agarase to elucidate structural features of the charged polysaccharides in agar. Carbohydr.Res.16,435-445 (1971).

8. Duckworth, M., Turvey, J.R.: The specificity of an agarase from a Cytophaga species Biochem. J. 113, 693-696 (1969).

9. Duckworth, M., Turvey, J.R.: Action of a bacterial agarase on agarose and alkali-treated porphyran. Biochem. J. 113, 687-697 (1969).

10. Turvey, J.R., Williams, F.L.: The agar-type polysaccharides from the red alga Ceramium rubrum. Carbohydr. Res. 49, 419-425 (1976).

11. Batey, J.F., Turvey, J.R.: The galactan sulfate of the red alga Polysiphonia Lanosa.Carbohydr. Res. 43, 133-143 (1975).

12. Craigie, J.S., Leigh, C.: Carrageenans and agar. Handbook of Phycological methods, Vol. II, Physiological and biochemical methods; edited by J.A. Hellibust and J.S. Craigie, Cambridge Univ. Press. N.Y., in press.

13. Izumi, K.: Structural analysis of agar-type polysaccharides by nmr spectroscopy. Biochim. Biophys. Acta 320, 311-317 (1973).

14. Hamer, G.K., Bhattacharjee, S.S., Yaphe, W.: Analysis of the enzymic hydrolysis of agarose by ^{13}C-n.m.r. spectroscopy. Carbohydr. Res. 54, C7-C10 (1977).

15. Bhattacharjee, S.S., Yaphe, W., Hamer, G.K.: ^{13}C-N.m.r. spectroscopic analysis of agar, κ-carrageenan and ι-carrageenan. Carbohydr. Res. in press.

16. Bhattacharjee, S.S., Hamer, G.K., Yaphe, W.: A study of agar and carrageenan by ^{13}C-n.m.r. spectroscopy. J. Phycology (supplement) 13, abstr. pap. 22 (1977).

17. Bhattacharjee, S.S., Hamer, G.K., Yaphe, W.: A study of agar and carrageenan by ^{13}C-n.m.r. spectroscopy. In press., Proc. Ninth Int. Seaweed Symp. Santa Barbara (U.S.A.)

18. Hamer, G.K., Bhattacharjee, S.S., Yaphe, W.: ^{13}C-n.m.r. spectroscopic analysis of the enzymic hydrolysis products of agarose. Abstr. CIC/ACS Conf. Montreal, 1977, Carb-11.

19. Yarotsky, S.V., Shashkov, A.S., Usov, A.I.: Analysis of ^{13}C-n.m.r. spectra of some red seaweed galactans. Bioorg. Chem. 3, No. 8, 1135-1137 (1977).

20. Young, K., Hong, K.C., Duckworth, M., Yaphe, W.: Enzymic hydrolysis of agar and properties of bacterial agarases. Proc. Int. Seaweed Symp., 7th University of Tokyo Press, Tokyo. pp 469-472 (1972).

21. Duckworth, M., Yaphe, W.: Thin-layer chromatographic analysis of enzymic hydrolysates of agar. J. Chromatography 49, 482-487 (1970).

SIMPLE LABORATORY METHOD FOR DETERMINATION OF GEL STRENGTH

K. Czapke
Morski Instytut Rybacki, Gdynia, Poland

The growing interest in seaweed as a raw material for industry makes the assistance of science and scientific research work necessary to insure an adequate development of this field and to shape it into a modern form.

The search for new seaweed species and their uses makes simple and time saving laboratory investigation methods desirable. Especially if the same species growing in different seawater habitats show considerable differences in their chemical composition and as a result the substances obtained from them have often quite different physical and chemical properties.

For this reason a simple laboratory method for determination of gel strength used in my own laboratory work, and found useful and convenient also by other scientists will be described and discussed.

The gelling substances derived from seaweeds, agar-agar being one of the best known ones, have a most important significance. Used in bacteriology and medicine it serves in defense against infectious deseases. Besides it also finds great application in modern food and confectionary industry.

One of the characteristic features of gelling properties is the gel strength, i. e. the resistance to the breaking of gel. This depends on the concentration of the gelling substance in the gel and on the chemical nature of that substance. The gel strength is theoretically defined as the weight in grammes necessary for breaking the surface of a gel of known concentration within an area of one square centimetre. Usually it is referred to the

Marine Algae in Pharmaceutical Science
edited by H. A. Hoppe, T. Levring, Y. Tanaka

concentration of one per cent of the gelling product although there are some divergluces from this rule.

For practical reasons and from a commercial point of view the gel strength is of special importance as it is essential for classifying gelling products during manufacture and application. The measurement of gel strength is therefore of considerable value both for scientific work and for industry.

Although the term gel strength is based on quite a simple theoretical assumption the methods and devices for measuring it are sometimes intricate and somewhat inconvenient.

One method for example (2), requires rather large samples of about 200 millilitres, comparatively large laboratory glass fittings, and employs a considerable amount of quicksilver as a means to produce sufficient pressure, to break the gel. The quicksilver is poured from a bottle into a beaker at a constant rate until the gel surface is broken by a cylindrical plunger on which the beaker is placed.

The relatively large samples are not desirable as they are labour, time, and space consuming. Furthermore the use of quicksilver has to be avoided for health reasons, due to the well known dangerous and poisonous effects of its vapours.

My proposed method is free from these disadvantages. It employs a simple device for carrying out the measurement and in the conventional set of gramme weights for producing the pressure necessary to break the gel.

The device consists of few simple components. A common test-tube, a little piston made of light material, fitted with two small plates at both ends also of light material, a usual test-tube stand, and a set of commonly used weigths from 1 gramme to about 300 grammes. The test-tube serves as a container for the sample the gel strength of which is to be measured. It is an ordinary glass test-tube about 1.5 centimetres in diameter and about 15 cm in lenth. The piston is a little thin rod, round in cross section, about 0.5 cm in diameter and about 10 cm in length, provided at the lower end with a disc of an area amounting exactly to 1 square centimetre. The upper end of the piston is fitted with a round plate circ. 3 cms in diameter provided with an upwards bent rim circ. 2 or 3 cms high. The weight of the piston with plates is to be exactly known in whole numbers.

In order to perform a measurement of the gel strength the piston is to be put lightly into the test-tube and be set cautiously onto the sample surface taking care not to break it. At the same time the piston is to be kept in vertical position by means of three little claps clamped onto the test-tube outlet. Each of the clasps has a very little protrusion by means of which it lossely touches the piston shaft to enable free movement downwards. This last arrangement prevents breaking the gel surface by leaning of the piston against the test-tube wall. The piston shaft has to stand out at least 1 cm above the test-tube outlet. Now the weights are put onto the upper plate of the piston until the gel surface is broken under the pressure of the load of both the weights and the piston. The result of the measurement is the sum of both loads expressed as grammes per square centimetre for 1 % concentration of the gel solution.

For convenience the measurement procedure can be performed with the sample test-tube placed in the test-tube stand.

The measurements may also be carried out with samples contained in a beaker, Petri dish or like vessel. In this case the piston has to be put into a short glass tube shorter in length than the piston itself. The glass tube is fastened 1 cm above the gel surface by means of a common laboratory stand and clamp. In any event, as already mentioned, the piston shaft has to stand out at least 1 cm above the test-tube outlet or the end of the glass tube, to enable free vertical movement downwards.

While preparing the gel samples for measurement some precautions have to be taken. The concentration of the solution must be exactly known. The volume of the sample must amount to about 10 millilitres, i. e. large enough to fill the test-tube sufficiently to let the piston shaft stand out above the test-tube outlet. The samples have to be prepared at least a day before the measurements take place and should stand at room temperature carefully stoppered in order to prevent forming a stronger gel concentration in the upper layer of the sample through evaporation.

The degree of accuracy of this method, like other of its kind, amounts to 1 gramme per square centimetre and is suitable for all scientific and industrial purposes.

Two questions deserve to be touched upon while discussing the above method as there are some differences of opinion on these points. The first one involves the temperature at which the measurements ought to be performed. Some methods prefer 20° C while others do not. It seems resonable to take the laboratory room temperature 20° C as a standard temperature since it has been accepted in analytical work for about two centuries and is obligatory as well in the calibration of all chemical apparatuses.

The second question is whether the area of the piston employed can differ from the dimensions mentioned above and without reducing the results to a unit measurement. Such a custom does not suit the terms of proper exchange of information. For this reason seems advisable to accept the square centimetre as a standard unit for the dimension of the piston as well as for expressing gel strength, in accordance with general usage in the sciences.

The device described above for gel strength measurement may be easily constructed. Only few simple utensils, commonly found in every conventional laboratory, are needed.

The test-tube and test-tube stand belong to the basic equipment of any laboratory, as do the weights. The only part which must be specially constructed is the piston. It can be assembled from a short, thin piece of glass tube ca. 10 cms in length and ca. 0.5 cm in diameter, a small rubber or cork stopper, and a piece of tin, aluminium or plastic plate.

The stopper should be chosen so that its larger area is exactly 1 square centimetre. Otherwise it is to be adjusted to this dimension. The smaller area of the stopper is to be cut thin so as to fit into the glass tube, and to serve as connection means with the tube by way of stoppering.

An upper plate of suitable size and shape is to be provided with a hole in the centre which serves to connect the plate with the glass tube by means of a hand-made rubber or cork "tack". This plate may be stuck with adhesive to the tack-head if necessary.

The centring clasps can be made of short, thin, narrow plate pieces bent so as to allow being clamped onto the test-tube outlet and adjusted to keep the piston loosely in vertical position.

As the growth of the seaweed industry in the next 50 years is expected to be far greater than that during the first half of this century (3), every contribution to simplify and unify laboratory work, like the method discussed above, may be welcomed.

1. Chapman, V. J.: Seaweeds and their Uses, Methuen Co. LTD., London, 1970

2. Zikieiev, B. W.: Pierierabotka Wodnogo Nierybnogo Syria, Pishchepromizdat, Moskow, 1950

3. Woodward, F. N.: Proceedings of the Fifth International Seaweed Symposium, Pergamon Press, Oxford, 1966

USE OF THE LAMINARIA TENT IN OBSTETRICAL PRACTICE

Konstantin Feochari
Clinical Investigation Service, St. Mary's Hospital, Quebec H3T 1M5, Canada

With modern technological advances increasing at a preponderant rate, the concomitant problems of industrialized urbanization, environmental pollution, overpopulation, and disruption of the natural balance between numerous ecosystems have come about. A grave consequence resulting from this state of affairs is the concurrent depletion of many natural resources, which in a great many instances involves the selfish and wasteful exploration by man. The ocean remains one of the final natural resources to be conquered by man, to provide the necessary nutritional products to feed a world where food shortages are inevitable in the very near future.

Perhaps a not so obvious asset of the ocean's benefit to mankind resides in the utilization of many natural products from the sea in clinical application in medicine. A perfect example of this is the use of the Laminaria tent in obstetrical practice for the purpose of cervical dilatation. Although Laminaria tents were in practical use as early as 1860, technical difficulties resulting from improper handling and insertion procedures rendered their use to almost nil in America by the early twentieth century (1,2). However, because of recent advances in antibiotic therapy as well as in sterilization procedures, the use of the Laminaria tent in clinical medicine today is once again regaining a strong foothold. Thus, despite the marvels of modern medicine, clinicians must once again resort to the superficial "simplicity" of the seaweed Laminaria digitata as an effective, safe, and reliable cervical dilator.

The seaweed Laminaria digitata belongs to the Laminariaceae family, which contains several of the largest oceanic seaweeds (3). Members of this family inhabit predominantly waters of temperate and cold temperatures of the northern and southern hemispheres. The coasts of the Northern At-

Marine Algae in Pharmaceutical Science
edited by H. A. Hoppe, T. Levring, Y. Tanaka

lantic and Pacific Ocean as well as those of the Arctic Ocean and White Sea contain approximately thirty species of Laminaria (4). The Laminariaceae family exhibits a distinctive alternation of heteromorphic phases. The leafy fronds are either simple or divided and contain considerably sized sporophytes (asexial plants). The stipe supporting the fronds ranges considerably in length, which may be so short as to be hardly visible to the naked eye (genus Hedophyllum) or may be several feet long, as is typical of the Laminaria species. Upon germination , the morphologically distinct spore patches on the fronds give rise to the microscopic gametophytes (sexual plants) which are easily distinguishable from the common asexual sporophytes (3).

Laminaria digitata is a large brown seaweed (tangle) with a cylindrically shaped flexible stipe, which variably measures from 0.5 to 1.5 inches in diameter and 1 to 5 feet in length. The stipe is flattened distally and is smooth in texture because this species does not possess the epiphytic stipe growths which are so common to its neighbour, Laminaria hyperborea. The thick stipe expands to form a flat, shallow dome-shaped holdfast which firmly anchors the plant to the underlying rocky substratum. Apically, the stipe gradually narrows and becomes compressed and then spreads out into a flat and leathery lamina. At full maturity the lamina is strap-shaped, divided, and measures two or more feet across the breadth of the frond (3,5).

Laminaria digitata is a perennial species. In its initial developmental stage, the lamina begins to divide into morphologically distinct strap-like segments, which appear to vary in number in relation to their environmental niche. Specimens from sheltered localities have significantly few segments in comparison to those from relatively open and exposed sites. With increasing algal maturity, a vast number of small, slightly thickened confluent patches of unilocular sporangia appear on both sides of the lamina. These sporangia contain the spores which eventually develop into the microscopic gametophytes that are essential for species propagation. New frondal growth occurs at the base of the lamina at the beginning of early spring (3). The old darkish brown frond remains attached to the stipe until it finally atrophies and is shed into the sea.

The shed fronds accumulate in great masses and are washed up on shore by the surf. Cast fronds thrown up on the shoreline are initially clear green but with time climatic factors bleach them to a dirty white color.

Laminaria digitata often occur in abundant stands on the rocky shores of the Atlantic and Pacific coasts and the alga may often be seen partially exposed at low tide (4,6,7). The seaweed occupies the upper limit of the sublittoral zone in colder waters adjacent to its neighbour below, Laminaria hyperborea (3).

The chemical composition of Laminaria digitata and its variation in different geographical and seasonal conditions has been well documented (8, 9,10). There has been an extensive amount of research done on the carbohydrate constitution of Laminaria digitata (11). Alginic acid, laminaran, fucoidan, D-mannitol, D-glucose, and D-galactose have all been extracted from the alga and analyzed (4,12). Alginic acid is the main carbohydrate that is present in this species and accounts for approximately 15 to 40% of the total plant constitution on a dry weight basis. The laminaran content (a reserve carbohydrate in storage form in brown algae) of the entire plant is 2% in April, while in July it accounts for approximately 21% of the alga composition (13). Laminaran is present neither in the stipe nor in the basal parts of the seaweed but is found exclusively in the frond. Considerable work has also been performed on the chemical constitution of the water-soluble laminaran, its production by the plant and the antilipemic activity of laminaran sulfate (14,15,16). In January Laminaria digitata contains a mannitol concentration of 5% which gradually increases to 21% in June (17). In contrast, the algal alginic acid content decreases from a value of 25% in January to 19% in July (11). In addition to the various studies done on the carbohydrate constitution of Laminaria digitata, research has also been done on the protein and amino acid content of the alga and their combined nutritional value. The total algal protein content has been estimated to be 12% in March which thereafter declines to a value of 5% by August. The presence of carotene, carotinoids, niacin, iodine, potassium, fatty acids, and antimicrobial substances have also been described to occur in Laminaria digitata (4).

The shed fronds that are washed on shore are collected and processed into a variety of algal products, including algal meal. The alga is commomly employed to make fodder, which is subsequently used as an additional nutritional additive for cattle fodder (4). Being an important initial raw material used in the production of alginates and alginic acid, Laminaria digitata is also harvested directly in the ocean in addition to being gathered on the shores (18). Experiments have been performed dealing with the optimum frequency and season of harvesting as well as on the methodology of drying the alga; this refers specifically to the non-thermal drying of the seaweed (19). Other applications of Laminaria digitata include its role in the manufacture of glue (20) and the use of the dessicated stipe in the production of Laminaria tents for cervical dilatation.

The cervix uteri has been a major obstacle blocking the entry to the uterus which has faced obstetrician-gynecologists for centuries. In various gynecological conditions and in the different stages of gestation, the practicing clinician must often perform certain diagnostic or therapeutic intrauterine techniques. Often these maneuvers require sufficient cervical dilatation before facilitative uterine entry is achieved. These obstetrical procedures range in form from insertion of intrauterine devices into the uterus to the induction of abortion up to the fourteenth week of pregnancy. The problem that the cervix has presented to obstetricians as a major uterine blocking agent may be more appreciated by the host of materials and substances that were employed by nineteenth century physicians to overpower and dilate the persistingly stubborn closed cervix. Springs, bougies, dilators, roots, stems, catheters, sponges, screws, pliers, bags, and balloons have all been utilized at one time or another as cervical dilating agents (21). In addition the cervix has been heated, greased, shocked, hormonized, narcotized, and paralyzed in an even more futile attempt to open it up. However, common sense prevails that many of these crude and often "laymen" methods of cervical dilatation should manifest various complicative side effects. As expected, each of the above mentioned dilating methods is accompanied by numerous complications, which include laceration, hemorrhage, perforation, infertility, hematoma, infection or damage to other tissues and organs (21). Thus, despite the widespread theoretical and analytical advances of modern medicine, there is as yet no single exclusive method for cervical dilatation that is safe, reliable, and efficient.

The preferred use of Laminaria tents over other cervical dilating agents in Germany, Japan, Switzerland and elsewhere is evident from the fact that they have been employed in these countries for over one hundred years (22). In America, interest in the Laminaria tent flourished in the 1860's and 1870's with the development of special instrumentation for effective cervical insertion and experimentation with various tent sizes and shapes, which allowed insertion into abnormally shaped uteri (2). As Dr. J. Braxton Hicks quoted in 1869 (1); "Of all the materials used for dilating the cervix uteri, there are none so cleanly efficient and convenient as those made from the Laminaria digitata. Although now largely employed by many, they are not so well known as they deserve to be; and therefore, allusion to the subject may not be altogether useless, nor some practical hints unacceptable to some readers." With time however, the use of the Laminaria tent became more unfavourable due to the increasing frequency of uterine perforations which resulted in infection, especially in tents fabricated from land plants containing Clostridia spores (21). Even though oceanic bacteria are relatively nonpathogenic and nonvirulent for man, inefficient packaging and improper handling of the harvested seaweed at that time as well as increasing pollution of shorelines from industrialized coastal cities rendered Laminaria tents as potentially infectious agents (21,23). The growing link between Laminaria tents and abortions, which were lawfully declared a criminal offense in the Unites States in the late 1800's, resulted in the almost complete obliteration of the tent in America by the early twentieth century. The disrepute of the tent grew even stronger as a result of the development of other cervical dilator gadgetry which claimed to be safer and more efficient than the tent.

Laminaria tents are produced from the root of the seaweed Laminaria digitata (seatangle). The stipe is firstly dried and thereafter molded into an elongated stick shape form, measuring 5.5 to 6.0 centimeters in length with a diameter of 0.3 to 0.5 centimeter (2). A strong silk thread attached to one end permits the properly inserted tent to be easily grasped and removed from the cervix. The dessicated stipe is hygroscopic, that is when placed in the warm and moist environment of the cervix, it is capable of expanding from three to five times its original preinsertion diameter. Since the process of Laminaria expansion is gradual (spanning over 12 to 24 hours), there is decreased pain associated with cervical dilation and at the same time the cervix becomes soft and flexible. Further cervical dilation (if necessary) is thus more easily and less painfully achieved. Laminaria tents are generally employed for the purpose of cervical dilatation after which one step artificial abortion is performed. In young primigravidas whose cervix is narrow and stiff, gradual cervical dilatation with the Laminaria tent permits sufficient dilatation to allow the products of conception to be removed by suction curettage or by ovum forceps. After thorough vaginal disinfection, the patient is anesthesized and the cervix is grasped with a vulsella (24). Several small size Laminaria tents are inserted side by side into the cervix. The insertion of several smaller sized tents instead of one thick one usually prevents the difficulty in their removal. In addition, a usual procedure is to pass a sterilized gauze pad through the loops of the silk thread, tie the corners, and pack the gauze gently in the vagina (2). This procedure helps to anchor the tents firmly in the cervix and also prevents their migration towards the uterine fundus. The insertion of the tents into the closed cervix requires some technical expertise; the tent should be gently pushed to the point just beyond the internal os so that the ends protrude from both the internal and external os. Insertion of the tent(s) should be performed with utmost gentleness and caution since perforation of the uterus is a constant threat, particularly if the uterus is sharply flexed at the internal os (24). With the tents placed in the cervix, it is not obligatory for the patient to remain in the hospital. The patient may be sent home with specific instructions to avoid bathing, douching, intercourse and to return to the hospital within a period of 24 hours. Upon removal of the tent(s), the cervix should be sufficiently dilated and softened to permit the entry of a finger, ring forceps, small-sized ovum forceps, or other obstetrical instruments easily into the uterus. One step artificial abortion is then performed by removal of the fertilized ovum with forceps or by suction currettage.

However good the efficacy of Laminaria tent as a cervical dilating agent is, this does not exclude the fact that various complications may arise with their use, especially in the hands of a physician inexperienced in their use. As one obstetrician recounts (24), "I have twice seen disaster follow the use of tents. The last occasion was when one of my junior assistants pushed the tent through the cervical wall. The patient became infected and died of sepsis. This mistake is very easily made." The three main complications associated with the use of Laminaria tents include the following (2): (i) infection leading to possible sepsis due to perforating the uterus or failing to remove the tent at the appropriate time, (ii) cervical wall damage by the initial misplacement of the Hegar dilator with subsequent insertion of the tent(s) into the injured tissue, (iii) difficulty in tent removal from the cervix due to migration towards the uterine fundus. In situ, the intenal os is the point which exerts the greatest pressure on the tent as it dilates and an indentation on the tent is usually observed at this location (25). According to physical laws, the areas of least resistance will thus be distally towards the external os and within the uterine cavity itself. An inserted tent with a slight flaring of the portion placed near the external os will tend to swell at a faster rate than the uniform portion at the internal os. This will result in the tent being gradually extruded into the vagina. In the case of a tent with a uniform diameter, a rapid intrauterine expansion leads to diminished resistance posed by the cervix. This effect may result in the gradual extrusion of the tent into the uterine cavity. In a study done by Hanson et al. (25), three patients in which a uniform diameter tent was inserted into the cervix (medium-size Type A) had complications associated with the tent. The tents gradually migrated so far into the uterine cavity that upon examination they were not visible and could only be palpated at various depths in the cervical canal. Attempts to remove the migrated tents with forceps resulted only in the crumbling of the tents and these patients had to undergo operative procedures for their removal. A dissection of a small bladder flap with anterior cervical trachelotomy about midway to the internal os was performed and the exposed tents were subsequently removed from within the uterus (25). Hanson claims that uniform diameter tents should not be employed for cervical dilatation since there is always danger of tent migration into the intrauterine environment. The use of a slightly flared Laminaria tend (Type B) with the flared portion being placed distally (in the region closer to the external os) would prove more satisfactory, since cervical extrusion would result in the direction of the vagina and not the uterus.

The other major complication associated with the use of Laminaria tents, namely infection, is today almost nonexistent due to advances in modern sterilization procedures in both the handling and cervical insertion of the tents. Bacteriological studies (21) show that new unused tents treated with gamma irradiation and subsequently sterilized by ethylene oxide did not show any microbial growth upon incubation. Sterilization of unused tents with 99% alcohol allowed only the growth of viable E. coli. Through vaginal-cervical antiseptical cleansing prior to tent insertion also diminishes the possibility of infection. Clinical evidence shows that strict adherance to the above mentioned sterilization procedures should render the infection rate equal to or even less than that observed following the use of other presently used methods of cervical dilatation.

Perhaps the most comprehensive study on the use of the Laminaria tent in modern clinical practice has been performed by Newton (21). The Laminaria tent was used as the primary method for cervical dilatation in 500 patients undergoing abortion by suction curettage. All patients were less than 12 weeks pregnant and a followup examination was performed for each patient after the abortion. The most significant finding of this study was the reliable effectiveness of the Laminaria tent as a cervical dilator. In some cases dilatation was observed 3 to 4 hours after insertion, but maximal effects were usually observed 24 hours after insertion. The total complication rate was 8.0%, of which 2.2% of the patients developed a fever that was attributable either to the use of the tent or to the technical procedures involved in performing the abortion. Uterine perforations resulting in infection were not observed in any of the patients and significant cervical trauma was observed in only one patient. The data from this study relating to the complications encountered with the use of the tent compare favourably with data from other similar series in which cervical dilatation prior to abortion was achieved by methods other than by Laminaria tent. A concluding comment by Newton favouring the use of the tent is appropriate at this point: "The inconvenience of an extra clinic visit for placement of Laminaria tent, which takes only two to three minutes, is far outweighed by the time saved in cervical dilatation and management of complications resulting from use of other methods of cervical dilatation. Its ease of insertion, effectiveness, and apparent safety justify the consideration of Laminaria tent whenever cervical dilatation is required (21)."

The use of the Laminaria tent is once again regaining its long lost popularity as an effective cervical dilator in many countries, including America. Part of its success resides in the fact that many of the complicative factors which initially resulted in the disrepute and disuse of the tents can today be greatly minimized or completely obliterated. Technical expertise in cervical insertion can help reduce the danger of uterine perforation and infection. Various shapes and sizes of tents can be used to best suit the individual patient's needs, as for example curved tents in patients with flexed uteri. Treatment of new unused Laminaria tents with gamma irradiation followed by ethylene oxide sterilization has significantly lowered the rate of infection concomitant with their use. Besides cervical dilatation prior to abortion, Laminaria tents have also been employed to correct uterine malpositions, to dilate stenotic vaginas, to ripen the cervix in pregnancy, and for dilatation in patients with a hypoplastic cervix or with cervical atrophy (1,2,21). The use of the Laminaria tent has indeed passed the test of time. The gap between designation of the tent as a crude, antiquitous cervical dilator and the recognition of it as a modern medical device has finally been closed.

REFERENCES

(1) Hicks, J.B.: On seatangle tent. Practitioner 1, 83 (1869).

(2) Manabe, Y.: Laminaria tent for gradual and safe cervical dilatation. Amer. J. Obstet. Gynec. 110, 743 (1971).

(3) Carola, I.D.: British seaweeds-The Kew series, Ed. Eyre and Spottiswoode, 1963.

(4) Levring, T., Hoppe, H.A., and Schmid, O.J.: Marine algae-A survey of research and utilization, Cram, de Gruyter, Hamburg, 1969.

(5) Brennan, A.T.: Inst. f. Ind. Res. and Stand., Dublin, 1950.

(6) Wort, D.J.: The seasonal variation in chemical composition of Macrocystis integrifolia and Nereocystis luetkeana in British Columbia coastal waters. Canad. J. Bot. 33, 323 (1955).

(7) Sanford, F.B.: Fishery Leaflet 469, U.S. Dept. of the Interior, Washington, D.C. (1958).

(8) Primo, C.: Second Int. Seaweed Symp. 1955, quoted from Levring, T. et al. (4).

(9) Butler, M.R.: Fish. Res. Board Can. Ms. Repts. Exptl. Stat. 47 (1930).

(10) Haug, A. and Jensen, A.: Second Int. Seaweed Symp. 1955, quoted from Levring, T. et al. (4).

(11) Jensen, A.: Norsk. Inst. f. Tang-og Tareforskning Rep. 10 (1956), quoted from Levring, T. et al. (4).

(12) Smith, F. and Montgomery, R.: The chemistry of plant gums and mucilages, Reinhold, New York, 1959.

(13) Haug, A.: Teknisk Ukeblad, Oslo; WFA (4) 39 (1954), quoted from Levring, T. et al. (4).

(14) Friedlaender, M.H.G. et al.: Molecular weight and hydrodynamic properties of laminarin. Biochem. Biophys. Acta 14, 136 (1954).

(15) Mookerjea, S. and Hawkins, W.W.: The antilipemic activity of laminarin sulfate. Can. J. Biochem. Physiol. 36, 261 (1958).

(16) Hegnauer, R.: Chemotaxonomie der Pflanzen, Birkhaeuser, Stuttgart (1962), quoted from Levring, T. et al. (4).

(17) Jensen, A. and Haug, A,: Norsk Inst. f. Tang-og Tareforskning Rep. 14 (1956), quoted from Levring, T. et al. (4).

(18) Humm, H.J.: The brown algae, in Marine Products of Commerce, Ed. Tressler and Lemon, Reinhold, New York, 1951.

(19) Reid, H.C. and Jackson, Ph.: J. Sci. Food Agric. 7, 4 (1956).

(20) Maass, H.: Alginsaeure und Alginate, in Chemie und Technik, Ed. Strabanbov, Verlagsgesellschaft mbH., Heidelberg, 1959, quoted from Levring, T. et al. (4).

(21) Newton, B.W.: Laminaria tent: Relic of the past or modern medical device. Am. J. Obstet. Gynec. 113, 442 (1972).

(22) Watteville, H. de : Proc. Roy. Soc. Med. 62, 828 (1969).

(23) Longaker, H.L.: Personal communication, 1971.

(24) Munro, K.: Operative obstetrics, 7th ed. 1964.

(25) Hanson, F.W., Niswander, K.R. and Trelford, J.D.: Cervical migration of Laminaria tents. Am. J. Obstet. Gynec. 114, 835 (1972).

MEDICINAL AND PHARMACEUTICAL UTILISATION OF PURIFIED "AGAR AGAR" EXTRACTED FROM GELIDIELLA ACEROSA OF INDIAN SHORES

B. V. Gopal

Cellulose Products of India Ltd.,
RA. 9, Tirunagar,
Madurai, 625 006, India

Introduction

The industrial production of Agar Agar in India was started around 1965. The primary red-purple seaweed, Gelidiella acerosa belonging to the Rhodophyceae family, was first located in the Gulf of Mannar south at the eastern sea coast of the Indian Peninsula during the year 1964 by a survey team of M/s. Cellulose Products of India Ltd., Ahmedabad. The secondary weed Gracilaria Edulis also grows abundantly in the same region. Unlike Gelidium seaweeds growing around the other countries of the world, Gelidiella acerosa is a small plant growing on coral substratum to a maximum length of 10 cms in its natural habitat of turbulent waters at depths ranging from 1 to 3 meters. The plant has stringlike stiffness. Harvesting is carried out by primitive methods of hand picking using goggles for underwater visibility during the mature season from December to March. The weeds thus harvested are hauled to the shore in wet condition in country sailboats, sun dried and hand sorted to eliminate calcarious incrustrations and other contaminating sea weeds for packing in gunny bags and transportation to the processing plants.

Marine Algae in Pharmaceutical Science
edited by H. A. Hoppe, T. Levring, Y. Tanaka

Manufacture

The large scale manufacture of Agar Agar is of recent origin in the country (by M/s. Cellulose Products of India Ltd., Ahmedabad-Madurai in the year 1970). The edible (food grade) quality of Agar is manufactured mostly using a mixture of Gelidiella and Gracilarie in extraction while the bacterilogical quality is produced from Gelidiella, the primary weed.

The sea weeds varying in quality and purity from batch to batch, are carefully mixed and processed under controlled conditions to remove the impurities at each stage (with emphasis on bacteriological controls) to derive a final Agar product to meet the stringent specifications of food and Pharmacopia specifications for bacteriological work.

The various stages of the manufacturing processes are:

1. The seaweeds are dried, dusted and subjected to cold strong acid pretreatment followed by thorough water washings to remove the sea salts, corals and other contaminating weeds.
2. For purified grades of Agar, the washed weeds are further pre-treated with hot alkali solutions to obtain Agarose rich Agar having better gelling properties.
3. In view of the inferior gel strength of the Agar in the weeds the washed weeds are generally extracted at atmospheric pressure by boiling till the Agar goes completely into solution in the extract.
4. The extract with insoluble impurities is filtered adopting stage filteration by using filter media.
5. The filtered extract is then filled in shallow trays and allowed to gel.
6. This crude gel in trays is frozen for long periods at subzero (^{o} C) temperatures; the Agar film separates out as the water freezes.

7. The Agar film is separated by thawing the ice; soluble impurities are then removed by treated water washings.
8. The Agar film thus obtained is bleached using hypochlorite solutions, chemically treated for removal of residual impurities, sterilized for bacterial control and dreied in hot air ovens under controlled temperatures.
9. At all stages of the process bacterial controls are adopted by sterilising the equipment employed.
10. The dried Agar strips are packed to size if edible; the pharmaceutical grade Agar is cut, ground and blended for uniformity of quality, to be packed finally in containers for marketing as powder.
11. For obtaining an even more pure grade of Agarose-rich Agar with best gelling characteristics, further chemical treatments are carried out using sequestering agents ro remove the Agaropectines in cold water soluble forms.

Uses

Agar with characteristic gelling (around 40° C at room temperature to form a clear transparent and stiff elastic gel), melting (above 85° C with reversible property), emulsifying, suspending, absorbing, lubricating and colloidal properties has multifarious uses. They are:

1. Agar as a gelling Agent

   The food grade quality of Agar in the form of strips, also familiarly known as China grass / Faluda Grass is used by the common man in the country for preparation of edible gellies, deserts, puddings etc. with milk, sugar, chocolate and flavours. Agar in powder form is used by confectionaries.

2. Agar as an emulsifying Agent

2.1. Agar finds its place in preparation of cosmetic and pharmaceutical creams.

2.2. Edible emulsions in medicine in view of its negligible sugar and protein value.

2.3. In photography to replace gelatine for special application for hyperthin, high-resolution emulsions.

3. Agar as a suspending and coagulating Agent.

3.1. For colloidal graphite.

3.2. In radiology.

3.3. In analytical coagulation of calcium sulphate, Arsenic sulphide, Ferric hydroxide and Barium sulphate.

3.4. In general in science as a colloid for studying osmosis, dialysis, hydrations, bound water, peptization, flocculation etc. and for forming artificial membranes simulating natural cell walls, immobilising agent in photomicrography.

3.5. Purified Agarose-rich Agar, having an elastic texture and higher gel stability of gel even in low concentration finds its use as a suspending agent in immunoelectrophoresis.

4. Agar in medicine and bacteriological culture

4.1. Agar is used as a perfect laxative in medicine. Agar with its innert characteristics is a chief constituent in medicinal emulsions for laxative preparations. It is the opinion of certain physicians that Agar is the only perfect non habit forming laxative without any harmful after effects.

4.2. Antibiotics and vaccines are manufactured with the aid of Agar.

4.3. As an innert medium for culturing microorganisms in bacteriological studies, public health laboratories, research laboratories and pharmaceutical industries. Orchid culture is considered easy with Agar medium. There is no substitute for Agar in preparation of solidified culture media in view of the following unique properties. Agar forms a clear solution when boiled in water at 95° C at 1.5 and 2 % concentrations. The solution sets to form clear gel plates below 40° C. This ideal setting temperature helps sometimes in mixing nutrient culture fluids and introducing spores of bacteria in the liquid stage below 45° C. This ideal setting temperature sometimes helps in mixing nutrient culture fluids and introducing spores of bacteria in the liquid stage below 45° C which are not harmful to them. The solution sets to form a firm and resilient gel below 40° C which does not melt till the temperature exceeds 85° C. The spores of bacteria can also be streaked into the firm surface of the gel (which does not break) and allowed to propagate at required incubating temperatures (45° C to 80° C) depending on each species of bacteria or fungi. 1.5 % concentration of Agar is recommended for alkaline medias while higher concentration of above 2 % is required for acidic media. Agar at required concentration is easily dissolved and sterilised simultaneously by autoclaving at 15 lbs pressure. The solution obtained being clear is normally used without filtration but may be filtered using bits of "Chardin" papers. The clean gel plates thus obtained are considered ideal for single spore cultures.

4.4. Agar is used as a constituent in medicinal tablets and capsules.

5. Agar is used in dentistry and criminology for making impressions.

6. Agar can be used to lubricate dies in drawing tungsten and tantalum filaments.

7. Agar is also known for its use in agriculture as an insecticide activator and carrier and in coating nitrogen bacteria culture.

8. Agar is said to have been used beneficially as a flash inhibitor in sulphur mining explosives and in improving density, brightness and adhesion of copper plate on aluminium.

9. Agar also finds its place in died finishing of paper and textiles of superfine quality.

10. Agar is advantageously used in the canning of preserved and cooked fish.

11. Ion exchange properties of Agar provide future vistas to explore.

(By courtesy of M/s. Cellulose Products of India Ltd., Ahmedabad-Madurai, India).

Reference

George Smith, use of Agar Agar for preparing culture media
George Smith & Harold Rastrick Arnold; London Edward Arnold & Co, Ltd., An Introduction to Industrial Mycology.

STUDIES ON PTEROCLADIA CAPILLACEA (GMEL.) BORN. ET THUR.

Part I

Phytochemical Investigations

Kasim C. Güven and Erden Güler

Department of Pharmacy and Technology, Faculty of Pharmacy, University of Istanbul, Turkey

In recent years many papers have been published on phytochemical screening of marine algae. The chemical constituents of Gelidium sp. have also been studied by various workers. In1859 Payen (20) first isolated agar from Gelidium amansii; in 1962 Katayama (9) summarized essential oil contents of some algae including Gelidium sp. Different sterol compounds (cholesterol, 24 methylene cholesterol, chalinasterol, ß-sitosterol, 22-dehydrocholesterol, fucosterol) were isolated from G. corneum (6), G. amansii, G. japonidum, G. subcostatum (12) and G. Sesquipedale (15). The nitrogenous compounds as total nitrogen, soluble nitrogen and protein of G. amansii (28, 29), G. crinale (2), G. latifolium, G. elegans, Gelidium sp. (30), Pterocladia capillacea (21) and seasonal variation of protein concentrations in G. spinulosum, G. sesquipedale, G. pusillum, G. attenuatum and G. pusillum var. pulvinatum (24) and the free amino acids of G. amansii (18), C. cartilagineum (13), G. latifolium (7) and G. pusillum (14) have also been reported.

The other compounds isolated were: lipid and fat from G. sesquipedale (11) and G. crinale (2), polysaccharides and cellulose from Gelidium sp. (12, 22) holocellulose from the cell membrane of G. amansii (1), galactosyl glycerol, floridoside, mesoinositol, laminitol and mytilitol from G. pristoides (19) and G. cartilagineum (13), catalase from G. amansii (16, 17),

Marine Algae in Pharmaceutical Science
edited by H. A. Hoppe, T. Levring, Y. Tanaka

xantohphyll, cytochrome, and chlorophyll a (I) from G. corneum (6), G. amansii (10), G. pusillum, G. spinulosum and G. sesquipedale (3), caroten, thiamine, riboflavin, nicotinamide, choline and folic acid from G. corneum (6), G. crinale (2), G. sesquipedale (25), G. cartilagineum (13) and G. amansii (8), calcium, phosphorous, cobalt, zinc, silver and iodine from G. crinale (2) and some other Gelidium sp. (12, 27, 31).

A previous study publsihed from this laboratory described Vitamin B_{12} content of G. capillaceum (5).

In this paper agar, volatile oil, sterol, free amino acid and protein contents of Gelidium capillaceum were reported.

Experimental Pterocladia capillacea was collected from Turkish coasts on Marmara Sea (Eskihisar-Tütünçiftlik, Mudanya, Tirle, Kapanca) and Aegean Sea (Kuşadasi, Gökova, Marmaris) from July through September in the years 1971 - 1976.

The alga was dried and milled to a coarse powder before the following studies were carried out.

1. Agar

1.1 Extraction of agar

Powdered alga was macerated with different alkaline solutions; a - 12 % ammonia b - 5 % sodium hydroxide c - 5 % calcium oxide d - a mixture of a and b in the ratio of 1:1, heated in an autoclave or oven at different temperatures (90°, 100°, 110° C) for 1 - 3 hr. e - Algae was autoclaved with four portions of distilled water at 110° C for 4, 4, 6 and 6 hr respectively according to Tseng (1952) (26). The hot jelly extract obtained was filtered through linen and after coagulation at room temperature was put into a freezer for two days. The frozen material was left to thaw at room temperature and filtered. After bleaching with 7 % sodium hypochloride solution agar flakes were separated by filtering on linen and washed with distilled water several times and then agar gel was lyophilized.

Yield of agar according to the applied methods were: a - 10 %, b - 4 %, c - 6.4 %, d - 6.8 %, e - 15 %.

1.2 Analysis of agar

Sulphate content and ash residue of agar were determined gravimetrically and were found to be 0.7 and 2.4 % respectively.

Agar was hydrolized with N hydrochloric acid for 3 hr at 100° C in a sealed tube. The hydrolysate was evaporated to remove hydrochloric acid, the residue was dissolved in distilled water and applied to a column (2 x 25 cm) of MB-3 Amberlit H^{+}, OH^{-} (Fluka AG) resin and eluted with deionized water. The eluate was applied on Whatman No 41, 43 or S & S 2043a papers and developed in ethyl acetate-acetic acid-water (3:1:1) solvent system. The reagents used were:

a - 1 ml of saturated silver nitrate solution and 200 ml acetone,

b - 10 g sodium hydroxide and methanol to make 500 ml.

Sugar components of this agar were detected as galactose, xylose and rhamnose by comparison with the reference samples.

2. Essential oil

The volatile oil of G. capillaceum was obtained by steam distillation of the fresh alga. The distillate was saturated with sodium chloride and extracted with ether. The ether extract was distilled at 35° C. Yield 0.06 %. The contents of this volatile oil were determined by gas chromatographic method[+] and eugenol, geraniol, geranyl formate, geranyl acetate, terpi-

---

[+] Packard Becker Gas Chromatograph Model 419

neol, citronellol, nonanol, isobutyric acid, eucalyptol, α and ß pinen were identified.

3. Alkaloid

General alkaloid extraction technique was used. The alga was extracted with methanol for 48 hr in a Soxhlet, then the extract was distilled and the residue was acidified with 0.1 N hydrochloric acid. The aqueous phase was filtered and washed with ether to remove nonalkaloidal parts. The acidic solution was separated and an ammonia solution was added to make slightly alkaline and extracted with ether. The ether extract was distilled and the residue was applied on precoated silicagel plates and developed in butanol-acetic acid-water (3:1:1) solvent system. After spraying alkaloid reagents (iodoplatinate or Dragendorff) a fleck was observed on the chromatogram. The yield was very poor. The studies for the identification of this compound are continuing.

4. Sterol

For the isolation of sterol compound the method of Smith et al. (1973) (23) was used. The alga was extracted with acetone under reflux for 4 hr and the acetone extract was distilled. The residue was saponified with 10% potassium hydroxide in ethanol for 2 hr. The nonsaponified parts were extracted several times with ether, then the ether extract was washed with distilled water and the etheral part was distilled. The residue was crystallized from methanol. This crystallized compound was purified by preparative thin-layer chromatography using 20 x 20 cm plates coated silicagel and developing solvent system benzen-ether (1:1). The sterol band was visualized under UV light at 254 nm or by spraying 0.5 % antimony trichloride in chloroform or 50 % sulfuric acid reagents. After visualisation under UV light the sterol band was scraped off and eluted several times with ether. The extract was distilled and white crystals were obtained in a yield of 0.0021 %, m. p. 140° C.

C. H. analysis[+]; C 82.86 %, H 11.42 %. The i. r. spectrum[++] showed characteristic peaks for Δ 5 sterols at 805 and 845 cm^{-1}.

The mass spectrum[+++] of this compound showed an apparent ion peak at m/e 386. Other peaks were observed at m/e 371 ($M-CH_3$); 368 ($M-H_2O$); 353 [$M-(CH_3+H_2O)$] ; 301; 275; 273 (M-side chain); 255 [M-(side chain+ H_2O)] ; 247; 231 [M-(side chain and part of ring D+H)] and 213.

The i. r. and mass spectra, elementary analysis as well as tlc mobility, m. p. and m. m. p. were identical with those of an authentic sample of cholesterol.

5. Free amino acids and protein

5.1 Free amino acids

For extraction of free amino acids the powdered alga was macerated with phenol-acetic acid-distilled water (PAW) (1:1:1) for 24 hr. After filtration of PAW solution, protein was precipitated by adding 10 % trichloracetic acid solution and centrifugated. The supernatant was separated and condensed in rotary evaporator and the residue was dissolved in 5 ml of distilled water and applied to a column (2 x 25 cm) of Dowex 50 x 8 H^+ (Fluka AG.). The column was then washed with distilled water and the amino acids were eluted with 2 N and 4 N ammonia solutions respectively. The eluates were distilled in a rotary evaporator and the residue was dissolved in citrate buffer (pH 2.2). Quantitative determination of the amino

[+] Ilse Beetz Mikroanalytisches Laboratorium, Kronach, West-Germany

[++] Perkin Elmer 577 Grating IR Spectrophotometer

[+++] Mass Spectrophotometer, Jeol, J. M. S.-D 100, Japan

acids were made by the aid of an automatic amino acid analyzer[+]. The results are shown in Table 1.

Table 1 - Free amino acids of Pterocladia capillacea (mcg/g)

Lysine	1.54
Histidine	0.96
Arginine	0.82
Aspartic acid	3.29
Glutamatic acid	12.21
Glycine	1.17
Alanine	1.60
Valine	2.57

5.2 Protein

5.2.1 Extraction and fractionation

For the isolation and fractionation of the protein part of alga, the previously proposed method in this laboratory was used (4).

The powdered alga was macerated with sodium carbonate solution (2 %) for 24 hours. The extract was filtered and acidified with N hydrochloric acid and 3 % barium chloride solution was added. The precipitate was separated by centrifugation and treated with sodium carbonate solution (1 %), then centrifugated and filtered. The filtrate was neutralized with 0.1 N hydrochloric acid. Yield of protein was 1.51 %. The isoelectric point of this protein extract was 1.5. For fractionation, the protein solution was applied to a column (3 x 50 cm) of Sephadex G-25 Fine (Pharmacia) and eluted with distilled water with a flow rate of 1.8 ml/min and each 5 ml portions were collected. The absorbance of each eluate was read at 360 nm using a Beckman DB Spectrophotometer. The fractionation curve is shown in Fig. 1.

[+] Beckman Automatic Amino Acid Analyzer Model 121

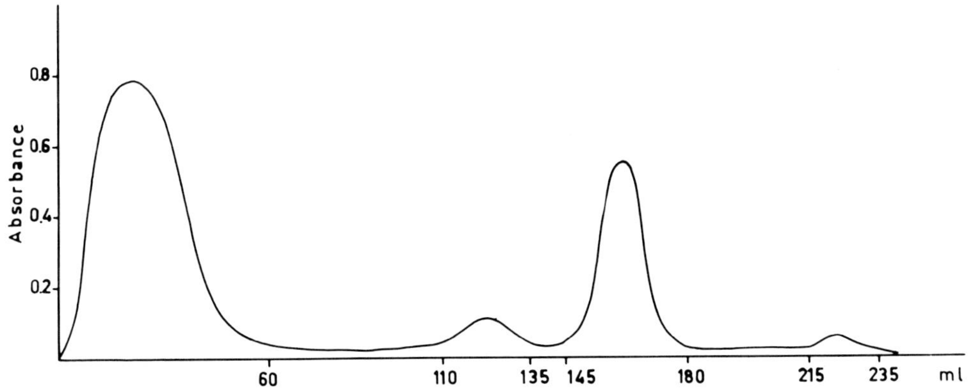

Fig. 1 - Fractionation curve of Pterocladia capillacea protein.
Fraction I: 0-60 ml, Fraction II: 110-135 ml,
Fraction III: 145-180 ml, Fraction IV: 215-235 ml

5.2.2.1. Amino acid contents of protein fractions

The protein fractions were hydrolyzed with 6 N hydrochloric acid for 24 hr at 110° C in sealed tubes. The excessive acid was removed in a rotary evaporator. The residues were dissolved in citrate buffer (pH 2.2) and were analysed qualitatively by two-dimentional thin-layer and paper chromatographic techniques. For this purpose cellulose MN 300 adsorbant film and S & S No 2043a papers were used. The developing solvents were 1) n. butanol-acetic acid-water (3:1:1) and 2) phenol-water (3:1) and the reagent was 0.5 % ninhydrin in acetone. For visualization the chromatograms were heated at 110° C for 2 - 3 min. The quantitative amino acid contents of these fractions were determined using a Beckman Model 121 Automatic Amino Acid Analyzer. The results are shown in Table 2.

5.2.2.2. Carbohydrate contents of protein fractions

For the identification of the monosaccharide contents of protein fractions the method mentioned above for agar analysis was used. The monosaccharides identified by comparison with authentic samples were galactose, arabinose and fucose in Fraction I, galactose and arabinose in Fraction II and III.

Table 2 - Amino acid contents of Pterocladia capillacea protein fractions (mcg/mg)

Amino acids	Fraction I	Fraction II	Fraction III	Fraction IV
Lysine	17.98	1.60	0.14	1.54
Histidine	2.14	0.15	-	Trace
Arginine	13.09	0.52	Trace	1.27
Aspartic acid	66.94	3.85	0.17	2.70
Threonine	27.15	0.71	0.47	1.01
Serine	17.02	0.94	0.42	1.66
Glutamic acid	62.53	3.23	1.02	4.51
Proline	18.99	0.92	0.46	1.64
Glycine	29.72	5.17	0.33	2.70
Alanine	29.04	2.22	0.62	1.87
Cystine	238.85	11.53	0.96	10.57
Valine	1.69	-	-	-
Methionine	0.85	0.14	-	-
Isoleucine	22.29	1.70	-	1.18
Leucine	35.94	2.22	-	1.84
Tyrosine	8.69	Trace	-	0.32
Phenylalanine	81.93	0.66	-	0.77

The pharmacological activities as lipolytic, hypoglycemic and anticoagulant effects of these fractions are to be reported in Part II.

This study was supported partly by the Technical and Scientific Research Council of Turkey.

Discussion

The extraction of agar from Gelidium sp. was reported by various authors (12, 20, 22). In this work the agar content of the closely related Pterocladia capillacea was studied and different extraction methods were applied. The yield of agar and its monosaccharide content was also determined. The monosaccharides detected are similar to the earlier findings (22).

Volatile oil amount of Gelidium sp. was reported by Katayama (9) as 0.13 % of dry alga while we found that this yield was 0.06 % of fresh alga. The constituents of this essential oil were also determined.

An alkaloid was isolated and identified from a marine alga in this laboratory for the first time. In this work the existence of alkaloid in G. capillaceum was demonstrated.

The main steroid product, cholesterol,has been isolated from different marine algae. We extracted cholesterol from this algae.

Free amino acid (7, 13, 14, 18) and protein (2, 21, 24, 28-30) contents of Gelidium sp. were reported by various authors. In this study Pterocladia capillacea was examined and protein part of this alga was fractionated. Then amino acid contents of each fraction were determined. In addition to these findings galactose, arabinose and fucose were identified as the monosaccharide contents of Fractions I - III.

Summary

In this study the extraction and analysis of agar and essential oil, the existence of alkaloid, free amino acid and protein contents of Pterocladia capillacea were reported. In addition to these findings the protein part of this alga was fractionated and the amino acid and carbohydrate contents were determined.

References

1. Akahane, T. and T. Fuse. 1968. Polysaccharides (holocellulose fraction) of the cell membrane of [the alga] Gelidium amansii. Nippon Nogei Kagaku Kaishi 42 (12): 740-742; C. A. 70, 64432 (1969).

2. Alfimov, N. N. 1963. Biochemistry of some macrophytes of the Black Sea. Botan. Zh. 48: 132-135; C. A. 60, 5885 (1964).

3. Camba, J. S. 1964. Effect of light intensity and temperature on the concentration of chlorophyll in certain benthic marine algae. Compt. Rend. 259 (7): 1432-1435; C. A. 61, 15033 (1964).

4. Güven, K. C. and B. Tekinalp. 1971. About Sargassum vulgare Agardh. Botanica Marina XIV: 63-64.

5. Güven, K. C. and E. Güler. 1976. Vitamin B_{12} content of Gelidium capillaceum Kütz. Ibid. XIX: 395-396.

6. Heilbron, I. M. 1942. Some aspects of algal chemistry. J. Chem. Soc. I: 79-89.

7. Impellizzeri, G., S. Mangiafico, G. Oriente, M. Piattelli, S. Sciuto, E. Fattorusso, S. Magno, C. Santacroce and D. Sica. 1975. Amino acids and low-molecular-weight carbohydrates of some marine red algae. Phytomchemistry 14: 1549-1557.

8. Kanazawa, A. 1961. The vitamin B complex in marine algae. I. On vitamin contents. Kagoshima Daigaku Suisan Gakubu Kiyo 10: 38-69; C. A. 58, 5986 (1963).

9. Katayama, T. 1962. Volatile constituents. In: Lewin, R. A. (Ed.), Physiology and Biochemistry of Algae, Academic Press, New York, N. Y., pp. 467-473.

10. Kato, S. 1959. Algal cytochrome of c-type. J. Biochem. 46: 629-632.

11. Laur, M. H. 1963. Identification and separation by thin-layer chromatography on silicic acid of the lipid constituents of Rhodophyceae. Compt. Rend. 257 (7): 1501-1502; C. A. 60, 852 (1964).

12. Levring, T., H. A. Hoppe and O. J. Schmid. (Eds.) 1969. Marine Algae, Cram, De Gruyter & Co., Hamburg.

13. Lindberg, B. 1955. Methylated taurines and choline sulfate in red algae. Acta Chem. Scand. 9: 1323-1326.

14. Madgwick, J. C. and B. J. Ralph. 1972. Free amino acids in Australian Marine Algae. Botanica Marina XV: 205-209.

15. Meunier, H., S. Zelenski and L. Worthen. 1969. Comparison of the sterol content of certain Rhodophyta. In: Youngken, H. W. (Ed.), Food-Drugs From The Sea Proceedings, Marine Technology Society, Washington, D. C., pp. 319-325.

16. Mitsuzo, T. 1953. Studies on the catalase in marine algae. II. On the optimum pH of the catalase in marine algae. Bull. Japanese Soc. Sci. Fish. 19 (7): 798-802; Biol. Abst. 30, 2804 (1956).

17. Mitsuzo, T. 1953. Studies on the catalase in marine algae. III. On the optimum temperature in marine algae. Ibid. 19 (7): 803-808; Biol. Abst. 30, 2805 (1956).

18. Miyazawa, K., K. Ito and F. Matsumoto. 1969. Aminosulfonic acids in six species of marine algae. Nippon Suisan Gakkaishi 35 (12): 1215-1219; C. A. 72, 87161 (1970).

19. Nunn, J. R. and M. M. Holdt. 1955. Galactosyl glycerol from Gelidium pristoides and Gracilaria confervoides. J. Amer. Chem. Soc. 77 (9): 2551-2552.

20. Payen, M. 1859. Sur la gélose et les nids de salangane. Compt. Rend. Acad. Sci. 49: 521-530.

21. Pellegrini, M. 1970. Contribution à l'étude chimique des algues Méditerranéennes. Fractions azotées de quelques Rhodophycées. Bull. Mus. Hist. Natur. Marseille 30: 189-204.

22. Percival, E. and R. H. McDowell. (Eds.). 1967. Chemistry and Enzymology of Marine Algal Polysaccharides, Academic Press, London.

23. Smith, L. L., A. K. Dhar, J. L. Gilchrist and Y.Y. Lin. 1973. Sterols of the brown alga, Sargassum fliutans. Phytochemistry 12: 2727-2732.

24. Torregrosa, R. E. 1964. Seasonal variations in chemical composition, extraction and characteristics of agar from some algae (Gelidium) of the South-Atlantic coast of Spain. Invest. Pesquera 26: 165-194; C. A. 62, 3111 (1965).

25. Torres-Pombo, J., J. Seoane-Camba and I. Ribas. 1969. Extraction of choline sulfuric ester from Gelidium sesquipedale. In: Margalef, R. (Ed.), Proc. Int. Seaweed Symp., 6th, Subsecretaria De La Marina Mercante, Madrid, pp. 611-614.

26. Tseng, C. K. 1952. Agar. In: Kirk, R. E. and D. F. Othmer (Eds.), Encyclopedia of Chemical Technology, Vol. 1, The Interscience Encyclopedia, Inc., New York, N. Y., pp. 232-238.

27. Tsuruga, H. 1968. Some long-lived radionuclides in marine organisms on the Pacific coast of Japan. J. Radiat. Res. 9 (2): 63-72; C. A. 70, 26662 (1969).

28. Yamada, N. 1961. Studies on the manure for seaweed. I. On the change of nitrogenous component of Gelidium amansii Lmx. cultured with different nitrogen sources. Bull. Japanese Soc. Sci. Fish. 27 (11): 953-957; Biol. Abst. 41, 9055 (1963).

29. Yamada, N. 1964. Fertilizer for seaweed. II. Effect of phosphorus on the nitrogen absorption of an agar seaweed (Gelidium amansii) and influence of light intensity on this effect. Nippon Suisan Gakkaishi 30 (11): 908-911; C. A. 64, 2709 (1966).

30. Yamada, N. and Y. Iwahashi. 1964. Fertilizer for seaweed. IV. Fertilizing yellowed Gelidium beds. Ibid. 30 (12): 986-992; C. A. 64, 2709 (1966).

31. Wong, K. M., V. F. Hodge and T. R. Folsom. 1971. Concentration of plutonium, cobalt and silver radionuclides in selected Pacific seaweeds. Nucl. Sci. Abst. 25: 55151; C. A. 76, 137557 (1972).

STUDIES ON PTEROCLADIA CAPILLACEA (GMEL.) BORN ET THUR.

Part II

Pharmacological, Antibacterial and Antifungal Investigations

K. C. Güven and E. Güler

Department of Pharmacy and Technology, Faculty of Pharmacy, University of Istanbul, Turkey

E. Aktin and H. Koyuncuoğlu

Department of Neurology, Department of Pharmacology, Faculty of Medicine, University of Istanbul, Turkey

In recent years several comprehensive accounts of pharmacological activities of marine algal products have been published by Baslow (3), Levring-Hoppe-Schmid (19), Chapman (5) and Jackson (16). In the sixteenth century certain algae were probably used for the treatment of goitre. Marine algae have been used in the treatment of ascariasis, in Basedow's disease, in kidney and lung diseases, in stomach troubles as well as in scabies. Laminaria has been used in surgery, and the other main products of algae such as carrageenan, agar, alginic acid in pharmacy (3, 5, 16, 19). Mannitol and sodium alginate were administered as blood plasma substitute.

The main problems that hinder working with marine algae are the technical difficulties of isolation and characterization of the active compounds, limited amounts of available algae and the difficulties in the collection of the samples.

Although some pharmacological studies have recently been published, the knowledge of the bioactive substances isolated from algae is rather scanty.

Marine Algae in Pharmaceutical Science
edited by H. A. Hoppe, T. Levring, Y. Tanaka

Many papers have previously been published from this laboratory describing lipolytic, hypoglycemic, anticoagulant and fibrinolytic activities of protein parts of different algae (1, 2, 8 - 14).

Pharmacological activities of Gelidium sp. have been studied by various authors. G. latifolium has been reported to have antilipemic properties (2). Ether and alcoholic extracts of G. amansii have shown a slight inhibition against M. tuberculosis while acetone extract of G. capillaris has exhibited considerable inhibition toward M. avium and M. tuberculosis (17). Extracts of G. filicinum had limited value against gram negative and positive organisms but showed considerable activity against M. tuberculosis H 37 Rv and M. smegmatis (20). Water extract of G. cartilagineum var. robustum was reported to be toxic in mice (15). It was also reported that agar extracted from this species has antiviral properties (4, 7). Agar of G. amansii, after decomposition, was reported to be useful as antiulcer, antipepsin and antiinflammation agent (21). G. cartilagineum was used for the treatment of lung diseases and scrofula in Japan (5) and G. filicinum for miscarriage in Hawaii (22).

In this paper Pterocladia capillacea was studied pharmacologically and the lipolytic, hypoglycemic and anticoagulant activities of the protein fractions were reported. Antibacterial and antifungal properties of the essential oil obtained from the same alga were also studied.

Experimental Pterocladia capillacea (Gmel.) Born. et Thur. was collected from the coasts of Marmara and Aegean Sea in July through September 1971 - 1976. The extraction and fractionation technique of protein from G. capillaceum was reported in Part I of this study.

1. Determination of lipolytic activity

Forty normal albino rats of Wistar strain, weighing 200 - 250 g were used. The rats were anesthetized by i. p. injection of pentobarbital (50 mg/kg). Fractions were dissolved in normal saline and the pH was adjusted to 7.4 before injection into a tail vein. Two control groups consisting of 5 and 10 rats for each fraction were used and the animals were administered saline (1 ml/kg) or heparin (100 IU/kg) respectively. Fifteen min. after the injection, 4.8 ml blood was withdrawn by an intracardiac puncture into a syringe containing 0.2 ml of a 10 % sodium citrate solution, centrifuged at 2500 rpm for 15 min. and plasma was removed. For the measurement of lipolytic activity, 0.5 ml of the assay system consisting of 1 ml plasma, 2 ml phosphate buffer (pH 7.4) and 1 ml 1 % soybean oil emulsion (Intralipid 10 % Vitrum, Deutsche Kali G. m. b. H.) was used and the free fatty acids (FFA) were determined before and one hour after incubation at 37° C in a waterbath. FFA determination was carried out by the modified Duncombe's method (Güven, Bergisadi and Güler, unpublished). The lipase activity was calculated as the difference in FFA content in the assay system before and after 1 h incubation and expressed in terms of microunits. One microunit of lipase activity equals the production of 1 μ mole FFA/hr under the given conditions.

Statistical analyses of the results were made and p values calculated.

2. Determination of hypoglycemic activity

For this purpose 40 rabbits weighing 2 - 3 kg were used and saline (1 ml/kg) and regular insulin (0.1 IU/kg) were administered for control. The fractions, insulin and saline were tested on the same rabbits according to a circulation schedule after a fortnightly intervals.

The animals were fasted overnight. The substances were administered via an ear vein and blood samples for glucose level determinations were obtained by an ear puncture before and 30, 60, 120, 180 and 240 minutes after the injections. The difference between fasting level and the lowest value after injection was considered as the maximum blood sugar fall.

3. Determination of anticoagulant, antithrombin and fibrinolytic activities

For this purpose citrated fresh human plasma obtained from 10 normal persons was used. As a control system 0.5 IU/ml heparin and buffer (pH 7.35) were used and anticoagulant, antithrombin and fibrinolytic activities were determined by the following methods respectively:

3.1 Recalcification time

A mixture of citrated plasma and sample was recalcified and the clotting time measured (23).

3.2 Plasma thrombin time

A solution of 3 U/ml thrombin[+] in barbital-saline buffer was added to the mixture of citrated plasma and sample and clotting time measured (24). Heparin was used in the concentration of 0.2 IU/ml.

3.3 Euglobulin lysis time (ELT)

The mixture of citrated plasma and acidic water (pH 5.2) was put in a refrigerator (+4° C) for 30 min and centrifuged. After discarding the supernatant thoroughly, ELT was determined (6, 26). Heparin, in the concentration of 1 IU/ml, was used as control. The time when the clot was no more a uniform unit was accepted as the lysis time.

[+] Topostasin Roche per os

4. Determination of antibacterial and antifungal activities

For this purpose the disc method with nutrient agar medium was used. The test organisms were Staphylococcus aureus, Klebsiella, Esherichia coli, Pseudomonas aeruginosa, Bacillus subtilis and Candida albicans. After applying different amounts of the essential oil to the center of the Petri dishes containing agar medium the plates were incubated at 37° C for 24 hr. The effect of the essential oil on the growth of the microorganisms was tested after the incubation period. This was undertaken by measuring the diameter of the area around the center of the Petri dishes completely free from any fungal growth.

Table 1 - Lipolytic activity of protein fractions I - III

Substance	Lipolytic activity $\pm$SE (uU/ml plasma)	p
Saline (1 ml/kg)	$0.76 \pm 0.08^{+}$	-
Heparin (100 IU/kg)	$9.00 \pm 0.46^{++}$	0.001
Fraction I (20 mg/kg)	$6.74 \pm 0.46^{++}$	0.001
Fraction I (5 mg/kg)	$3.62 \pm 0.97^{+}$	0.02
Fraction II (20 mg/kg)	$0.60 \pm 0.17^{+}$	0.5
Fraction III (20 mg/kg)	$0.76 \pm 0.25^{+}$	0.9

$^{+}$ mean of 5 experiments
$^{++}$ mean of 10 experiments

Results

1. Lipolytic activity

Lipolytic activity of the protein fractions compared with saline and heparin are shown in Table 1. According to these findings only Fraction I showed significant activity.

Table 2 - Blood glucose levels $\pm$SE after the administration of Fraction III

Time (min)		0	30	60	120	180	240
Substance							
Saline[+] (1 ml/kg)		104.8 $\pm$ 5.37	103.7 $\pm$ 4.90	100.9 $\pm$ 4.53	102.6 $\pm$ 2.81	104.3 $\pm$ 5.08	104.9 $\pm$ 4.40
Insulin[++] (0.1 IU/kg)		91.4 $\pm$ 2.54	85.1 $\pm$ 2.25	73.7 $\pm$ 2.60	82.4 $\pm$ 3.07	93.8 $\pm$ 3.01	94.3 $\pm$ 1.11
	$p <$	0.05	0.001	0.001	0.001	0.10	0.02
Fraction III[+++] (20 mg/kg)		93.9 $\pm$ 3.13	107.2 $\pm$ 2.78	84.9 $\pm$ 3.97	72.7 $\pm$ 3.25	90.5 $\pm$ 4.68	110.0 $\pm$ 4.20
	$p <$	0.10	0.90	0.01	0.001	0.05	0.50
Fraction III[+++] (10 mg/kg)		96.6 $\pm$ 5.17	107.1 $\pm$ 3.13	83.3 $\pm$ 6.30	96.8 $\pm$ 6.41	90.5 $\pm$ 6.47	99.5 $\pm$ 4.90
	$p <$	0.30	0.90	0.05	0.50	0.20	0.50
Fraction III[+++] (5 mg/kg)		57.6 - 4.00	93.4 - 7.29	63.8 - 2.41	75.7 - 7.67	81.8 - 2.95	102.4 - 4.10
	$p <$	0.001	0.30	0.001	0.01	0.001	0.90

[+] means of 20 experiments

[++] menas of 30 experiments

[+++] means of 10 experiments

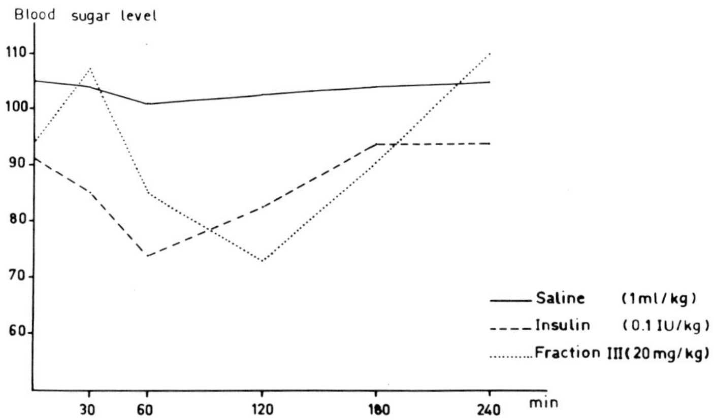

Fig. 1 - Graphical representation of hypoglycemic activity of Fraction III compared with that of insulin and saline

2. Hypoglycemic activity

The results of hypoglycemic activity of the protein fractions are shown in Table 2 and Fig. 1. Fraction III showed hypoglycemic activity but the lowest blood sugar level was achieved two hours after the administration while for insulin this time was one hour. Fraction I and II exhibited no activity.

3. Anticoagulant, antithrombin and fibrinolytic activities

3.1 Recalcification time

The results are shown in Table 3, Fig. 2, 3 and 4. Total extract and Fraction I were found to be active. It is seen that Fraction I is substantially more active as compared to total extract.

3.2 Plasma thrombin time

The results are shown in Table 4. In Fig. 5 and t graphical representations of the plasma thrombin time of total extract and Fraction I as compared with that of heparin are given. Fraction I was found more active in comparison to other fractions.

Table 3 - Recalcification time for heparin, total extract and fractions. Each value is the mean of 10 experiments.

Sample	Doses (in 1 ml plasma)	Recalcification time (sec) $\pm$ SE	p <
Buffer	0.02 ml	102.0 $\pm$ 2.27	-
Heparin	0.5 IU	371.5 $\pm$ 8.83	0.001
Total extract	0.5 mg	197.0 $\pm$ 11.94	0.001
	1 mg	359.4 $\pm$ 17.93	0.001
	1.5 mg	564.6 $\pm$ 13.37	0.001
Fraction I	0.05 mg	169.9 $\pm$ 8.25	0.001
	0.15 mg	326.8 $\pm$ 10.95	0.001
	0.25 mg	493.8 $\pm$ 9.37	0.001
Fraction II	5 mg	275.3 $\pm$ 10.64	0.001
Fraction III	3 mg	208.7 $\pm$ 8.01	0.001
	4 mg	331.1 $\pm$ 9.63	0.001
	5 mg	469.8 $\pm$ 17.76	0.001

3.3 Euglobulin lysis time (ELT)

The results are shown in Table 5 and Fig. 7. Since the fraction doses could not be adjusted according to heparin values for comparison, total extract and the fractions were all assayed at the same dosage. As can be seen, Fraction I had a marked activatory effect on the fibrinolytic system while Fractions II and III had almost no effect.

4. Antibacterial and antifungal activities

The diameter of inhibition zones of volatile oil on tested microorganisms is shown in Table 6. As can be seen the essential oil has shown activity against the organisms tested.

This study was supported in part by the Technical and Scientific Research Council of Turkey.

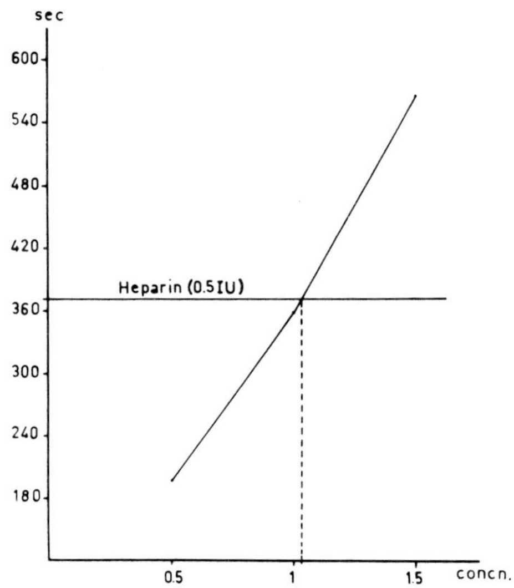

Fig. 2 - Graphical comparison of the recalcification time for heparin and total extract

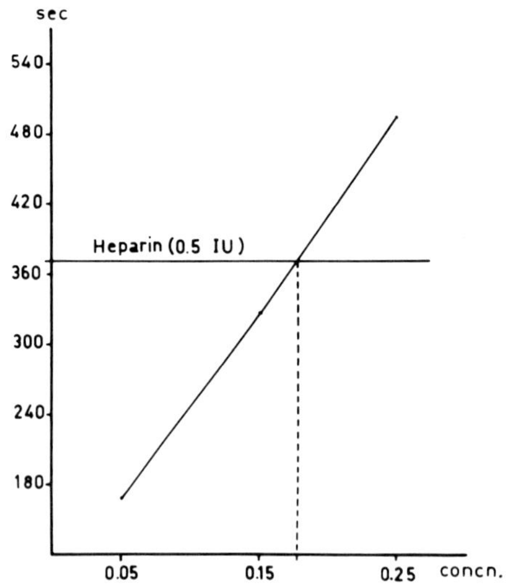

Fig. 3 - Graphical comparison of the recalcification time for heparin and Fraction I

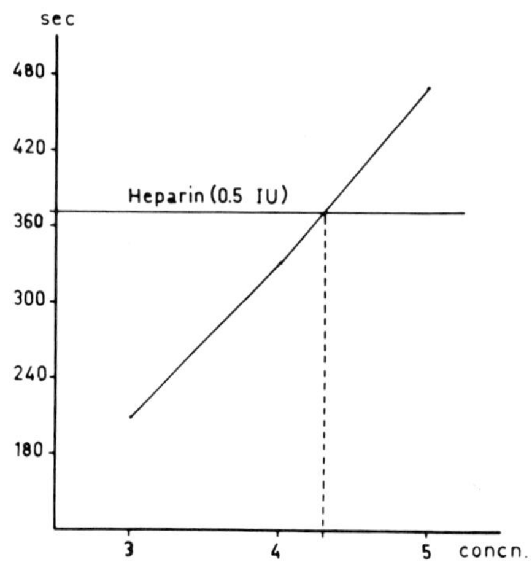

Fig. 4 - Graphical comparison of the recalcification time for heparin and Fraction III

Table 4 - Plasma thrombin time of heparin, total extract and fractions. Each value is the mean of 10 experiments.

Sample	Doses (in 1 ml plasma)	Plasma thrombin time (sec) $\pm$ SE	p <
Buffer	0.02 ml	21.2 $\pm$ 0.49	-
Heparin	0.2 IU	44.3 $\pm$ 3.48	0.001
Total extract	0.35 mg	31.3 $\pm$ 2.76	0.01
	0.50 mg	50.2 $\pm$ 3.14	0.001
	0.65 mg	73.3 $\pm$ 5.63	0.001
Fraction I	0.025 mg	24.4 $\pm$ 0.88	0.01
	0.05 mg	45.6 $\pm$ 2.44	0.001
	0.10 mg	72.7 $\pm$ 6.67	0.001
Fraction II	5 mg	27.7 $\pm$ 1.70	0.001
Fraction III	5 mg	29.6 $\pm$ 1.62	0.001

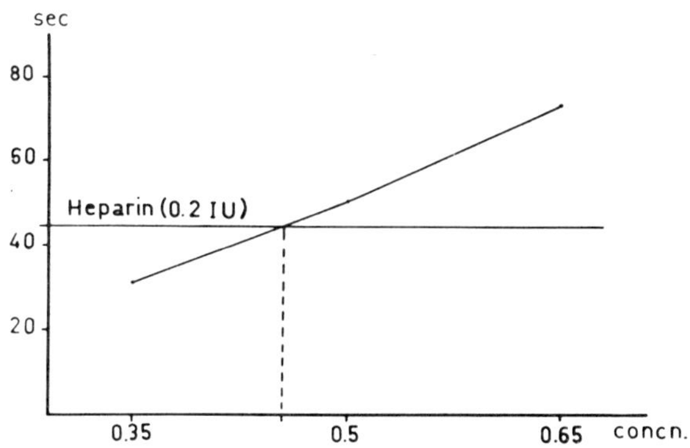

Fig. 5 - Graphical comparison of the plasma thrombin time of heparin and total extract

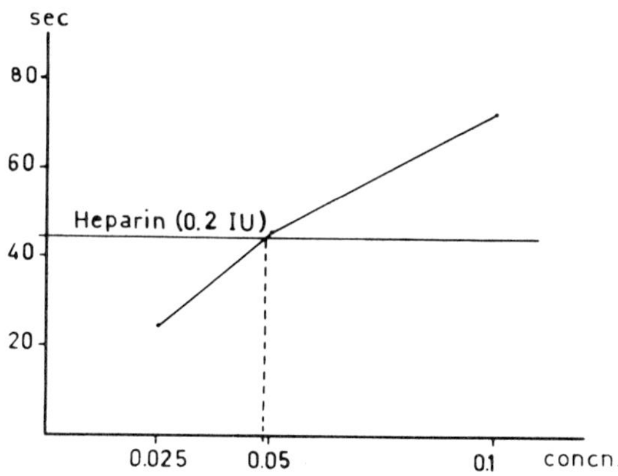

Fig. 6 - Graphical comparison of the plasma thrombin time of heparin and Fraction I

Table 5 - Euglobulin lysis time of heparin, total extract and fractions. The values are the means of 10 experiments for each.

Sample	Doses (in 1 ml plasma)	ELT (min) ± SE	p <
Buffer	0.1 ml	249.8 ± 4.86	-
Heparin	1 IU	245.6 ± 3.77	0.9
Total extract	0.01 mg	227.5 ± 5.94	0.02
	0.10 mg	178.1 ± 8.34	0.001
	0.20 mg	151.3 ± 6.93	0.001
Fraction I	0.01 mg	175.8 ± 9.18	0.001
	0.10 mg	90.1 ± 3.51	0.001
	0.20 mg	69.4 ± 4.70	0.001
Fraction II	0.01 mg	246.4 ± 5.93	0.9
	0.10 mg	238.6 ± 4.61	0.2
	0.20 mg	243.4 ± 7.45	0.5
Fraction III	0.01 mg	247.0 ± 5.41	0.9
	0.10 mg	247.3 ± 6.61	0.9
	0.20 mg	244.7 ± 7.54	0.9

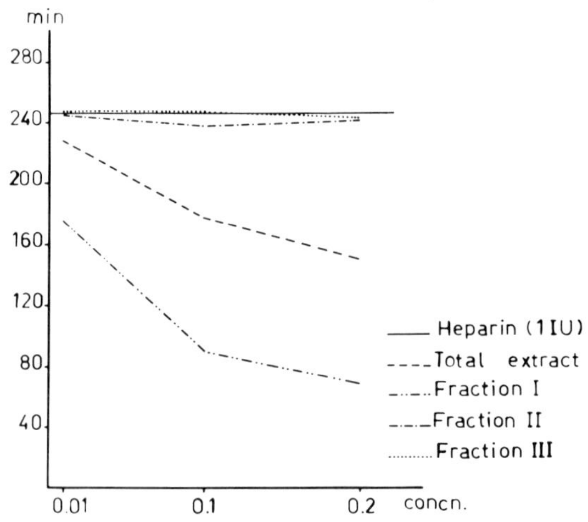

Fig. 7 - Graphical plotting of the mean values for euglobulin lysis time of total extract and fractions

Table 6 - Diameter of the inhibition zones produced by volatile oil on agar plates seeded with different test microorganisms. 0.5 µl volatile oil did not inhibit the growth of the organisms.

Test organisms / Sample	1 µl	5 µl
Staphylococcus aureus	6 mm	14 mm
Klebsiella	3 mm	22 mm
Esherichia coli	2 mm	20 mm
Pseudomonas aeruginosa	6 mm	14 mm
Bacillus subtilis	4 mm	26 mm
Candida albicans	4 mm	12 mm

Discussion

It has been reported by various authors that different compounds such as fucoidin, sargalin and carrageenan which are extracted from marine algae exhibited lipolytic, hypoglycemic and anticoagulant activities respectively (3, 5, 16, 19).

In some papers previously published from this laboratory, it was shown that the protein part of several marine algae (Sargassum vulgare, Polysiphonia subulifera, Gracilaria verrucosa, Gelidium latifolium, Phyllophora nervosa, Cystoseira barbata, Corallina rubens) and their fractions had lipolytic, hypoglycemic and anticoagulant effects (1, 2, 8 - 14).

In an earlier paper we studied the lipolytic activity of an extract obtained from G. latifolium and compared it with heparin (2). It was shown that this extract releases the same substance as heparin does, that is, lipoprotein lipase.

In the present study Pterocladia capillacea was investigated. Pharmacological studies showed that Fraction I had lipolytic, anticoagulant and fibrinolytic effects while Fraction III was observed to have hypoglycemic activity.

Fucoidin and carrageenan are sulphated polysaccharides. The former had lipolytic activity, while the latter was shown to have anticoagulant effect. On the other hand sargaline, a phenol compound, has hypoglycemic activity. It is interesting to note that all the fractions obtained from G. capillaceum were protein. Fraction I contained three, Fraction II and III two monosaccharides, as well.

The research work on the elucidation of the mechanism of these different pharmocological activities is continuing.

The antibacterial and antifungal activities of the essential oil obtained from Pterocladia capillacea are similar to the results published by various authors (18, 25).

Summary

Pharmacological investigations of protein fractions obtained from Pterocladia capillacea showed that Fraction I has lipolytic activity while Fraction III revealed hypoglycemic effect. On the other hand, Fraction I exhibited anticoagulant and fibrinolytic activities.

The antibacterial and antifungal activities of the essential oil extracted from this alga were also determined.

References

1. Aktin, E. and K. C. Güven. 1965. Studies on antilipemic properties of the algae collected from Turkish coasts. Qualitas Plantarum et Materia Vegetabiles 12: 210-215.

2. Aktin, E. and K. C. Güven. 1969. Studies on the lipase released by the substances obtained from Gracilaria verrucosa (Huds.) and Gelidium latifolium (Greville). Arzneim. Forsch. 19: 190-193.

3. Baslow, M. H. (Ed.) 1969. Marine Pharmacology, The Williams & Wilkins Co., Baltimore.

4. Bhakuni, D. S. and M. Silva. 1974. Biodynamic substances from marine flora. Botanica Marina XVII: 40-51.

5. Chapman, V. J. (Ed.) 1970. Seaweeds and their Uses, Methuen & Co. Ltd., London.

6. Copley, A. L., S. Niewiarowski and J. Maréchal. 1959. A micromethod of euglobulin fibrinolysis in plasma of human subjects and small laboratory animals. J. Lab. Clin. Med. 53: 468-473.

7. Gerber, P., J. D. Dutcher, E. V. Adams and J. H. Sherman. 1958. Protective effect of seaweed extracts for chicken embryos infected with influenza B or mumbs virus. Proc. Soc. Exptl. Biol. and Med. 99 (3): 590-593; Biol. Abst. 33, 22399 (1959).

8. Güven, K. C. and E. Aktin. 1962. A study on the antilipemic and anticoagulant activities of Phyllophora nervosa (D. C.) Grev. Eczacilik Bülteni 4: 199-206.

9. Güven, K. C. and E. Aktin. 1964. Studies on antilipemic and anticoagulant properties of algae collected from Turkish coasts. Botanica Marina VII: 1-4.

10. Güven, K. C. and B. Tekinalp. 1971. About Sargassum vulgare Agard. Botanica Marina XIV: 63-64.

11. Güven, K. C., A. Bora, E. Aktin, C. Öker, H. Hatemi and E. T. Çetin. 1972. Investigations on Phyllophora nervosa (D. C.) Grev. Ibid. XV: 46-48.

12. Güven, K. C., E. Aktin, H. Koyuncuoğlu and N. Bergişadi. 1974. Lipolytic and hypoglycemic compounds obtained from Cystoseira barbata J. Ag. Arzneim. Forsch. 24: 144-147.

13. Güven, K. C., Ş. B. Ulutin, E. Mutluay and O. N. Ulutin. 1974. Anticoagulant-antithrombin and fibrinolytic actions of extract of marine alga Corallina rubens L. Haemostasis 2: 260-268.

14. Güven, K. C., E. Mutluay, E. Aktin and H. Koyuncuoğlu. 1975. Chemical and pharmacological investigations on Corallina rubens. Planta Medica 27: 5-13.

15. Habekost, R. C., I. M. Fraser and B. W. Halstead. 1955. Observations on toxic marine algae. J. Washington Acad. Sci. 45 (4): 101-103; Biol. Abst. 30, 20545 (1956).

16. Jackson, D. F. (Ed.) 1964. Algae and Man, Plenum Press, New York, N. Y.

17. Kamimoto, K. 1956. Antibacterial substances extracted from seaweeds. II. Effect of extracts from seaweeds on the growth of some acid fast bacteria. Nippon Saikingaku Zasshi 11: 307-313; C. A. 51, 18138 (1957).

18. Katayama, T. 1964. Volatile constituents of algae. XX. Pharmacological action of volatile constituents and biochemical significance of acrylic acid. Kagoshima Daigaku Suisan Gakubu Kiyo 13: 58-72; C. A. 62, 10822 (1965).

19. Levring, T., H. A. Hoppe and O. J. Schmid (Eds.) 1969. Marine Algae, Cram, De Gruyter & Co., Hamburg.

20. Maurer, C. C. 1964. Antibacterial substances in Chilean marine algae. Anales Fac. Qim. Farm., Univ. Chile 16: 114-121; C. A. 64, 13122 (1966).

21. Murakami, M., K. Kuriaki, Y. Kawashima and S. Tachikawa. 1972. Sulfated agar polysaccharide. Japan 72: 20007; C. A. 77, 52304 (1972).

22. Nagata, K. M. 1971. Hawaiian medicinal plants. Economic Botany 25: 245-254.

23. Owen, C. A., F. D. Mann, M. M. Hurn and J. M. Stickney. 1955. Evaluation of disorders of blood coagulation in the clinical laboratory. Am. J. Clin. Path. 25: 1417-1426.

24. Ratnoff, O. D. and J. E. Colpy. 1954. An accelerating property of plasma for the coagulation of fibrinogen by thrombin. J. Clin. Invest. 33: 1175-1182.

25. Roos, H. 1957. Untersuchungen über das Vorkommen antimikrobieller Substanzen in Meeresalgen. Kieler Meeresforschungen 13: 41-58.

26. Ulutin, Ş. B. 1970. Idrarin fibrinolitik aktivitesini ölçmek için yeni bir metod. I. Ü. Cerrahpaşa Tip Fakültesi Dergisi 1: 34-38.

Lipid Constituents of the Red Alga Ceramium rubrum. A Search for Antimicrobial and Chemical Defense Substances.

T. Noguchi[a], M. Ikawa[b], J.J. Uebel[c], and K.K. Andersen[c]
Laboratory of Marine Biochemistry[a], Faculty of Agriculture, University of Tokyo, Bunkyo-ku, Tokyo 113, Japan, and Departments of Chemistry[c] and Biochemistry[b], University of New Hampshire, Durham, New Hampshire 03824, USA.

INTRODUCTION

Recently we reported that of 21 species of red algae and 8 species of brown algae tested only the red alga Ceramium rubrum strongly inhibited the growth of Bacillus subtilis (1). An active substance was isolated in crystalline form and identified as elemental sulfur. When 8 species of red algae and 8 species of brown were analyzed for free sulfur content, it was found that C. rubrum contained considerably more free sulfur than any of the others tested (0.008% of dry weight vs. 0.0015% or less for most of the others). Since organic polysulfides which break down to release sulfur have been found in the brown alga Dictyopteris (2) and cyclic polysulfides have been found in the red alga Chondria californica, an alga which exhibits antimicrobial activity against Vibrio anguillarium (3), a reinvestigation of the lipid constituents of C. rubrum was undertaken.

MATERIALS AND METHODS

Materials. C. rubrum was collected from 2 different locations. Sample I was collected in the Great Bay estuary off Adams Point in New Hampshire, and Sample II was collected off the ocean beach at Wallis Sands, New Hampshire.

Fractionation of Sample I. 830 g (wet wt.) of C. rubrum were extracted 3 times with 2 l of 95% ethanol each time. The combined extracts were concentrated to dryness in vacuo and the residue extracted 3 times with 300 ml portions of n-pentane. The pentane extract was washed 3 times with water, dried over $MgSO_4$, and concentrated in vacuo to dryness. The

Marine Algae in Pharmaceutical Science
edited by H. A. Hoppe, T. Levring, Y. Tanaka

TABLE 1. FRACTIONATION OF C. RUBRUM LIPIDS (SAMPLE I)

Fract. no. (tube no.)	Eluting agent	Wt. (mg)	Components
1 (5-20)	Hexane-100%	25.2	Sulfur
	Hexane:benzene-95:5		Hydrocarbons
	Hexane:benzene-90:10		
2 (41-46)	Hexane:benzene-70:30	6.4	Hydrocarbons (?)
			Fatty acid esters(?)
3 (47-50)	Hexane:benzene-70:30	70.3	Fatty acid Et esters
	Hexane:benzene-50:50		
4 (51-57)	Hexane:benzene-50:50	292.7	Fatty acid Et esters
	Hexane:benzene-10:90		
5 (58-63)	Hexane:benzene-10:90	192.4	Di-(2-ethylhexyl)-
	Benzene-100%		phthalate (DEHP)
6 (64-74)	Benzene-100%	44	Triglycerides
	Benzene:ether-90:10		
7 (75-83)	Benzene:ether-90:10	169	Triglycerides
(dark	Benzene:ether-60:40		Phaeophytin a(minor)
green)	Benzene:ether-20:80		Sterol (minor)
8 (84-103)	Benzene:ether-20:80)	29.6	Free fatty acids
	Ether-100%		
9 (104-107)	Methanol-100%	63.5	Free fatty acids
Total amount recovered		893.1 (92%)	

residue (0.97 g) was dissolved in a small volume of hexane and chromatographed on a 2.8 x 27 cm column containing 40 g of Silica Gel (J.T. Baker Chem. Co.) under hexane and 15 ml fractions collected. The solvent sequence is given in Table 1. Each fraction was concentrated and fractions showing similar thin-layer chromatographic (TLC) patterns were combined.

Fractionation of Sample II. 3.5 kg (wet wt) of *C. rubrum* were extracted with 7 l of ethyl acetate. The mixture was centrifuged and the residue extracted 2 times with 2.4 l portions of chloroform:methanol (3:1). The extracts were combined and concentrated to dryness *in vacuo*, yielding 10.74 g of lipids. 4.97 g of the residue were dissolved in 30 ml of

TABLE 2. FRACTIONATION OF C. RUBRUM LIPIDS (SAMPLE II)

Fract. no. (tube no.)	Eluting agent	Wt. (mg)	Components
1 (14-17)	Hexane-100%	57.7	Hydrocarbons
2 (18-26)	Hexane-100%	68.7	Hydrocarbons
	Hex:benz-90:10		Sulfur
3 (27-45)	Hex:benz-90:10	16.6	Hydrocarbons
4 (46-67)	Hex:benz-90:10	9.6	Unsat. hydrocarbons
	Hex:benz-70:30		
5 (68-72)	Hex:benz-70:30	0	
6 (73-93)	Hex:benz-70:30	32.6	Fatty acid derivs.
	Hex:benz-50:50		
	Hex:benz-30:70		
7 (94-124)	Hex:benz-30:70	78.3	Triglycerides
	Hex:benz-10:90		
8 (125-142)	Hex:benz-10:90	40.6	Triglycerides
	Benzene-100%		DEHP (minor)
9 (143-184)	Benzene-100%	260.8	DEHP
	Benz:ether-80:10		
10 (185-204)	Benz:ether-90:10	1375.6	Triglycerides
	Benz:ether-60:40		Free fatty acid (major)
			Sterol
			Phaeophytin a
			Chlorophyll a
11 (205-226)	Benz:ether-20:80	525.1	Free fatty acid (major)
	Ether-100%		Phaeophytin a
12 (227-265)	Ether-100%	271.5	Polar fatty acid derivs.
	Methanol-100%		Phaeophytin a (minor)
			Sterol (?)
13 (266-277)	Methanol-100%	1587.7	Polar fatty acid derivs.
			Phaeophytin a (minor)
14 (278-290)	Methanol-100%	208.8	Phaeophytin a (minor)
			Aromatic compds.
Total amount recovered		4533.6 (91%)	

hexane and chromatographed on a 4 x 35.5 cm column containing 200 g of Silica Gel (Baker) under hexane and 15 ml fractions collected. The solvent sequence is given in Table 2. The fractions were concentrated and fractions showing similar TLC patterns were combined.

Identification of Components. Components in the fractions were identified by the following methods: (a) TLC - Thin layer chromatography was done using Silica Gel 60F-254 (EM Laboratories) precoated plates, which were developed with hexane:ether (95:5) or benzene:ether (60:40). Components were detected with iodine vapor or UV; (b) NMR - The nuclear magnetic resonance spectrum was obtained using the JEOL MN-100 spectrometer. Samples were dissolved in deuterochloroform containing 1% tetramethylsilane; (c) IR - The infrared spectrum was obtained using a Perkin-Elmer Model 337 grating infrared spectrophotometer. Samples were dissolved in carbon tetrachloride or chloroform; (d) GLC - Gas liquid chromatography was performed on a Barber-Colman Series 5000 gas chromatograph using a DEGS on Anachrom A glass column. Fatty acids and their derivatives were esterified or transesterified to methyl esters by refluxing in boron trichloride-methanol; (e) UV-Vis - The ultraviolet and visible spectrum was obtained on a Beckman DB-G recording grating spectrophotometer; (f) GC-MS - The gas chromatography mass spectrometry was performed at the biomedical research mass spectrometer facility at the Massachusetts Institute of Technology, Cambridge, Mass.

RESULTS

Extractions. Two methods of extraction were employed in isolating the lipids. The Great Bay sample (I) was extracted with ethanol and the more non-polar lipids extracted from the ethanol extract with pentane. The yield amounted to 0.12% of the wet weight. The ocean sample (II) was extracted with ethyl acetate, followed by chloroform:methanol, similar to the procedure of Wratten and Faulkner (3). The yield in this case amounted to 0.31%.

Fractionation by Column Chromatography. The general nature of the constituents in the various fractions from silica gel column chromatography are listed in Tables 1 and 2.

Sulfur. Free sulfur was detected by MS in both fresh samples of *C.*

TABLE 3. HYDROCARBONS OF C. RUBRUM (SAMPLE I)

No. of C atoms & degree of unsat.	% of total hydrocarbons	
	This study	Youngblood et al. (4)
15:0	8 (15:0 + 15:1)	1.3
15:1		-
16:0	-	0.2
17:0	72	96
17:1	3	2.2
17:2	-	0.1
21:x	17	-

rubrum and was obtained in crystalline form from Fraction 2 of Sample II. This result with fresh seaweed confirms our previous findings of the occurrence of free sulfur in dried C. rubrum.

Hydrocarbons. GC indicated the presence of 3 major hydrocarbon fractions. These were identified by GC-MS and are shown in Table 3. The overwhelming major component was $C_{17:0}$. The C_{21} fraction appeared to be a complex mixture of multiply unsaturated C_{21} hydrocarbons, which could not be well resolved. The amount of hydrocarbons in each case amounted to no more than 3% of the total lipids. Our results are essentially in agreement with those of Youngblood et al. (4) (see Table 3) except that they did not report the presence of any C_{21} hydrocarbons.

Fatty Acid Derivatives. Fatty acid derivatives accounted for the bulk of the lipid extracts in both samples, amounting to about 69% of the Sample I extract and 79% of the Sample II extract. Sample I yielded mostly ethyl esters of fatty acids. This was due to the fact that ethanol was used in the extraction procedure and is to be looked at as an artifact of the isolation procedure. In the case of Sample II, where no ethanol was involved in the extraction, no significant amount of ethyl esters could be detected. Triglycerides were a major lipid component. This class was identified by the characteristic multiplet (AB of an ABX) at 4.2 ppm due to the methylene protons of the glycerol moiety of triglycerides (5). Significant amounts of free fatty acids were also found in both samples. Sample II gave fractions which appeared to contain large amounts of polar fatty acid derivatives which have not yet been identified.

TABLE 4. FATTY ACID COMPOSITION OF C. RUBRUM LIPIDS (SAMPLE I)

No. of C atoms	This report*	Klenk et al. (6)*	No. of C atoms	This report*	Klenk et al. (6)*
10:0	-	-	18:3	-	2.1
12:0	3.8	-	18:4	-	1.0
12:1	4.9	-	19:0	-	0.4
14:0	8.2	5.6	20:0	-	-
14:1	-	+	20:1	-	+
15:0	-	0.6	20:2	-	-
16:0	43.4	38.6	20:3	-	0.9
16:1	10.6	7.3	20:4	3.8	4.8
16:2	0.9	0.5	20:5	16.2	17.2
16:3	-	0.6	21:0	-	-
16:4	-	0.6	22:0	-	-
17:0	-	0.8	22:1	-	-
18:0	0.9	1.3	22:5	}1.7	0.5
18:1	5.3	13.6	22:6		1.4
18:2	0.4	1.2	* % of total fatty acids		

The fractions in Table 1 containing fatty acids were examined by GLC after converting the fatty acids to their methyl esters. By taking into account the form in which the fatty acid occurred in each fraction, the amount of each fatty acid in each fraction was determined, which were then summed from all the fractions. The principal fatty acid was palmitic acid. Of interest was the presence of appreciable amounts of the highly saturated C_{20} and C_{22} fatty acids. Our results are in essential agreement with those of Klenk *et al.* (6) (see Table 4).

Since the hydrocarbons found in the seaweed probably arise from the decarboxylation of fatty acids, one might expect some relationship between these two classes. One of the most noticeable features is the predominance of $C_{17:0}$ hydrocarbon and the virtual absence of $C_{18:0}$ fatty acid. This would indicate that there is a deliberate mechanism to get rid of excess $C_{18:0}$ fatty acid. As chain lengths increase to 18 and higher, the acids tend to become highly unsaturated. Therefore the C_{21} hydrocarbons would be expected to be more highly unsaturated than the lower hydrocarbons as appears to be the case.

Plasticizer. The plasticizer di-(2-ethylhexyl)-phthalate (DEHP) made up a significant part of the lipid fraction in both samples, 20% in I and 5% in II. It was identified by its NMR and IR spectra. Although the solvents used in this study were all redistilled and kept in glass, the source of the plasticizer is still not certain. There are numerous reports in the literature of the occurrence of phthalate esters in microorganisms, plants, and animals (7).

Pigments. The principal pigment was phaeophytin a, a degradation product of chlorophyll a. It was identified through its UV-vis spectrum. Although its amount was small it imparted a very dark green to almost black color to some of the fractions.

Sterols. Sterol could be demonstrated by TLC. These were minor constituents and were not further investigated.

Aromatic compounds. Fraction 14 of Sample II showed a strong downfield peak in its NMR spectrum in the aromatic proton region. This fraction is under further investigation.

DISCUSSION

Any substance which an organism accumulates or produces which is toxic or disagreeable to other organisms would tend to give the organism containing it an ecological advantage over other organisms. Free sulfur could be considered such a substance. It has been widely used as an antifungal and a plant and animal antiparasitic agent and it also is antibacterial (1, 8, 9). Since the free sulfur content of C. rubrum is elevated as compared to other seaweeds in the area (1), it probably makes this seaweed a less desirable target for other organisms and serves as a chemical defense substance. We were not able to show the presence of organic polysulfides, which can break down to release sulfur, as has been noted in other seaweeds (2, 3).

The fatty acids, which can be released from triglycerides by the action of lipases, should also be taken into account in any discussion of chemical ecology. Of the saturated fatty acids C_8-C_{12} acids have been shown to have the greatest antibacterial activity (10, 11), the C_4-C_{12} acids the greatest antifungal activity (12, 13) and C_7-C_{12} acids the greatest anti-*Chlorella* activity (14). Of special interest are the unsaturated fatty acids where the double bond is terminally located away from the carboxyl group. C_3 acrylic acid is especially active against *Chlorella* (14) and undecylenic acid also displays strong anti-*Chlorella* activity (14). Acrylic acid is also an antibacterial agent that has been demonstrated in a number of Rhodophyta, Phaeophyta, Chlorophyta, and Chrysophyta (15). The presence of some C_{12} fatty acid and the presence of considerable quantities of highly unsaturated C_{20} acids has been demonstrated in *C. rubrum*. Although the latter have not been tested for biological activity, it has been shown that linoleic acid and linolenic acid do inhibit the growth of *Chlorella* (unpublished results).

ACKNOWLEDGEMENT

This study was made possible by a grant from the Leslie S. Hubbard Marine Program Fund to the University of New Hampshire.

REFERENCES

1. Ikawa, M., Thomas, V.M., Jr., Buckley, L.J., Uebel, J.J.: Sulfur and the toxicity of the red alga *Ceramium rubrum* to *Bacillus subtilis*. J. Phycol. 9, 302-304 (1973).
2. Moore, R.E.: Bis-(3-oxoundecyl) polysulfides in *Dictyopteris*. Chem. Comm., 1168-1169 (1971).
3. Wratten, S.J., Faulkner, D.J.: Cyclic polysulfides from the red alga *Chondria californica*. J. Org. Chem. 41, 2465-2467 (1976).
4. Youngblood, W.W., Blumer, M., Guillard, R.L., Fiore, F.: Saturated and unsaturated hydrocarbons in marine benthic algae. Marine Biol. 8, 190-201 (1971).
5. Bovey, F.A.: Nuclear magnetic resonance spectroscopy, Academic Press, N.Y., p. 105ff (1969); See also, High resolution NMR spectral catalog, Varian Associates, Palo Alto, Calif., Spectrum no. 368 (1962).
6. Klenk, E., Knipprath, W., Eberhagen, D., Kool, H.P.: Über die ungesättigten Fettsäuren der Fettstoffe von Süsswasser- und Meeresalgen. Hoppe-Seyler Z. Physiol. Chem. 334, 44-59 (1963).
7. Peakall, D.B.: Phthalate esters: Occurrence and biological effects.

Residue Rev. 54, 1-41 (1975).

8. Weld, J.T., Gunther, A.: The antibacterial properties of sulfur. J. Exptl. Med. 85, 531-542 (1947)
9. Libenson, L., Hadley, F.P., McIlroy, A.P., Wetzel, V.M., Mellon, R.R.: Antibacterial effect of elemental sulfur. J. Infect. Dis. 93, 28-35 (1953).
10. Hassinen, J.B., Durbin, G.T., Bernhart, F.W.: The bacteriostatic effects of saturated fatty acids. Arch. Biochem. Biophys. 31, 183-189 (1951).
11. Kabara, J.J., Swieczkowski, D.M., Conley, A.J., Truant, J.P.: Fatty acids and derivatives as antimicrobial agents. Antimicrob. Agents Chemother. 2, 23-28 (1972).
12. Gershon, H., Parmegiani, R.: Organic fluorine compounds. II. Synthesis and antifungal properties of 2-fluoro fatty acids. J. Med. Chem. 10, 186-188 (1967).
13. Hoffman, C., Schweitzer, T.R., Dalby, G.: Fungistatic properties of fatty acids and possible biochemical significance. Food Res. 4, 539-545 (1939).
14. McGrattan, C.J., Sullivan, J.D., Jr., Ikawa, M.: Inhibition of Chlorella (Chlorophyceae) growth by fatty acids, using the paper disc method. J. Phycol. 12, 129-131 (1976).
15. Sieburth, J.M.: Antibacterial substances produced by marine algae. Develop. Indust. Microbiol. 5, 124-134 (1964).

(University of New Hampshire Sea Grant Publication No. UNH-SG-AB-107)

Contributions on the Content, Seasonality and Gelling Properties of the Phycocolloid from Three *Hypnea* Species from Hawaii.*

Keto E. Mshigeni
Department of Botany, University of Dar es Salaam
P.O. Box 35060, Dar es Salaam, Tanzania, East Africa.

ABSTRACT

Studies have been conducted on phycocolloids from three species of *Hypnea* Lamouroux (Rhodophyta, Gigartinales) from Hawaii. In *Hypnea cervicornis* J. Agardh the phycocolloid yields obtained (on a dry weight basis) were 35.2 ± 0.9, 30.1 ± 0.8, 34.1 ± 0.4 and 35.0 ± 1.0% in March, June, September and December, respectively. Corresponding values of 30.6 ± 0.7, 26.1 ± 0.5, 30.2 ± 0.4 and 30.2 ± 1.3% were obtained for *H. chordacea* Kuetzing, while in *nidifica* J. Agardh values of 45.5 ± 1.8, 32.9 ± 0.8, 29.3 ± 0.3 and 30.5 ± 0.8%, were obtained. Infrared spectra of the phycocolloids were determined and found to have absorption peaks typical of kappa carrageenan. The gelling properties of the phycocolloids were also studied. In *H. cervicornis* the gel strength of unmodified phycocolloid ranged from 282 ± 17 g/cm^2 in June to 378 ± 32 g/cm^2 in September. That of *H. chordacea* ranged from 350 ± 17 g/cm^2 in June to 530 g/cm^2 in December, while that of *H. nidifica* ranged from 298 ± 17 g/cm^2 in June to 615 ± 29 g/cm^2 in December. Treatment with solutes significantly increased the gel strength and gave rise to values of up to 1000 g/cm^2 and beyond. The viscosities of the phycocolloids were also investigated and found to be significantly lowered by treatment with lime. The economic significance of these findings is discussed.

---

* Extracted from a thesis submitted to the Graduate Division of the University of Hawaii in partial fulfillment of the requirements for the degree of Doctor of Philosophy in Botanical Sciences in Fall, 1974.

Marine Algae in Pharmaceutical Science
edited by H. A. Hoppe, T. Levring, Y. Tanaka

INTRODUCTION

The increased demand for seaweed colloids as thickeners, as gelling agents, as emulsion stabilizers and suspending agents in food-, pharmaceutical- and other modern industries has stimulated research on the content, nature and gelling properties of phycocolloids from a wide variety of seaweeds.

In the genus *Hypnea* Lamouroux (Rhodophyta, Gigartinales) the first known discussion of the industrial potential of any of its species is that of Reed (1906) who reported on the suitability of the phycocolloid from *Hypnea nidifica* J. Agardh as a culture medium for micro-organisms. This species is also used by the Hawaiians for curing stomach troubles (Reed, 1906) and hence seems to be of pharmaceutical significance. A second species studied in this connection is *H. musciformis* (Wulfen) Lamouroux which has also been reported (Humm, 1942) to produce usable "agar". Indeed the gel from this species has been found (Ferguson, 1946) to be an excellent bacteriological "agar" having a very low viscosity and a setting point of just about 30°C. In some Asian countries (Levring *et al.*, 1969), *H. musciformis* is also used as a vermifuge. This further suggests that the species has potential use in pharmaceutical industry. *H. spicifera* (Suhr) Harvey is a third species in the genus which has been studied and considered to be a good source of "agar" (Fox and Stephens, 1943).

The phycocolloid from *Hypnea musciformis* has been reported to have stronger gelling properties than any other known algal colloid (Humm and Williams, 1948) and its gelling properties have been found to be greatly modified by treatment with solutes (Humm and Williams, 1948; Humm, 1951). These early reports on the gelling properties of *H. musciformis* phycocolloids were subsequently substantiated by other workers. Hamilton and Carroll (1962), for example found that the gel strength of the species was higher than that of the well known carrageenan producer, *Chondrus crispus* (L.) Stackhouse. Rama Rao and Krishnamurthy (1968) also found that the gel strength of the phycocolloid from *H. musciformis* was greatly increased by solutes such as potassium chloride.

The strong gel-forming property of **the** phycocolloid from *Hypnea* has led to its being referred to as a "gelan" (Stoloff and Silva, 1957; Stoloff, 1962). However, in the literature, some authors also refer to this phycocolloid as an "agar" (e.g., Humm, 1942; Fox and Stephens, 1943; Humm, 1944; Ferguson, 1946; Zaneveld, 1959; Desikachary, 1965), as a "kappa carrageenan" (Yaphe and Baxter, 1955; Yaphe, 1959), or simply as "hypnean" (see Levring *et al.*, 1969). This nomenclatural inconsistence reflects the lack of understanding on the nature of the gel from *Hypnea*.

Although Yaphe and Baxter's (1955) and Yaphe's (1957, 1959) research using specific enzymes to digest the phycocolloid from *Hypnea musciformis* has revealed the phycocolloid to be largely a kappa carrageenan, the nature of the phycocolloid for the majority of *Hypnea* species is totally unknown. Yet, it cannot be assumed that all the other 30 or so species of *Hypnea* reported in the literature would also have kappa carrageenan. In *Eucheuma* J. Agardh, for instance, some species (e.g., *E. spinosum* J. Agardh) contain iota carrageenan whereas others (e.g., *E. striatum* J. Agardh) contain kappa carrageenan (Stancioff and Stanley, 1969). It is therefore desirable for other hitherto neglected species to be investigated.

The method employed to establish the nature of *Hypnea musciformis* phycocolloid (Yaphe and Baxter, 1955; Yaphe, 1959) was the use of the kappa carrageenan specific enzyme, kappa carrageenase obtained from the marine bacterium, *Pseudomonas carrageenevora*. This enzyme digested the phycocolloid from the alga into sulphated galactose. Subsequent studies on the phycocolloid from *H. spicifera* (Clingman and Nunn, 1959) and *H. japonica* Tanaka (Hong *et al.*, 1972) have also confirmed that the phycocolloid from *Hypnea* is very similar to kappa carrageenan.

In their studies of infrared spectra of a wide variety of red algal colloids, Stancioff and Stanley (1969) published the characteristic absorption peaks of kappa, iota and lambda carrageenans. However, no one has yet established that IR absorption peaks of *Hypnea* phycocolloids are the same as those of kappa carrageenan.

Another aspect of study which is desirable for the phycocolloid industry is a quantitative estimation of gel content. In the genus *Hypnea* scattered reports of this nature appear in the literature. However, these were confined to only a few species. In Indonesia a value of 30% has been reported for *H. musciformis* (Zaneveld, 1959) while in India the values reported for this species range from 30-35% according to Rama Rao and Krishnamurthy (1968) and from 53-57% in the publication of Desikachary (1965), all on a dry weight basis. These differences in the yields of phycocolloid reported in the literature may be a reflection of differences in the techniques of phycocolloid extraction and the duration of boiling the seaweeds, or may be due to seasonal variations in colloid content. In Brazil, for example, the phycocolloid content of *H. musciformis* was 48% in March while in October the yield was only 35% (Humm and Williams, 1948).

There undoubtedly appears to be a seasonal variation in phycocolloid content in some seaweeds. In the temperate latitudes, most of the seaweeds investigated, e.g., *Gelidium robustum* (= *G. cartilagineum* sensu Cooper and Johnstone, 1944) and *Chondrus crispus* (Black *et al.*, 1965), contain the highest levels of phycocolloids in summer and the lowest in winter. Similar results were obtained by Dawes *et al.* (1973) on three species of Eucheuma from Florida. On the tropical shores of India, in contrast, studies on Hypnea musciformis revealed the highest values of phycocolloid in winter and early spring (December to March) and the lowest levels in summer, i.e., June (Rama Rao, 1970). It is not, however, known whether populations of Hypnea on other tropical shores show a similar behaviour with respect to phycocolloid seasonality.

The industrial value of a seaweed colloid depends upon its gelling properties (see Levring et al. 1969; Chapman, 1970). Data of this nature have been accumulating for a few species of Hypnea, e.g., H. musciformis, H. spicifera and to a lesser extent H. cervicornis J. Agardh (Levring et al., 1969). Further studies on additional species are of commercial importance. To this end, therefore, the following investigation was conducted on three species from Hawaii, viz., H. cervicornis, H. chordacea and H. nidifica.

MATERIALS AND METHODS

Harvesting and Drying of Seaweeds

Thalli of Hypnea cervicornis, H. chordacea and H. nidifica were picked by hand at Diamond Head reef, Oahu Island, Hawaii (U.S.A.) during spring low tides and placed in plastic bags. These were taken to the laboratory and rinsed with fresh water to remove sea salt from the thallus surface. This was followed by careful sorting out to remove non-Hypnea adhering algae or rock debris. The algae were then placed in an oven at 50°C to dry.

Phycocolloid Extraction

The dried seaweeds were cleaned further by re-soaking 30.0 g sample in one litre of cold tap water placed in a 2.0 litre stainless steel beaker with gentle stirring for 15 minutes. After the removal of all the visible adhering particles the seaweed was redried.

The steps involved in phycocolloid extraction from 20.0 g samples of the clean and dry Hypnea thalli are summarised in Fig.1 and the apparatus used are shown in Figs. 2 and 3. After extraction the phycocolloid was dried to a constant weight and the values obtained were converted to percentages.

Handling the Extracted Phycocolloid

The dry phycocolloid from each species was milled into a powder using a Thomas mill and placed in a bottle with a lid. After labelling (to indicate the species, the dates of harvesting and the yield), the bottles were stored in a dry place until the phycocolloids were required for other studies.

Preparation of Phycocolloid Films for Infrared Spectroscopy

The techniques used in the preparation of films of the phycocolloids for infrared spectroscopy studies are essentially similar to those recently described by Semesi and Mshigeni (1977) involving the casting of phycocolloid films over mercury (Stancioff and Stanley, 1969). The films obtained were kept dry in a desiccator until needed for the determination of the IR spectra.

Determination of IR Spectra

Infrared spectra of the phycocollids were determined using an IR 10 Infrared Spectrophotometer available in the Department of Chemistry, University of Hawaii. The curve for each tested species was traced and re-drawn on a single graph sheet to enhance comparison.

Gel Strength Determination

Powdered phycocolloid from the tested species (3.6 g) and potassium chloride (0.36 g), were mixed and added to 160 ml hot distilled water in a tared beaker with vigorous agitation in a boiling water bath. The final weight was adjusted to 180 g using hot distilled water followed by vigorous stirring.

The hot gel was poured into a crystallizing dish and then left to cool in a 25°C water bath for one hour.

The solid gel was inverted in the crystallizing dish and then placed on the pan of the scale of a Marine Colloids Gel Tester (Model GT - 1) (Fig. 4) available in the laboratory of Professor Maxwell S. Doty, University of Hawaii. After placing the crystallizing dish with the gel to be tested in position (Fig. 4) the plunger was released and the force indicator on the scale of the pan was carefully watched. The point where the gel surface was ruptured by the plunger was shown by the snapping back of the pointer. The reading on the scale when this occurred was taken as the gel strength of the phycocolloid.

Viscosity Determinations

The viscosity of the phycocolloid for each species was determined by dispersing 2.7 g of the extract in a 250 ml Berzelius beaker (Fig. 5) in 170 ml hot distilled water in a water bath, with strong agitation. After thoroughly dispersing the phycocolloid, the final volume was brought to 180 g using boiling distilled water.

A spindle of the viscometer was preheated to 75°C in a hot water bath, attached to a Brookfield viscometer, and then immersed in the hot phycocolloid. When the temperature of the phycocolloid fell to 75°C, viscosity readings on the viscometer were taken.

Modifications of the Phycocolloids by Treatment with Lime

The procedure for modifying the phycocolloid with lime involved the extraction of the gel as in Fig. 1 except that instead of using 2% sodium hydroxide, lime was added (0.2 g lime per g of clean dry phycocolloid). The effect of lime on gel strength and viscosity were determined for each of the tested species.

Effects of Potassium Chloride Concentration on Gel Strength

In this study the effect of potassium chloride concentration on gel strength was investigated using a concentration gradient ranging from 0.0 - 3.6% potassium chloride in 2% gels, i.e., containing 3.6 g phycocolloid per 180 g of hot solution.

RESULTS

Phycocolloid Content

The results of this study are illustrated in Fig. 6. In *Hypnea cervicornis* the yield of phycocolloid exceeded 30% of the dry weight throughout the year. The highest yields were obtained in spring and winter months, and the lowest values occurred in summer. Although the phycocolloid yield of *H. chordacea* was consistently lower than that of *H. cervicornis*, the pattern of seasonal change closely resembled that of the latter species.

Hypnea nidifica (Fig. 6) showed the widest seasonal variations in its phycocolloid content. In March a mean yield of 45.5 $\pm$ 1.8% was obtained. By June and September the yields had declined to 32.9 $\pm$ 0.8 and 29.3 $\pm$ 0.3%, respectively. In December, however, the values had begun to increase again.

Infrared Spectra

The results of this study are shown in Fig. 7. The IR spectral peaks of all the three species closely resembled those shown (Stancioff and Stanley, 1969) to be characteristic of kappa carrageenan.

Gel Strength

The results on the gel strengths of the phycocolloids from the three tested species (Fig. 8) showed that *H. nidifica* and *H. chordacea* consistently had stronger gels than *H. cervicornis*. Each species, however, showed a decline in the gel strength during summer.

Gel Viscosity

The results on gel viscosity experiments are summarised in Fig. 9. *Hypnea cervicornis* consistently showed the highest gel viscosity, followed by *H. nidifica* and *H. chordacea*. This was quite a reverse of what was found in the gel strength experiments.

Effects of Lime on the Gelling Properties of the Phycocolloids

The results of this study are summarised in Fig. 10. It was observed that lime treatment appreciably improved the gel strength. As shown in Fig. 11, however, treating with lime markedly lowered the gel viscosity.

Effects of Potassium Chloride Concentration on Gel Strength

The results of this study are summarised in Fig. 12. No gelation occurred in the complete absence of potassium chloride. With the addition of this solute, however, the phycocolloid could gel. This continued until an optimum value was reached beyond which the gel strength began to decline again (Fig. 12). With 3.6% potassium chloride gelation was zero again.

DISCUSSION

As summarised in Fig. 6, all the three species of *Hypnea* showed appreciable yields of phycocolloid. The finding that the highest yields generally occurred in winter and spring months while the lowest levels occurred in summer is similar to that reported by Rama Rao (1970) for *H. musciformis* in India. Thus from gel content, the best harvesting season of *Hypnea* for phycocolloid production in the tropics appears to be during the cooler months. This is different from what is known for temperate algae such as *Gelidium robustum* (Cooper and Johnstone, 1944) and *Chondrus crispus* (Black *et al.*, 1965). Thus the algae of the tropical and temperate latitudes seem to show different patterns of phycocolloid seasonality.

It is known that during tropical summers, algae of the intertidal and upper subtidal habitats are subjected to high light intensities and consequent thallus bleaching (Mshigeni, 1974). Bleached thalli would be expected to show a reduced capacity for photosynthesis and hence a lower rate of phycocolloid and indeed overall carbohydrate production. Furthermore, the high tropical summer temperatures undoubtedly cause increased respiratory rates and hence a decline in the total carbohydrate level and therefore also of phycocolloid. The combined effects of light and temperature may thus be the major causal factors for the declined levels of phycocolloids during summer for the tropical species.

Several other factors are, however, also known to affect phycocolloid yields. The fertility of the sea in the growth habitat may, for example, significantly improve vegetative growth but not necessarily phycocolloid production. Indeed Neish and Shacklock (1971), working on *Chondrus crispus*, observed that thalli growing rapidly in nitrate-enriched sea water yielded a lower level of carrageenan than those growing in non-enriched sea water. Dawes *et al.* (1973) made similar observations on *Eucheuma* in Florida and suggested that if nitrogen supplies are limiting, intermediates in the photosynthetic carbon cycle may be diverted from protein- to carbohydrate synthesis. If nitrate levels are high, Dawes *et al.* (1973) suggest, the intermediates of photosynthesis are diverted towards protein synthesis. This would be expected to cause an increase in the protein levels and a decrease in carrageenan levels. In view of what has been discussed above, therefore, the unusually low level phycocolloid for *Hypnea nidifica* in December (Fig. 6) may be explained along these lines. Although no investigations were made on the possible seasonal changes of nitrate levels in the study site, it would be expected that because of run-offs draining into the sea associated with winter rains, nitrate levels in the shoreline Hawaiian waters are probably highest in winter (December). Thus in our attempts to understand the possible causal factors for the observed seasonal changes in phycocolloid levels, we also need to consider the possible effect of the prevailing field conditions.

As shown in Fig. 7, the IR spectra of all the three species showed a broad absorption band at 1240 cm^{-1} which is typical of all sulphated polysaccharides, a sharp absorption peak at 930-940 cm^{-1} indicating the presence of 3,6-anhydrogalactose, and another sharp peak at 845 cm^{-1} known to be due to the presence of galactose 4-sulphate (Stancioff and Stanley, 1969). Note the absence of a single broad absorption band at 800-860 cm^{-1} which is exhibited by lambda carrageenan and of a sharp absorption peak at 805 cm^{-1} known to be present in iota carrageenan (Stancioff and Stanley, 1969). The present findings thus suggest that all the tested Hawaiian species of *Hypnea* produce a carrageenan of the kappa type. These results support those by Yaphe and Baxter (1955) and Yaphe (1957, 1959) based on the enzymatic hydrolysis of the phycocolloid from *H. musciformis*.

The observed seasonal change in the gel strengths as summarised in Fig. 8, is in close agreement with what Rama Rao (1970) reported for *Hypnea musciformis* in India. The causal factors for the observed summer decline in gel strength are not understood. However, the fact that the lowest values of gel strength were obtained during the months when the carrageenan levels were also lowest, suggests that the causal factors involved may be related to those discussed for seasonal variations in phycocolloid content.

With respect to the findings on gel viscosity (Fig. 9), the author is unaware of any published values of *Hypnea* gel viscosity which would enhance comparisons. In their studies on the viscosity of the phycocolloid from *H. musciformis*,

Humm and Williams (1948) discussed their results only in terms of "relative viscosity" and this renders comparisons difficult. However, from the present study, it seems that the phycocolloid from *H. cervicornis* would be more useful than that from *H. chordacea* and *H. nidifica* where a high viscosity gel is desired.

The results of the experiments on the effects of lime and potassium chloride on the gelling properties of *Hypnea* phycocolloids (Figs. 10-12) are in close agreement with what is known in the literature (Stancioff and Stanley, 1969; Humm and Williams, 1948; Rama Rao and Krishnamurthy, 1968) for *Hypnea* and other seaweeds. Thus solutes such as calcium hydroxide and potassium chloride can be used to improve the gelling properties of *Hypnea* phycocolloids to give a gel of desired gel strength and viscosity. However, this improvement is only within certain working limits (Fig. 12).

The present study thus sheds further light on the view that *Hypnea* is a potential source of carrageenan for industry. All the three tested species showed appreciable quantities of carrageenan and, very probably, the same may apply to *Hypnea* populations on other tropical and subtropical shores. Similar studies on other species from other regions are thus desirable. Together with these, information on the standing crops of the species (e.g., see Mshigeni, 1977) and on their reproduction and regeneration mechanisms (see Mshigeni, 1976) should be sought.

ACKNOWLEDGEMENTS

I wish to express my deep gratitude to Dr Maxwell S. Doty and Dr G.A. Santos both of the Botany Department, University of Hawaii (U.S.A.) for advice, encouragement and criticisms, the Rockefeller Foundation for financial support, my wife Grace for untiring help and unlimited co-operation throughout the study, and Mrs Z. Kassam for typing the manuscript.

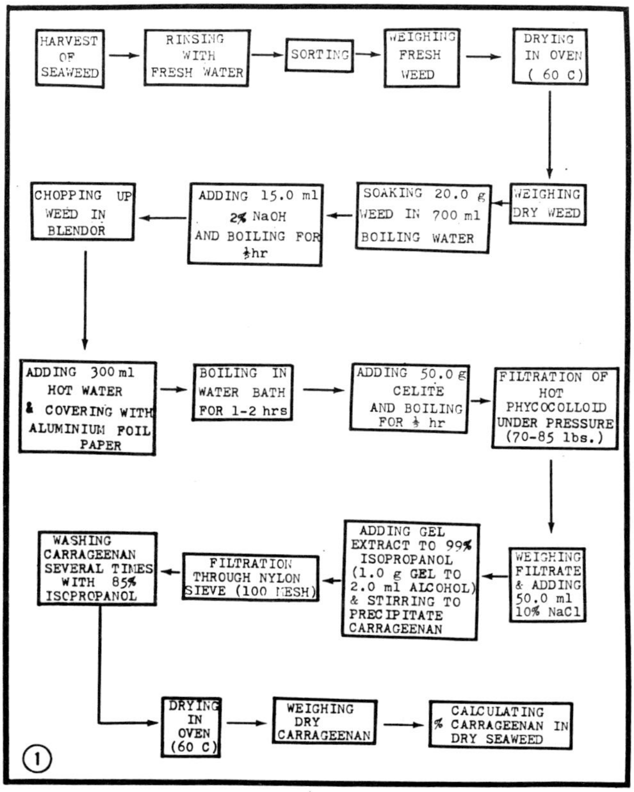

Fig. 1. Outline of the steps followed in phycocolloid extraction.

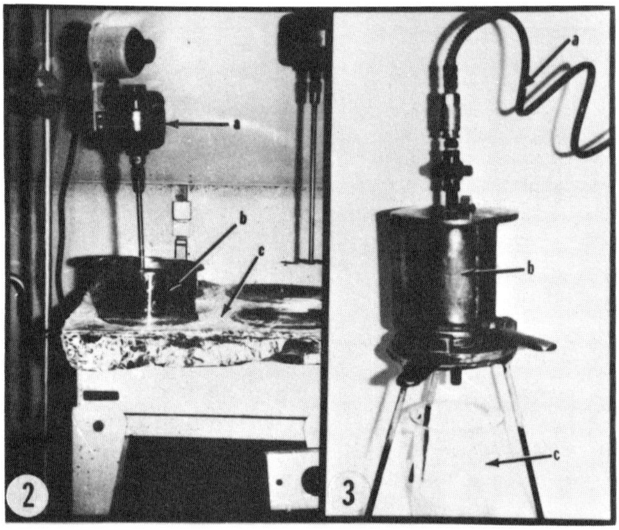

Fig. 2. Phycocolloid extraction apparatus. Note electric stirrer a for stirring up the seaweed; the stainless steel beaker b in which the seaweed is placed; and the electrically heated water bath c in which the seaweed is boiled.

Fig. 3. Phycocolloid filtration apparatus. Note attachment a connected to a high pressure (70-85 lbs/sq. in.) source the thick-walled "bomb" filter b in which the boiled seaweed and celite 545 (filter aid) are placed, and beaker c for receiving the filtrate.

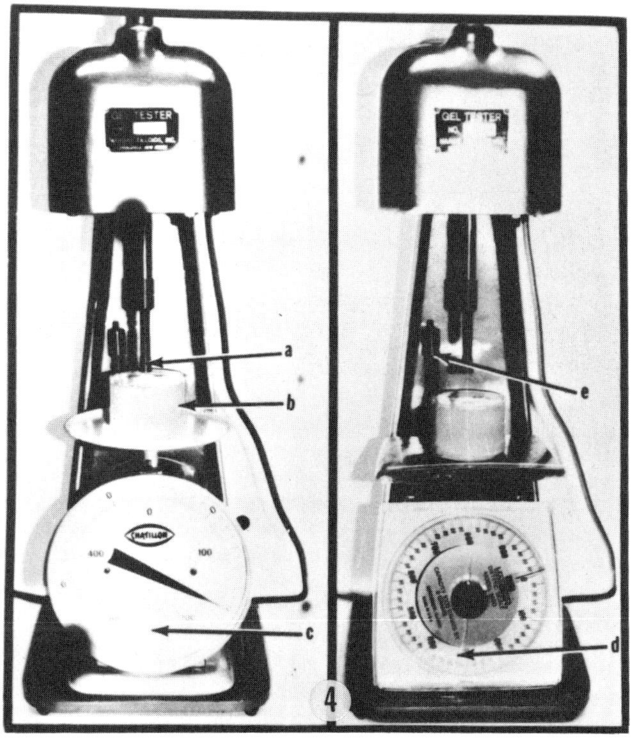

Fig. 4. Gel strength apparatus. Note plunger _a_ of the Marine Colloids Gel Tester, the crystallising dish _b_ containing the gel to be tested; the scale _c_ for testing weak gels, the scale _d_ for testing stronger gels, and a wider plunger _e_ for weak gels.

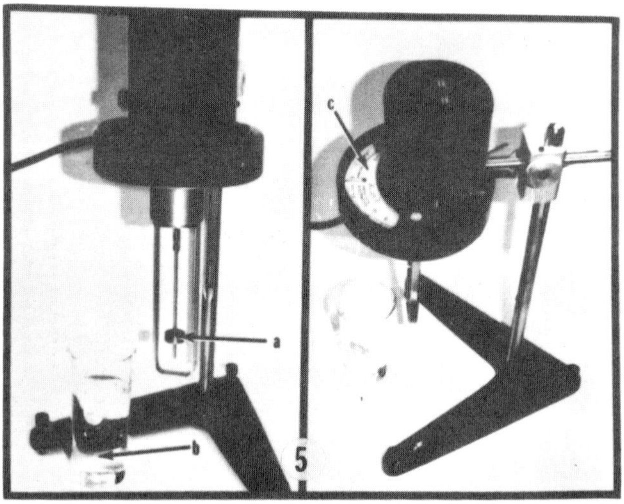

Fig. 5. Gel viscosity apparatus. Note the plunger of the viscometer a, the Berzelius beaker b in which the gel to be tested is placed and the scale c from which viscosity readings are taken.

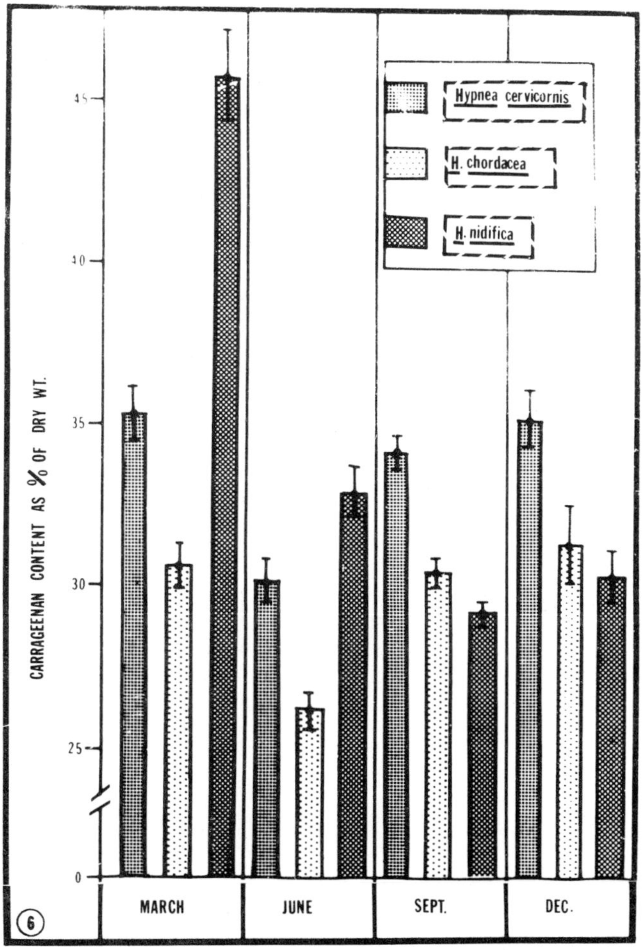

Fig. 6. Graphic presentation of the data on <u>Hypnea</u> phycocolloid yields.

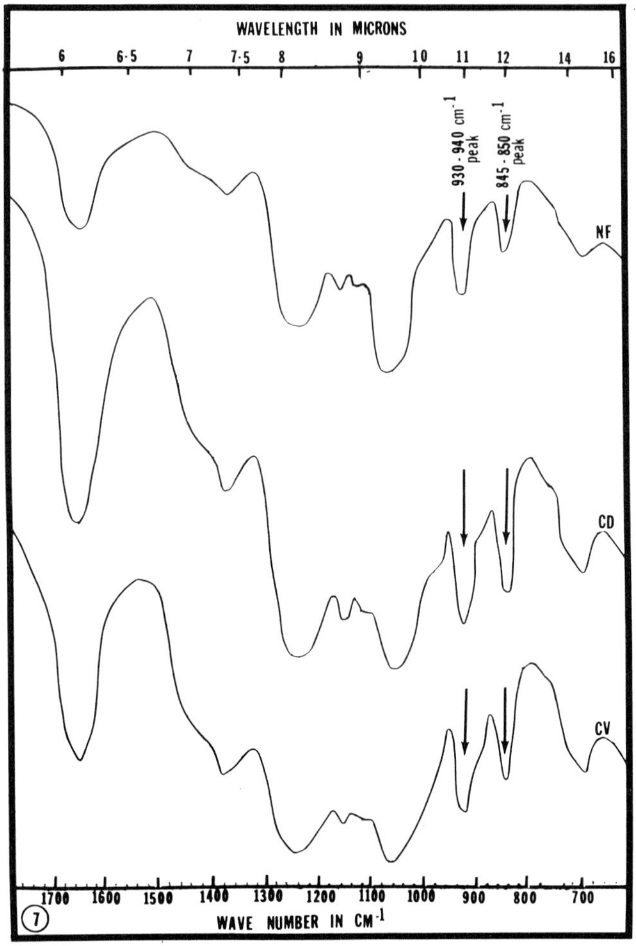

Fig. 7. Infrared spectral curves of phycocolloids from *Hypnea cervicornis* (CV), *H. chordacea* (CD) and *H. nidifica* (NF). Note the sharp absorption peaks (arrows).

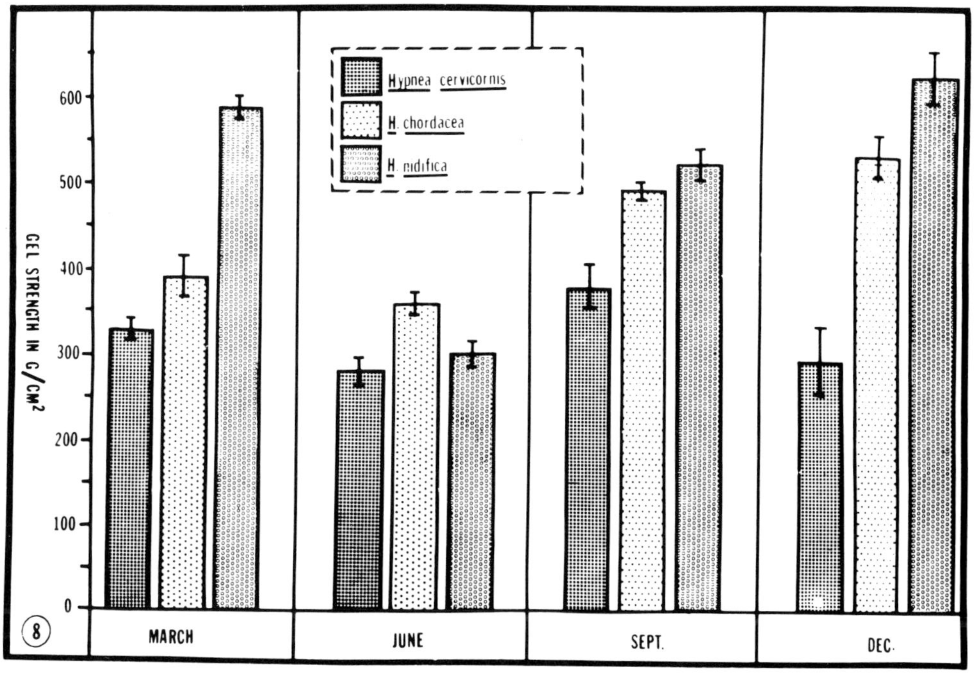

Fig. 8. Gel strengths of phycocolloids from *Hypnea cervicornis*, *H. chordacea* and *H. nidifica* extracted in different seasons.

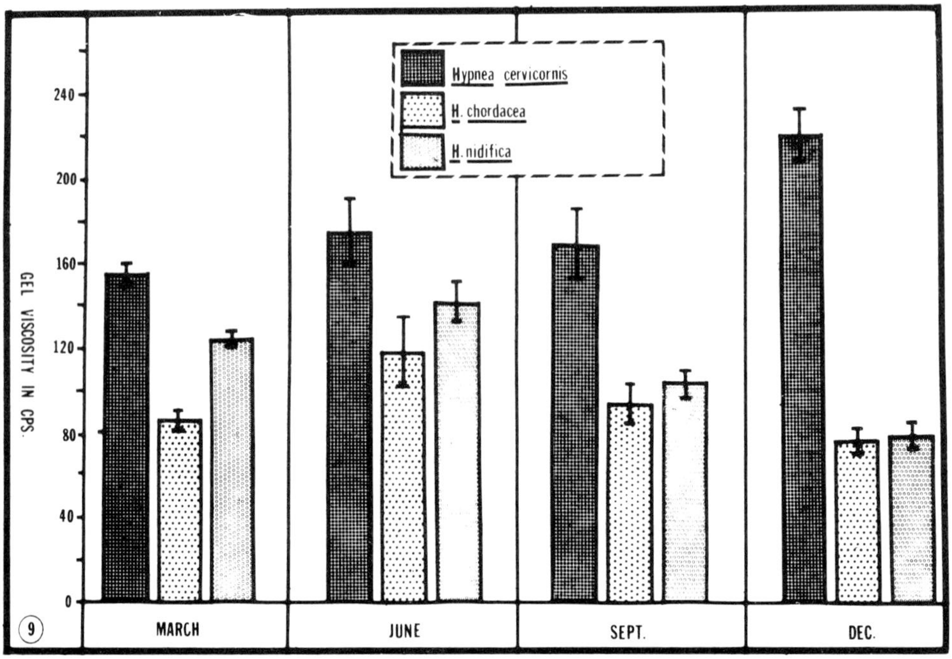

Fig. 9. Gel viscosities of phycocolloids from *Hypnea cervicornis*, *H. chordacea* and *H. nidifica* extracted in different seasons.

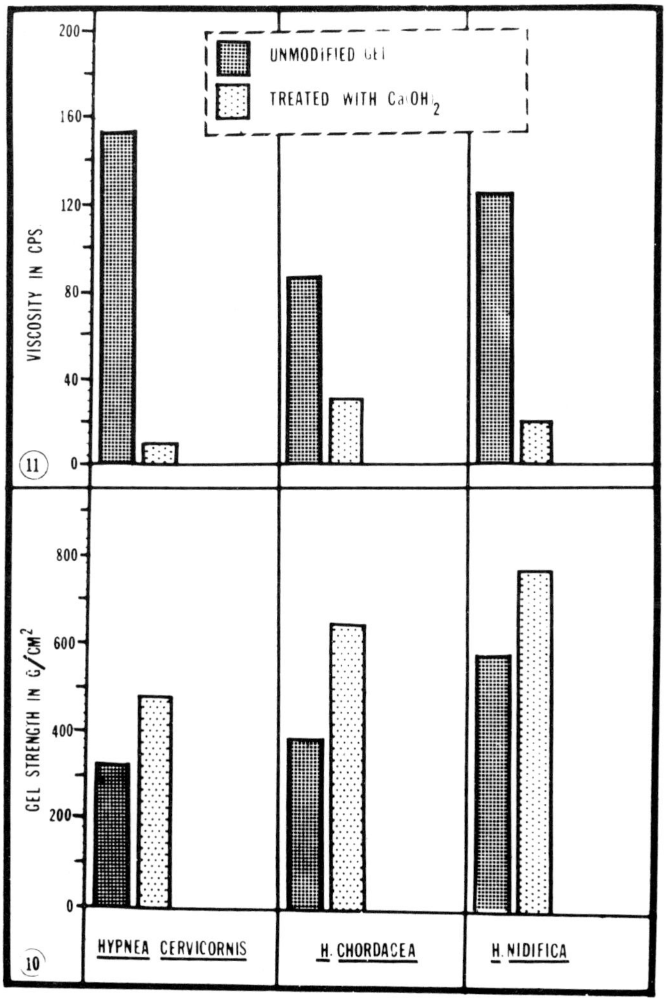

Fig. 10. Effects of lime on the strength of gels from Hypnea cervicornis, H. chordacea and H. nidifica.

Fig. 11. Effects of lime on the viscosity of the phycocolloids from Hypnea cervicornis, H. chordacea and H. nidifica.

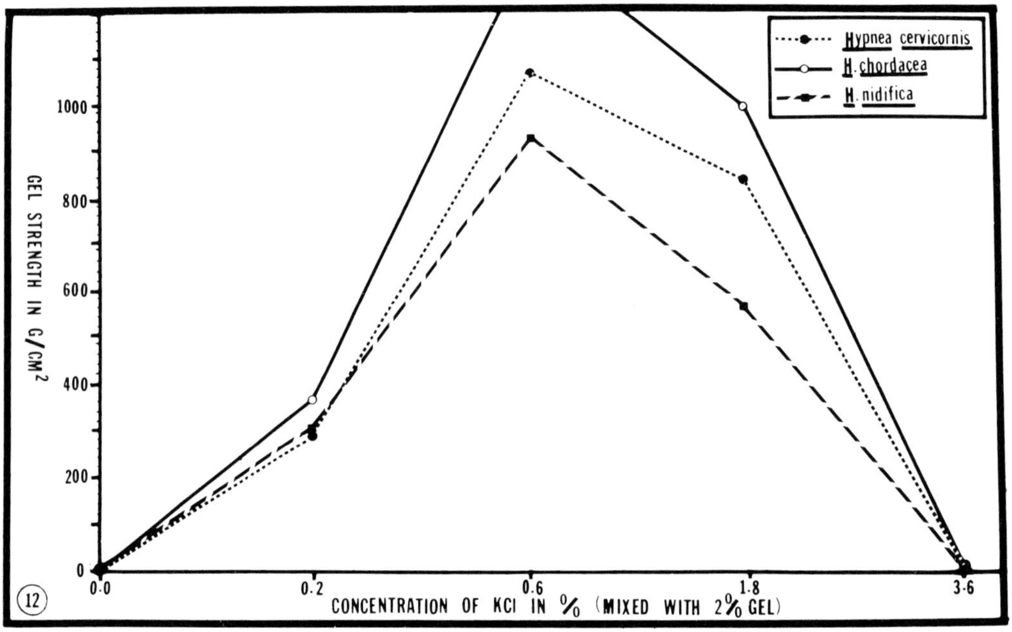

Fig. 12. Effects of potassium chloride concentration on gel strength in Hypnea cervicornis, H. chordacea and H. nidifica. Note with 0.6% KCl the strength of the gel from H. chordacea exceeded the capacity of the gel tester used.

REFERENCES

Black, W.A.P., W.R. Blackmore, J.A. Colquohoun and E.T. Dawar: The evaluation of some marine algae as a source of carrageenan and its kappa and lambda components. J. Sci. Fd. Agric. 16, 543-585 (1965).

Chapman, V.J.: Seaweeds and their uses. Methuen, London, 304 pp (1970).

Clingman, A. and J.R. Nunn: Red seaweed polysaccharides. Part III. Polysaccharide from Hypnea spicifera. J. Chem. Soc. London, Febr. 1959, 493-498 (1959).

Cooper, N.C. and G.R. Johnstone: The seasonal production of agar in Gelidium cartilagineum, a perennial red alga. Amer. J. Bot. 31, 638-640 (1944).

Dawes, C.J., J.M. Lawrence, D.P. Cheney and A.C. Mathieson: mss. Ecological studies of Florida Eucheuma .... Seasonal variation of carrageenan, total carbohydrate, protein and lipid. 30 typescript pages (1973).

Desikachary, T.V.: Seaweed resources in India. In Krishnamurthy, V. (ed.), Proc. Seminar on Sea, Salt and Plants, Bhavnagar, India, 7-24 (1965).

Ferguson, W.E.J.: Agar in Australia. Counc. Sci. Industr. Res. Bull. 203, 1-43 (1946).

Fox, F.W. and E. Stephens: Agar from South African Seaweeds. S. Afr. J. Sci. 39, 147-149 (1943).

Hamilton, R.D. and J.J. Carroll: A comparison of polysaccharides of Hypnea musciformis and Chondrus crispus. Nature 196, 1201 (1962).

Hong, K.C., L.R. Cruess and S.C. Skoryna: The effect of Hypnea japonica extracts on bone metabolism in the rat. Proc. Intl. Seaweed Symp. 7, 566-568 (1972).

Humm, H.J.: Seaweeds at Beaufort, North Carolina, as a source of agar. Science 96, 230-231 (1942).

Humm, H.J.: Agar resources of the South Atlantic and East Gulf coasts. Science 100, 209-212 (1944).

Humm, H.J.: The red algae of economic importance: agar and related phycocolloids. In Tressler, D.K. and J.M. Lemon (ed.), Marine products of commerce. Reinbold, N.Y., 47-93 (1951).

Humm, H.J. and L.G. Williams: A study of agar from two Brazilian seaweeds. Amer. J. Bot. 35, 287-292 (1948).

Levring, T., H.A. Hoppe and O.J. Schmid: Marine algae. A survey of research and utilization. Cram, de Gruyter and Co., Hamburg. 421 pp (1969).

Mshigeni, K.E.: The biology and ecology of selected Hawaiian species in Hypneaceae (Rhodophyta, Gigartinales). Ph.D. dissertation, University of Hawaii, 739 pp (1974).

Mshigeni, K.E.: Studies on the reproduction of selected species of Hypnea (Rhodophyta, Gigartinales) from Hawaii. Bot. Mar. 19, 341-346 (1976).

Mshigeni, K.E.: Seasonal changes in the standing crops of three Hypnea species (Rhodophyta, Gigartinales) in Hawaii. Bot. Mar. 20, 303-306 (1977).

Neish, A.C. and P.F. Shacklock: mss. Greenhouse experiments on the propagation of strain T_4 of Irish moss. Technical Report 14 of the Atlantic Regional Laboratory, Nat. Res. Council of Canada, Halifax, N.S., Canada. 25 typescript pages (1971).

Rama Rao, K.R.: Study on the growth cycle and phycocolloid content in Hypnea musciformis (Wulfen) Lamouroux. Bot. Mar. 13, 163-165 (1970).

Rama Rao, K.R. and V. Krishnamurthy: Study of the preparation and the properties of the phycocolloid from Hypnea musciformis from Veraval, Gujarat coast. Bot. Mar. 11, 129-133 (1968).

Reed, M.S.: The economic seaweeds of Hawaii and their food value. Ann. Rept. Agr. Expt. Station, 1906, 61-88 (1906).

Semesi, A.K. and K.E. Mshigeni: Studies on the yield and infrared spectra of phycocolloids from Chondrococcus hornemannii (Lyngbye) Schmitz and Sarconema filiforme (Sonder) Kylin from Tanzania. Bot. Mar. 20, 271-275 (1977).

Stancioff, D.J. and N.F. Stanley: Infrared and chemical studies on algal polysaccharides. Proc. Intl. seaweed Symp. 6, 595-609 (1969).

Stoloff, L.: Algal classification : an aid to improved industrial utilization. Econ. Bot. 16(1), 86-94 (1962).

Stoloff, L. and P.C. Silva: An attempt to determine possible taxonomic significance of the properties of water-extractable polysaccharides in red algae. Econ. Bot. 11(4), 327-330 (1957).

Yaphe, W.: The use of agarase from Pseudomonas atlantica in the identification of agar of marine algae (Rhodophyta). Can. J. Microbiol. 3, 987-993 (1957).

Yaphe, W.: The determination of kappa carrageenan as a factor in the classification of the Rhodophyceae. Can. J. Bot. 37, 751-757 (1959).

Yaphe, W. and B. Baxter: The enzymic hydrolysis of carrageenan. Appl. Microbiol. 3, 380-383 (1955).

Zaneveld, J.S.: The utilization of marine algae in tropical South and East Asia. Econ. Bot. 13(2), 89-131 (1959).

Studies on the littoral ecology and ecophysiology of the carrageenophytes, *Hypnea musciformis* (Wulfen) Lamouroux and *H. valentiae* (Turner) Montagne in Tanzania

K.E. Mshigeni and W.R. Mziray
Department of Botany, University of Dar es Salaam,
P.O. Box 35060, Dar es Salaam, Tanzania.

ABSTRACT

Investigations on the littoral ecology and ecophysiology of *Hypnea musciformis* and *H. valentiae* (Rhodophyta) from Oyster Bay, Dar es Salaam, Tanzania, have been conducted. *Hypnea valentiae* was found to have a wider range of distribution in the intertidal belt than *H. musciformis*. This species also performed better and grew faster than *H. musciformis* in the temperature-, salinity-, or illuminance tolerance experiments. Both species, however, showed inability to withstand conditions of desiccation even for a duration of one hour. Standing crop seasonality of these species was also investigated. Both showed maximum values of standing crops during August to October and minimal values during December to February and May and June. The coolest months of the year (May to July) also coincided with the season with the lowest frequency of mid-day extreme low-tide emersion. The implications of these findings on the ecology and agronomy of the seaweeds are discussed.

INTRODUCTION

Recently it has been realised that species of *Hypnea* Lamouroux (Rhodophyta, Gigartinales) contain relatively large quantities of carrageenan, a phycocolloid which has many applications in pharmaceutical and other industries owing to its desirable gelling properties (Yaphe and Baxter, 1955; Yaphe, 1959). This realisation has stimulated interest in research on many

Marine Algae in Pharmaceutical Science
edited by H. A. Hoppe, T. Levring, Y. Tanaka

aspects of the biology of the genus. These include studies on spore germination and early development of representative species (Mshigeni, 1976a; 1976b), studies on its reproductive phenomena (Isaac and Hewitt, 1953; Mshigeni, 1976c) and field cultivation (Mshigeni, 1976d). In these studies it has been revealed that in the tropics, species of *Hypnea* so far investigated reproduce throughout the year and that the tetrasporic phase is, by far, the dominant one. It has further been shown that the spores of the species investigated so far show the mediate discal pattern of spore germination (Inoh, 1947). Under favourable environmental conditions in the tropics, the sporelings can develop into adult fronds in only three months (Mshigeni, 1976d).

Most investigations on *Hypnea* have, however, been of floristic nature only. Nevertheless, such reports are helpful in attempting to depict the global distribution pattern of 30 or so species of the genus. In the Indian Ocean such reports include the works of Schmitz (1895), Weber-van Bosse (1928), Kylin (1938), Boergesen (1943, 1954), Isaac and Hewitt (1953), Dawson (1954), Zaneveld (1955), Durairatnam (1961), Isaac (1967) and Jaasund (1976). In the Atlantic Ocean the works of Taylor (1928, 1960), Lawson (1957), Chapman (1963), Conover (1964) and Bodard (1968), are worth of mention. In the Pacific floristic accounts by Chamberlain (1880), Okamura (1907), Setchell and Gardner (1924), Setchell (1926), Kylin (1941), Tanaka (1941), Taylor (1945) and Trono (1968) are amongst the ones documented in the literature. Thus the genus *Hypnea* shows cosmopolitan distribution on tropical and subtropical shores.

Reports pertinent to the ecology of *Hypnea* species have also been published. A few of these have been focused on population changes in different months of the year. Lawson (1957), for example, studied *H. musciformis* populations on the shores of Ghana and found that there were significant seasonal variations in the populations of the seaweed. The variations showed correlation with the behaviour of the tides. Rama Rao (1972) conducted similar studies in India and likewise concluded that there were significant seasonal variations in *H. musciformis* populations and that these were largely controlled by tidal behaviour. However, these conclusions were not substantiated by any experimental data to demonstrate the performance of the species in question under gradients of desiccation, temperature, salinity, etc., approximating those experienced by the algae during tide-induced emersion in the field.

In a recent study on *Hypnea* populations in Hawaii, Mshigeni (1977a) also revealed that intertidal populations of *H. cervicornis*, *H. chordacea* and *H. nidifica* similarly showed seasonal changes in abundance and that these variations were explainable in terms of seasonal changes in tide behaviour. The populations of these species were most abundant and also extended highest up the shore during the season with lowest frequency of mid-day extreme low tides and minimal quantities during the season when mid-day tide-induced emersion was of highest frequency (Mshigeni, 1977a).

It is known that light, temperature, salinity and desiccation largely control intertidal seaweed populations. With respect to the species of *Hypnea*, however, these have merely been suggested as possible controlling factors. Very few actual experiments (e.g., Biebl, 1962) have been conducted to demonstrate how the various species in the genus perform under gradients of these parameters. Mshigeni (1976e, 1976f), however, conducted experiments to test the growth response of *H. cervicornis* and *H. chordacea* to light-, salinity- and temperature gradients on the shores of Hawaii. In this study it was revealed that *H. cervicornis* which extends higher up the shore was more tolerant to all the tested parameters than *H. chordacea* whose littoral distribution extended lower down the intertidal belt.

One of the objectives of the present study was, therefore, to conduct comparative studies of similar nature on two other species of Hypnea which have, hitherto, never been investigated along similar lines. Investigations of this nature are likely to yield fruitful information which can easily find application in the increased commercial production of species of Hypnea for carrageenan industry. The two selected species are *Hypnea musciformis* (Wulfen) Lamouroux and *H. valentiae* (Turner) Montagne. These were selected on the basis of their relative abundance on the shores of Tanzania (Jaasund, 1976) and their larger size in comparison with other species. They are thus the most attractive *Hypnea* species which could be potential sources of carrageenan in Tanzania.

In the execution of seaweed growth response experiments, one can use either fragments of adult plants as experimental material (e.g., Whitford and Kim, 1966; Humm and Kreuzer, 1975; Dawes, 1974), or young stages of the seaweeds in question. In the present study the experimental work was conducted using germlings raised from carpospores. Details on the techniques employed in the capturing of the spores are as published in earlier publications by the senior author (Mshigeni, 1976a; Mshigeni and Lorri, 1977a, 1977b). Young stages of seaweeds are generally believed to be more sensitive to environmental parameters such as salinity, illuminance, desiccation and temperature than adult plants (Mshigeni, 1974). It is thus clearly desirable to know the physiological tolerance of the young stages of economic seaweeds since information of this nature is of primary importance in revealing factors controlling their viability and growth rates. Furthermore, such knowledge is useful in explaining certain field observations on the seasonal, geographical and littoral distribution patterns of seaweeds in general.

MATERIALS AND METHODS

FIELD STUDIES

Selection of Study Site

Prior to the conduction of the field studies, many shores in the vicinity of Dar es Salaam, Tanzania, were surveyed and *Hypnea*- rich habitats were located. Densest population stands were found at Oyster Bay. This was therefore selected as the study site.

The species richly occurring at Oyster Bay are *Hypnea valentiae*, *H. musciformis* and *H. pannosa*. *H. cornuta* and *H. spinella* are also represented but these are more sparsely distributed. Other associated benthic algae include several species of *Ulva*, *Enteromorpha*, *Caulerpa* and *Halimeda* in the Chlorophyta; several species of *Turbinaria*, *Cystoseira*, *Dictyota* and *Sargassum* in the Phaeophyta and several species of *Gracilaria*, *Laurencia*, *Spyridia* and *Acanthophora* (Rhodophyta).

Identification of the *Hypnea* Species

The species of *Hypnea* collected were identified using the works of Boergesen (1943, 1954), Tanaka (1941), Taylor (1960), Dawson (1954, 1961), Weber-van Bosse (1928) and Jaasund (1976). Herbarium specimens of *Hypnea* collected from various parts of the world and deposited in the seaweed herbaria at Uppsala and Lund (Sweden) were also examined by the senior author to confirm the identifications. The two species *H. musciformis* and *H. valentiae* are morphologically distinct from each other. In the first place, the former is characterised by the possession of swollen curved hook-like structures at the tips of the main axes. *H. valentiae* fronds lack such structures. Secondly, fronds of *H. musciformis* are densely covered with short lateral branchlets whereas those of *H. valentiae* are much less densely branched (its branchlets being more lax and much longer). Thirdly, adult *H. musciformis* fronds are generally shorter (below 15 cm) than those of *H. valentiae* (mostly above 15 cm). Finally, *H. musciformis* fronds primarily show epiphytic growth habit whereas those of *H. valentiae* are much less commonly found as epiphytes. Photographic illustrations of these seaweeds (Fig. 2, 3) have, therefore, been included in the report to illustrate some of these differences.

It should, however, be borne in mind that there are taxonomic and nomenclatural confusions on the taxon referred to in the literature as *Hypnea musciformis*. Kuetzing (1868), Kylin (1938) and Tanaka (1941), for example, illustrated other species of *Hypnea* which also bear hook-like structures on the ends of the thallus axes (e.g., *H. ecklonii*, *H. episcopalis*, *H. alopecuroides*, *H. intricata*, *H. tenuis*, *H. arenaria* and *H. japonica*). Whether these are possibly mere morphological variants of *H. musciformis* is yet to be investigated.

The name *Hypnea valentiae* has also been reported to embrace several species such as *H. nidifica*, *H. cornuta*, *H. hamulosa* etc. (Hauck, 1887; Dawson, 1961). In other opinions (e.g., Jaasund, 1976), each of these species is a distinct taxon. Thus there is clearly a need for a monograph to review the status of the genus. Suffice to mention here is that the information given above is hoped to enable the reader to understand what species the authors are talking about.

Littoral Distribution Studies

The littoral distribution pattern of the selected species was investigated using quantitative sampling procedures along a 160 m transect stretching from the botton of the cliffs on the shore to about the spring-tide low water mark. A 0.4 x 0.4 m wooden quadrat was used throughout.

The sampling quadrat was placed on the ground starting from

0, 10, 20, 30, 40.........to 160 m seaward from the botton of the cliffs. Where the quadrat fell all the Hypnea musciformis and H. valentiae fronds were hand-picked. These were placed in an appropriately numbered collecting bag. All the bags were subsequently transported to the laboratory and contents carefully sorted out, bag after bag, to remove adhering rock debris and other contaminants. The Hypnea plants from each bag were then weighed fresh, dried in an oven at 60°C and when a constant weight was reached, weighed again to determine the dry weight. From the weights obtained standing crop values in g/m^2 were calculated by dividing the weight values by 0.16, the area in m^2 covered in each quadrat sample. The data showing standing crop variation as a function of the distance seaward from the shore for the month of July, 1976 are summarised in Fig. 4.

Standing Crop Seasonality Studies

In an attempt to reveal whether the standing crops of the two Hypnea species vary seasonally or not, the procedure of sampling was executed at monthly intervals and continued for a period of slightly over one year. The data presented in this report cover the period May 1976 through June 1977 (a total of 14 months).

For each single month, all the values for each sampling zone, 0, 10, 20, 30 m etc. seaward from the shore were lumped together and then an overall average monthly value was calculated. The values (each representing an average of 17 samples) obtained were then expressed in terms of mean standing crop as function of the month of the year, as summarised graphically in Fig. 5.

Data on the Dar es Salaam Tides

Since tidal factors have been shown to exert significant influences on Hypnea populations elsewhere, the data on the tides of Dar es Salaam (which are published by the East

African Harbours Corporation) were analysed. It is known that extreme low tides occurring around mid-day are most lethal to tropical intertidal seaweeds (Mshigeni, 1976f). Hence the number of days with mid-day (1000 to 1500 hours) extreme low-tides (0.0 m and below) were counted from predicted tide tables. This was done for each month during the 14-months period. The data so obtained are summarised in Fig. 6. It was hoped that these data would be useful in attempting to explain the observed phenomena on various aspects of the biology of these species.

LABORATORY STUDIES

Cultural Techniques

The plants used in the laboratory experiments were 10-week old laboratory-raised carpospore germlings of *Hypnea valentiae* (mean size 2.6 mm ± 1.6) and *H. musciformis* (mean size 1.6 ± 0.5 mm). These had been raised in a chamber supplied with a 12L : 12D photoregime from fluorescent cool-white light tubes, an illuminance of 1000 lux, and a temperature of 28 ± 2°C. The cultural medium used was millipore-filtered seawater enriched with nutrients as described by Schreiber (1927). The millipore filters (0.45 µm pore diameter) served to remove contaminants such as spores of other benthic algae or phytoplankton such as diatoms. The media were changed every three days.

Light Response Experiments

Three levels of light each obtained by differential screening were tested, viz., 500, 1500 and 3000 lux. These were supplied from cool-white fluorescent light tubes.

Twenty sporelings of each species were carefully transferred from the stock population of the sporelings into each of three replicates of pyrex glass culture vessels. Each vessel contained 300 ml of the culture medium. Three replicates of culture vessels were placed under each light regime tested for each species. An automatic lighting system was set up using a time switch to make the photoperiod correspond to that in the field during the research period (12L : 12D).

The culture vessels were covered with an appropriate lid to minimise evaporation of water and hence the concentration of the culture medium (while at the same time allowing a small opening for aeration). The plants were then left under the described light conditions for three weeks. After this period, their sizes were measured. This was done either directly against a squared graph paper for the larger-sized H. valentiae sporelings, or by a skiagraphic technique in a dark room (Mshigeni, 1977b) for the smaller-sized H. musciformis sporelings. Mean sporeling height for each species and its confidence limits (P=0.05) was then determined statistically for each illuminance level tested (Snedecor and Cochran, 1967).

Salinity Response Experiments

In this study 11 salinity levels were tested, viz. 5, 10, 15, 25, 35, 45, 55, 65, 70, 75, 80‰ S. Higher salinities (above 35‰) were prepared by adding calculated amounts of dry salt (obtained by evaporating seawater) to normal seawater and stirring till the salt dissolved. Salinities below 35‰ were obtained through dilutions with distilled water.

Aliquots of each salinity level (250 ml) were transferred into two growth containers. The groups of 20 sporelings of each species were placed in each container for all three replicates of the salinity levels. The sporelings were left under each salinity treatment for a period of 24 h. After this period, the seawater from each container was carefully decanted and the sporelings rinsed with the normal growth medium. Each container was subsequently supplied with 300 ml of the normal growth medium at 35‰, and subjected to normal cultural conditions again.

Since a given seaweed species can tolerate only a certain specific range of salinity, it would be expected that the treatment of the sporelings under excessively high or excessively low salinities would be lethal (Biebl, 1962). Therefore the sporelings which would be killed by the salinity treatments were expected to completely succumb after a few additional days of laboratory culture. Measurements to determine the percentages of sporelings which had survived the salinity treatments were thus made three weeks after the salinity treatments.

Temperature Tolerance Experiments

Four temperature levels were tested in this experiment: $20^{o}C$, $28^{o}C$ (room temperature), $38^{o}C$ and $47^{o}C$. Apart from the $28^{o}C$ temperature level all the other levels were obtained in calibrated Gallenkamp incubators.

For each species, samples of 20 germlings were transferred into 20 test vessels each containing 300 ml of the culture medium. For each temperature tested, the duration of the temperature treatment was 1, 2, 4, 6 and 10 h. before returning the plants to room temperature. The room-temperature treatment was left to continue under the room conditions. The culture medium was changed every three days.

After two weeks, the growth response of the sporelings was assessed. This was done by counting the number of sporelings still alive and expressing the results as a percentage of the total. The data were then summarized graphically.

Desiccation Tolerance Experiments

Fourteen test containers were obtained. To each of these 10 ml of the cultural medium was added. Groups of 20 sporelings of *H. musciformis* were transferred to each of seven of the test vessels. The same procedure was followed for *H. valentiae* sporelings. 300 ml of the growth medium was then added to one container of each species to serve as the controls. All the growth medium from the remaining vessels was quickly pipetted out. The lot was subsequently taken up to the roof of the 3-storey Botany Department building where the sporelings were exposed to the air for a period of 1, 2, 3, 4, 5 and 6 h. After each time interval, a representative container for each species was supplied with 300 ml of the culture medium and returned to the room conditions. The control samples were left submerged in the growth medium in the open air for all the six hours. After this period the medium was changed for a fresh one and then returned to room conditions. At each interval, atmospheric conditions of temperature and relative humidity of the area where the experiment was conducted, were measured using a suitable thermometer and hygrometer.

Since the lethal effects of desiccation were more noticeable and more drastic than those of the other factors earlier tested, the growth responses were measured after only one week following the date of desiccation. As in the previous experiments, the growth response was measured in terms of the percentage of the sporelings that survived the desiccation treatment.

RESULTS

FIELD STUDIES

Description of the Study Area

The profile of the area where the field study was conducted is shown in Fig. 1. It is seen that the area just beneath the cliffs slopes gradually to an intertidal basin at about 40-60 m seaward from the cliff, which contains a permanent pool of water with mean depth of 20 to 40 cm during the spring low tides. Beyond the basin there is a rocky outcrop which is subjected to low-tide emersion. This gently slopes towards the infralittoral fringe.

The infralittoral fringe portion of the shore experiences a moderate degree of wave action. The force and the direction of the currents in East Africa varies seasonally, depending upon whether the North-east or the South-east monsooon is prevailing (Isaac, 1968). The hottest season of the year (December to February) is dominated by the north-east monsoon whereas the coolest season (May to July) is dominated by the south-east monsoon.

Littoral Distribution

The distribution of *Hypnea valentiae* and *H. musciformis* as a function of distance seaward from the shore is summarised in Fig. 4. From these results it is clear that the littoral distribution of *H. valentiae* extends higher up the shore than *H. musciformis*. The latter species is more abundantly represented along the infralittoral fringe.

Standing Crop Seasonality

The data on the mean standing crops (g/m^2) of *Hypnea musciformis* and *H. valentiae* in different months of the year are summarised in Fig. 5. It was found that in both species the highest values of standing crops occurred during the months of August to October, and the minimal values occurred during December to February and May and June.

Data on the Tides

The results of this study show that the highest frequency of extreme mid-day low tides occurs during the months of December to February, and lowest frequency is observed during May to August (Fig. 6).

LABORATORY EXPERIMENTS

Light Responses

The morphological appearance of the sporelings of *Hypnea valentiae* and *H. musciformis* at the end of the light treatment period is shown in skiagraphs presented in Figs. 7 to 12. A graphic presentation of their measurements is shown in Fig. 13.

It was found that the sporelings of *Hypnea valentiae* grew significantly faster (P=0.05) than those of *H. musciformis* under all the light levels tested (Fig. 13). In the case of *H. valentiae* the growth rate of the sporelings at 3000 lux was significantly faster than that of the other two levels tested. For both species growth was significantly slower at 500 lux than in the upper illuminance levels tested.

Salinity Responses

The results of this study are summarised in Fig. 14. It was found that *Hypnea musciformis* sporelings could tolerate a salinity range of 25‰ to 75‰ S. They all died at 5 to 15‰ and 80‰. For *H. valentiae* the upper limit salinity tolerance was not obvious from this experiment since they tolerated even the highest salinity level tested (i.e., 80‰S). Like *H. musciformis*, however, the sporelings could not, in general terms, tolerate salinities below 25‰ S. The recorded 20% survival of *H. valentiae* sporelings at 10‰S merits a brief comment. Although the sporelings in question were still alive, they were obviously in the process of disintegration as most of the thalli had lost their characteristic red pigmentation.

Temperature Tolerance Responses

The results of this study are summarised in Fig. 15. It was found that the sporelings of *Hypnea valentiae* were able to tolerate a temperature of 20°C for a duration of up to 10h whereas those of *H. musciformis* failed to survive at this temperature even at the 1 h duration of exposure. Surprisingly, the latter species showed a poor survival rate (50%) at the 1 h duration exposure. The response of the exposure treatment appeared to be critical at the temperature of 38°C.

The sporelings of *Hypnea valentiae* showed full survival for durations of 1, 2, 4 and 6 h exposure to 38°C temperature but completely succumbed after 10 h exposure at this temperature. Sporelings of *H. musciformis*, in contrast, succumbed even at the 6 h duration. At 2 h exposure, only 5% of the sporelings survived while they fully tolerated the same temperature for 4 h. At 47°C, sporelings of *H. musciformis* could not survive even for 1 h treatment. Those of *H. valentiae* were able to tolerate this temperature for only 1 h. Beyond this all the sporelings died.

Desiccation Tolerance

The results of the desiccation tolerance experiment (Table 1) showed that both *Hypnea musciformis* and *H. valentiae* sporelings succumbed under all the levels of desiccation tested (i.e., 1, 2, 3, 4, 5 and 6 h- exposure to the air). The controls, however (i.e., never desiccated), showed 100% survival.

DISCUSSION

As shown in Fig. 4, *Hypnea valentiae* has a wider range of littoral distribution at Oyster Bay than *H. musciformis*. From the observation that this species extends higher up the shore (where the habitat experiences prolonged emersion at low tide) than *H. musciformis*, it would be expected that *H. valentiae* would show better performance under conditions of high salinity, high temperature and prolonged desiccation, etc. than *H. musciformis*. In general terms this postulation is in agreement with the results of the laboratory experiments.

The results of the salinity experiments (Fig. 14) show that in both species the lower limit of tolerance is around 25‰ S.

Although Hypnea valentiae sporelings showed 20% survival at 10‰ salinity, the survivors consisted of individuals which were essentially progressively dying. Indeed, when they were examined several days later, they had all died. Thus the two species of Hypnea cannot tolerate estuarine habitats in the immediate vicinity of river mouths where salinity values between 0 and 10‰ occur.

With respect to the responses of the species at high salinities, Hypnea musciformis exhibited its upper limit tolerance at about 75‰ S. H. valentiae, in contrast, showed possibilities of tolerating much higher salinities since, even at 80‰, 85% of the sporelings were still alive and healthy-looking.

The results of this study closely agree with those reported by Conover (1964) who showed the lower limit salinity tolerance of Hypnea musciformis from Texas lagoons to be 28‰ S. The upper limit of 75‰ S observed in this study for H. musciformis from Tanzania, however, far surpasses Conover's (1964) results of 33‰ S. These differences may be attributed to genetic differences, to habitat differences or to differences in the experimental procedures. Near the Equator, (e.g., Dar es Salaam) salinities in the tide pools where this species often grows, are generally high during the periods of low-tide emersion due to the much higher temperatures which accelerate the rate of water evaporation from the pools. Thus in time, tropical populations of H. musciformis may have tended to be well adapted to the prevailing higher salinities, thus extending the salinity tolerance range as compared to populations on sub-tropical shores (e.g., Texas lagoons).

Other findings reported in the literature on Hypnea musciformis do support the idea that geographical differences may influence tolerance limits of benthic algae. Biebl (1962), for example, showed that mangrove populations of H. musciformis from Puerto Rico tolerated a salinity value of 40% seawater (approximately

14‰ S). Furthermore, Mshigeni (1976e), showed that sporelings of intertidal species H. cervicornis and H. chordacea from Hawaii, were capable of growing at salinity levels as low as 15‰ S for up to 20 days. It thus appears that the geographical habitat within which an algal species is growing, is an important factor in determining its range of salinity tolerance and, very likely, its tolerance to other environmental parameters. In the same line of argument, it has been shown that certain genera of marine algae may have one or more representatives (e.g., Bangia fuscopurpurea) which can be transferred from the marine habitat to fresh water via a number of spore generations (Hartog, 1972).

The results of the temperature tolerance experiments reveal some points which are noteworthy. Firstly, the existence of an upper limit temperature tolerance (Fig. 15) beyond which the sporelings of both species could not survive, and secondly the fact that the duration of the subjection of the sporelings to the more extreme levels of temperature was critical to survival. Thus 75% of the sporelings of Hypnea valentiae could, at least, survive a one-hour duration exposure to 47°C while those of H. musciformis could not. Furthermore, although H. valentiae sporelings could tolerate 1 h exposure to 47°C, they totally failed to tolerate 10 h exposure to 38°C. In this connection, Biebl (1972) suggested that the duration of heating was an important question when studying heat resistance. He reported, for instance, that Ceramium pacificum could survive 5- minute heating up to 36°C, 30 minute heating to 34°C, but could not withstand 12 h heating only up to 24°C.

Generally sporelings of Hypnea valentiae were more tolerant to a wider range of temperature than those of H. musciformis (Fig. 15). Sporelings of the former species fully tolerated 20°C for as long as 10 h while those of the latter succumbed at this treatment. The response of H. musciformis to 20°C temperature treatment is of interest since Biebl (1962) had reported a lower limit tolerance of 9°C for mangrove populations of this species from Puerto Rico. The big difference between Biebl's and the present results could possibly stem from genetic-, habitat- or geographical differences. Furthermore, the differences could be explained along the argument that adult plants (which were used in Biebl's experiments) are generally more tolerant to environmental parameters than sporelings.

A few anomalies should also be noted from the present results. The survival of only 5% at 2 h exposure to 38°C in the case of *H. musciformis* could be due to factors other than the temperature treatment (Fig. 15). This is because similar sporelings had tolerated a temperature of 38°C for a duration of 4 h.

With respect to the desiccation response, the results of this study (Table 1) showed clearly that the sporelings of both *Hypnea valentiae* and *H. musciformis* stand no chance of surviving when completely emersed, even for an hour, while resting on a vessel that approximates a bare rock substratum. The controls which remained permanently immersed, in contrast, showed 100% survival. One would therefore postulate that sporelings of the two species of *Hypnea* which could, by chance, start their development on a bare intertidal rock substratum in the field, are likely to be killed by relatively short periods of emersion during low tides. Fortunately, however, observations in the field in fact show that under most conditions, the sporelings of benthic algae will usually be shaded by larger (mostly adult) plants. Furthermore, they experience periodic wave-splash which wets the emersed algae, hence supplying them with a thin film of moisture that keeps replenishing evaporated water. In addition, even the relative humidity within the intertidal belt on the shore is certainly much higher than the recorded 57 to 64% relative humidity at the roof of the Botany Department building, University of Dar es Salaam on the day the desiccation experiment was conducted. Thus the rates of thallus dehydration in the field would be expected to be slower than those under experimental conditions described above. As such it would appear that the effects of desiccation in the field condition would be somewhat milder than was the case in the present investigation.

The results of this study therefore suggest that the nature of microenvironments at which seaweeds thrive in the field need closer scrutiny than the literature reports. Further research on desiccation is called for since this appears to be one of the most neglected topics on the subject.

In the case of the light response experiments, it should be noted that the illuminance levels tested are much lower than the prevailing conditions in the field. A scrutiny of the results obtained in these experiments indeed shows (Figs. 7-13) that there was a general tendency for an increased rate of growth from 500 lux to 3000 lux. One would thus expect the optimum levels of light to be higher than the maximum level tested. Nevertheless, it should also be noted that *Hypnea musciformis* was shown by Whitford and Kim (1966) to be indifferent to light intensity and performed equally well under dim (300 ft-c) and bright (3,000 to 5,000 ft-c) light conditions.

Another finding of the present study worth emphasizing is the fact that the sporelings of *Hypnea valentiae* grew significantly faster than those of *H. musciformis* under all the light levels tested (Figs. 7-13). This, combined with the fact that it had a wider range of salinity- and temperature tolerance and that it attains a larger vegetative size (Fig. 2, 3) than *H. musciformis*, suggests that other things being equal, *H. valentiae* is a better candidate than the latter species when cultivating the seaweeds for animal feeds or for carrageenan production.

The finding that the maximum value of standing crops for both Hypnea musciformis and H. valentiae (Fig. 6) is during the period August to October is of interest. In view of the fact that the study site is in the tropics where direct insolation is known to be catastrophic to many intertidal seaweeds, one would expect to find the highest survival of intertidal species during the months June and July (the coolest season). This would include the survival of sporelings which are especially sensitive to factors such as high salinity, high temperature or dessication which prevail in intertidal habitats. However, since it may take three months or more (Mshigeni, 1976e) for freshly germinated spores of some frondose species to attain adult size, the effects of the favourable conditions for algal growth prevailing in June and July would be expected to be noticed on the standing crops of the species in question two or more months later. This probably helps to explain why the maximum values of standing crops (Fig. 6) were found in August to October.

In addition to the favourable cool conditions during May to July, this is (Fig. 5) also the season with minimal frequency of extreme ('zero' and 'minus') low tides occurring around mid-day (1000 and 1500 hours), when the effects of exposure to direct insolation are most lethal to intertidal seaweeds in the tropics (Mshigeni, 1976f). The low frequency of occurrence of the extreme mid-day low tides also means that a larger percentage of plants of the species in question would survive during the months June and July, and this could also possibly explain the observed standing crop maxima during August to October. Thus tidal factors also seem to be involved in determining the observed seasonal changes in the Tanzanian populations of the two Hypnea species in a way similar to the findings of Lawson (1957) on the shores of Ghana, Rama Rao (1972) in India and Mshigeni (1977a) in Hawaii.

With respect to the standing crop minima during December to

February, one could perhaps partly explain this finding from the point of view of the fact that December and January are the hottest months in Dar es Salaam, hence large proportions of intertidal populations of Hypnea are probably killed by tide-induced exposure to direct insolation during this season. In the case of the minima for the months March to May, the decline in the standing crops could perhaps be explained in terms of algal killing by low salinity exposure since this is the rainy season. Yet, another postulation is the possible involvement of strong monsoon currents in tearing off the algae in question from their substrata.

As aforementioned, the shores of East Africa experience strong north-east monsoons during December to February and south-east monsoons during May to July (Isaac, 1968). During these months large quantities of erect frondose algae are torn off from the substrata by the excessive force of monsoon currents and cast ashore as drift material. Indeed it is during this season that representatives of sublittoral genera such as Titanophora, Halymenia, Eucheuma, etc. are cast ashore in largest quantities. Thus excessive water movement may also be a factor contributing to the decline of Hypnea musciformis and H. valentiae populations during the months of December to February (north-east monsoon) and May to July (south-east monsoon).

The results of the present study thus seem to have important agronomic implications: In the first place, when raising crops of the seaweeds through mariculture, illuminance levels of 3000 lux and above are recommended. Secondly, care should be taken not to subject the sporelings to salinities below 25‰ S. Thus, although estuarine habitats may have rich nutrient levels brought down to the sea from terrestrial habitats, crops of the seaweeds should be raised sufficiently far away from river mouths to avoid killing by low salinity exposure. Thirdly the crops of the seaweeds should be raised in habitats which are not subjected to prolonged emersion at low tide. Thus the tidal- and microclimatic conditions of the farming site must be investigated first. Finally, from what has been discussed on the seasonal variations of the standing crops of the seaweeds, it would be recommended that the best season for raising commercial crops of Hypnea in Tanzania is during the months of July to October. Investigations of this nature are important for any seaweed of potential pharmaceutical significance.

ACKNOWLEDGEMENT

This work was made possible through a research grant to the senior author from the University of Dar es Salaam Research and Publications Committee and the International Foundation for Science (Stockholm, Sweden: Agreement No.G103). To these he wishes to extend his deepest gratitude. The technical assistance of M. Fundi, S. Abdallah, H. Suleiman, B. Mponda and Miss Minja, the kind assistance of Mrs Z. Kassam who typed the manuscript, and the warm encouragment from Professor Dr Tore Levring, are also very gratefully acknowledged. I also wish to thank Mr Göran Michanek for Providing some literature materials which were valuable in the discussions, and UNESCO for providing a travel grant which enabled me to present this paper in the Special Symposium: Marine Algae in Pharmaceutical Science during the IXth International Seaweed Symposium at Santa Barbara, California.

TABLES

Table 1. Percentage survival of sporelings of *Hypnea musciformis* and *H. valentiae* after exposure to gradients of desiccation.

DURATION OF EXPOSURE (HOURS)	TEMPERATURE AROUND SPORELINGS (°C)*	ATMOSPHERIC RELATIVE HUMIDITY (%)*	PERCENTAGE SURVIVAL	
			H. mus.	*H. val.*
0	34	64	100	100
1	32	60	0	0
2	35	57	0	0
3	36	57	0	0
4	36	58	0	0
5	35	62	0	0
6	34	64	0	0

* Measured at the end of the treatment period.

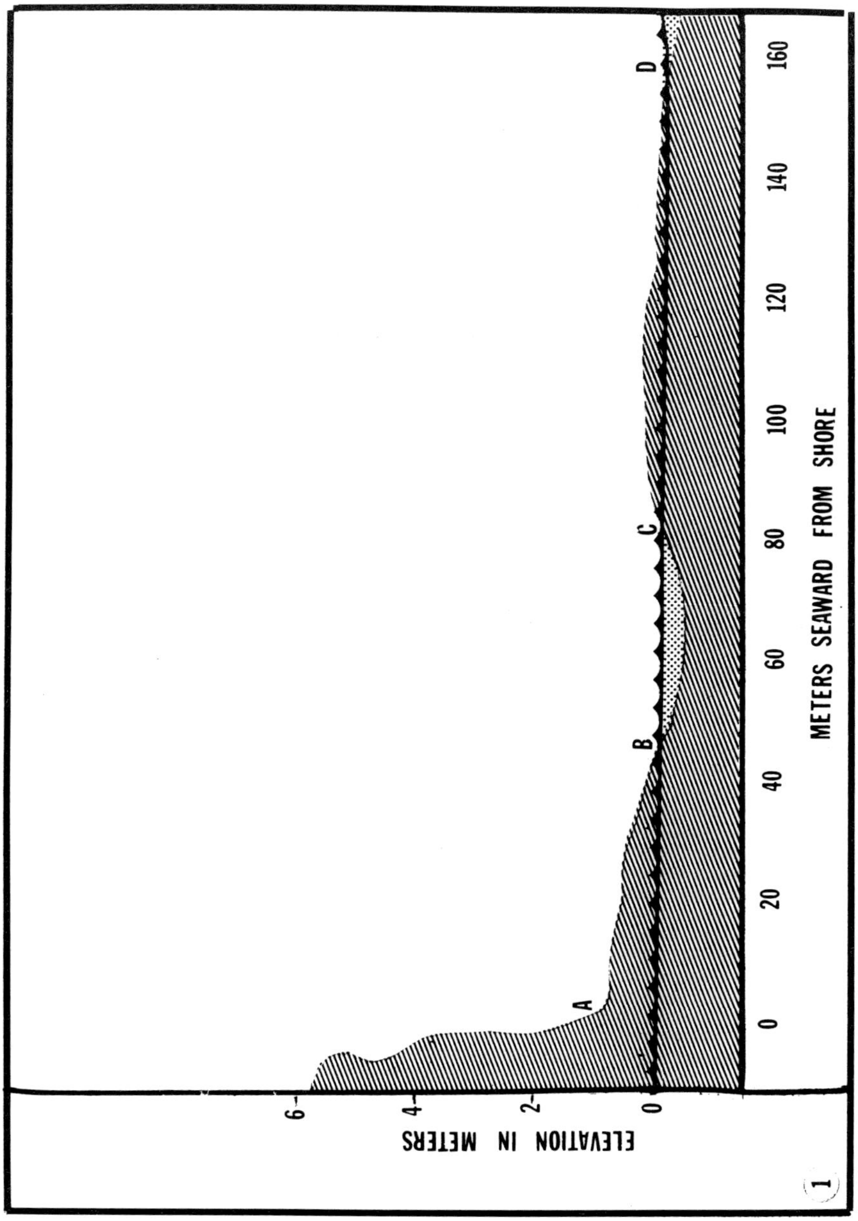

Fig. 1. Profile of the study area at Oyster Bay, Dar es Salaam. Note: A is zone just beneath the cliff experiencing longest duration of emersion at low tide; BC is intertidal basin permanently immersed at low tide; CD is lower intertidal area emersed at low tide, and D is area around spring tide low water mark experiencing frequent wetting due to wave action.

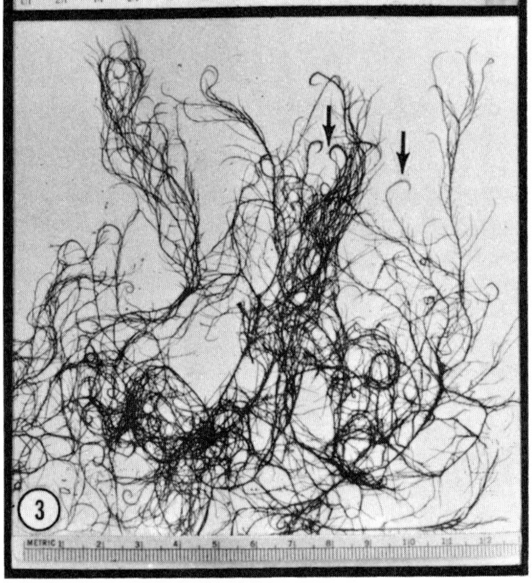

Fig. 2. Hypnea valentiae showing branching habit.

Fig. 3. Hypnea musciformis showing branching habit. Note characteristic crozier hooks (arrows) on ends of main branches.

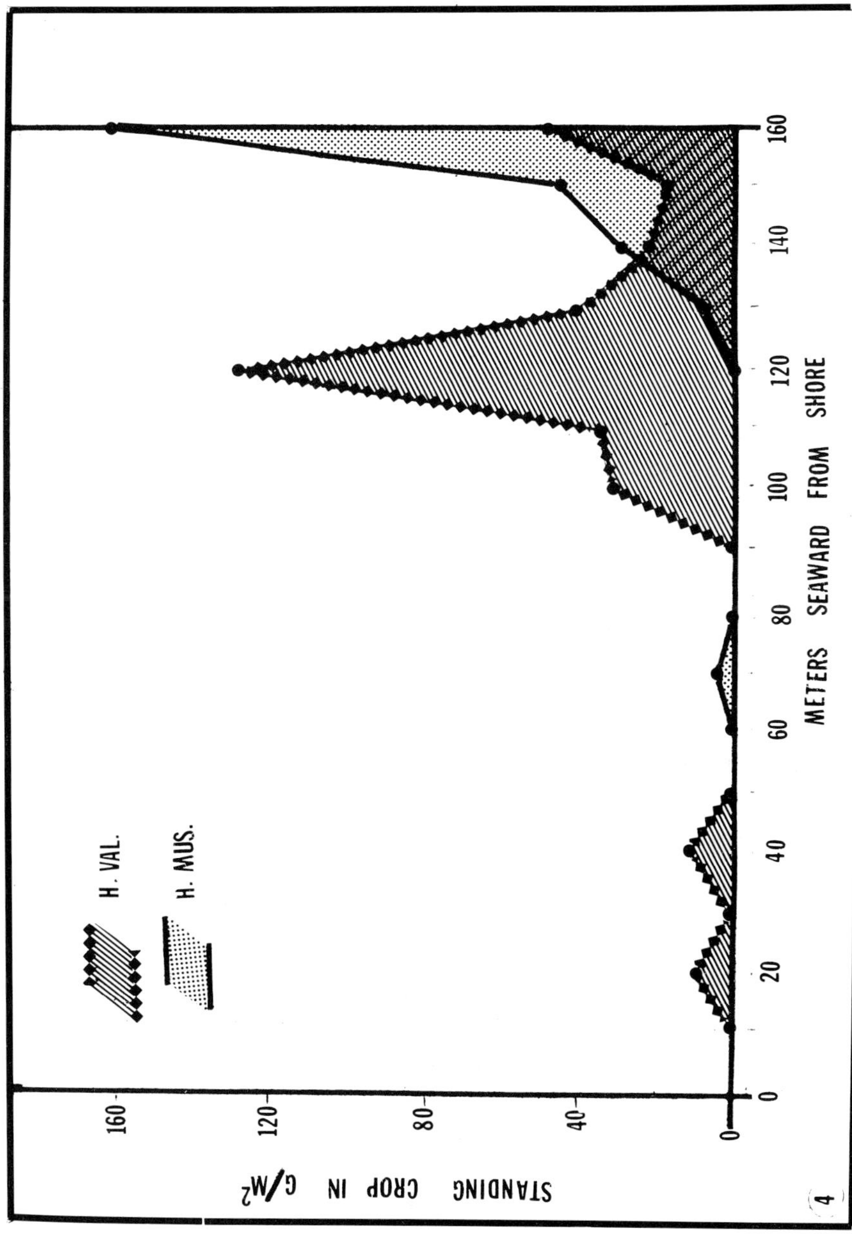

Fig. 4. Standing crop variation of <u>Hypnea valentiae</u> (<u>H. val.</u>) and <u>H. musciformis</u> (<u>H. mus.</u>) as a function of distance seaward from the shore (July, 1976).

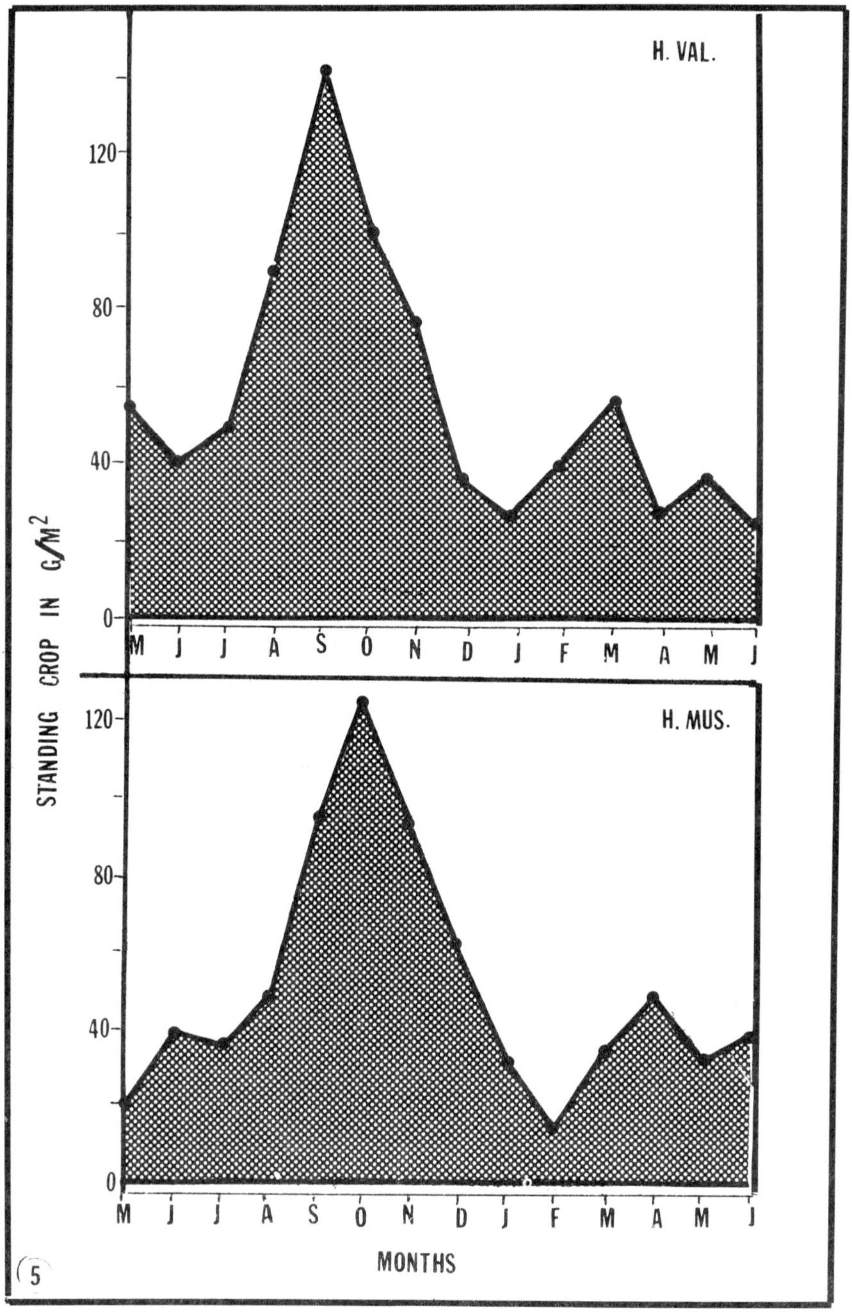

Fig. 5. Mean standing crop variation of <u>Hypnea valentiae</u> (<u>H. val.</u>) and <u>H. musciformis</u> (<u>H. mus.</u>) as a function of months of the year (May, 1976 to June, 1977).

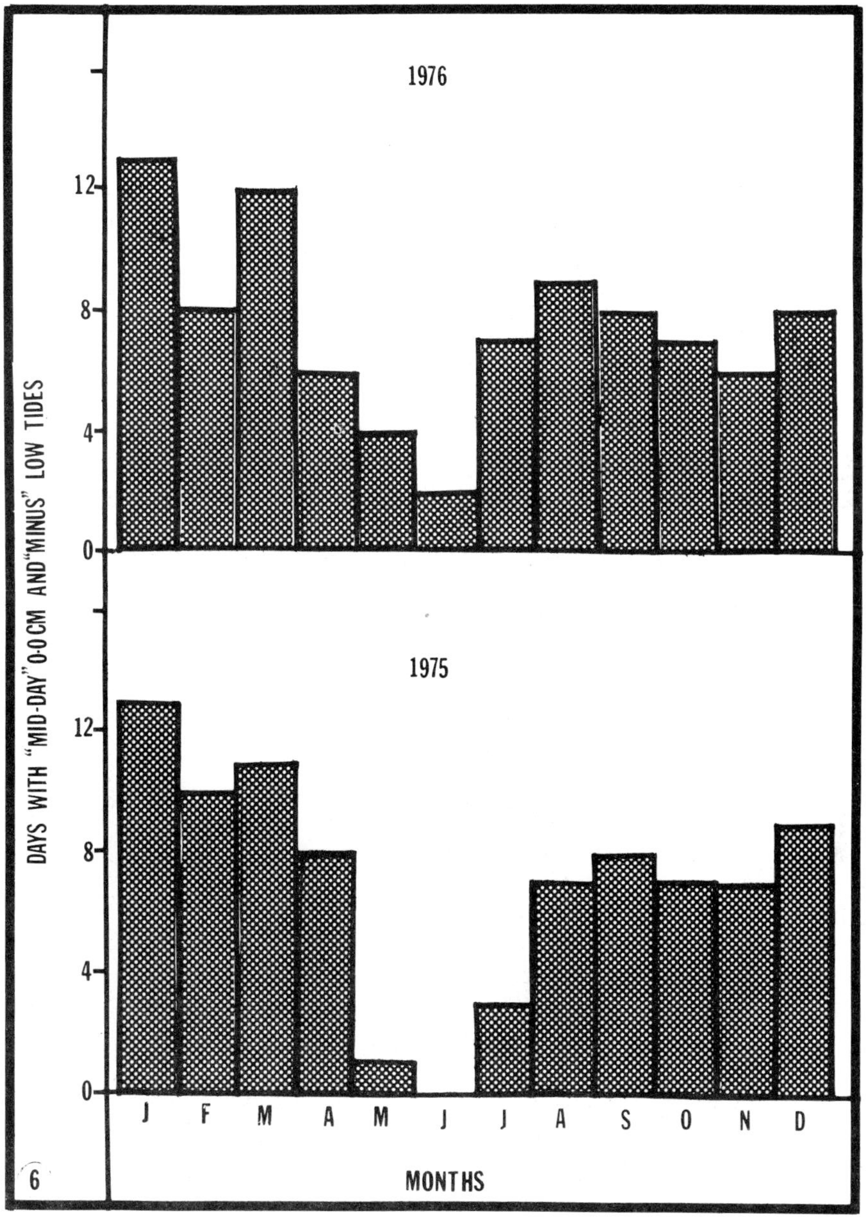

Fig. 6. Days with predicted extreme (0.0 cm and "minus") low tides in Dar es Salaam occurring around 1000 and 1500 hours as a function of months of the years 1975 and 1976.

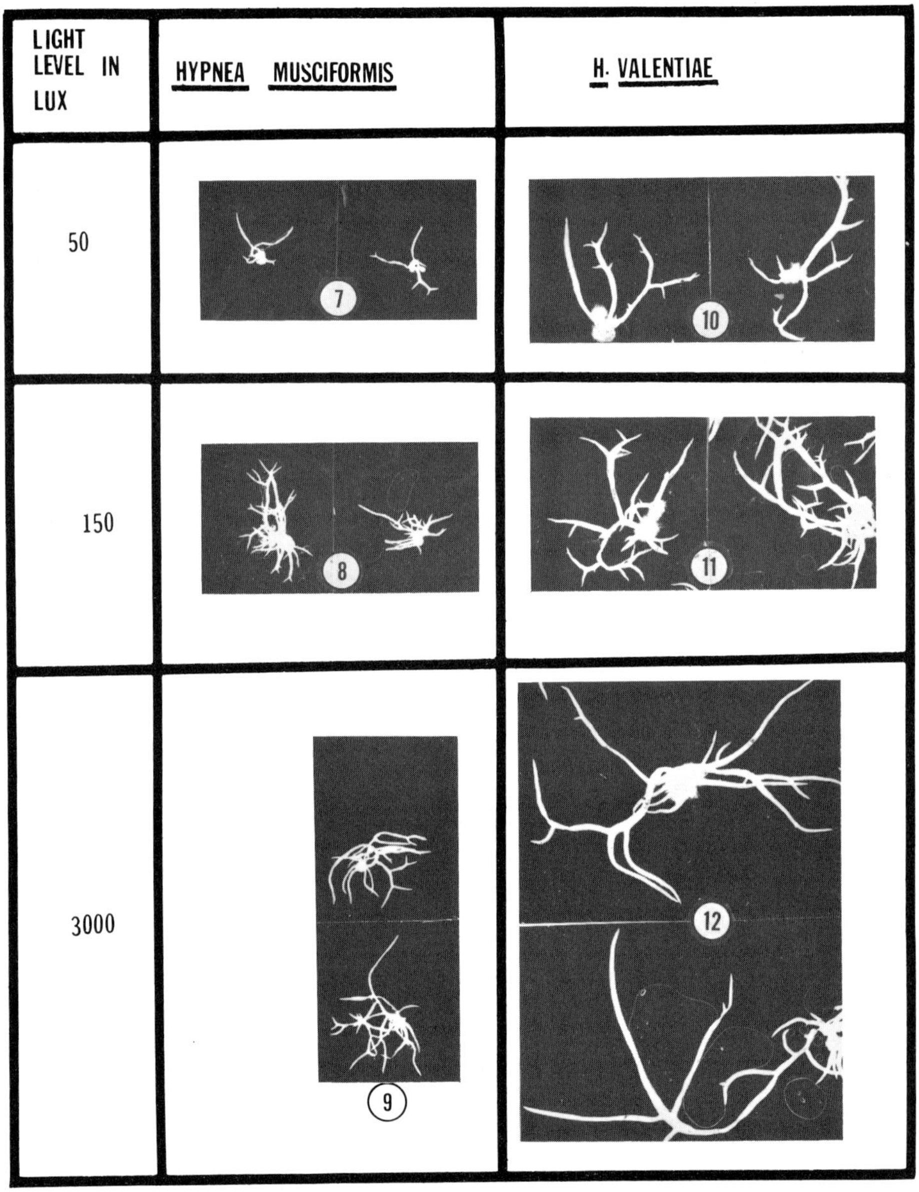

Figs. 7-9. Skiagraphs showing appearance of Hypnea musciformis sporelings grown at three different levels of illuminance.

Figs. 10-12. Skiagraphs showing appearance of Hypnea valentiae sporelings grown at three different levels of illuminance.

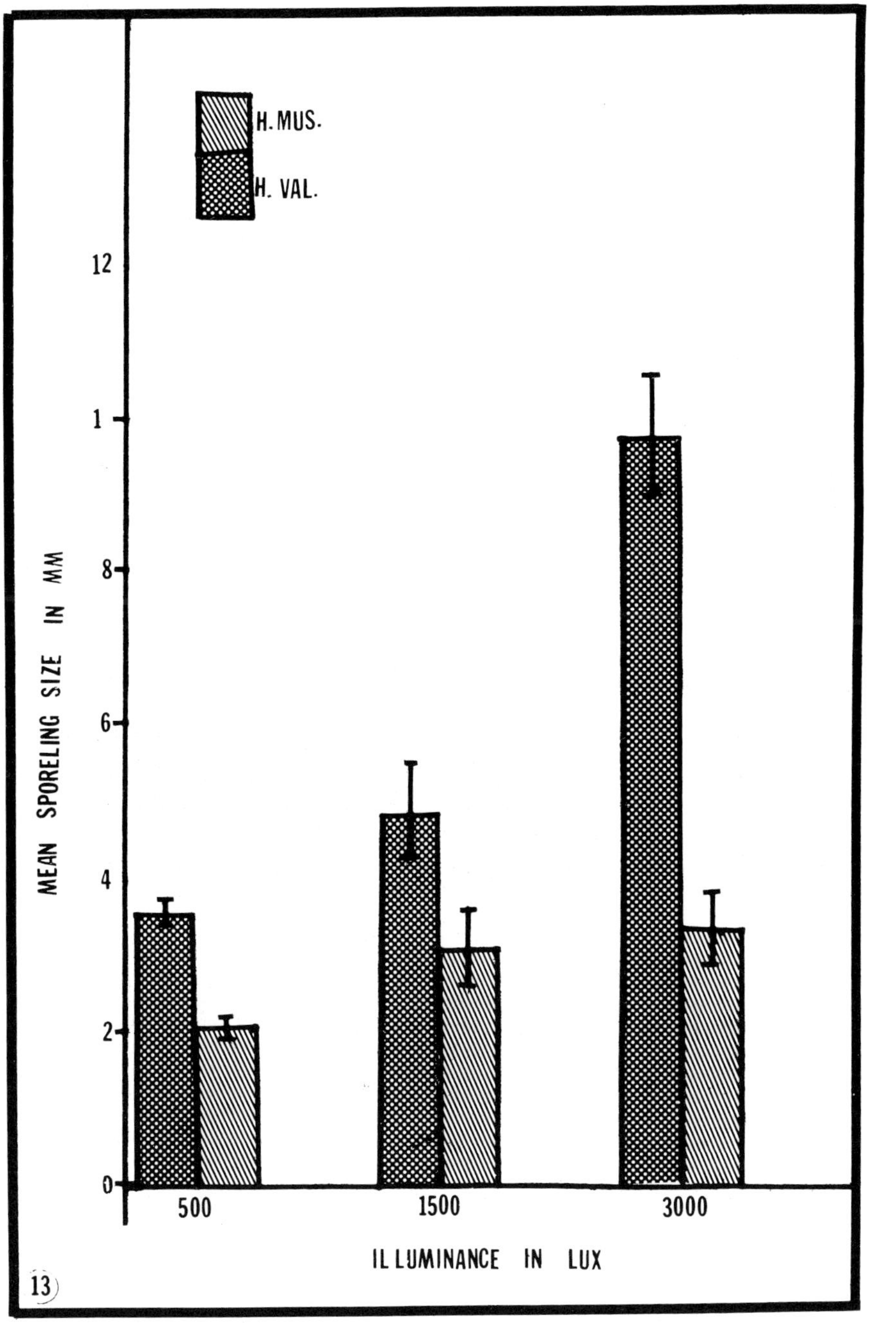

Fig. 13. Growth responses of _Hypnea valentiae_ (_H. val._) and _H. musciformis_ (_H. mus._) sporelings as a function of illuminance (lux).

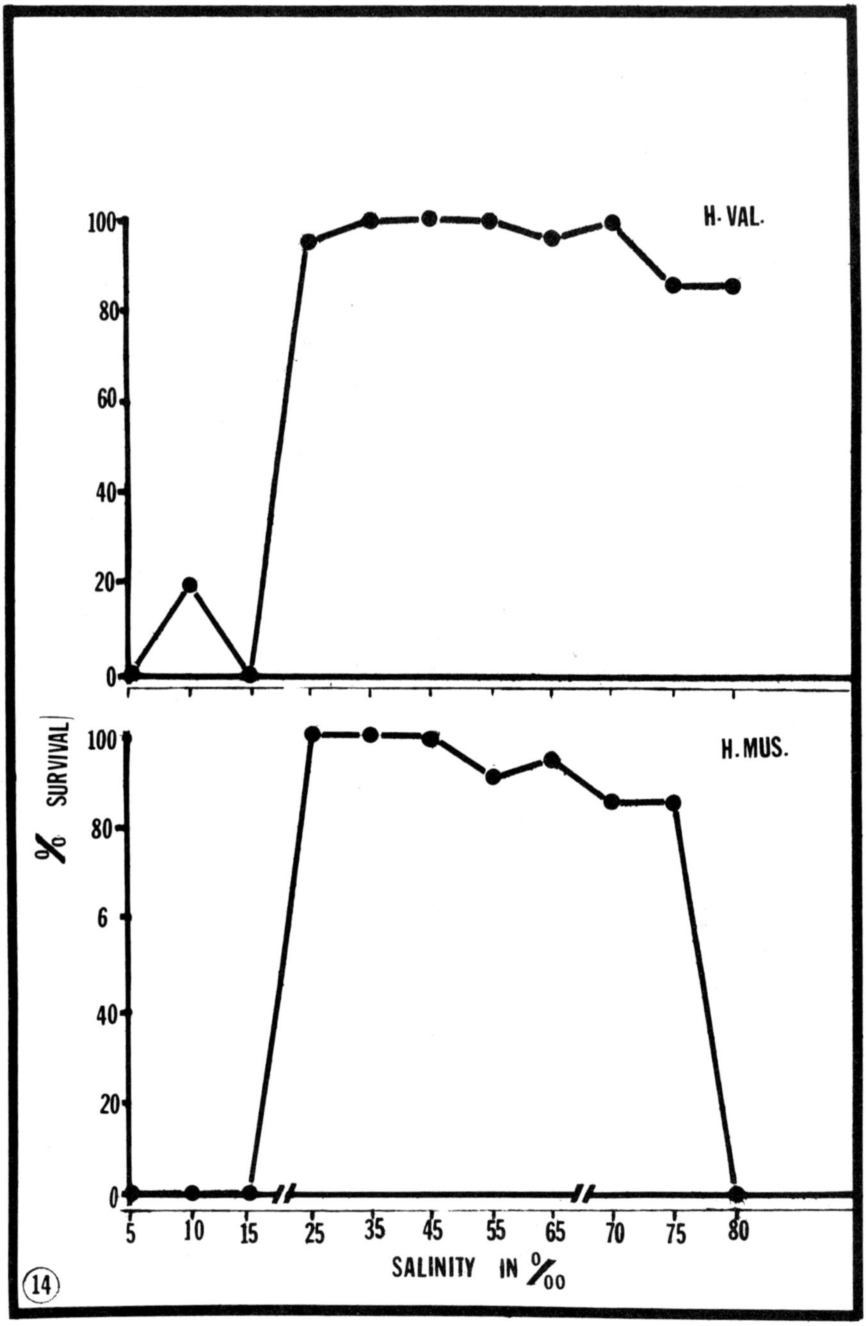

Fig. 14. Growth responses of *Hypnea valentiae* (*H. val.*) and *H. musciformis* (*H. mus.*) sporelings as a function of salinity.

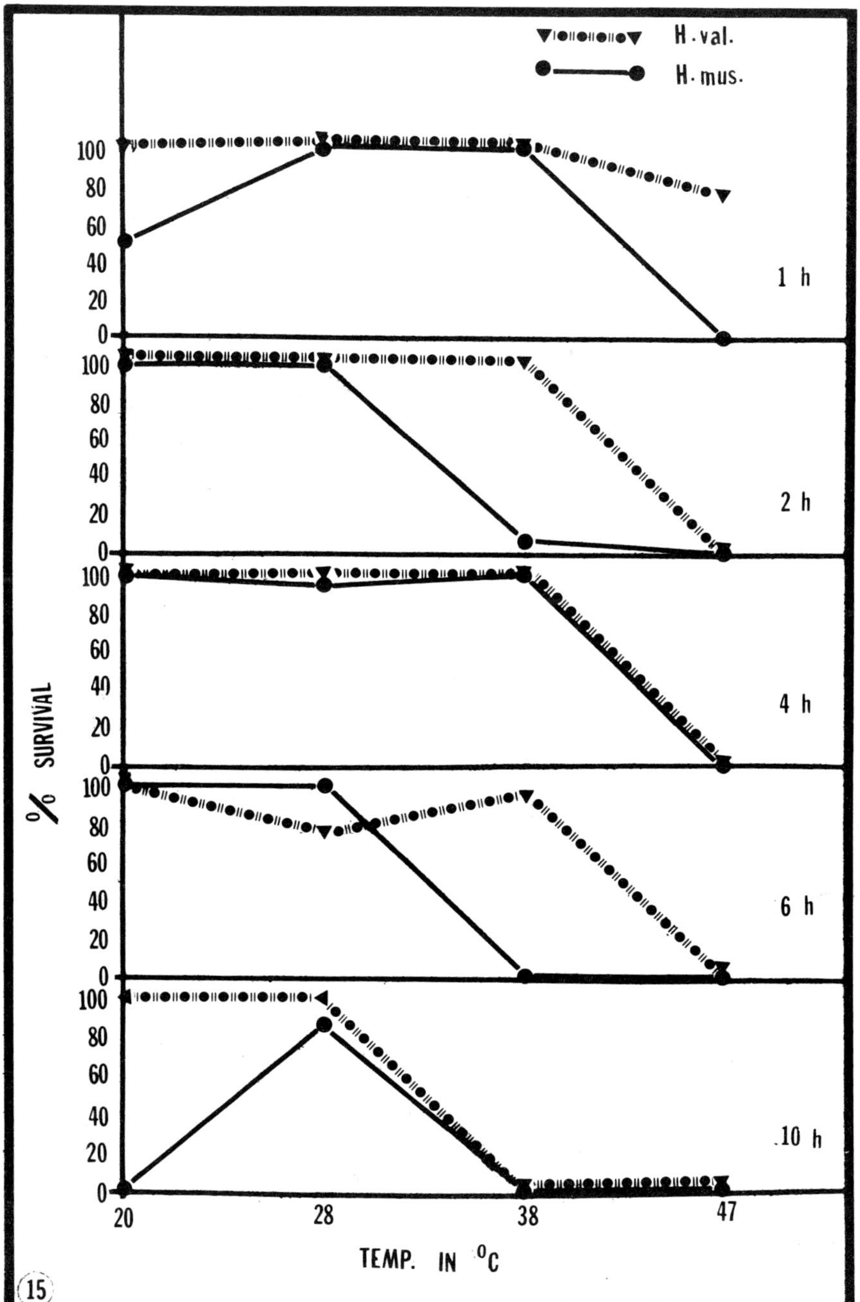

Fig. 15. Growth responses of Hypnea valentiae (H. val.) and H. musciformis (H. mus.) sporelings as a function of temperature.

REFERENCES

Biebl, R.: Protoplasmisch-okologische Untersuchungen an Mangrovealgen von Puerto Rico. Protoplasma 55, 573-606 (1962).

________.: Temperature resistance of marine algae. Proc. Intl. Seaweed Symp. 7, 23-28 (1972).

Bodard, M.: Les Hypnea au Sénégal (Hypnéacées, Gigartinales). Bull. de I' I.F.A.N. 30, Sér. A (3), 811-825 + Figs. 1-30 (1968).

Boergesen, F.: Some marine algae from Mauritius. Pt. III. Rhodophyceae. Dansk Vidensk. Selsk. Biol. Middel. 19, 54-63 (1943).

________: Some marine algae from Mauritius. Dansk. Vidensk. Selsk. Biol. Meddel. 22, 23-28 (1954).

Chamberlain, J.E.: Algae of the Hawaiian Islands. Thrum's Hawaiian Almanac and Annual for 1861, 32-33 (1880).

Chapman, V.J.: The marine algae of Jamaica. Pt. II. Phaeophyceae and Rhodophyceae. The Institute of Jamaica, 112-116 (1963).

Conover, J.T.: The ecology, seasonal periodicity and distribution of benthic plants in some Texas lagoons. Bot. Mar. 7, 4-41 (1964).

Dawes, C.J.: On the mariculture of the Florida seaweed, Eucheuma isiforme. Florida Sea Grant Program Report, No.5, 1-10 (1974).

Dawson, E.Y.: Marine plants in the vicinity of the Institut Oceanographique de Nha Trang, Viet Nam. Pac. Sci., 8, 373-469 (1954).

________ : Marine red algae of Pacific Mexico. Part IV. Gigartinales. Pac. Nat. 2, 191-343 (1961).

Durairatnam, M.: Contributions to the study of the marine algae of Ceylon. Fisheries Research Sta., Dept. of Fisheries, Ceylon Bull. 10, 1-117 (1961).

Hartog, C.: The effect of salinity tolerance of algae on their distribution, as exemplified by *Bangia*. Proc. Intl. Seaweed Symp. 7, 274-276 (1972).

Hauck, F.: Ueber einige von J.M. Hildebrandt in rothem Meere und indischen Ocean gesammelte Algen. Hedwigia 26, 18-45 (1887).

Humm, H.J. and J. Kreuzer: On the growth rate of the red alga, *Hypnea musciformis*, in the Caribbean Sea. Carib. J. Sci. 15, 1-4 (1975).

Inoh, S.: *Kaiso no hassei*: Development of marine algae (in Japanese), Tokyo. 225 pp (1947).

Isaac, W.E.: Marine botany of the Kenya coast.I. A first list of Kenya marine algae. J.E. Afr. Nat. Hist. Soc. and Nat. Mus. 26, 75-83 (1967).

__________: Marine botany of the Kenya coast. III. General account of the environment, flora and vegetation. J.E. Afr. Nat. Hist. Soc. and Nat. Mus. 27, 7-28 (1968).

__________ and F. Hewitt: The morphology, geographical distribution and ecology of *Hypnea spicifera* (Suhr) Harvey. J.S. Afr. Bot. 19, 73-83 (1953).

Jaasund, E.: *Intertidal seaweeds in Tanzania. A field guide*. University of Tromsø. 160 pp (1976).

Kuetzing, F.: *Tabulae phycocogicae*, Bd. 18, i + 35 pp + 100 pls (1968).

Kylin, H.: Verzeichnis einiger Rhodophyceen von Sudafrika. Lunds Univ. Arsskr., N.F. Avd. 2, 34(8), 1-26 (1938).

__________: Californische Rhodophyceen. Lunds Univ. Arsskr., N.F. Avd. 2(37), 2-20 (1941).

Lawson, G.W.: Seasonal variation of intertidal zonation on the coast of Ghana in relation to tidal factors. J. Ecol. 45, 831-860 (1957).

Mshigeni, K.E.: The biology and ecology of selected Hawaiian species in the Hypneaceae (Rhodophyta, Gigartinales). Ph.D. dissertation, University of Hawaii. 739 pp (1974).

______________ : Developmental studies in _Hypnea cervicornis J. Agardh_ and _H. chordacea_ Kuetzing: spore germination. Bot. Mar. _19_, 217-221 (1976a).

Mshigeni, K.E.: Holdfast and shoot differentiation in sporelings of _Hypnea cervicornis_ J. Agardh (Rhodophyta, Gigartinales). Bot. Mar. _19_, 223-226 (1976b).

______________ : Studies on reproduction of selected species of _Hypnea_ (Rhodophyta, Gigartinales) from Hawaii. Bot. Mar. _19_, 341-346 (1976c).

______________ : Field cultivation of _Hypnea_ (Rhodophyta) spores for carrageenan: prospects and problems. Bot. Mar. _19_, 227-230 (1976d).

______________ : Effects of the environment on developmental rates of sporelings of two _Hypnea_ species (Rhodophyta Gigartinales). Mar. Biol. _36_, 99-103 (1976e).

______________ : The seasonality of mid-day 'zero' and 'minus' low tides on the tropical shores of Hawaii and their effects on intertidal seaweed populations. Experientia _32_, 1259-1260 (1976f).

______________ : Seasonal changes in the standing crops of three _Hypnea_ species (Rhodophyta, Gigartinales) in Hawaii. Bot. Mar. _20_, 303-306 (1977a).

______________ : A skiagraphic method of recording developmental patterns of frondose algae. Bot. Mar., in press (1977b).

______________ and W.S.M. Lorri: Spore germination and early stages of development in _Hypnea musciformis_ (Rhodophyta Gigartinales). Mar. Biol. _42_, 161-164 (1977a).

__________ and __________ : Spore germination and early stages of development in Hypnea valentiae (Turner) Montagne. Bot. Mar., in press (1977b).

Okamura, K.: Icones of Japanese Algae, Japan. Vol. 1, 47 (1907).

Rama Rao, K.: Role of critical tide factor in the vertical distribution of Hypnea musciformis (Wulfen) Lamouroux. Proc. Indian Nat. Sci. Acad. 38(B), 267-272 (1972).

Schmitz, F.: Marine Florideen von Deutsch Ostafrika. Engler Bot. Jahrb. 21, 137-177 (1895).

Schreiber, E.: Die Reinkultur von marinen Phytoplankton und deren Bedeutung fur die Erforschung der Produktions-fahigkeit des Meerwassers. Wiss. Meeresuntersuch., N.F. 16(10), 1-34 (1927).

Setchell, W.A.: Tahitan algae. Univ. Calif. Publ. Bot., 12, 100-101 (1926).

__________ and N.L. Gardiner: New marine algae from the Gulf of California. Proc. Calif. Acad. Sci. 12(29), 695-949 (1924).

Snedecor, G.W. and W.G. Cochran: Statistical Methods. 6th Edition, Iowa. 593 pp (1967).

Tanaka, T.: The genus Hypnea from Japan. Sci. Pap. Inst. Algol. Res. Hokkaido Univ. 2, 227-250 (1941).

Taylor, W.R.: Marine algae of Florida. Carnegie Inst. Washington Publ. 25, 155-157 (1928).

__________: Pacific marine algae of the Allan Hancock expeditions to the Galapagos Islands. Allan Hancock Pac. Expd. 12, 226-228 (1945).

__________: Marine algae of the eastern tropical and sub-tropical coasts of the Americas. The Univ. of Michigan Press, Ann. Arbor, 870 pp (1960).

Trono, G.C.: The taxonomy and ecology of the marine benthic algae of the Caroline Islands. Ph.D. dissertation, University of Hawaii, 235-240 (1968).

Weber-van Bosse, A.: Liste des algues du Siboga. IV. Rhodophyceae. Pt. 3. Gigartinales et Rhodymeniales. Siboga Exped. Monogr. 59d, 392-533 (1928).

Yaphe, W.: The determination of K-carrageenan as a factor in the classification of the Rhodophyceae. Can. J. Bot. 37, 751-757 (1959).

__________ and B. Baxter: The enzymic hydrolysis of carrageenan. Appl. Microbiol. 3, 380-383 (1955).

Zaneveld, J.S.: Economic marine algae of tropical South and East Asia and their utilization. Indo-Pacific Fish Council 3, 1-54 (1955).

Taxonomic Index

Subject Index

BOTANICA MARINA